ADVANCES IN

Computation, Modeling and Control of Transitional and Turbulent Flows

ADVANCES IN

Computation, Modeling and Control of Transitional and Turbulent Flows

Editors

Tapan K. Sengupta
Indian Institute of Technology, Kanpur, India

Sanjiva K. Lele
Stanford University, USA

Katepalli R. Sreenivasan
New York University, USA

Peter A. Davidson
Cambridge University, UK

World Scientific

NEW JERSEY • LONDON • SINGAPORE • BEIJING • SHANGHAI • HONG KONG • TAIPEI • CHENNAI • TOKYO

Published by

World Scientific Publishing Co. Pte. Ltd.

5 Toh Tuck Link, Singapore 596224

USA office: 27 Warren Street, Suite 401-402, Hackensack, NJ 07601

UK office: 57 Shelton Street, Covent Garden, London WC2H 9HE

British Library Cataloguing-in-Publication Data
A catalogue record for this book is available from the British Library.

ADVANCES IN COMPUTATION, MODELING AND CONTROL OF TRANSITIONAL AND TURBULENT FLOWS

ISBN 978-981-4635-15-8

Printed by FuIsland Offset Printing (S) Pte Ltd Singapore

Preface

The *IUTAM Symposium on Advances in Computation, Modeling and Control of Transitional and Turbulent Flows* was held in Goa, India, during December 15–18, 2014. The aim of the symposium was to bring together renowned experts from fluid dynamics, computational science, dynamical systems and control to share results from the latest developments in the field and to foster further progress by interaction and exchange of ideas. Following the long-standing IUTAM tradition the symposium was organized with a single track of technical sessions so as to provide ample opportunity for scientific dialogue and discussion, and especially to expose young scientists to the cutting-edge of research.

The scientific program of the symposium included 10 technical sessions presenting 56 contributed papers, a poster session with 11 posters, a panel discussion, and 8 invited keynote lectures delivered by internationally renowned scholars. A review panel selected these papers and posters from a larger list of abstracts submitted. The symposium attracted participation by delegates from 19 countries with approximately half of the contributed papers from India.

Authors of the papers presented at the symposium, including keynote speakers, were invited to submit a technical paper to these Proceedings, based on their presentations at the symposium. The papers received were reviewed by an expert review panel and edited for this volume. We hope that the papers capture the scientific excitement of the symposium. Several keynote speakers and most other speakers contributed to the volume, in some cases by including results updated after the symposium. The full scientific program of the symposium is included at the end of this volume.

We hope that these technical papers provide a glimpse of the new ideas and results discussed at the symposium and that they will foster further research and collaboration. The reader may note the broad range of themes represented in the papers, from careful assessment of numerical algorithms, fundamental studies of transition and instabilities, flow control, aeroacoustics and environmental flows, to dynamo theory and astrophysics. Included are papers on the need for simplified models capable of describing flow physics in diverse regimes or conditions.

The choice of Goa as the venue was not accidental: the famous D. D. Kosambi, an originator of the so-called Kurhunen-Loeve expansion, or the Proper Orthogonal Decomposition, was born here and his birthday is celebrated as the science day in the state of Goa. A short sketch of his life and times is included as an appendix. The choice of the subject for the conference was due to its timely interest in the scientific firmament, including many from the host country. Everyone agreed that the richness and beauty of the fluid phenomena discussed in these proceedings paralleled the stunningly enthralling Goa, the meticulous organization of the symposium, and the warm hospitality of the local organizing committee.

Color Plates

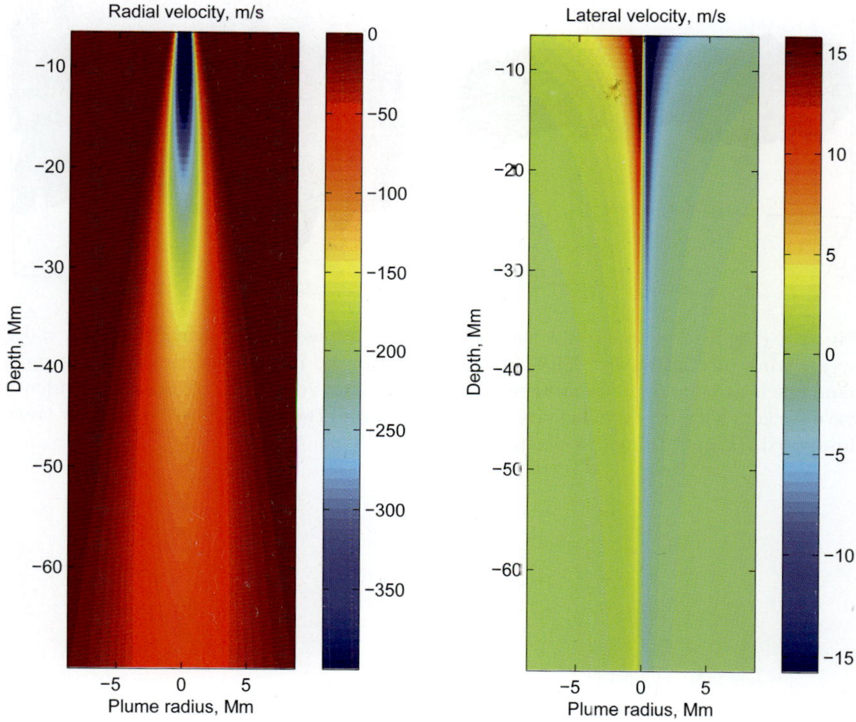

Fig. 1. (Figure 3 of paper 2) Velocity profile obtained from a model of plumes[23]. The plume gradually grows in size as it descends (left panel, radial velocity) and the degree of entrainment (right panel, horizontal or lateral velocity). Despite the increasing degree of entrainment, the diameter of the plume grows slowly. If the plumes entrains sufficiently, it will destabilize and mix into the ambient, resulting in overturning convection. However, in the weakly mixing mechanism described here, the increasing density stabilizes descending plumes, preventing them from breaking up (Courtesy, M. Rieutord).

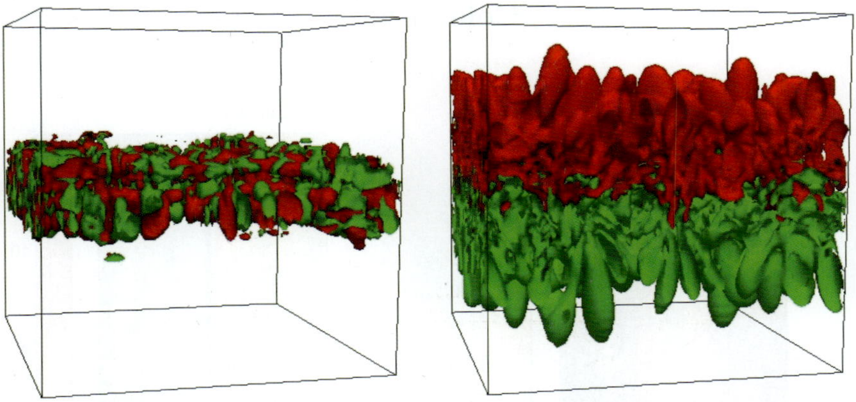

Fig. 2. (Figure 2 of paper 3) A slab of turbulence spreads in a rotating fluid ($Ro = 0.1$) by emitting inertial waves. The left-hand panel is the initial condition and the right-hand one is at $\Omega t = 6$. Red marks negative helicity and green positive helicity. Evidently wave packets with negative helicity travel upward and those with positive helicity travel downward. (Figure from Davidson, 2013[2].)

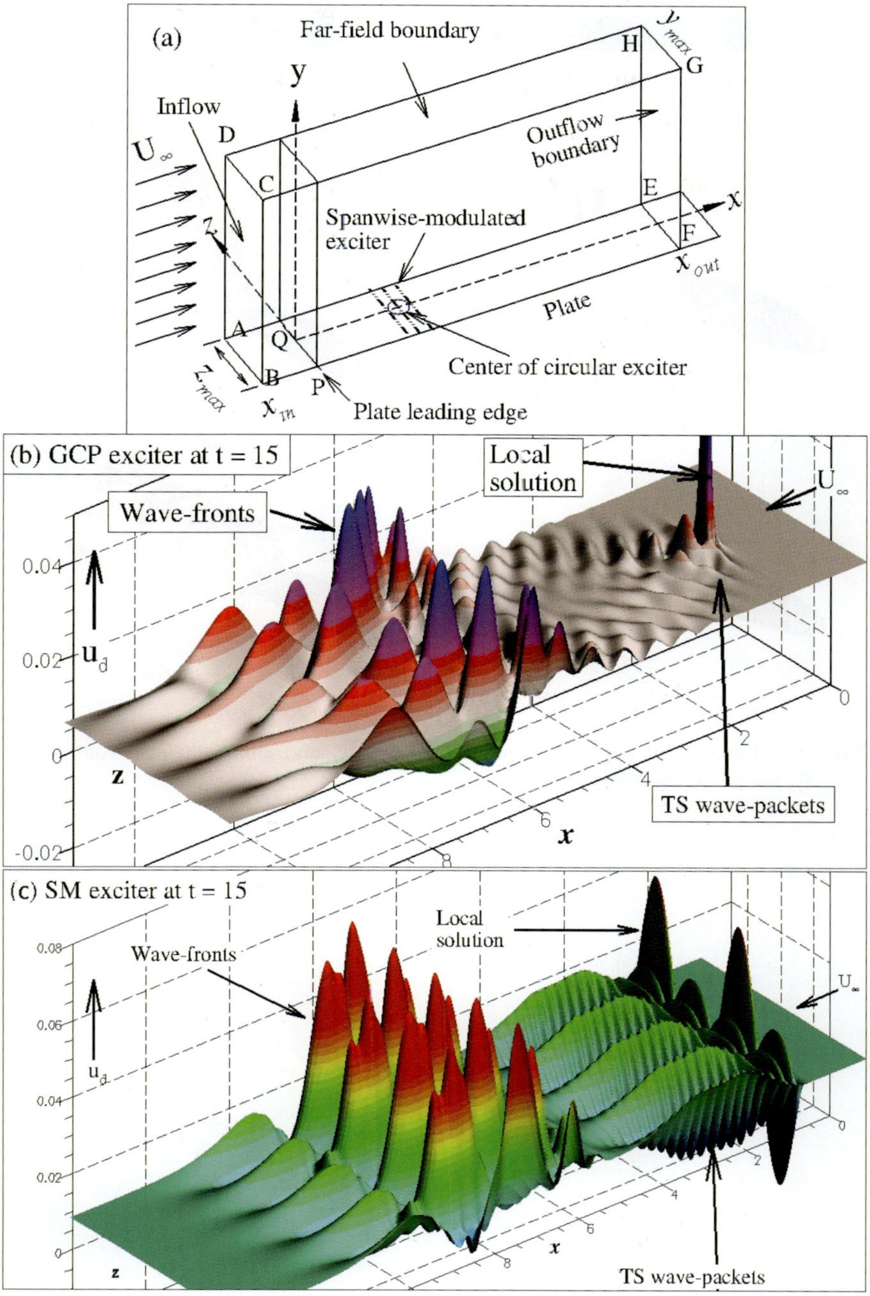

Fig. 3. (Figure 2 of paper 5) Perspective plot of u_d at a fixed height shown for (a) GCP and (b) SM exciters. The frequency of excitation is $F = 2\pi\nu f/U_\infty^2 = 0.5 \times 10^{-4}$; ν is kinematic viscosity and f is frequency in Hz.

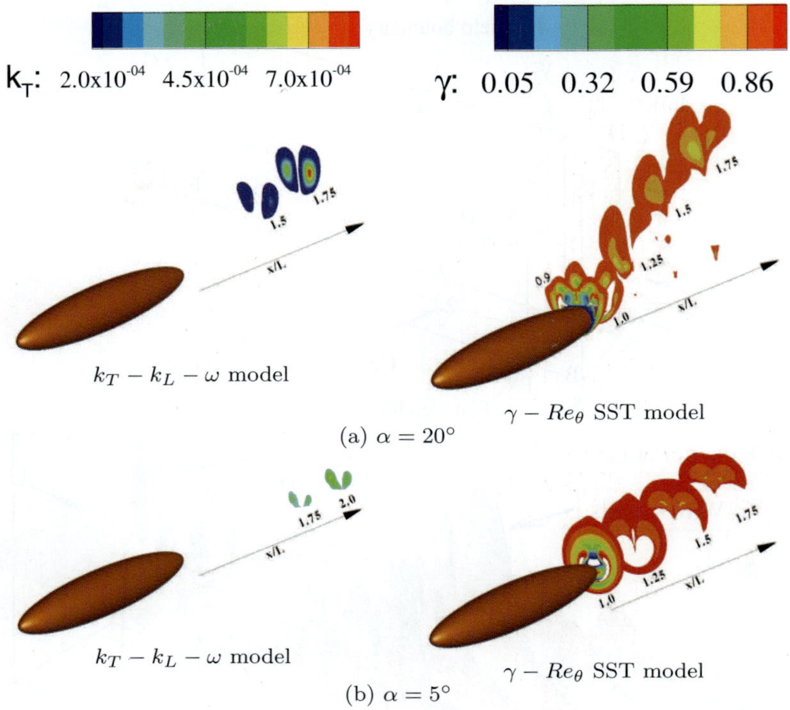

Fig. 4. (Figure 7 of paper 9) Kinetic energy (left) and gamma (right) contours at $\alpha = 20°$ for prolate spheroid.

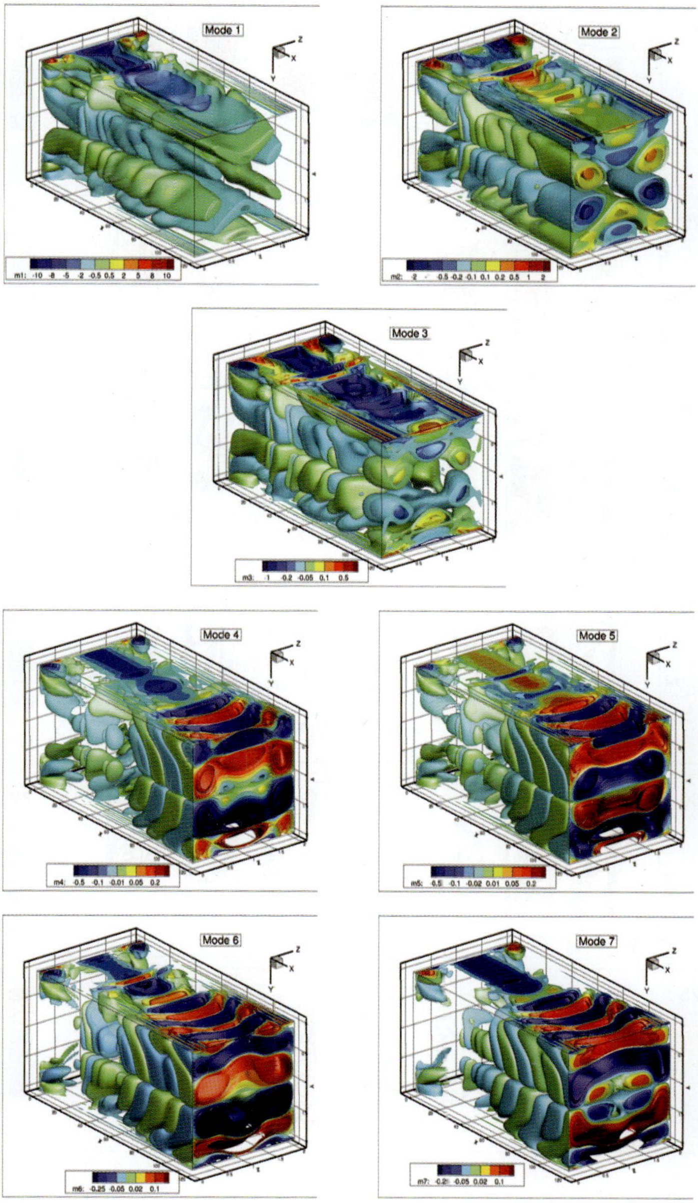

Fig. 5. (From paper 10) Iso-surfaces of mode 1, 2, 3, 4, 5, 6 and 7.

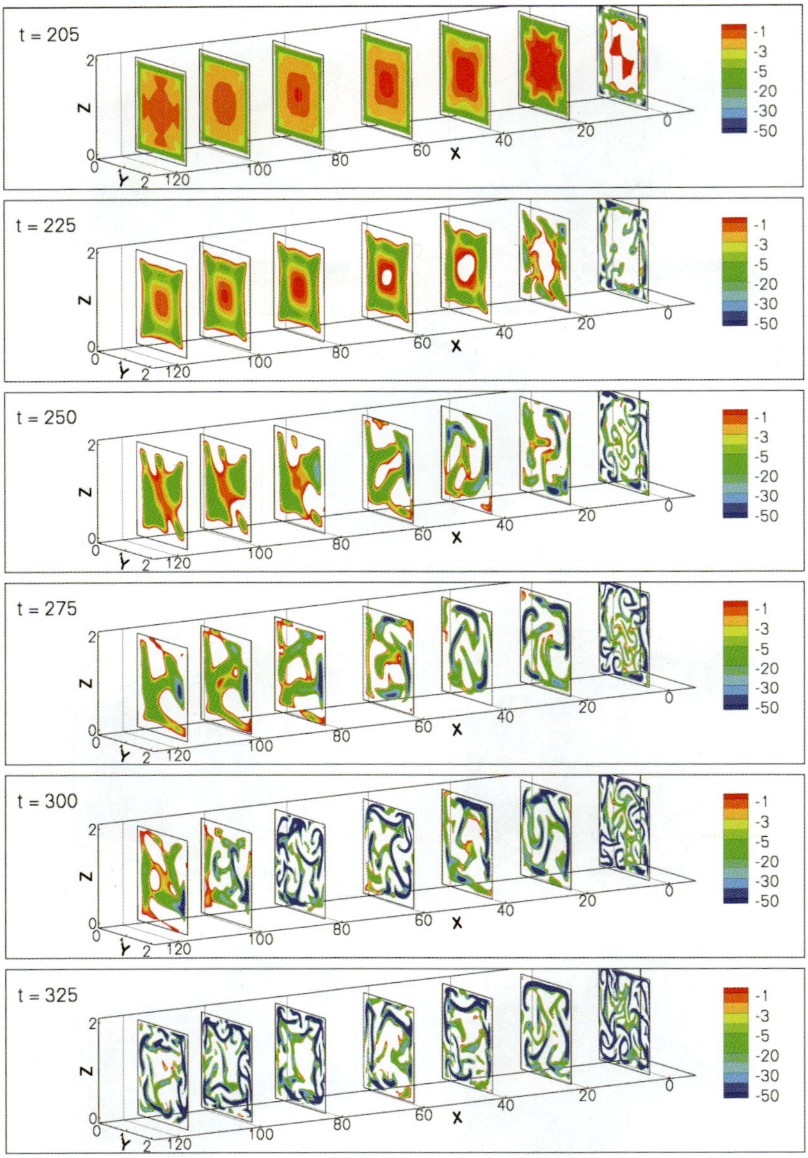

Fig. 6. (From paper 10) Disturbance energy (E_d) contours at different planes and indicated times.

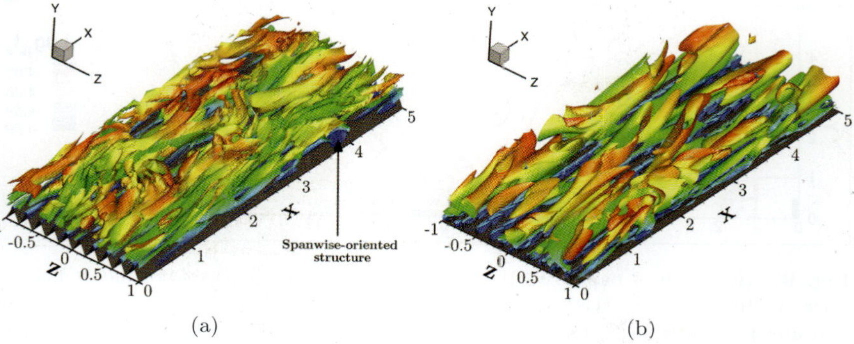

Fig. 7. (Figure 4 of paper 13) Near-wall vortical structures deduced using the Q-criterion, where Q is measured in wall units: (a) V-groove riblets, and (b) Thin blade riblets. The arrow in Fig. 4(a) points out one of the spanwise-oriented structures responsible for causing the flow reversal at the corresponding location depicted in Fig. 3(a).

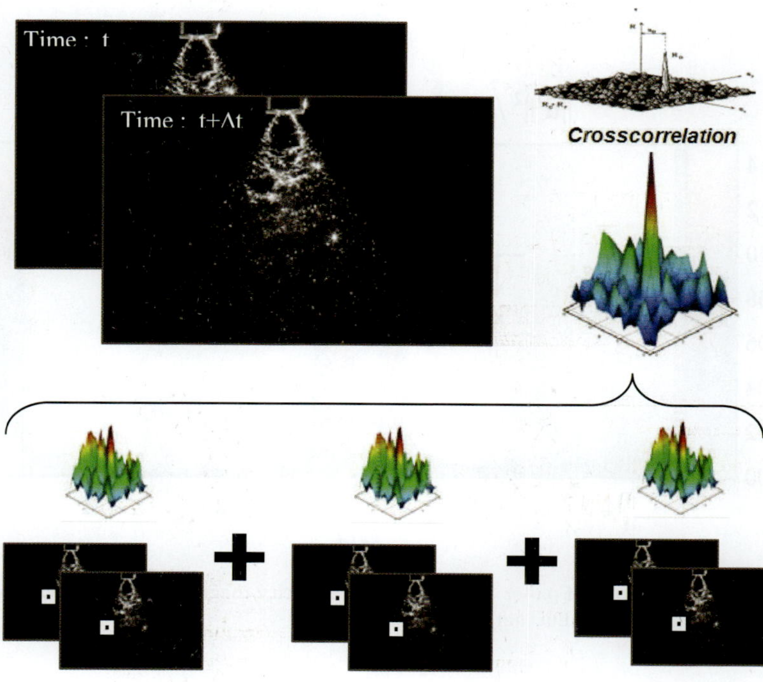

Fig. 8. (Figure 5 of paper 21) Ensemble correlation technique for PIV.

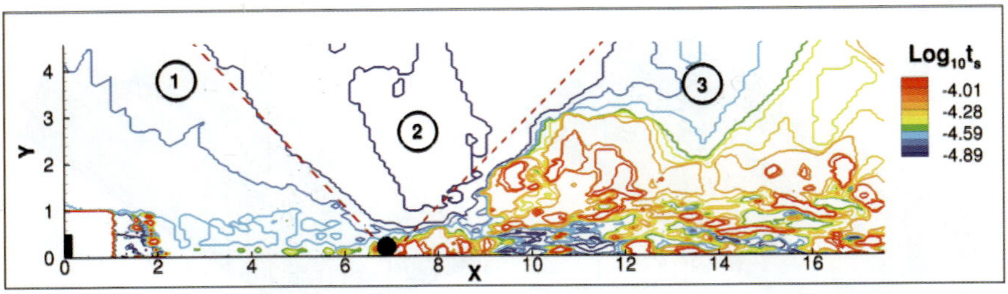

Fig. 9. (Figure 9 of paper 36) Integral timescale of the perturbation boost field for Case A obtained from the autocorrelation of pressure perturbation boost. The dot indicates the source location. The dashed lines demarcate three major timescale regions in the nearfield as identified by the circled numbers.

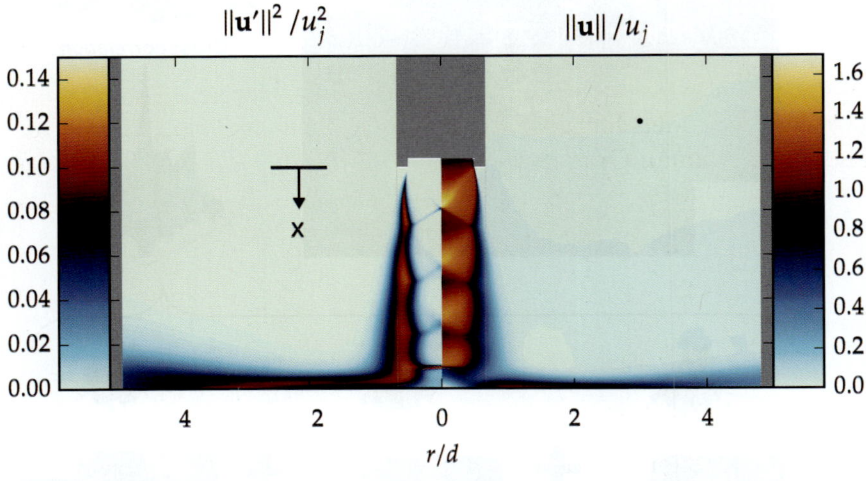

Fig. 10. (Figure 3 of paper 38) Mean of the velocity magnitude (right half) and mean square of the velocity fluctuation (left half).

Contents

Keynote Speakers

Contributed Papers

Large-Eddy Simulation of the Navier-Stokes Equations: Deconvolution, Particle Methods, and Super-Resolution

A. Leonard

Graduate Aerospace Laboratories, California Institute of Technology,
Pasadena, CA 91105, USA
**E-mail: leonard@caltech.edu*

The technique of large-eddy simulation (LES) continues to play an important role in the numerical simulation of fluid dynamic processes in engineering and scientific applications. This paper will review and discuss a few of the various schemes for LES applied to the incompressible Navier-Stokes equations and to the scalar advection-diffusion equation. In particular, subgrid models based on deconvolution will be discussed. An interesting connection between the tensor diffusivity model, a particular version of a deconvolution model, and Lagrangian particle methods for the vorticity transport equation and for the scalar convection-diffusion equation will be explored. In addition, the possibility of using super-resolution, i.e. recovering fine-scale information knowing only coarse-scale information, in LES will be investigated.

Keywords: large-eddy simulation, deconvolution, Lagrangian particle methods, super-resolution

1. Introduction

In the numerical simulation of fluid dynamic phenomena in engineering and science, the method of large-eddy simulation (LES) plays an increasingly prominent role. This is so because the range of scales involved in many applications may be too large for direct simulation methods, on the one hand, or the flow too complex to be treated accurately by the Reynolds-Averaged Navier-Stokes (RANS) equations or by unsteady RANS, on the other.

Generally in LES a considerable fraction of the energy-containing eddies must be accounted for explicitly, leaving only the effect of the subgrid eddies on the larger computed scales to be modeled. Thus even LES can be too costly for some applications. If this is the case there are still a number of options available. For example, one could develop an advanced subgrid model, reducing the number of computed scales that are necessary. Or in flows that involve turbulent boundary layers one could use the technique known as Detached Eddy Simulation[1] in which LES is replaced by RANS in the boundary layers. Finally, in climate modeling, the atmosphere-ocean system is too complex to be treated by LES at the present time although so-called "eddy-permitting" simulations of idealized problems of the climate system are beginning to appear. Thus, for example, we see also efforts to develop low-order models for the climate based on stochastic differential equations[2].

In this paper we focus on a particular model, the tensor diffusivity model, for the subgrid stresses in LES based on approximate deconvolution as described below. In Sec. 2 we rederive the model and discuss its properties, including the forward and backward scattering of turbulent energy into and out of the unresolved small scales, respecttively. Possibilities for supplementing the model with an additional dissipation term are then discussed. Lagrangian particle methods are reintroduced in Sec. 3 because the leading-order truncation error of this method is the tensor diffusivity model for two-dimensional vorticity transport equation. For three-dimensional vorticity transport and for scalar advection, the particle velocity must be modified slightly from the local velocity. In Sec. 4 we explore the potential of super-resolution as a subgrid model. In super-resolution one recovers fine-scale information knowing only coarse-scale information and some knowledge of the structure of the fine scales. Summary and conclusions are presented in Sec. 5.

2. Approximate Deconvolution

In approximate deconvolution one performs a partial inversion of the filter operation used in defining the filtered products that appear in the large-eddy equations. One thus recovers a portion of the subgrid stresses. Stolz & Adams[3], for example, have used such an approach.

To develop the particular deconvolution approach of interest here we start with the filtered product of two scalar functions, each a function of d spatial variables in a d-infinite domain given by

$$\overline{uv} = \int \frac{\exp\left(\frac{-|\mathbf{x}-\mathbf{x}'|^2}{\sigma^2}\right) u(\mathbf{x}')v(\mathbf{x}')d\mathbf{x}'}{(\sqrt{\pi}\sigma)^d}, \tag{1}$$

where we have employed Gaussian filtering. The task is then to approximate the above filtered product in terms of functionals of $\overline{u}$ and $\overline{v}$. For example we can derive a series expansion in powers of σ^2 as follows. Let $\hat{g}(\mathbf{k})$ be the Fourier Transform of $g(\mathbf{x})$ given by

$$\hat{g}(\mathbf{k}) = \int e^{i\mathbf{k}\cdot\mathbf{x}} g(\mathbf{x}) d\mathbf{x}. \tag{2}$$

Substituting the inverse transforms of $u(\mathbf{x})$ and $v(\mathbf{x})$ into (1) we have

$$\overline{uv} = \left(\frac{1}{2\pi}\right)^{2d} \int\int\int \frac{\exp(\frac{-|\mathbf{x}-\mathbf{x}'|^2}{\sigma^2}) e^{-i(\mathbf{p}+\mathbf{q})\cdot\mathbf{x}'} \hat{u}(\mathbf{p})\hat{v}(\mathbf{q}) d\mathbf{p} d\mathbf{q} d\mathbf{x}'}{(\sqrt{\pi}\sigma)^d}, \tag{3}$$

and integrating over $\mathbf{x}'$ we have

$$\begin{aligned}\overline{uv} &= \left(\frac{1}{2\pi}\right)^{2d} \int\int e^{-|\mathbf{p}+\mathbf{q}|^2\sigma^2/4} e^{-i(\mathbf{p}+\mathbf{q})\cdot\mathbf{x}} \hat{u}(\mathbf{p})\hat{v}(\mathbf{q}) d\mathbf{p} d\mathbf{q} \\ &= \left(\frac{1}{2\pi}\right)^{2d} \int\int e^{-\mathbf{p}\cdot\mathbf{q}\sigma^2/2} e^{-i(\mathbf{p}+\mathbf{q})\cdot\mathbf{x}} e^{-|\mathbf{p}|^2\sigma^2/4} \hat{u}(\mathbf{p}) e^{-|\mathbf{q}|^2\sigma^2/4} \hat{v}(\mathbf{q}) d\mathbf{p} d\mathbf{q}.\end{aligned} \tag{4}$$

Expanding the factor $e^{-\mathbf{p}\cdot\mathbf{q}\sigma^2/2}$ in a Taylor series and performing the inverse transforms, (4) gives the desired result,

$$\overline{uv}(\mathbf{x}) = \sum_{n=0}^{\infty} \left(\frac{\sigma^2}{2}\right)^n \frac{1}{n!} \frac{\partial^n \overline{u}}{\partial x_{i_1} \partial x_{i_2} ... \partial x_{i_n}} \frac{\partial^n \overline{v}}{\partial x_{i_1} \partial x_{i_2} ... \partial x_{i_n}}, \tag{5}$$

where repeated indices are summed. This formula was apparently first derived by Yeo[4] (see also Yeo & Bedford[5]) using Taylor series expansions in real space. Formally (5) is a convergent series because $\overline{u}$ and $\overline{v}$ are C_∞ functions. Note that if either $\overline{u}$ or $\overline{v}$ is linear in all its variables then all the terms $n \geq 2$ are zero. The stretched vortex case discussed below illustrates this point. In general the number of terms that are useful, i.e. accurately computable, will be limited by the numerical method used to compute the spatial derivatives and the behavior of u and v at high spatial wavenumbers. See, in particular, the discussion in Carati *et al.*[6].

As an example, consider the solution

$$\begin{aligned} u(x) = v(x) &= +1 \quad x < 0 \\ &= -1 \quad x > 0 \end{aligned} \tag{6}$$

to the inviscid Burgers' equation on the infinite domain. We have that $\overline{u^2} = 1$ but $\overline{u} = -erf(x/\sigma)$ so that

$$\frac{d^n \overline{u}}{dx^n} = (-1)^n \frac{2}{\sqrt{\pi}\sigma^n} e^{-x^2/\sigma^2} H_{n-1}(x/\sigma), \tag{7}$$

where the H_n's are the Hermite polynomials. As n increases,

$$e^{-x^2/2\sigma^2} H_{n-1}(x) \sim \cos(\sqrt{2(n-1)}x/\sigma - (n-1)\pi/2) \tag{8}$$

so that, for this case, the wavelength of the terms in the series (5) decreases as $1/\sqrt{n-1}$. As a consequence any term with n greater than than N_{max} given by the Nyquist criterion:

$$N_{max} = 1 + \left[\frac{1}{2} \left(\frac{\pi\sigma}{\Delta} \right)^2 \right] \tag{9}$$

should not be used, where Δ is the grid spacing for the simulation.

To illustrate the result of using the expansion (5) in a transport equation we repeat an analysis[7] of the filtered advection-diffusion equation with diffusivity κ for the scalar Ψ,

$$\frac{\partial \overline{\Psi}}{\partial t} + \nabla \cdot \overline{\mathbf{u}\Psi} = \kappa \nabla^2 \overline{\Psi}. \tag{10}$$

This is, of course, not a closed equation because of the appearance of the term $\overline{\mathbf{u}\Psi}$. However, applying the result of (5) to (10) and keeping only the first two terms in the expansion, we find

$$\frac{\partial \overline{\Psi}}{\partial t} + \nabla \cdot \overline{\mathbf{u}}\,\overline{\Psi} = -\frac{\sigma^2}{2} \overline{S}_{ij} \frac{\partial^2 \overline{\Psi}}{\partial x_i \partial x_j} + \kappa \nabla^2 \overline{\Psi}, \tag{11}$$

where $\overline{S}_{ij}$ is the strain-rate tensor of the filtered velocity field,

$$\overline{S}_{ij} = \frac{1}{2} \left(\frac{\partial \overline{u}_i}{\partial x_j} + \frac{\partial \overline{u}_j}{\partial x_i} \right). \tag{12}$$

Thus $\overline{S}_{ij}$ plays the role of a diffusion tensor for the filtered scalar field. (In general, approximating the term $\overline{u_i \Phi}$ in the transport equation for Φ by the first two terms of the equivalent expansion gives the tensor diffusivity model[7–9] or TDM for that filtered product.) To help understand the effect of this subgrid model we can transform, at any particular $\mathbf{x}$, to the principle coordinates of $\overline{S}_{ij}$ say $\mathbf{x}'$. We find that the first term on the RHS of (11) becomes

$$-\frac{\sigma^2}{2} \overline{S}_{ij} \frac{\partial^2 \overline{\Psi}}{\partial x_i \partial x_j} = -\frac{\sigma^2}{2} \left(\overline{\alpha} \frac{\partial^2 \overline{\Psi}}{\partial x_1'^2} + \overline{\beta} \frac{\partial^2 \overline{\Psi}}{\partial x_2'^2} + \overline{\gamma} \frac{\partial^2 \overline{\Psi}}{\partial x_3'^2} \right) \tag{13}$$

where the eigenvalues of $\overline{S}_{ij}$ satisfy

$$\begin{aligned} &\overline{\alpha} \geq \overline{\beta} \geq \overline{\gamma} \\ &\overline{\alpha} + \overline{\beta} + \overline{\gamma} = 0, \end{aligned} \tag{14}$$

so that one has negative diffusion along the stretching direction(s), x_1' and possibly x_2'. Therefore implementation of this model could be a delicate matter as will be discussed later.

The TDM has number of favorable characteristics. In an experimental study, Liu *et al.*[10] found good correlation between subgrid stresses and the TDM for the turbulent jet as did Sullivan *et al.*[11] for the atmospheric boundary layer. In numerical simulations of forced homogeneous isotropic turbulence, Borue & Orszag[9] found a significant improvement in the correlation when comparing the TDM predictions to those of the Smagorinsky model that assumes gradient diffusion with a turbulent viscosity ν_T given by

$$\nu_T = (c_s\Delta)^2(2\overline{S}_{ij}\overline{S}_{ij})^{1/2}, \tag{15}$$

where c_s is the Smagorinsky constant, typically $c_s \approx 0.2$, and Δ is the grid spacing used in the simulation. One of their comparisons is shown in Fig. 1 where the correlation coefficient between the model predictions and the actual subgrid stresses is 0.27 for the Smagorinski model and 0.88 for the tensor diffusivity model. The TDM stresses are not material-frame indifferent but neither are the actual subgrid stresses. However, as shown by AL & Winckelmans[12], the divergence of the TDM stresses is material-frame indifferent, as it should be.

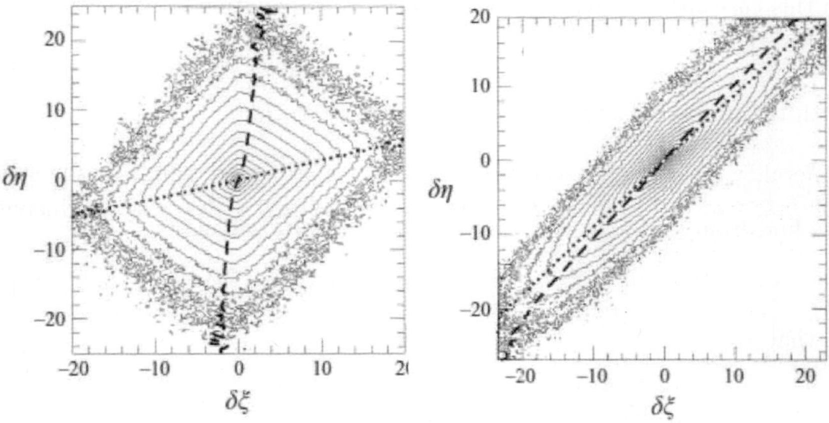

Fig. 1. Joint PDF's between the Smagorinski eddy viscosity tensor (left) and the tensor diffusivity (right) for their $j \neq i$ components ξ and the same components of the actual subgrid-scale stress tensor η (from Borue & Orszag[9]).

By considering the structure of the triple correlation function in homogeneous, isotropic turbulence AL[8] estimated the turbulence energy dissipation rate due to the TDM as $\epsilon_{TDM} \approx 0.6\epsilon \pm 0.2\epsilon$, where ϵ is the total dissipation rate. Winckelmans *et al.*[13] indeed found in isotropic decay that the TDM alone did produce energy dissipation without numerical instability as long as the skewness of the velocity derivative remains negative. However, they found that additional dissipation was needed to achieve good results for the rate of turbulence decay. Supplementing TDM with a Smagorinsky model worked reasonably well. Other authors have explored the use of the TDM and have proposed the elimination of the energy backscatter from the TDM as a way to provide the required additional dissipation[14,15]. Referring to Eq. (13), we see that this procedure would effectively eliminate the term in the TDM proportional to $\overline{\alpha}$ and possibly to the term proportional to $\overline{\beta}$. However the following application of the TDM seems to contraindicate the use of this latter strategy, i.e., we shall see that, in this special case, the TDM provides the exact backscatter and also the exact forward scatter.

Burgers' vortex

Consider the dynamics of a Burgers' vortex subjected to a time-dependent axisymmetric velocity field $\mathbf{U}(\mathbf{x}, t) = (U_r, U_\theta, U_z) = (-b(t)r\mathbf{e}_r/2, 0, b(t)z\mathbf{e}_z)$ producing an unsteady strain rate on the vortex. In this case the 2D inviscid vorticity transport equation reduces to

$$\frac{\partial \omega}{\partial t} + U_r \frac{\partial \omega}{\partial r} = \omega \frac{\partial U_z}{\partial z}, \tag{16}$$

with the unfiltered ω having a Gaussian distribution:

$$\omega(r, t) = \frac{\Gamma}{\pi \sigma_u^2(t)} e^{-r^2/\sigma_u^2(t)}. \tag{17}$$

We find from (16) that the unfiltered width $\sigma_u(t)$ satisfies

$$\frac{d\sigma_u^2(t)}{dt} = -b(t)\sigma_u^2(t) \tag{18}$$

so that

$$\sigma_u^2(t) = \sigma_{u0}^2 e^{-\int_0^t b(t')dt'} \tag{19}$$

where σ_{u0} is the initial width of the Gaussian in (17). Thus the exact filtered solution will also be Gaussian with width σ_f given by

$$\sigma_f^2 = \sigma_u^2 + \sigma^2. \tag{20}$$

Because the curl operator commutes with the the filtering operation (1) we have

$$\overline{\nabla \times (\boldsymbol{\omega} \times \mathbf{u})} = \nabla \times (\overline{\boldsymbol{\omega} \times \mathbf{u}}) = \nabla \times (\overline{\boldsymbol{\omega}} \times \overline{\mathbf{u}}) + \frac{\sigma^2}{2} \nabla \times \left(\frac{\partial \overline{\boldsymbol{\omega}}}{\partial x_i} \times \frac{\partial \overline{\mathbf{u}}}{\partial x_i} \right) + O(\sigma^4) \tag{21}$$

so that the filtered vorticity equation with the TDM is found to be

$$\frac{\partial \overline{\omega}}{\partial t} + U_r \frac{\partial \overline{\omega}}{\partial r} = \overline{\omega} \frac{\partial U_z}{\partial z} - \frac{\sigma^2}{2} \frac{\partial U_r}{\partial r} \frac{1}{r} \frac{\partial}{\partial r} \left(r \frac{\partial \overline{\omega}}{\partial r} \right), \tag{22}$$

where we have used the fact that $\overline{U}_r = U_r$ and that $\overline{U}_z = U_z$. If $\overline{\omega}$ has an initial Gaussian distribution then the solution to (22) above remains Gaussian with a width $\sigma_t(t)$ satisfying

$$\frac{d\sigma_t(t)^2}{dt} = -b(t)(\sigma_t(t)^2 - \sigma^2). \tag{23}$$

Comparing (23) with (20), we see that $\sigma_t(t) = \sigma_f(t)$ for all time, i.e., *the TDM yields the exact filtered solution for the Burgers' vortex with arbitrary stretching rate $b(t)$.*

The Smagorinsky model applied to an axisymmetric vortex produces a turbulent viscosity ν_T given by (15) with

$$2\overline{S}_{ij}\overline{S}_{ij} = 3b^2(t) + \left[r \frac{\partial}{\partial r} \left(\frac{\overline{u}_\phi}{r} \right) \right]^2, \tag{24}$$

where $\overline{u}_\phi$ is the filtered azimuthal velocity component,

$$\overline{u}_\phi = \frac{1}{r} \int_0^r \overline{\omega}(r', t) r' dr'. \tag{25}$$

Note that the second term on the RHS above produces dissipation even when $b(t) = 0$. And, because of this term, an initially Gaussian vortex will not remain Gaussian. To proceed further with this example let us assume that the stretch rate is large enough so that

$$b^2(t) >> \left[r \frac{\partial}{\partial r} \left(\frac{\overline{u}_\phi}{r} \right) \right]^2 = O\left(\frac{\Gamma^2}{\sigma_s^4} \right), \tag{26}$$

and, hence, an initially Gaussian vortex will remain approximately Gaussian using the Smagorinsky model. Here $\sigma_s(t)$ is the width of the Gaussian for this model.

With the approximation (26) ν_T depends on time only with the result that $\sigma_s(t)$ satisfies

$$\frac{d\sigma_s^2(t)}{dt} = -b(t)\sigma_s^2(t) + 4\nu_T. \tag{27}$$

By choosing $(c_s\Delta)^2 = \sigma^2/4\sqrt{3}$ we find that (27) becomes

$$\begin{aligned} \frac{d\sigma_s^2(t)}{dt} &= -b(t)(\sigma_s^2(t) - \sigma^2) \quad (b(t) > 0) \\ &= -b(t)(\sigma_s^2(t) + \sigma^2) \quad (b(t) < 0), \end{aligned} \tag{28}$$

i.e., $\sigma_s(t)$ satisfies the exact dynamics of $\sigma_f(t)$ and $\sigma_t(t)$ during positive stretching ($b(t) > 0$) but is excessively enlarged when $b(t) > 0$. In other words, energy is forward scattered to the subgrid scales during contraction of the vortex ($b(t) < 0$) rather than backscattered as it should be. Figure 2 illustrates a particular case.

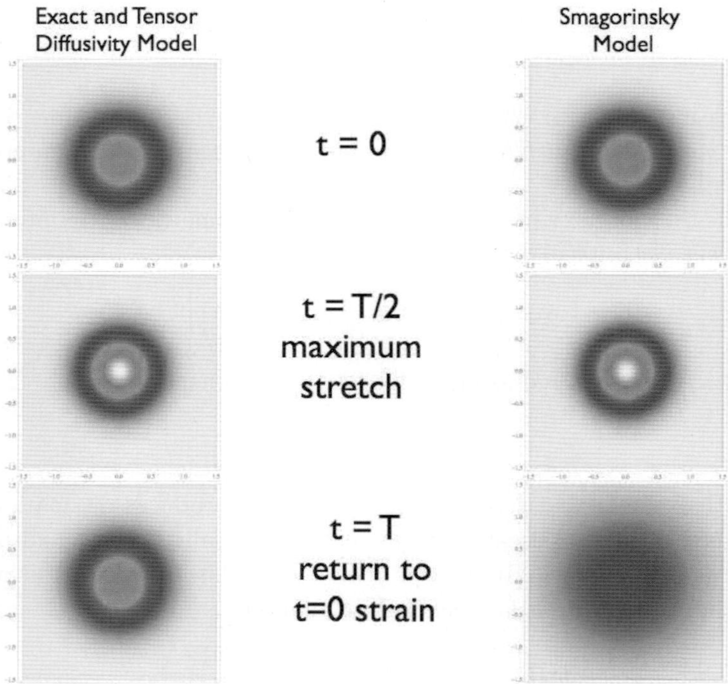

Fig. 2. The filtered vorticity field of a Burgers' vortex. Left side: exact and tensor diffusivity model. Right side: Smagorinsky model with the approximation (26). Filter width $\sigma^2 = 1/4$, initial unfiltered width $\sigma_{u0}^2 = 1/12$, maximum stretch $= \exp\left(\int_0^{T/2} b(t)dt\right) = 5$. Net stretch $= 1$ at $t = T$.

Our conclusion then is that the tensor diffusivity model is a worthy component of a subgrid model when its implementation does not lead to spurious effects due to its anti-diffusive nature and should be supplemented with a dissipative term, most likely $O(\sigma^4)$ yet to be determined.

3. Lagrangian Particle Methods

In this section we outline the use of Lagrangian particle methods for the scalar and the vorticity transport equations and consider the truncation error of these schemes. (See, e.g., Cottet & Koumoutsakos[16] and Winckelmans[17].) We will see that there is an interesting connection to the tensor diffusivity model (TDM) and will make use of it.

For a concrete example, consider the 2D inviscid vorticity transport equation for the vorticity field $\boldsymbol{\omega}(\mathbf{x},t) = \omega(\mathbf{x},t)\mathbf{e}_z$:

$$\frac{\partial \omega}{\partial t} + u_j(\mathbf{x},t)\frac{\partial \omega}{\partial x_j} = 0, \tag{29}$$

where it is understood that the velocity field is computed from the Poisson equation $\nabla^2 \mathbf{u} = -\nabla \times \boldsymbol{\omega}$. Let the the vorticity field be represented by a collection of particles thus

$$\omega(\mathbf{x},t) = \sum_k \alpha_k \gamma(\mathbf{x} - \mathbf{x}_k(t)), \tag{30}$$

where γ is the particle distribution function chosen here to be Gaussian with width σ,

$$\gamma(\mathbf{z}) = \frac{\exp(-|\mathbf{z}|^2/\sigma^2)}{\pi\sigma^2}. \tag{31}$$

For the 2D inviscid case the α_k's remain constant in time and if we move the particles with the velocity at the center of each particle:

$$\frac{d\mathbf{x}_k(t)}{dt} = \mathbf{u}(\mathbf{x}_k(t),t), \tag{32}$$

the above scheme (30) and (32) solves the following equation exactly:

$$\frac{\partial \omega}{\partial t} + \sum_k \alpha_k u_j(\mathbf{x}_k(t),t)\frac{\partial \gamma(\mathbf{x} - \mathbf{x}_k(t))}{\partial x_j} = 0. \tag{33}$$

Thus the truncation error ϵ of the scheme is given by

$$\epsilon = \sum_k \alpha_k \left[u_j(\mathbf{x}_k(t),t) - u_j(\mathbf{x},t)\right] \frac{\partial \gamma(\mathbf{x} - \mathbf{x}_k(t))}{\partial x_j}. \tag{34}$$

Using a Taylor series expansion for $\mathbf{u}(\mathbf{x}_k(t),t)$,

$$\mathbf{u}(\mathbf{x}_k,t) = \mathbf{u}(\mathbf{x},t) + (\mathbf{x}_k - \mathbf{x})_m \frac{\partial \mathbf{u}(\mathbf{x},t)}{\partial x_m} + O(|\mathbf{x}_k - \mathbf{x}|^2), \tag{35}$$

and the repeated use of the relation

$$\mathbf{z}\gamma(\mathbf{z}) = -\frac{\sigma^2}{2}\nabla\gamma(\mathbf{z}), \tag{36}$$

we find that

$$\epsilon = \frac{\sigma^2}{2}\frac{\partial u_l}{\partial x_m}\frac{\partial^2 \omega}{\partial x_l \partial x_m} + \frac{\sigma^2}{4}\nabla^2 u_l \frac{\partial \omega}{\partial x_l} + O(\sigma^4). \tag{37}$$

Because ω is a 2D vorticity field the second term on the RHS above is zero so that the modified 2D transport equation for vorticity is found to be

$$\frac{\partial \omega}{\partial t} + u_j(\mathbf{x}, t)\frac{\partial \omega}{\partial x_j} + \frac{\sigma^2}{2}\frac{\partial u_l}{\partial x_m}\frac{\partial^2 \omega}{\partial x_l \partial x_m} = O(\sigma^4). \tag{38}$$

Thus the Lagrangian method for the 2D vorticity equation produces implicitly the tensor diffusivity model[14,18]. It is important to note that we have chosen Gaussian particles for the Lagrangian method with width σ and the TDM is based on a Gaussian filter with the same width σ.

For the scalar advection equation and for the 3D vorticity transport equation the second term on the RHS of (37) does not vanish in general but (A) modifying the particle velocity from (22) to

$$\frac{d\mathbf{x}_k(t)}{dt} = \mathbf{u}(\mathbf{x}_k(t), t) - \frac{\sigma^2}{4}\nabla^2 \overline{u}(\mathbf{x})|_{\mathbf{x}=\mathbf{x}_k} \tag{39}$$

and (B) modifying the velocity gradient as follows for the dynamics of the vector strengths of the 3D vortex particles

$$\frac{d\boldsymbol{\alpha}_k}{dt} = \overline{\omega}_j(\mathbf{x}_k)\left[\frac{\partial \overline{\mathbf{u}}}{\partial x_j}(\mathbf{x})|_{\mathbf{x}=\mathbf{x}_k} - \frac{\sigma^2}{4}\frac{\partial \nabla^2 \overline{\mathbf{u}}}{\partial x_j}(\mathbf{x})|_{\mathbf{x}=\mathbf{x}_k}\right] \tag{40}$$

yields the TDM for both the scale advection equation and the 3D vorticity transport equation to order $O(\sigma^4)$. These modifications are related to the fact that an unfiltered scalar ϕ maybe partially recovered from the filtered field $\overline{\phi}$ by

$$\phi = \overline{\phi} - \frac{\sigma^2}{4}\nabla^2 \overline{\phi} + O(\sigma^4). \tag{41}$$

As mentioned in the previous section, the TDM needs to be supplemented with a dissipative term. It is possible that such dissipation could be achieved implicitly by the use of particle redistribution schemes, which, in general, are required by Lagrangian methods to maintain adequately small spacing between particles in all directions. See Fig. 2 of Winckelmans[17] that suggests that the use of a third-order redistribution scheme may supply the additional dissipation and yield good results at least for the 2D vorticity transport equation.

4. Super-Resolution

Super-resolution in our application would be the process of reconstructing the unfiltered field from knowing only the filtered field (Candes & Fernandez-Granda[19]). This would seem to be an ideal candidate for building a subgrid model. By reconstructing u from $\overline{u}$ and v from $\overline{v}$ one could filter the product uv to determine $\overline{uv}$. It turns out that this is possible, under special circumstances, when one knows the spatial structure of the elements that make up the unfiltered field.

The following is a simple 1D example from Ref. 19. Suppose $f(x)$ is periodic in x with period 1 and is composed of N δ functions:

$$f(x) = \sum_{j=1}^{N} a_j \delta(x - x_j). \tag{42}$$

The Fourier transform $\hat{f}(k)$ is

$$\begin{aligned} \hat{f}(k) &= \int_0^1 e^{-i2\pi kx} f(x)dx \\ &= \sum_{j=1}^{N} a_j e^{-i2\pi kx_j} \end{aligned} \tag{43}$$

In this example the coarse information that is known are the $n = 2f_c + 1$ lowest frequency components of $\hat{f}(k)$ we denote by $\hat{\overline{f}}(k)$ Or, equivalently, we know $\overline{f}(x)$ the inverse transform of $\hat{\overline{f}}(k)$. This is illustrated in Figs. 3 and 4. *But also known is the fact that $f(x)$ is a collection of δ functions.* What is not known is N, the number of δ functions, their strengths a_j (possibly complex), and their locations x_j. Recovery of the exact $f(x)$ is then possible if the minimum separation between the elements Δ satisfies $\Delta \geq 2/f_c$.

The unique solution $f(x)$ is then given by

$$\underset{f}{min} \, ||f||_{TV} = \underset{f}{min} \sum_{j=1}^{N} |a_j| \text{ subject to given } \hat{\overline{f}}(k), \tag{44}$$

where TV stands for Total Variation. In Ref. 19 it is shown how to transform the minimization problem (44) into a convex programming problem in $(n+1)^2/2$ variables.

We now present a similar example closer to a subgrid modeling problem in fluid mechanics. Consider a 2D vorticity field composed of N point

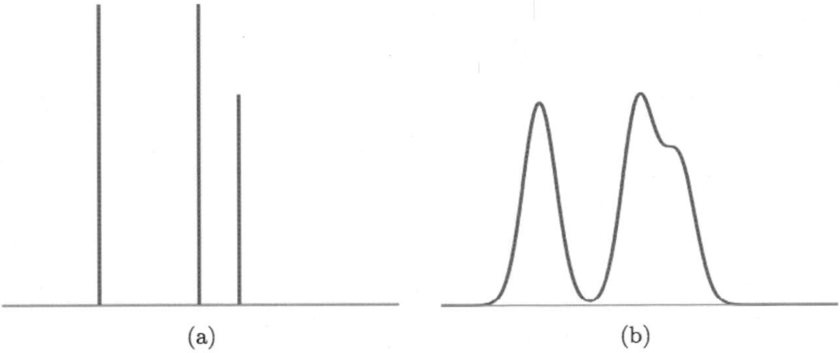

Fig. 3. (a) Schematic of the unfiltered 1D function $f(x)$, (b) filtered function $\overline{f}(x)$ (from Ref. 19).

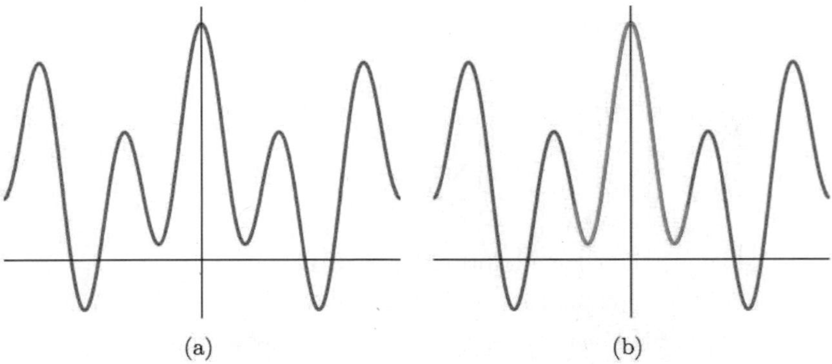

Fig. 4. Real part of the (a) Fourier-transformed function $\hat{f}(k)$, (b) Fourier transform of the filtered function $\hat{\overline{f}}(k)$ (from Ref. 19).

vortices:

$$\omega(\mathbf{x}, t) = \sum_{j=1}^{N} \Gamma_j \delta(\mathbf{x} - \mathbf{x}_j(t)) \tag{45}$$

which, assuming an inviscid flow, move according to point-vortex dynamics. (See (47) below.) The subgrid problem would then be to determine N, the Γ_j's, and the $\mathbf{x}_j$'s from the filtered field $\overline{\omega}$ given by

$$\overline{\omega}(\mathbf{x}, t) = \sum_{j=1}^{N} \Gamma_j \frac{e^{-|\mathbf{x}-\mathbf{x}_j(t)|^2/\sigma^2}}{\pi\sigma^2}. \tag{46}$$

Figure 5 illustrates this problem. Note that, based on the constraint concerning the 1D example discussed above, the minimum spacing between vortices will have to be greater than some constant times the filter width σ for a unique solution to be recoverable.

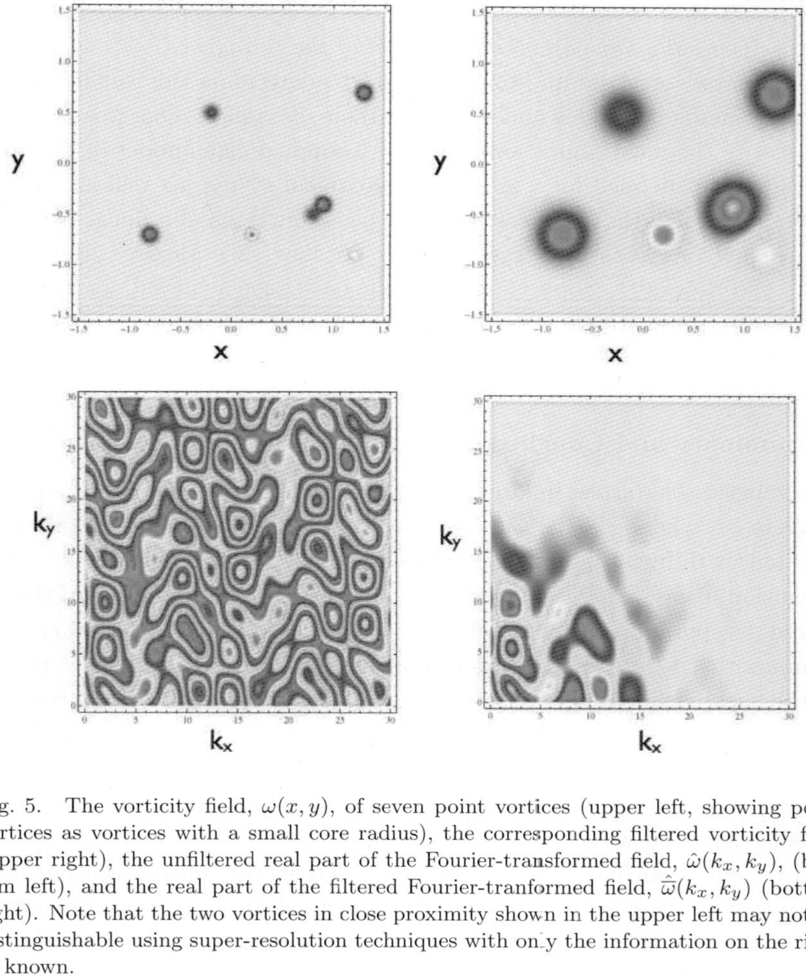

Fig. 5. The vorticity field, $\omega(x, y)$, of seven point vortices (upper left, showing point vortices as vortices with a small core radius), the corresponding filtered vorticity field (upper right), the unfiltered real part of the Fourier-transformed field, $\hat{\omega}(k_x, k_y)$, (bottom left), and the real part of the filtered Fourier-tranformed field, $\hat{\bar{\omega}}(k_x, k_y)$ (bottom right). Note that the two vortices in close proximity shown in the upper left may not be distinguishable using super-resolution techniques with only the information on the right as known.

However, also note that in this case – inviscid 2D flow with point vortices – one could avoid the subgrid modeling problem altogether by simply tracking the N point vortices with the $2N$ equations of motion,

$$\frac{d\mathbf{x}_i(t)}{dt} = \frac{1}{2\pi}\sum_{j\neq i}^{N}\frac{(\mathbf{x}_j(t)-\mathbf{x}_i(t))\times\mathbf{e}_z\Gamma_j}{|\mathbf{x}_j(t)-\mathbf{x}_i(t)|^2}. \tag{47}$$

Thus, in general, if a turbulent field could be well-represented by a small[a] *number of known components of the field and if their mutual dynamics could also be determined then one can bypass the need for subgrid modeling. One would essentially have a low-order model of the turbulence.*

Of course, now the questions are how to determine this small number of components or modes and what are their dynamics. One possibility is to identify the most important so-called Koopman eigenmodes of the flow. These modes, defined for a nonlinear dynamical system as well as a linear system, can play the same role that the eigenfunctions of a linear operator can, namely serve as basis functions for the flow. See Mezić[20], Rowley *et al.*[21], and Rowley[22]. Some enticing results have been achieved, for example, in the case jet in crossflow[21] but much more needs to be done to establish a low-order model of turbulence.

5. Summary and Conclusions

The subgrid-scale stress in large-eddy simulation may be written as a power series in σ^2, where σ is the width of the filter used to define the large scales. The tensor diffusivity model, corresponding to the first two terms in this expansion, has several useful features. It provides a good fraction of the required dissipation in turbulent flows with the additional dissipation coming from another term, yet to be determined, that is probably $O(\sigma^4)$. It also yields backscatter of energy into the large scales during local contraction or thickening of vortex tubes, as illustrated by the Burgers' vortex example presented above. Care must be taken, however, in implementing this model because of its anti-diffusive property along principal axes of the strain-rate tensor associated with positive extension.

In Lagrangian particle methods discussed in this paper, the particles carry vorticity times volume in the case of the vorticity transport equation and scalar density times volume in the case of the scalar advection-diffusion equation. The leading order truncation error for these methods is precisely the term corresponding to the tensor diffusivity model, as long as the particles are transported with the local velocity at the center of each particle

[a]With today's turbulence simulations using $(n000)^3$ grid points only $n000$ components would be "small" in comparison.

for the 2D vorticity equation and with given $O(\sigma^2)$ correction for the 3D vorticity equation and for the scalar equation. Likewise, for the 3D vorticity equation, the velocity gradient tensor for the vortex stretching term must be evaluated an $O(\sigma^2)$ distance away from the center of the particle.

The possibility of using super-resolution, i.e. recovering fine-scale information knowing only coarse-scale information, to construct a subgrid model was investigated. This is a promising approach that needs much further investigation but requires *a priori* knowledge of the structure of the elements that make up the turbulent field.

References

1. P. R. Spalart, Detached-Eddy Simulation, *Annu. Rev. Fluid Mech.* **41**, 181 (2009).
2. A. J. Majda, C. Franzke and B. Khouider, An applied mathematics perspective on stochastic modeling for climate, *Phil. Trans. R. Soc. A* **366**, 2429 (2008).
3. S. Stolz and N. A. Adams, An approximate deconvolution procedure for large-eddy simulation, *Phys. Fluids* **11**, 1699 (1999).
4. W. K. Yeo, A generalized high pass/low pass filtering procedure for deriving and solving turbulent flow equations, PhD thesis, Ohio State University, (1988).
5. W. K. Yeo and K. W. Bedford, *Closure-free turbulence modeling based upon a conjunctive higher order averaging procedure*, in *Computational Methods in Flow Analysis*, eds. H. Niki and M. Kawahara, (Okayama University of Science, 1988), pp. 844–851.
6. D. Carati, G. S. Winckelmans and H. Jeanmart, On the modeling of the subgrid-scale and filtered-scale stress tensors in large-eddy simulation, *J. Fluid Mech.* **441**, 119 (2001).
7. A. Leonard, Large-eddy Simulation of Chaotic Convection and Beyond, AIAA Paper 97-0204 (1997).
8. A. Leonard, Energy Cascade in Large-eddy Simulations of Turbulent Fluid Flows, *Adv. Geophys.* **18 A**, 237 (1974).
9. V. Borue and S. A. Orszag, Local energy flux and subgrd-scale statistics in three-dimensional turbulence, *J. Fluid Mech.* **366**, 1 (1998).
10. S. Liu, C. Meneveau and J. Katz, On the properties of similarity subgrid-scale models as deduced from measurements in a turbulent jet, *J. Fluid Mech.* **275**, 83 (1994).
11. P. P. Sullivan, T. W. Horst, D. H. Lenschow, C.-H. Moeng and

J. C. Weil, Structure of subfilter-scale fluxes in the atmospheric surface layer with application to LES modeling, *J. Fluid Mech.* **482**, 101 (2003).

12. A. Leonard and G. S. Winckelmans, *A Tensor Diffusivity Subgrid model for LES*, in *Direct and Large-Eddy Simulation III*, eds. P. R. Voke, N. D. Sandham and L. Kleiser, (ERCOFTAC, vol. 7, Kluwer, 1988), pp. 147–162.
13. G. S. Winckelmans, A. A. Wray, O. V. Vasilyev and H. Jeanmart, Explicit-filtering large-eddy simualtion using the tensor-diffusivity model supplemented by a dynamic Smagorinsky term, *Phys. Fluids* **13**, 1385 (2001).
14. G.-H. Cottet, Artificial Viscosity Models for Vortex and Particle Methods, *J. Comput. Phys.* **127**, 299 (1996).
15. G. Balarac, J. Le Sommer, X. Meunier and A. Vollant, A dynamic regularized gradient model of the subgrid-scale scalar flux for large eddy simulations, *Phys. Fluids* **25**, 075107 (2013).
16. G.-H. Cottet and P. Koumoutsakos, *Vortex Methods, Theory and Practice* (Cambridge University Press, 2000).
17. G. S. Winckelmans, *Chapter 5 Vortex Methods*, in *Encyclopedia of Computational Mechanics*, eds. E. Stein, R. de Borst and T. J. R. Hughes, Volume 3:*Fluids*. (John Wiley & Sons, 2004), pp. 129–153.
18. A. Leonard, Vortex Methods for Flow Simulation, *J. Comput. Phys.* **37**, 289 (1980).
19. E. J. Candes and C. Fernandez-Granda, Towards a Mathematical Theory of Super-Resolution, *Comm. Pure Appl. Math.* **67**, 906 (2014).
20. I. Mezić, Analysis of Fluid Flows via Spectral Properties of the Koopman Operator, *Annu. Rev. Fluid Mech.* **45**, 357 (2013).
21. C. W. Rowley, I. Mezić, S. Bagheri, P. Schlatter and D. S. Henningson, Spectral analysis of nonlinear flows, *J. Fluid Mech.* **641**, 115 (2009).
22. C. W. Rowley, These Proceedings.

Convective Transport in the Sun

Shravan M. Hanasoge
Tata Institute of Fundamental Research, Mumbai 400005, India

Laurent Gizon
Max-Planck-Institut für Sonnensystemforschung, Katlenburg-Lindau, Germany
Georg-August-Universität, Institut für Astrophysik, Göttingen, Germany

Katepalli R. Sreenivasan
Department of Physics, Department of Mechanical Engineering and the Courant Institute of Mathematical Sciences, New York University, New York, USA
Email: krs3@nyu.edu

The convective envelope of the Sun exhibits many interesting fluid dynamical phenomena. Some of them are outlined here: differential rotation, weak poleward meridional circulation, thermal transport, helioseismic constraints on velocities, Reynolds stresses, momentum transport and the plume models for the large-scale motion. We consider these problems briefly from a fluid dynamics perspective and comment on the impact of the numerous observations of the Sun on our understanding of these properties.

Keywords: Sun's interior motion, helioseismic constraints on velocities, observations and modeling

1. Introduction

The transport of solar luminosity from the core outwards is accomplished by radiation up to approximately $0.7R_\odot$, where $R_\odot \sim 6.96 \times 10^8$ m is the solar radius. At this radius, the opacity of solar plasma increases to an extent that photon scattering becomes significant, rendering radiative transport relatively ineffectual. A thick thermal boundary layer thus forms (purely radiative transport transitions to purely convective transport over a few density scale heights) and energy transport from this region out to the solar photosphere (surface) is largely accomplished by thermal convection. The parameters governing fluid motion in the solar convection zone lie in an extraordinary regime: Reynolds number, $Re \sim 10^{10}$, Rayleigh number, $Ra \sim 10^{24}$, thermal Prandtl number, $Pr \sim 10^{-6}$, and the magnetic

Prandtl number, $Pr_m \sim 10^{-6}$. (Since huge property variations occur across the depth of the convection region, these numbers are only representative.) Wide ranges of spatial and temporal scales that may be expected from these parameters suggest that direct numerical simulations of solar convection will not be feasible in the foreseeable future. Thus a possible path for progress lies in enunciating a phenomenology that is consistent with observations and can explain large-scale phenomena such as differential rotation and meridional circulation (to be explained below). Because the Sun has been observed and studied extensively, it is an excellent object on which to test hypotheses pertaining to convection and turbulence at high Reynolds numbers in a rotating and stratified fluid. We mention a few relevant issues here; a longer discussion of these and other related issues can be found in forthcoming article in 2016 in Annual Review of Fluid Mechanics, by the same authors[1].

2. Differential Rotation

A celebrated inference of helioseismology (i.e., the seismology of the Sun) is differential rotation: measurements of global modes of oscillation of the Sun reveal that the equator rotates at a faster rate than the poles, as depicted[2] in Figure 1. The oscillations of the Sun are of small amplitude compared to the sound speed of the medium and linear theory describes their behavior with great accuracy[3]. The contours of constant rotation are approximately radially directed, and not axially aligned Taylor-Proudman cylinders[2]. The latitude-dependent specific angular momentum of the Sun is given by $\mathcal{L} = \Omega^2 r^2 \sin^2 \theta$, where θ is spherical co-latitude, r is radius and $\Omega = \Omega(r, \theta)$ is the rotation rate. For solid body rotators, the angular momentum associated with the equator ($\theta = \pi/2$) is greater than at the poles ($\theta = 0$); and for a solar-like rotator, the equator-to-pole angular momentum difference is even larger. The mechanism governing the establishment of this form of differential rotation is an outstanding question in solar and stellar hydrodynamics.

3. Meridional Circulation

It was discovered[5] that a weak equator-to-pole meridional circulation persists at the surface or photosphere ($\sim 20\,\mathrm{m\,s^{-1}}$). The circulation transports parcels of high angular momentum from the equator towards the poles (at the near-surface layers) and back towards the equator at deeper layers of the Sun. The overall action is to erase (or moderate) the observed rotational

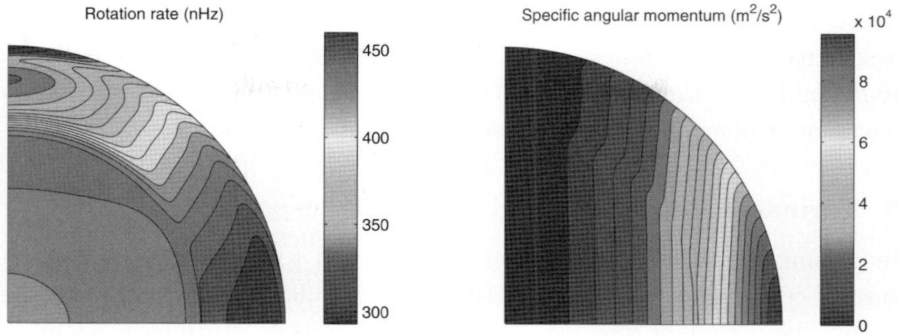

Fig. 1. Angular velocity (left panel) and the specific angular momentum (right panel) shown as a function of latitude and radius, as inferred by helioseismology using data taken by the Michelson Doppler Imager[4]. The outer envelope is seen to rotate differentially and rotation contours are not parallel to the rotation axis as Taylor-Proudman balance might suggest. Courtesy, H.M. Antia.

shear, spin the pole up relative to the equator, thereby creating anti-solar differential rotation. An argument more general than this can be made: *any* circulation along the meridian would spin the pole up relative to the equator[6,7].

While this meridional circulation may well straddle the entire depth of the convection zone (i.e., from about $0.7R_\odot$ outwards), the deepest layer which it plumbs is a subject of some controversy[8]. Thus, it is fair to conclude at this stage that the return depth of meridional circulation is not conclusively known. This is a challenging problem to model.

4. Reversal of the Solar Magnetic Field and Sunspots

The Sun supports a global magnetic field that reverses polarity every (roughly) 11 years. The regularity of pole reversal is quite in contrast to Earth's stochastic reversals (which are also far less frequent). At the beginning of the cycle (the minimum), the field is largely poloidal while a toroidal field is observed on the surface during the course of the solar cycle (maximum). Understanding the dynamo of the Sun will prove insightful in appreciating field generation mechanisms in a variety of stars (especially those that are Sun-like) across our galaxy, for which there are now an abundance of observations[9]. While a number of mechanisms to explain the formation of large-scale magnetism have been proposed[10], one widely regarded and compelling theory is the Babcock-Leighton flux-transport dynamo[11], which relies on rotational shear associated with differential rotation to enable the poloidal-toroidal field transformation. Much as an elastic band resists stretching, so too would the magnetic field act to diminish the shear associated with differential rotation. However, seismic observations show a remarkably constant differential rotation, leaving us with the fundamental question of how differential rotation is set up and sustained.

5. Turbulent Convection and Energy Transport

Increasingly, seismic investigations have turned towards inferring the nature of coherent convective structures in the bulk of the solar convection zone. The dominant view has been that vigorous overturning motions (in the sense of mixing-length theory) are ubiquitous in the bulk and form the primary coherent structures associated with convection. This flow field would likely possess significant Reynolds stresses which would support the observed large-scale fluid circulations in the Sun. It would also be consistent with mixing length theory which posits that hot parcels of fluid rise, exchange their heat with their new surroundings and fall back as cold downflows. We term this sequence of actions as "overturning motions".

However, seismic and other investigations have attempted for decades to record these large-scale motions with no success[12]. These failed efforts, combined with results from high-resolution surface convection simulations, led to alternate theories based on coherent descending plumes from the surface, in mass balance with weak up-flow from the deep convection zone[13]. These convective flows would possess much lower turbulent kinetic energy in the bulk, consistent with the apparent absence of large-scale coherent overturning convection. The lower kinetic energy implies that the total amount of thermal flux would be much smaller than in the overturning convection case. Motions with such low kinetic energy are associated with weak Reynolds stresses, which means that there are no obvious means of sustaining large-scale solar fluid circulations. Recently (see below), seismic upper bounds were placed on the convective power spectrum[14] at a depth of $r/R_\odot = 0.96$ (28 Mm or 28,000 km below the photosphere). An interesting problem is to investigate whether the descending plumes are in line with these seismic constraints and whether they can generate the appropriate amount of radially outward heat flux.

6. Helioseismic Constraints

The Sun supports a spectrum of resonant acoustic modes that are excited in the immediate sub-photosphere by convection in these layers. Although the excitation mechanism is nonlinear and there is likely to be coupling with turbulence near the photosphere, acoustic waves may be effectively treated as small-amplitude linear oscillations about the solar background[3]. These waves propagate in the interior and emerge at the surface, causing Doppler shifts of the observed absorption lines. The rate of displacement of plasma at the photosphere due to the impinging of waves may thus be recorded, e.g., as is done at approximately 16 million spatial points by the Helioseismic and Magnetic Imager HMI[15] onboard the Solar Dynamics Observatory satellite.

Wave excitation in the Sun is a stochastic process. The raw observed wavefield velocities are well modeled as a zero-mean random process, and are seismically uninformative without an exact knowledge of the full distribution of sources. Consequently, the measurements that are used are two-point correlation functions, which have non-zero expectation values[16]. Wave travel times are typically measured by fitting the measured cross-correlation functions with reference wavelets, and these in turn can be formally interpreted in terms of Green's functions of the medium[17]. In particular, waves are advected by flows associated with the convective cells through which they propagate. This shifts the travel times of waves; and interpreting the shifts allows for the inference of convective flow velocities. Wave travel times form the input to an inverse problem, i.e., determining the properties of the medium that are able to explain observed changes in wave travel times.

To make these inferences, nearly a trillion raw observations of the photospheric wave field were analyzed[14] using methods of time-distance helioseismology[16]. Because of the formidable size of this observational dataset, downsampling and averaging form an important part of the analysis[18,19]. A primary goal has been[14] to characterize flow amplitudes of large scales of convection. Failing to detect large-scale convective cells, the work has been able to place upper bounds (see Figure 2) on the flow amplitudes. The derived constraints[14] suggest significant departures from the conventional understanding of heat transport in the convection zone. Both mixing-length theory[20] and numerical simulations[21–23] require flows of much larger magnitude to transport heat outwards than the constraints allow. Indeed, this result suggests the prevalence of an intrinsically different mechanism of convection than is currently surmised. A problem of some seriousness is the thermal-transport mechanism, on which we speculate in the following section; a much more difficult issue to address is how angular momentum is redistributed in order to produce the observed differential rotation.

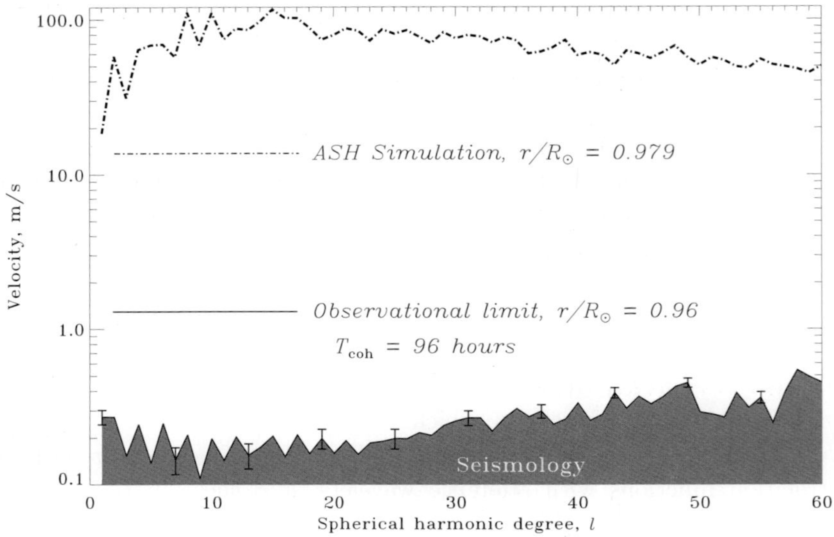

Fig. 2. Seismic upper bounds [14] on large-scale *lateral* (horizontal) convective-velocity amplitudes in the solar interior at the depth $r/R_\odot = 0.96$. These constraints are compared with spectra derived at a similar depth from global convection simulations ASH [21]. To obtain observational constraints, we assume a convective coherence time of 96 hours, consistent with estimates from the mixing length theory [20]. Assuming that the convection simulations are producing inflated amplitudes of convective velocities [7] it is possible to estimate on the basis of back-of-the-envelope calculations that a minimum convective velocity of 30 ms^{-1} (over a broad range of spatial wavenumbers) is required at a depth of $r/R_\odot = 0.96$ to support the angular momentum transport in the Sun.

7. Thermal Transport

Cold down-flow plumes formed at intergranular lanes at the photosphere could play a central role in determining the nature of deep interior convection. The stability of these descending plumes controls the extent of thermal mixing and consequently the kinetic energy of transport flows. It has been suggested [24,25] that these plumes travel all the way to the base of the convection zone, i.e., the timescale over which the cold plumes break up is longer than the travel time through the convection zone. Density and pressure increase rapidly with depth in the convection zone, implying that the plumes that are seeded at the photosphere, hemmed in by the high pressure and density of the fluid into which they enter, will decrease very slowly in size despite entraining fluid from their surroundings. The up-flows on the other hand, which form at the thick boundary layer at the base of the convection zone[26], need move only at very low speeds due to the enormous heat-carrying capacity of the high-density fluid at that depth. If descending plumes are indeed stable, the theory of overturning convection does not strictly apply since there is no transference of heat from one parcel of fluid to another. In this view, vigorous overturning convection occurs only in the vicinity of the thin upper boundary layer, where it is driven by the rapid sub-photospheric decrease in opacity. Density drops by some 3 orders in magnitude over a few hundred kilometers in the immediate sub-photosphere, resulting in rapidly thinning plasma and a region where free streaming radiation once again dominates energy transport.

8. Plume Model of Rieutord

One possible approach is to build a model of cold entraining plumes descending amid a warm diffuse upflow[24] into layers of higher density. A typical plume described in this theory is depicted in Figure 3. Their calculations show plumes that are about a factor of 10 smaller than the local seismic wavelength, and have much larger vertical velocities than lateral flow speeds. This may be attributed to weak entrainment since, if the entrainment rate were to be substantial, the plume would fall into the regime of well-mixed, overturning convection. Thus we assume an ensemble of plumes descending from the surface with predominantly radial velocities. Their relatively small size, compared to the local imaging acoustic wavelength, suggests that they may elude easy detection.

We estimate below the power spectrum of some $N \gg 1$ plumes randomly distributed on spherical surfaces. The flow field is statistically axisymmetric and so we may place individual plumes at the pole and rotate them to a random location (θ_i, ϕ_i). Since the laterally directed entrainment flows are substantially weaker than the radial flow, we set them to zero in the formulation below. Axisymmetry implies that the flow field depends only on the distance between some point to the axis of the plume. The spherical-harmonic coefficients of such a structure depend only on the degree ℓ and not on the azimuthal order m.

Denote the spherical-harmonic coefficients that describe the radial velocity of a descending plume at some latitude-longitude location (β_i, γ_i), respectively,

$$u_{\ell m} = a_\ell Y^*_{\ell m}(\beta_i, \gamma_i). \tag{1}$$

Since plumes may be at any given location on the spherical surface, the locations β_i, γ_i may be modelled as independent and identically distributed

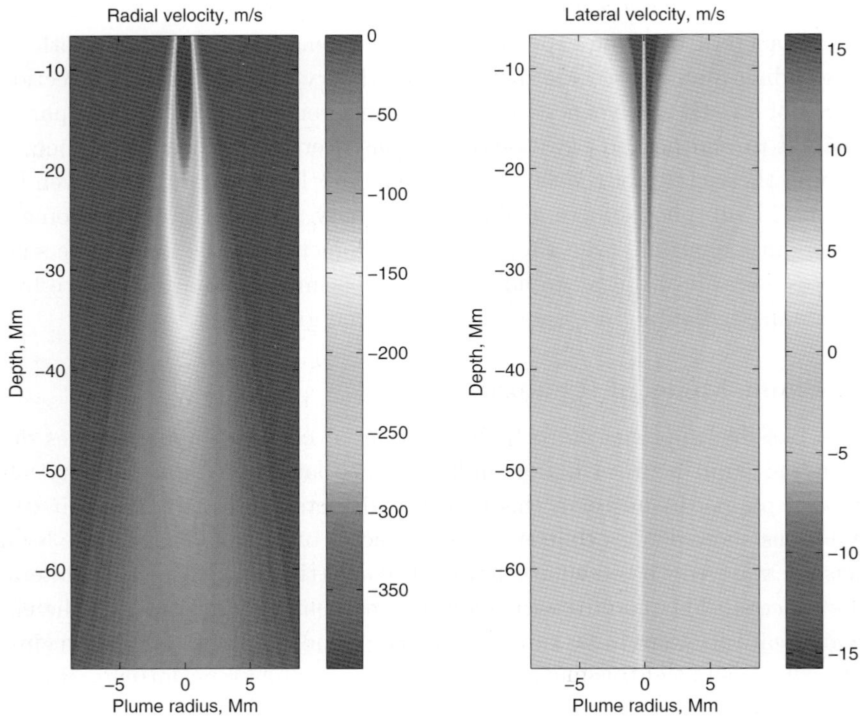

Fig. 3. Velocity profile obtained from a model of plumes [24]. The plume gradually grows in size as it descends (left panel, radial velocity) and the degree of entrainment (right panel, horizontal or lateral velocity). Despite the increasing degree of entrainment, the diameter of the plume grows slowly. If the plumes entrains sufficiently, it will destabilize and mix into the ambient, resulting in overturning convection. However, in the weakly mixing mechanism described here, the increasing density stabilizes descending plumes, preventing them from breaking up (Courtesy, M. Rieutord).

uniform random variables over the ranges $\gamma \in [0, 2\pi)$ and $\beta \in [0, \pi)$. The power spectrum is given by

$$\mathcal{P}_\ell = \sum_m |u_{\ell m}|^2 = |a_\ell|^2 \sum_{i,j=1}^{N} \sum_{m=-\ell}^{\ell} \langle Y_{\ell m}(\beta_i, \gamma_i)\, Y^*_{\ell m}(\beta_j, \gamma_j)\rangle. \tag{2}$$

Because the (i, j) represent statistically uncorrelated angle pairs (β_i, γ_i) and (β_j, γ_j), we have

$$\langle Y_{\ell m}(\beta_i, \gamma_i)\, Y^*_{\ell m}(\beta_j, \gamma_j)\rangle = \langle |Y_{\ell m}(\beta_i, \gamma_i)|^2\rangle \delta_{ij}. \tag{3}$$

Finally, using a standard spherical-harmonic identity,

$$\mathcal{P}_\ell = \sum_{i,j=1}^{N} \delta_{ij} \sum_{m=-\ell}^{\ell} \langle |Y_{\ell m}(\beta_i, \gamma_i)|^2 \rangle = \frac{2\ell+1}{4\pi} \sum_{i,j=1}^{N} \delta_{ij} = \frac{2\ell+1}{4\pi} N, \qquad (4)$$

we arrive at an estimate for the power spectrum associated with N plumes

$$\mathcal{P}_\ell = \frac{2\ell+1}{4\pi} N |a_\ell|^2. \qquad (5)$$

From this result, we can estimate the typical wave-travel-time response, by applying a multiplicative calibration function $\tau_\ell = \mathcal{C}_\ell \mathcal{P}_\ell$, where τ_ℓ is the wave travel-time response and $\mathcal{C}_\ell$ is a calibration function[14]. The larger the magnitude of flows, the larger the travel-time response, since advective flow due to thermal convection provides a phase-shift in waves. Seismic measurements intrinsically contain realization noise owing to the stochastic wave-excitation process described in section 6. Vigorous overturning convection is much larger in magnitude than the realization noise and would therefore be detectable[19]—which contradicts observations. The weakly mixing plume theory discussed here would in principle be in line with observations if the associated travel-time response is smaller in magnitude than realization noise. Indeed, we find that, as shown in Figure 4, the travel-time response is sufficiently small so that the weak-mixing would be seismically undetectable.

9. Reynolds Stresses and Momentum Transport

It is difficult to explain the maintenance of differential rotation, meridional circulation, and large-scale flow systems in the convection zone if the Reynolds stresses are small in comparison to current estimates. Differential rotation in the convective envelope is an $\ell \sim 2$ feature, with a relatively fast rotating equator. One theory has been that of a convection zone persisting in thermal wind balance, with a relatively warm pole. This is consistent with geostrophic turbulence, a consequence of weak Reynolds stresses as indicated by helioseismic bounds[14]. A latitudinal variation of entropy is capable of inducing solar-like differential rotation[27,28]. It has been estimated[27] that a very slightly warmer pole, i.e., by about 2-3 K, would be sufficient to set up solar differential rotation. A variation of the "efficiency" of convection as a function of inclination with respect to the rotation axis may result in a higher pole temperature aligned with the rotation axis[29,30]. Indeed, a faint temperature gradient, thermal in origin, has been observed at the photosphere of the Sun[31]. An alternative approach is to calculate the

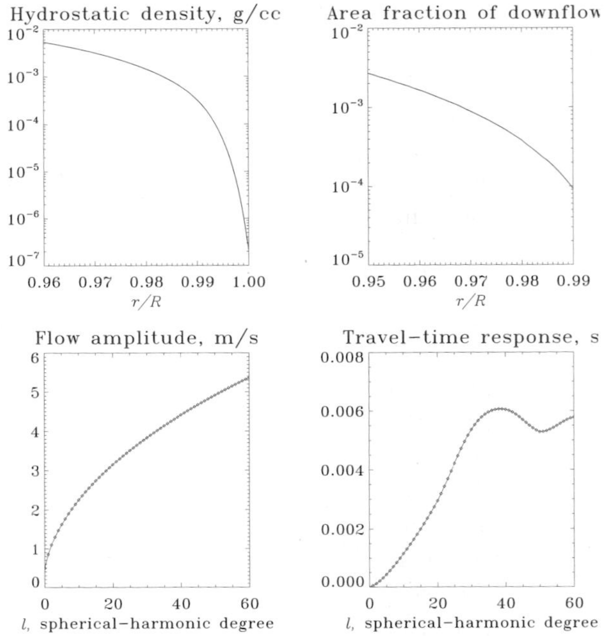

Fig. 4. Dominantly radial-heat-transport model of Section 8. Density increases by several orders in magnitude from the photosphere to interior layers (upper-left panel), acting to compress descending plumes. Entrainment overcomes this effect, resulting in an overall increase in the area occupied by the plumes (upper-right panel). The limit power spectrum of *radial* flows at $r/R_\odot = 0.96$ of an ensemble of cold descending plumes (lower-left panel) is so weak that it is practically helioseismically undetectable (lower-right panel; note also that Figure 2 shows *horizontal* or *lateral* flows, to which waves are more sensitive). We use a calibration function[14] to transform travel-time shifts to flow amplitudes. Thermal transport effected by these flows, rather than vigorous overturning convection, would be consistent with seismic constraints.

effect of centrifugal distortion at the base of the convection zone induced by the solid-body rotation of the Sun. The finding of the study[30] was that this distortion would be sufficient to set up a differential rotation.

It should be said that a latitudinal entropy gradient cannot alone account for the observed differential rotation. It has been pointed out[6,7] that a meridional circulation, regardless of its direction, would spin up the poles relative to the equator. Consequently Reynolds stresses must play some role in maintaining differential rotation and supporting meridional circulation. As already mentioned, there is definitive helioseismic evidence only for a poleward meriodional circulation in the near-surface layers; imaging the return circulation is an outstanding challenge that has thus far remained elusive[32]. Thus without seismic evidence of the variation of meridional flow amplitudes as a function of depth, the distribution of Reynolds stresses in the convection zone is not easy to assess.

The Sun exhibits cyclic magnetic activity, widely considered to be driven by the Babcock-Leighton dynamo mechanism[10,33,34] (while we focus on the Babcock-Leighton mechanism here, a variety of other models exist and the question of what governs the generation of large-scale magnetism is by no means settled[10]). The premise is that, at the beginning of the solar cycle, differential rotation at the base of the convection zone winds up the poloidal magnetic field, converting it into a toroidal field. As the cycle progresses, meridional circulation sweeps magnetic flux from various latitudes towards the poles and drags it to the base of the convection zone. This flux provides the starting poloidal field for the next solar cycle, thus closing the loop. If the conversion of poloidal into toroidal field occurs indeed due to differential rotation, the latitudinal rotational shear is subject to a weakening over the solar cycle owing to the elastic-band-like resistance of the magnetic field. Consequently, there must be continuous supply of angular momentum to the solar equator at the base of the convection zone to maintain rotational shear.

The putative discovery of large-scale fluid motions at the photosphere of the Sun[35] revealed that the velocity amplitudes are on the order of 8 m/s, smaller by a factor of 40 than flows associated with intermediate and other smaller scale convective structures[36]. Further, the photospheric convective spectrum shows a drop in power for $\ell < 120$, and at very low spherical-harmonic degrees, the power drops linearly with ℓ. Indeed, the reason for the lack of power on the largest scales is not well understood. We find that the non-mixing-length limit of thermal transport in the convection zone, based on the entraining-plume theory of Section 8[24], would be consistent with low kinetic energies and therefore in plausible agreement with the seismic constraints[14]. In light of the fact that the density at the depth $r/R_\odot = 0.96$ is some thousand times larger than at the photosphere, the measurements of large-scale motions on the order of 8 m/s at the surface[35] also imply very weak flow amplitudes at depth.

In contrast, constraints that were obtained using the method of ring-diagram analysis[37] suggest that convective velocity amplitudes in the solar interior are in agreement with ASH simulations; these contradictory inferences remain to be addressed.

10. Supergranulation

Supergranulation is a highly visible feature on the disk of the Sun, as shown in Figure 5. Although the phenomenon is believed to be a scale of convection, the intensity contrast between the centre and the edges of a supergranular structure is on the order of 1 K, barely a statistically significant measurement[38]. Consequently, the driving mechanism of supergranulation and its role in global solar dynamics is a matter of some debate. See[39] for a review of research into supergranulation that is now several decades in history.

The seismic inference of the internal structure of supergranulation provides an effective means of discriminating between theories. For instance, the depth variation of the flow system can indicate whether the mechanism is connected to ionisation or coherent interaction between small-scale granulation or magneto-convection, etc. Consequently, supergranulation has been a subject of central interest in helioseismology[40–51]. In recent years, the statistical study of ensembles of supergranules has gained popularity[52], which is based on the assumption that supergranules are statistically identical. Analyzing and averaging wave travel times over thousands of supergranules, precise estimates with very low error bars for the travel times associated with the average supergranule have been estimated[53]. Such precise travel times can ostensibly be used to create accurate models of supergranules.

However, seismic inferences have not produced a unified model that also successfully explains observed travel times. A number of these efforts are at odds with each other, and recent analyses[49,50,54] have arrived at a previously unanticipated result that supergranular flows peak at a depth of 2000 – 3000 km, reaching a maximum of 600 – 800 ms^{-1} in horizontal flow speed. In contrast, other studies[46,53] found that supergranular flows decrease in magnitude with depth. It has also been reported[42] that supergranules have wave-like properties, which are generally not well understood and additionally have induced controversy[55]. It is thus fair to say that supergranulation remains an open problem.

11. Conclusions

We have briefly discussed a number of phenomena related to the energy transport in the Sun. These phenomena are of considerable fluid dynamic interest and further insight on understanding of these phenomena must come from a convergence of fluid dynamics, helioseismology, numerical sim-

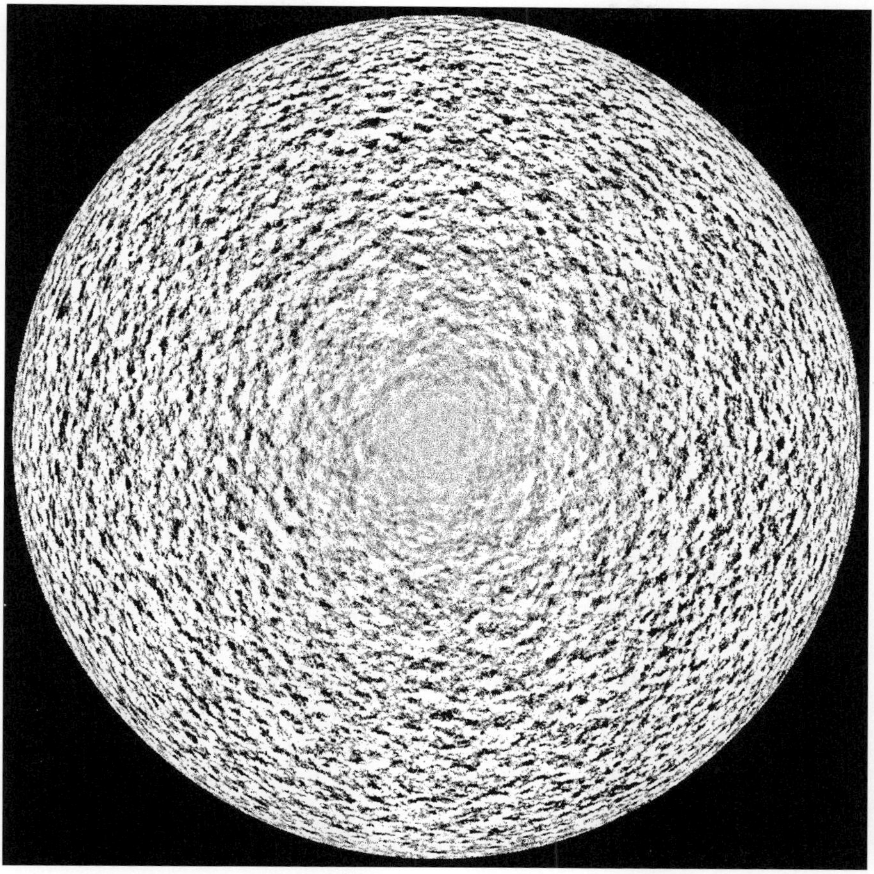

Fig. 5. Line-of-sight projected velocity field measured at the photosphere of the Sun by the Michelson Doppler Imager (Courtesy, M. Rieutord). Average rotation has been subtracted. The structure visible in this image is supergranulation, a flow system of horizontal size 35000-km in which horizontal velocities are about 300 m/s and vertical velocities are 4-10 m/s at the surface[36].

ulations and, of course, stellar models. Much progress has been made but it is not merely rhetorical to say that much remains to be explored.

12. Acknowledgments

This work is supported in part by the NYU-Abu Dhabi Center for Space Sciences. SH acknowledges a collaboration with the Max Planck Institute for Solar System Research through a Max Planck Partner Group established at the Tata Institute of Fundamental Research, Mumbai. LG acknowledges support from Deutsche Forschungsgemeinschaft SFB 963 "Astrophysical Flow Instabilities and Turbulence". This work utilizes HMI data that are courtesy of NASA/SDO and the HMI science team. We thank M. Rieutord for supplying Figure 5 and the model of the plume in Figure 3.

References

1. S. Hanasoge, L. Gizon and K. R. Sreenivasan, Seismic Sounding of Convection in the Sun, *ArXiv e-prints* (March 2015).
2. J. Schou, H. M. Antia, S. Basu, R. S. Bogart, R. I. Bush, S. M. Chitre, J. Christensen-Dalsgaard, M. P. di Mauro, W. A. Dziembowski, A. Eff-Darwich, D. O. Gough, D. A. Haber, J. T. Hoeksema, R. Howe, S. G. Korzennik, A. G. Kosovichev, R. M. Larsen, F. P. Pijpers, P. H. Scherrer, T. Sekii, T. D. Tarbell, A. M. Title, M. J. Thompson and J. Toomre, Helioseismic Studies of Differential Rotation in the Solar Envelope by the Solar Oscillations Investigation Using the Michelson Doppler Imager, *Astrophys. J.* **505**, 390 (September 1998).
3. J. Christensen-Dalsgaard, Helioseismology, *Reviews of Modern Physics* **74**, 1073 (November 2002).
4. P. H. Scherrer, R. S. Bogart, R. I. Bush, J. T. Hoeksema, A. G. Kosovichev, J. Schou, W. Rosenberg, L. Springer, T. D. Tarbell, A. Title, C. J. Wolfson, I. Zayer and MDI Engineering Team, The Solar Oscillations Investigation - Michelson Doppler Imager, *Solar Physics* **162**, 129 (December 1995).
5. T. L. Duvall, Jr., Large-scale solar velocity fields, *Solar Physics* **63**, 3 (August 1979).
6. M. S. Miesch and B. W. Hindman, Gyroscopic Pumping in the Solar Near-surface Shear Layer, *Astrophys. J.* **743**, p. 79 (December 2011).
7. M. S. Miesch, N. A. Featherstone, M. Rempel and R. Trampedach, On the Amplitude of Convective Velocities in the Deep Solar Interior, *Astrophys. J.* **757**, p. 128 (October 2012).
8. J. Zhao, R. S. Bogart, A. G. Kosovichev, T. L. Duvall, Jr. and T. Hartlep, Detection of Equatorward Meridional Flow and Evidence of Double-cell Meridional Circulation inside the Sun, *Astrophys. J.* **774**, p. L29 (September 2013).
9. D. Huber, M. J. Ireland, T. R. Bedding, I. M. Brandão, L. Piau, V. Maestro, T. R. White, H. Bruntt, L. Casagrande, J. Molenda-Żakowicz, V. Silva Aguirre, S. G. Sousa, T. Barclay, C. J. Burke, W. J. Chaplin, J. Christensen-Dalsgaard, M. S. Cunha,

J. De Ridder, C. D. Farrington, A. Frasca, R. A. García, R. L. Gilliland, P. J. Goldfinger, S. Hekker, S. D. Kawaler, H. Kjeldsen, H. A. McAlister, T. S. Metcalfe, A. Miglio, M. J. P. F. G. Monteiro, M. H. Pinsonneault, G. H. Schaefer, D. Stello, M. C. Stumpe, J. Sturmann, L. Sturmann, T. A. ten Brummelaar, M. J. Thompson, N. Turner and K. Uytterhoeven, Fundamental Properties of Stars Using Asteroseismology from Kepler and CoRoT and Interferometry from the CHARA Array, *Astrophys. J.* **760**, p. 32 (November 2012).

10. P. Charbonneau, Dynamo Models of the Solar Cycle, *Living Reviews in Solar Physics* **2**, p. 2 (June 2005).
11. M. Dikpati and P. A. Gilman, Simulating and Predicting Solar Cycles Using a Flux-Transport Dynamo, *Astrophys. J.* **649**, 498 (September 2006).
12. A. A. van Ballegooijen, On the surface response of solar giant cells, *Astrophys. J.* **304**, 828 (May 1986).
13. R. F. Stein and Å. Nordlund, Realistic Solar Convection Simulations, *Solar Physics* **192**, 91 (March 2000).
14. S. M. Hanasoge, T. L. Duvall, Jr. and K. R. Sreenivasan, Anomalously Weak Solar Convection, *Proceedings of the National Academy of Sciences* **109**, 11928 (2012).
15. J. Schou, P. H. Scherrer, R. I. Bush, R. Wachter, S. Couvidat, M. C. Rabello-Soares, R. S. Bogart, J. T. Hoeksema, Y. Liu, T. L. Duvall, D. J. Akin, B. A. Allard, J. W. Miles, R. Rairden, R. A. Shine, T. D. Tarbell, A. M. Title, C. J. Wolfson, D. F. Elmore, A. A. Norton and S. Tomczyk, Design and Ground Calibration of the Helioseismic and Magnetic Imager (HMI) Instrument on the Solar Dynamics Observatory (SDO), *Solar Physics* **275**, 229 (January 2012).
16. T. L. Duvall, Jr., S. M. Jefferies, J. W. Harvey and M. A. Pomerantz, Time-distance helioseismology, *Nature* **362**, 430 (April 1993).
17. L. Gizon and A. C. Birch, Time-Distance Helioseismology: The Forward Problem for Random Distributed Sources, *Astrophys. J.* **571**, 966 (June 2002).
18. T. L. Duvall, Jr., Nonaxisymmetric variations deep in the convection zone, in *GONG+ 2002. Local and Global Helioseismology: the Present and Future*, ed. H. Sawaya-Lacoste, ESA Special Publication, Vol. 517, February 2003.
19. S. M. Hanasoge, T. L. Duvall and M. L. DeRosa, Seismic Constraints on Interior Solar Convection, *Astrophys. J.* **712**, L98 (March 2010).
20. H. C. Spruit, A model of the solar convection zone, *Solar Physics* **34**,

277 (February 1974).
21. M. S. Miesch, A. S. Brun, M. L. De Rosa and J. Toomre, Structure and Evolution of Giant Cells in Global Models of Solar Convection, *Astrophys. J.* **673**, 557 (January 2008).
22. M. Ghizaru, P. Charbonneau and P. K. Smolarkiewicz, Magnetic Cycles in Global Large-eddy Simulations of Solar Convection, *Astrophys. J.* **715**, L133 (June 2010).
23. P. J. Käpylä, M. J. Mantere, G. Guerrero, A. Brandenburg and P. Chatterjee, Reynolds stress and heat flux in spherical shell convection, *Astronomy & Astrophysics* **531**, p. A162 (July 2011).
24. M. Rieutord and J.-P. Zahn, Turbulent plumes in stellar convective envelopes., *Astronomy & Astrophysics* **296**, p. 127 (April 1995).
25. H. Spruit, Convection in stellar envelopes: a changing paradigm, *Memorie della Societa Astronomica Italiana* **68**, p. 397 (1997).
26. J. Christensen-Dalsgaard, M. J. P. F. G. Monteiro, M. Rempel and M. J. Thompson, A more realistic representation of overshoot at the base of the solar convective envelope as seen by helioseismology, *Monthly Notices of the Royal Astronomical Society* **414**, 1158 (June 2011).
27. M. S. Miesch, Large-Scale Dynamics of the Convection Zone and Tachocline, *Living Reviews in Solar Physics* **2**, p. 1 (April 2005).
28. S. A. Balbus, A simple model for solar isorotational contours, *Monthly Notices of the Royal Astronomical Society* **395**, 2056 (June 2009).
29. L. L. Kitchatinov and G. Ruediger, Differential rotation in solar-type stars: revisiting the Taylor-number puzzle., *Astronomy & Astrophysics* **299**, p. 446 (July 1995).
30. S. A. Balbus and E. Schaan, The stability of stratified, rotating systems and the generation of vorticity in the Sun, *Monthly Notices of the Royal Astronomical Society* **426**, 1546 (October 2012).
31. M. P. Rast, A. Ortiz and R. W. Meisner, Latitudinal Variation of the Solar Photospheric Intensity, *Astrophys. J.* **673**, 1209 (February 2008).
32. D. C. Braun and A. C. Birch, Prospects for the Detection of the Deep Solar Meridional Circulation, *Astrophys. J.* **689**, L161 (December 2008).
33. M. Dikpati and P. Charbonneau, A Babcock-Leighton Flux Transport Dynamo with Solar-like Differential Rotation, *Astrophys. J.* **518**, 508 (June 1999).
34. M. Dikpati, Solar Dynamo Models, in *The Solar-B Mission and the Forefront of Solar Physics*, ed. T. Sakurai & T. Sekii, Astronomical

Society of the Pacific Conference Series, Vol. 325, December 2004.
35. D. H. Hathaway, L. Upton and O. Colegrove, Giant convection cells found on the sun, *Science* **342**, 1217 (2013).
36. S. M. Hanasoge and K. R. Sreenivasan, The Quest to Understand Supergranulation and Large-Scale Convection in the Sun, *Solar Physics* **289**, 3403 (September 2014).
37. B. J. Greer, B. W. Hindman, N. A. Featherstone and J. Toomre, Helioseismic Imaging of Fast Convective Flows throughout the Near-surface Shear Layer, *Astrophys. J.* **803**, p. L17 (April 2015).
38. N. Goldbaum, M. P. Rast, I. Ermolli, J. S. Sands and F. Berrilli, The Intensity Profile of the Solar Supergranulation, *Astrophys. J.* **707**, 67 (December 2009).
39. M. Rieutord, T. Roudier, F. Rincon, J.-M. Malherbe, N. Meunier, T. Berger and Z. Frank, On the power spectrum of solar surface flows, *Astronomy & Astrophysics* **512**, p. A4 (March 2010).
40. A. G. Kosovichev and T. L. Duvall, Jr., Acoustic tomography of solar convective flows and structures, in *SCORe'96 : Solar Convection and Oscillations and their Relationship*, eds. F. P. Pijpers, J. Christensen-Dalsgaard and C. S. Rosenthal, Astrophysics and Space Science Library, Vol. 225, December 1997.
41. T. L. Duvall, Jr. and L. Gizon, Time-Distance Helioseismology with f Modes as a Method for Measurement of Near-Surface Flows, *Solar Physics* **192**, 177 (March 2000).
42. L. Gizon, T. L. Duvall and J. Schou, Wave-like properties of solar supergranulation, *Nature* **421**, 43 (January 2003).
43. L. Gizon and A. C. Birch, Time-Distance Helioseismology: Noise Estimation, *Astrophys. J.* **614**, 472 (October 2004).
44. J. Zhao and A. G. Kosovichev, On the inference of supergranular flows by time-distance helioseismology, in *GONG+ 2002. Local and Global Helioseismology: the Present and Future*, ed. H. Sawaya-Lacoste, ESA Special Publication, Vol. 517, February 2003.
45. D. C. Braun, A. C. Birch and C. Lindsey, Local Helioseismology of Near-Surface Flows, in *SOHO 14 Helio- and Asteroseismology: Towards a Golden Future*, ed. D. Danesy, ESA Special Publication, Vol. 559, October 2004.
46. M. F. Woodard, Probing Supergranular Flow in the Solar Interior, *Astrophys. J.* **668**, 1189 (October 2007).
47. J. Hirzberger, L. Gizon, S. K. Solanki and T. L. Duvall, Structure and Evolution of Supergranulation from Local Helioseismology, *Solar*

Physics **251**, 417 (September 2008).

48. T. L. Duvall, Jr. and A. C. Birch, The Vertical Component of the Supergranular Motion, *Astrophys. J.* **725**, L47 (December 2010).
49. M. Švanda, Inversions for Average Supergranular Flows Using Finite-frequency Kernels, *Astrophys. J.* **759**, p. L29 (November 2012).
50. T. L. Duvall and S. M. Hanasoge, Subsurface Supergranular Vertical Flows as Measured Using Large Distance Separations in Time-Distance Helioseismology, *Solar Physics* **287**, 71 (October 2013).
51. D. E. Dombroski, A. C. Birch, D. C. Braun and S. M. Hanasoge, Testing Helioseismic-Holography Inversions for Supergranular Flows Using Synthetic Data, *Solar Physics* **282**, 361 (February 2013).
52. T. L. Duvall, Jr., A. C. Birch and L. Gizon, Direct Measurement of Travel-Time Kernels for Helioseismology, *Astrophys. J.* **646**, 553 (July 2006).
53. A. Birch, T. L. Duvall, L. Gizon and J. Jackiewicz, Helioseismology of the "Average" Supergranule, in *Bulletin of the American Astronomical Society*, June 2006.
54. T. L. Duvall, S. M. Hanasoge and S. Chakraborty, Additional Evidence Supporting a Model of Shallow, High-Speed Supergranulation, *Solar Physics* **289**, 3421 (September 2014).
55. M. P. Rast, J. P. Lisle and J. Toomre, The Spectrum of the Solar Supergranulation: Multiple Nonwave Components, *Astrophys. J.* **608**, 1156 (June 2004).

Rapidly-Rotating Turbulence and its Role in Planetary Dynamos

P. A. Davidson

Department of Engineering, University of Cambridge, Cambridge, CB2 1PZ, UK
E-mail: pad3@eng.cam.ac.uk

Columnar vortices aligned with the rotation axis frequently dominate the large scales in rapidly-rotating turbulence, both in laboratory experiments and in numerical simulations. We argue that, often, these columnar vortices are simply low-frequency inertial wave packets propagating away from a localised disturbance (a turbulent eddy or buoyant blob). An important feature of these low-frequency wave packets is that they transport negative helicity upward and positive helicity downward (relative to the rotation vector). This generation and subsequent spatial segregation of positive and negative helicity is potentially important for planetary dynamos.

Recent numerical simulations suggest that magnetic field generation in planets occurs predominantly within columnar vortices located outside the tangent cylinder, and these dynamos may be classified (at least approximately) as α^2. The simulations, which operate in a regime very far from that of the planets, also suggest that the helicity outside the tangent cylinder is predominantly negative in the north and positive in the south. In the more viscous and weakly forced simulations this helicity is often attributed to Ekman pumping at the mantle. However, Ekman pumping is less evident in the less viscous and more strongly driven simulations, and almost certainly plays no significant dynamical role in planetary interiors. We argue that, in the absence of Ekman pumping, the observed helicity distribution is a consequence of low-frequency helical wave packets initiated near the equatorial plane by buoyant plumes floating out towards the mantle. This provides a mechanism for maintaining a dynamo in the Earth, as well as in the gas giants.

Keywords: Rotating turbulence; dynamo; helicity

1. Waves and Columnar Vortices in Rapidly-Rotating Fluids

There are many cases in geophysics and astrophysics where a fluid is subject to a strong background rotation, say $\boldsymbol{\Omega} = \Omega \hat{\boldsymbol{e}}_z$. In such cases it often makes sense to transfer to a frame of reference rotating with the fluid. Of course, Newton's second law does not apply in such a non-inertial frame, but we can behave as if it does provided we add two fictitious forces to our equation of motion: the centrifugal force, which is irrotational and so merely augments the fluid pressure, and the Coriolis force, $2\boldsymbol{u} \times \boldsymbol{\Omega}$. In situations where free surfaces are not important it is usually sufficient to absorb the centrifugal force into the pressure and then work with an augmented pressure. Retaining the symbol p for the augmented pressure, the Navier-Stokes and vorticity equations in the rotating frame are:

$$\frac{D\boldsymbol{u}}{Dt} = -\frac{1}{\rho}\nabla p + 2\boldsymbol{u} \times \boldsymbol{\Omega} + \nu\nabla^2\boldsymbol{u} \tag{1}$$

$$\frac{D\boldsymbol{\omega}}{Dt} = (\boldsymbol{\omega} \cdot \nabla)\boldsymbol{u} + 2(\boldsymbol{\Omega} \cdot \nabla)\boldsymbol{u} + \nu\nabla^2\boldsymbol{\omega}. \tag{2}$$

The appearance of the Coriolis force in (1) and (2) defines the dynamics when departures from rigid-body rotation are weak, i.e. the Rossby number,

$$Ro = \frac{u}{\Omega l}, \tag{3}$$

is small. In particular, the fluid can sustain a form of incompressible internal wave motion, called inertial waves, the restoring force being the Coriolis force. It is also observed that, for $Ro << 1$, fluid motion is frequently dominated by columnar vortices aligned with the rotation axis, those vortices usually being termed Taylor columns. In fact, we shall see that these two phenomena, waves and columnar vortices, are closely related, being different manifestations of the same underlying dynamics. Moreover, both phenomena turn out to be central to the operation of planetary dynamos. Let us start, then, with Taylor columns.

1.1. *The Taylor-Proudman theorem*

When $Ro << 1$ and the influence of viscosity small, the vorticity equation simplifies to

$$\frac{\partial\boldsymbol{\omega}}{\partial t} = 2(\boldsymbol{\Omega} \cdot \nabla)\boldsymbol{u}. \tag{4}$$

If the motion is quasi-steady then $\partial\boldsymbol{\omega}/\partial t$ may be neglected and we obtain,

$$(\boldsymbol{\Omega} \cdot \nabla)\boldsymbol{u} = 0. \tag{5}$$

It appears that rapidly-rotating, quasi-steady motion is subject to the constraint that $\boldsymbol{u}$ is two-dimensional, in the sense that it is independent of the coordinate parallel to $\boldsymbol{\Omega}$. This is called the Taylor-Proudman theorem. Consider, for example, the experiment shown in Figure 1. A small object is slowly towed across the base of a rotating tank which is filled with water. As the object moves, the fluid cannot flow up and over the object, as we might have expected, since that constitutes a three-dimensional flow pattern. Rather, the column of fluid located between the object and the surface of the fluid moves across the tank, as if rigidly attached to the object. This column is known as a *Taylor column*, and it is a direct consequence of the need to maintain two-dimensional motion. Thus, for example, a fluid particle initially at point A will move across the tank, always centred above the object. It is natural to enquire how the fluid lying above the object knows to move with the object, and to answer this we must introduce the idea of inertial waves.

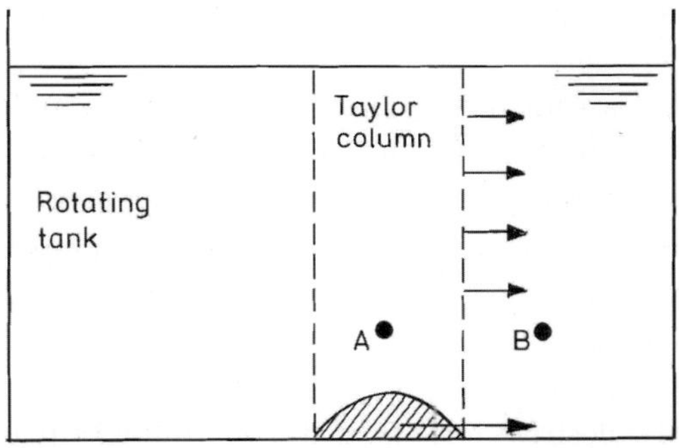

Fig. 1. A small object is slowly towed across the base of a rotating tank filled with water. As the object moves it carries with it a Taylor column. (Figure from Davidson, 2004.[1])

1.2. *Inertial waves and the formation of Taylor columns*

A rapidly-rotating, incompressible fluid can support internal wave motion. Application of the operator $\nabla \times \partial/\partial t$ to (4) yields the wave-like equation

$$\frac{\partial^2}{\partial t^2}(\nabla^2 \boldsymbol{u}) + 4(\boldsymbol{\Omega} \cdot \nabla)^2 \boldsymbol{u} = 0, \tag{6}$$

which allows for plane waves of the form $\boldsymbol{u} = \hat{\boldsymbol{u}} \exp[j(\boldsymbol{k} \cdot \boldsymbol{x} - \varpi t)]$. These are known as inertial waves and they have the dispersion relationship

$$\varpi = \pm \frac{2(\boldsymbol{k} \cdot \boldsymbol{\Omega})}{k}. \tag{7}$$

The group velocity of inertial waves, which is the velocity at which energy propagates away from a localised disturbance in the form of wave packets, is then

$$\boldsymbol{c_g} = \frac{\partial \varpi}{\partial k_i} = \pm \frac{2\boldsymbol{k} \times (\boldsymbol{\Omega} \times \boldsymbol{k})}{k^3} = \pm 2 \frac{k^2 \boldsymbol{\Omega} - (\boldsymbol{k} \cdot \boldsymbol{\Omega})\boldsymbol{k}}{k^3}. \tag{8}$$

Note that

$$\boldsymbol{c_g} \cdot \boldsymbol{\Omega} = \pm 2k^{-3}[k^2\Omega^2 - (\boldsymbol{k} \cdot \boldsymbol{\Omega})^2], \tag{9}$$

which tells us that the positive sign in (8) corresponds to wave energy travelling upward, while the negative sign corresponds to energy propagating downward.

Inertial waves are intrinsically helical, with a strong helicity $h = \boldsymbol{u} \cdot \boldsymbol{\omega}$. That is, equation (4), combined with the dispersion relationship (7), yields $\hat{\boldsymbol{\omega}} = \mp k \hat{\boldsymbol{u}}$, where $\hat{\boldsymbol{\omega}}$ is the amplitude of the vorticity. It follows that the vorticity and velocity fields are parallel and in phase, so that inertial waves have maximum helicity, with the positive sign in (9) corresponding to negative helicity, and the negative sign to positive helicity. Thus a wave packet with negative helicity will propagate upward ($\boldsymbol{c_g} \cdot \boldsymbol{\Omega} > 0$), while wave packets with positive helicity travel downward ($\boldsymbol{c_g} \cdot \boldsymbol{\Omega} < 0$). This is illustrated by the numerical simulation shown in Figure 2, where a slab of turbulence spreads in a rotating fluid at $Ro = 0.1$ by emitting inertial waves. The left-hand panel is the initial condition and the right-hand one corresponds to $\Omega t = 6$. Red is negative helicity and green positive helicity.

As we shall see, low-frequency inertial waves play a special role in many systems, particularly planetary dynamos. They are characterised by $\boldsymbol{k} \cdot \boldsymbol{\Omega} \approx 0$ and a group velocity of $\boldsymbol{c_g} = \pm 2\boldsymbol{\Omega}/k$, so that energy propagates along the rotation axis. To illustrate the point, consider the unbounded initial-value problem in which we slowly move a disc of radius R along the axis of rotation with a speed V, starting at time $t = 0$. Evidently low-frequency waves propagate in the $\pm\boldsymbol{\Omega}$ directions, carrying energy away from the disc at a speed $c_g = 2\Omega/k$. Since the smallest wavenumbers (largest wavelengths) travel fastest, and these have a magnitude of $k \approx \pi/R$, we find wave-fronts located a distance $(2/\pi)\Omega R t$ above and below the disc, as shown in Figure 3. So at time t the inertial waves generated by the disc fill a column of

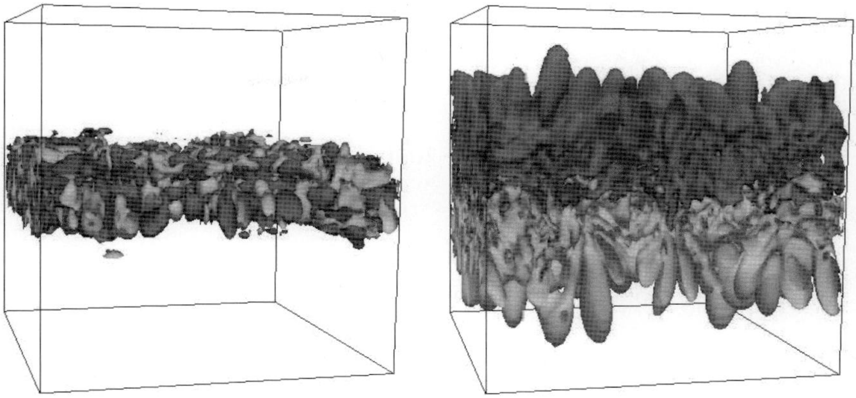

Fig. 2. A slab of turbulence spreads in a rotating fluid ($Ro = 0.1$) by emitting inertial waves. The left-hand panel is the initial condition and the right-hand one is at $\Omega t = 6$. Red marks negative helicity and green positive helicity. Evidently wave packets with negative helicity travel upward and those with positive helicity travel downward. (Figure from Davidson, 2013.[2])

radius R and half-length $l = (2/\pi)\Omega R t$, and these waves carry with them the information that the disc is moving. An exact solution of this initial-value problem shows that the fluid which lies within this column has the same axial velocity as the disc, while that lying outside the column is more or less quiescent in the rotating frame. In short, the inertial waves have created a transient Taylor column, whose length grows at the group velocity of low-frequency waves.

Returning to Figure 1, we can now understand how a Taylor column is maintained. When the object is towed slowly across the base of the tank it emits low-frequency inertial waves which travel upward at the group velocity. These reach the fluid surface on a time scale which is very short by comparison with the slow timescale of the movement of the object. As the inertial waves travel upward they carry with them the information that the object is moving, and in particular the information that tells the fluid within the column to move horizontally, in line with the towed object. Thus the moving Taylor column is maintained by a train of inertial wave packets radiated from the object, with the object itself acting rather like a radio antenna.

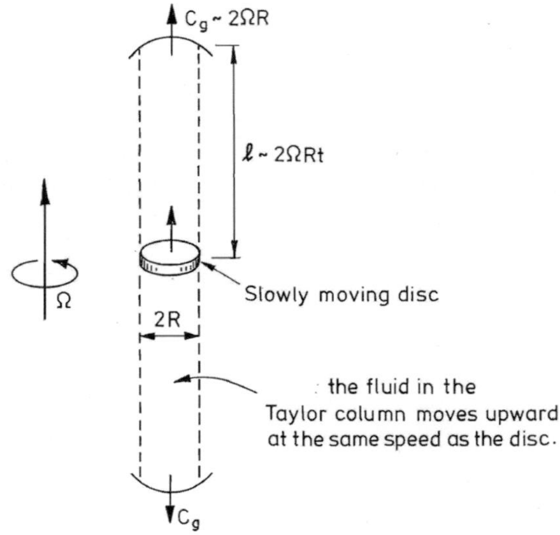

Fig. 3. Formation of a Taylor column by inertial waves generated by a slowly moving disc. (Figure from Davidson, 2001.[3])

2. The Formation of Columnar Vortices in Rotating Turbulence

2.1. *The spontaneous conversion of eddies into transient Taylor columns*

In Figures 1 and 3 we considered inertial waves driven by moving boundaries. In such cases there is an imposed time scale set by the movement of the boundary. For example, slowly-moving boundaries excite low-frequency waves and these propagate along the rotation axis. We now consider the initial-value problem consisting of a single eddy (blob of vorticity) sitting in a rapidly-rotating fluid. Here the situation is more complicated because there is no imposed time-scale. In such a situation the direction of propagation of wave energy is determined by the distribution of energy amongst the various $\boldsymbol{k}$-vectors in the initial velocity field. That is, when we Fourier transform the initial condition we apportion a certain amount of energy to each wavevector, $\boldsymbol{k}$, and that energy then disperses in the form of wave packets in accordance with (8). For an arbitrarily shaped eddy, then, we might expect the energy to disperse in all directions. It turns out, however, that there is still a tendency for wave energy to propagate preferentially along the rotation axis to form columnar structures. That is, spatially compact eddies spontaneously evolve into columnar vortices in which $(\boldsymbol{\Omega} \cdot \nabla)\boldsymbol{u}$ is small, *i.e.* they evolve into a kind of transient Taylor column[4].

To illustrate the point, consider the initial value problem consisting of a localised blob of vorticity sitting in an otherwise quiescent, rapidly-rotating fluid (Figure 4). Let the characteristic scale of the blob be δ, a typical velocity be u, and suppose $Ro = u/\Omega\delta << 1$. The subsequent motion consists of a spectrum of linear inertial waves whose energy disperses in all directions with a typical speed of $c_g \sim \Omega\delta$. However, as we shall see, this radiation of energy is subject to a powerful constraint which systematically favours dispersion of energy along the rotation axis.

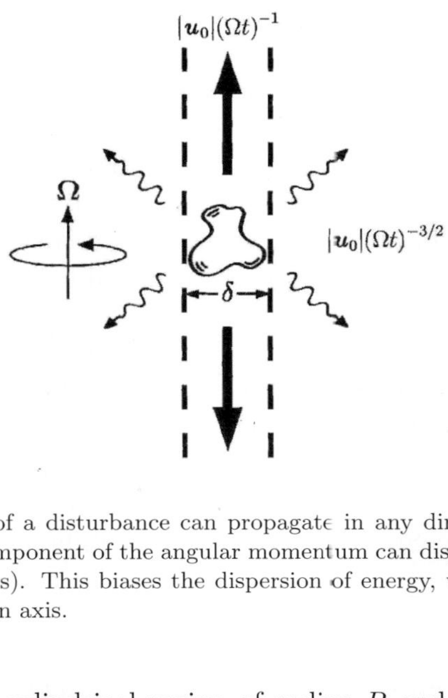

Fig. 4. The energy of a disturbance can propagate in any direction (wiggly arrows). However, the axial component of the angular momentum can disperse along the rotation axis only (solid arrows). This biases the dispersion of energy, with the energy density highest on the rotation axis.

Let V_R be the cylindrical region of radius R and infinite length that encloses the vorticity field $\boldsymbol{\omega}$ at $t = 0$ (the dotted lines in Figure 4). Then it is readily shown that the axial component of the angular impulse, $\frac{1}{3}\int_{V_\infty} [\boldsymbol{x} \times (\boldsymbol{x} \times \boldsymbol{\omega})]_z dV$, which is equal to the axial angular momentum held within V_R, $H_z = \int_{V_R} [\boldsymbol{x} \times \boldsymbol{u}]_z dV$, is conserved for all time[4]. We conclude that inertial waves cannot sustain a net radial flux of axial angular momentum, and so this component of angular momentum can disperse along the rotation axis only. It is easy to see that the constraint imposed by the conservation of H_z biases the dispersion of energy towards the rotation axis. For example, as the energy radiates to fill a three-dimensional volume whose size grows as $V_{3D} \sim (c_g t)^3 \sim (\delta\Omega t)^3$, we would expect that conservation of energy requires the velocity outside V_R to decay as $u \sim u_0(\Omega t)^{-3/2}$. However, inside the cylinder V_R the axial component of angular momentum is confined to a cylindrical region whose volume grows as $V_{1D} \sim c_g t\delta^2 \sim \Omega t\delta^3$. Conservation of H_z then suggests that the characteristic velocity inside V_R falls more slowly, as $u \sim u_0(\Omega t)^{-1}$. This is illustrated in Figure 4, and indeed these two temporal decay laws are readily confirmed by exact asymptotic analysis[4]. Thus the dominant influence of inertial wave radiation is to spread the energy of the initial vortex along the rotation axis.

An alternative way to understand why the fluid within V_R supports a higher radiation density is to note that the group velocity is perpendicular to $\boldsymbol{k}$, so that the energy associated with all horizontal wavevectors radiates along the rotation axis. Thus all the energy contained within a thin horizontal disc in $\boldsymbol{k}$-space propagates along a narrow cylinder in real space, and the process of channelling energy from a two-dimensional object in $\boldsymbol{k}$-space to a one-dimensional object (a narrow cylinder) in real space amplifies the radiation density[2]. In any event, whichever explanation is preferred, we conclude that an eddy which is initially confined to the region $|\boldsymbol{x}| < \delta$ disperses energy in all directions, but that the energy density is inevitably highest on the rotation axis, creating a columnar vortex above and below the initial disturbance, as shown in Figure 5. These columnar structures are manifestly a form of transient Taylor column. They are wave packets that propagate and spread at a rate dictated by the low-frequency group velocity, and so their length grows as $l_z \sim \delta\Omega t$, while their centres propagate as $z \sim \pm\delta\Omega t$. Such columnar wave packets are clearly evident in Figure 2.

2.2. *Columnar eddies in rapidly-rotating turbulence*

We now turn to turbulence. Consider the case where $Ro = u/\Omega l$ is small, l, being the integral scale. The short-term evolution of the large eddies is then more or less governed by the linear equation (4). Consider, for example, the case of freely-decaying turbulence. The large eddies will spontaneously emit inertial waves as discussed above, with a systematic bias to radiating energy along the rotation axis, which in turn elongates the turbulent eddies in the axial direction. In such a situation we can might tentatively neglect

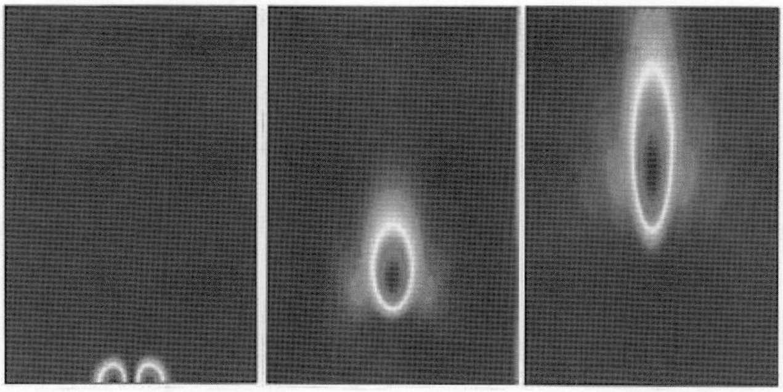

Fig. 5. A blob of vorticity spontaneously evolves intc a pair of columnar eddies through linear inertial wave propagation. Energy contours for increasing t are shown for $z > 0$. (Figure from Davidson, 2013.[2])

$\partial^2/\partial z^2$ in the Laplacian of (6), which then becomes

$$\frac{\partial^2}{\partial t^2}(\nabla_\perp^2 \boldsymbol{u}) + 4(\boldsymbol{\Omega}\cdot\nabla)^2\boldsymbol{u} = (non - linear\, terms\, of\, order\, Ro), \qquad (10)$$

where we have grouped together the weak nonlinear terms on the right. Here $\nabla_\perp^2$ represents the two-dimensional Laplacian defined in a plane normal to $\boldsymbol{\Omega}$. If we Fourier transform this equation in the transverse plane (but not in the direction of $\boldsymbol{\Omega}$) we have

$$\frac{\partial^2 \hat{\boldsymbol{u}}}{\partial t^2} = \left(\frac{4\Omega^2}{k_\perp^2}\right)\frac{\partial^2 \hat{\boldsymbol{u}}}{\partial z^2} + (non - linear\, terms\, of\, order\, Ro), \qquad (11)$$

where $\hat{\boldsymbol{u}}$ is the transformed velocity. The suggestion is that $\hat{\boldsymbol{u}}$, and hence turbulent energy, spreads by wave propagation along the $\boldsymbol{\Omega}$-axis, creating columnar structures in the process. Thus we might picture the large scales as an ensemble of columnar eddies immersed in a sea of small-amplitude inertial waves.

For the inhomogeneous case of turbulence spreading from a localised turbulent cloud, this cartoon seems to work well, as illustrated in Figure 6. The figure shows the numerical simulation of a horizontal slab of turbulence spreading in a rapidly-rotating fluid at $Ro = 0.1$[5]. The cloud grows by emitting low-frequency inertial waves which propagate along the rotation axis in the form of wave packets. The images correspond to the times $\Omega t = 0, 6$ and are iso-contours of axial kinetic energy coloured by helicity, with negative helicity marked blue and positive helicity red. The similarity to Figure 5 is striking and the fact that the helicity is (mostly) negative in the upper part of the flow, and positive in the lower regions, provides substantial support for the assertion that these are indeed inertial wave packets. This can be further confirmed by comparing the rate of spread of energy in the vertical direction with the group velocity of linear inertial waves, and the two match well in the simulations[5], and also in related laboratory experiments[4]. So there is little doubt that the columnar eddies observed in such cases are simply linear inertial wave packets, with non-linearity playing no significant role.

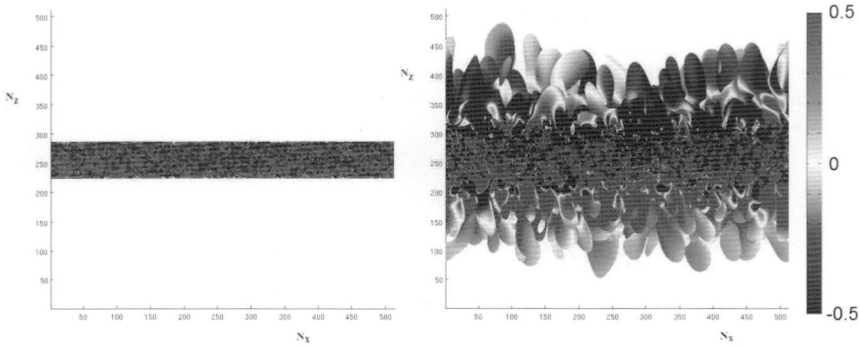

Fig. 6. Turbulence spreads in a rotating fluid by emitting inertial waves ($Ro = 0.1$). Left $\Omega t = 0$, right $\Omega t = 6$. Dark (light) colour marks positive (negative) helicity. The flow spontaneously converts into columnar vortices which have negative helicity above and positive helicity below.

The case of homogeneous, freely-decaying turbulence is, however, more complicated. Here there is evidence that non-linearity plays a significant role and that the initial value of Ro is important. Of course, in the linear approximation inertial waves exist without interaction. However, even when Ro is small, we cannot completely neglect the small but finite non-linear interactions of the inertial waves. The reason is that two small-amplitude waves with distinct frequencies can combine, via the weak non-linear term $\boldsymbol{u} \cdot \nabla \boldsymbol{u}$, to produce an inertial force at a third frequency which, in turn, can excite a third wave. Under the right conditions a so-called *resonant triad* can occur in which the natural frequency of the third wave matches that of the inertial forcing. The weak non-linearity then produces a significant response. The general trend for freely-decaying, homogeneous turbulence at moderate or small Ro is, like the inhomogeneous case above, towards a quasi-two-dimensional flow, dominated by large, columnar eddies (Figure 7). Curiously though, there is evidence that the behaviour is different depending on whether $Ro \sim 1$ (as in most laboratory experiments) or $Ro << 1$ (which typifies most numerical simulations).

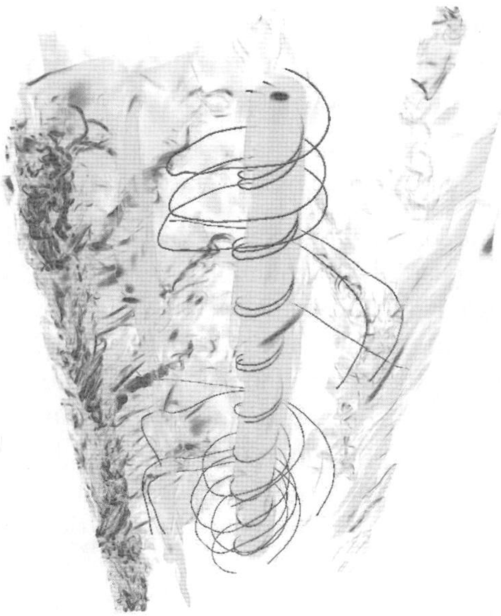

Fig. 7. Numerical simulation of rapidly rotating turbulence ($Ro \sim 0.06$), visualised by vorticity. (Courtesy of Pablo Mininni and Annick Pouquet.)

When $Ro \sim 1$ (or a little lower) we find that the parallel integral scale grows as $l_{\|} \sim l_{\perp}\Omega t$, exactly as in the inhomogeneous case, which is consistent with the notion that the large eddies elongate due to linear wave propagation[6,7]. (Here $l_{\|}$ and $l_{\perp}$ are integral scales in the parallel and perpendicular directions.) Within the large scales two sets of dynamics coexist. The columnar eddies interact non-linearly with each other on the slower time scale of $l_{\perp}/u$, while the inertial waves propagate on the faster time scale of Ω^{-1}, elongating the eddies. Of course, these large columnar eddies are immersed in a sea of smaller, disorganised vortices whose small size leaves them immune to the direct influence of the Coriolis force. Nevertheless, this filamentary vorticity tends to be spiralled up by the columnar eddies (Figure 7), and so the growing length-scale $l_{\|} \sim l_{\perp}\Omega t$ is enforced on it in a non-linear fashion. Thus linear and non-linear dynamics work hand in hand.

When $Ro << 1$ the behaviour of homogeneous turbulence is more complicated and, perhaps, less well understood. It seems that the growth in $l_{\|}$ is no longer fixed by the group velocity of inertial waves, $l_{\|} \sim l_{\perp}\Omega t$, but rather is determined by a longer non-linear time scale. The dynamics of such turbulence is still the subject of much discussion, and although there have been many attempts to understand this using resonant triad theory, a complete understanding remains elusive.

3. The Dispersion of Energy from Buoyant Eddies

In planetary cores the source of any convection or waves is buoyancy, and so we now turn to the generation of columnar structures from buoyant blobs. We start with a single blob and then move onto the case of a random sea of buoyant eddies, following the arguments in Davidson (2014)[8] in both cases. In the core of the Earth the Rossby number is small (around 10^{-6} based on the core radius) and so we continue to focus on the limit of small Ro. For reasons that will become apparent, we have in mind buoyant blobs slowly drifting out along the equatorial plane of the Earth, and so we shall take gravity to be perpendicular to the background rotation, $\boldsymbol{\Omega} = \Omega\hat{\boldsymbol{e}}_{\boldsymbol{z}}$.

Consider an isolated blob of buoyant material of scale δ. We model the fluid as Boussinesq and so the governing equation at low Ro is

$$\frac{\partial \boldsymbol{u}}{\partial t} = -\frac{1}{\rho}\nabla p + 2\boldsymbol{u} \times \boldsymbol{\Omega} + c\boldsymbol{g}, \;\; c = \rho^{'}/\rho, \tag{12}$$

where $\boldsymbol{g} = -g\hat{\boldsymbol{e_x}}$ is the local gravitational acceleration, $\rho^{'}$ is the density perturbation (a negative quantity) and ρ the background density. Equivalently, in terms of vorticity,

$$\frac{\partial \boldsymbol{\omega}}{\partial t} = 2(\boldsymbol{\Omega} \cdot \nabla)\boldsymbol{u} + \nabla c \times \boldsymbol{g}. \tag{13}$$

Now $\rho^{'}$ is governed by a simple advection-diffusion equation and so evolves on a slow time scale set by the magnitude of $\boldsymbol{u}$. Inertial waves, on the other hand, evolve on the fast timescale of Ω^{-1}. It follows that, at low Ro, we may treat $\rho^{'}$ as quasi-steady as far as the initiation of inertial waves is concerned. With $\rho^{'}$ independent of time, application of the operator $\nabla \times \partial/\partial t$ to 13 yields

$$\frac{\partial^2}{\partial t^2}(\nabla^2 \boldsymbol{u}) + 4(\boldsymbol{\Omega} \cdot \nabla)^2 \boldsymbol{u} = 2(\boldsymbol{\Omega} \cdot \nabla)(\boldsymbol{g} \times \nabla c), \tag{14}$$

and we see that the buoyancy term acts as a continual source of low-frequency inertial waves. These waves necessarily propagate in the $\pm\boldsymbol{\Omega}$

directions, carrying energy away from the buoyant blob. So, after a time t, we find wave-fronts located at $z \sim \pm\Omega\delta t$ above and below the blob, and within the cylindrical region defined by $z \sim \pm\Omega\delta t$ we have low-frequency inertial waves which have originated from the buoyant blob. We might expect these waves to create transient Taylor columns similar to those shown in Figure 6, and this is confirmed by (13) which, when the wave frequency is low, demands $(\boldsymbol{\Omega} \cdot \nabla)\boldsymbol{u} \approx 0$ outside the buoyant blob. We conclude that quasi-two-dimensional columnar vortices (i.e. transient Taylor columns) spontaneously emerge from the buoyant blob, much as they did from the impulsively started disc in Figure 3.

To determine the approximate structure of these transient Taylor columns we must consider the vertical 'jump conditions' across the buoyant blob after the initial passage of inertial waves. Since ρ' is quasi-steady and the inertial waves are of low frequency, (13) within the buoyant blob reduces to

$$2(\boldsymbol{\Omega} \cdot \nabla)\boldsymbol{u} + \nabla c \times \boldsymbol{g} \approx 0, \tag{15}$$

or equivalently,

$$2(\boldsymbol{\Omega} \cdot \nabla)\boldsymbol{\omega} \approx \boldsymbol{g}\nabla^2 c - (\boldsymbol{g} \cdot \nabla)\nabla c. \tag{16}$$

From (15) we find that the integrated vertical jump conditions across the blob are $\Delta u_x \approx 0$, $\Delta u_y \approx 0$, and

$$\Delta u_z \approx -\frac{g}{2\Omega}\int (\partial c/\partial y)dz, \tag{17}$$

while (16) yields the vorticity jump condition $\Delta\omega_z \approx 0$. From the jump condition for ω_z we see that a cyclonic columnar vortex below the buoyant blob corresponds to a cyclonic vortex above the blob, while an anticyclonic columnar vortex below corresponds to a anticyclonic vortex above. Moreover, for a Gaussian-like buoyant blob, (17) tells us that Δu_z is positive for $y < 0$ and negative for $y > 0$. It follows that u_z, which is antisymmetric about the plane $z = 0$, diverges from $z = 0$ for $y < 0$ (i.e. is positive for $z > 0$ and negative for $z < 0$) and converges to $z = 0$ for $y > 0$ (i.e. is negative for $z > 0$ and positive for $z < 0$). When combined with the requirement that upward propagating inertial waves have negative helicity, while downward propagating waves have positive helicity, we conclude that the inertial wave dispersion pattern consists approximately of[8]:

(1) a pair of cyclonic and anticyclonic columnar vortices above the blob;
(2) a matching pair of cyclonic and anticyclonic vortices below the blob;

(3) the cyclones above and below are located at the same value of y (same azimuthal angle in the case of the Earth), while the anticyclones are also located at coincident azimuthal angles;
(4) the anticyclones are located at negative y (smaller azimuthal angle in the case of the Earth), and the cyclones at positive y (larger azimuthal angle).

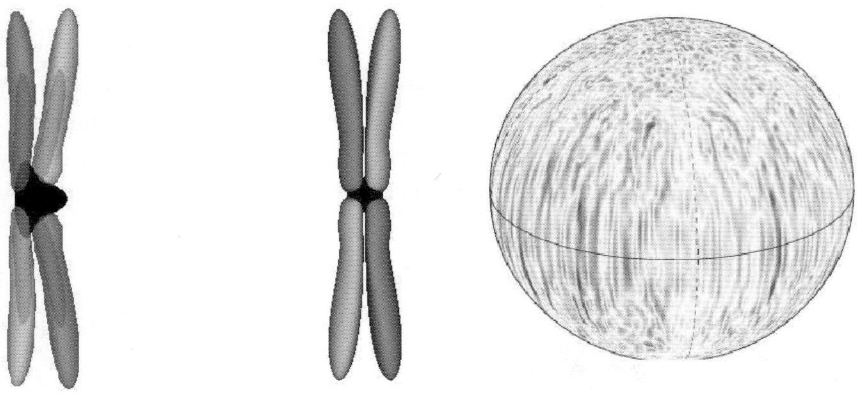

(a) (b) (c)

Fig. 8. (a) A buoyant blob floats radially outward ($Ro = 0.1$) while emitting low-frequency inertial waves. Positive (negative) axial velocity is dark (light). The resulting velocity field comprises four columnar vortices: a cyclone/anticyclone pair above and below the blob[9]. (b) As for (a), but with $Ro = 0.01$. (c) Typical numerical simulation of a dynamo showing alternating cyclone/anticyclone pairs[10].

This general structure is illustrated in Figure 8(a, b), which shows a numerical simulation of the velocity field generated by a buoyant blob (of initially Gaussian profile) which slowly migrates horizontally under the influence of gravity while generating low-frequency inertial waves. The images shown, corresponds to a time $\Omega t = 8$ after the release of the buoyant blob and the Rossby number is $Ro = 0.1$ in (a) and $Ro = 0.01$ in (b). The dispersion pattern is as anticipated above and might be compared with the typical flow pattern seen in numerical dynamo simulations, which consists of thin, alternating cyclones and anticyclones aligned with the rotation axis (Figure 8(c)).

Figure 9 shows what happens when we have a random distribution of such buoyant anomalies concentrated in a horizontal layer and drift-

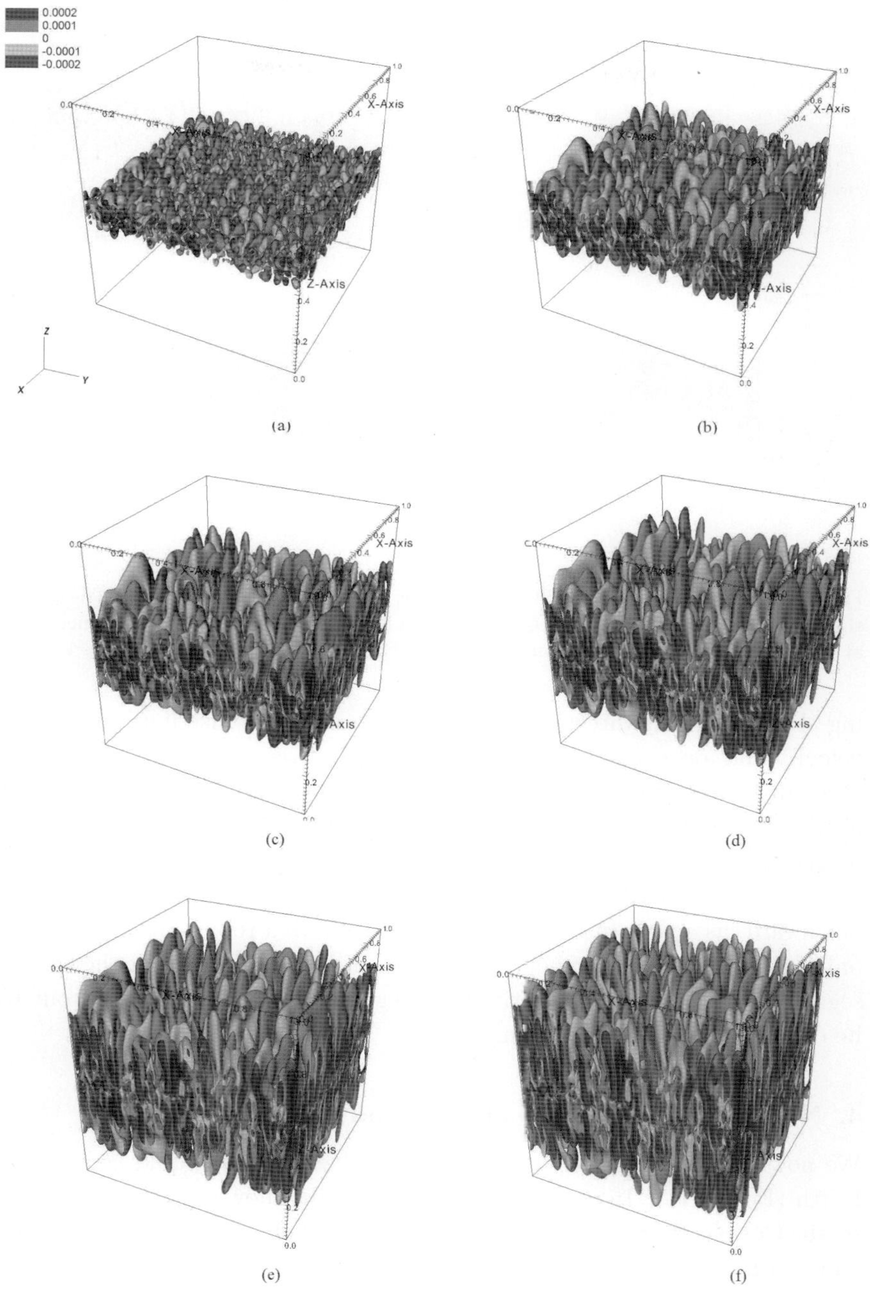

Fig. 9. Dispersion of wave energy from a sea of random buoyant blobs. The panels show iso-surfaces of u_z at $\Omega t = 2, 4, 6, 8, 10, 12$. (Figure from Davidson & Ranjan, 2015.[9])

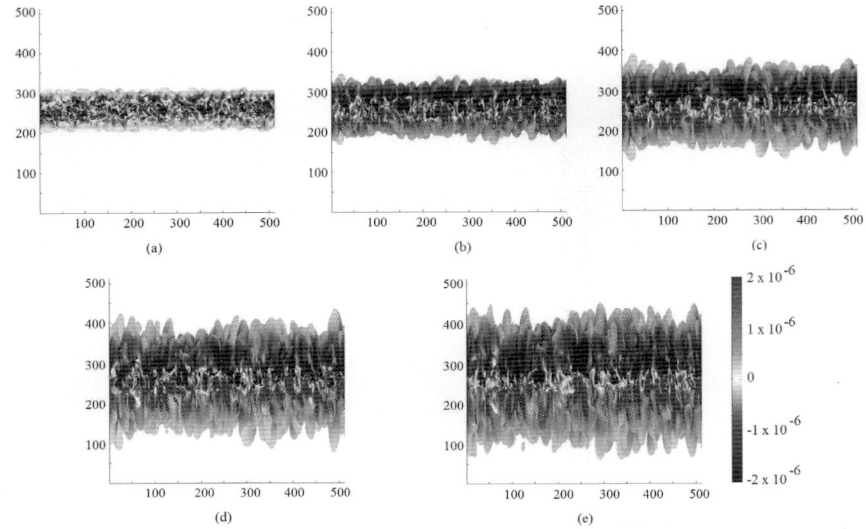

Fig. 10. The same initial condition as Figure 9, but coloured by helicity. The panels show iso-surfaces of $u_z^2/u_{zmax}^2 = 0.025$ coloured by helicity (dark positive, light negative) at $\Omega t = 2, 4, 6, 8, 10$. (Figure from Davidson & Ranjan, 2015.[9]).

ing slowly in the x direction[9]. The images show the distribution of axial velocity and, once again, we see that the inertial waves disperse in the form of axially elongated wave packets, arranged as cyclone-anticyclone pairs. It is striking that the overall flow pattern is highly reminiscent of that seen in dynamo simulations (Figure 8(c)). Equally important is the observation that the wave-packets carry negative helicity upward and positive helicity downward, as shown in Figure 10.[9] Thus we have a robust mechanism not only for the generation of helicity by buoyant blobs, but also for the spatial segregation of regions of positive and negative helicity. This turns out to be crucial for planetary dynamos.

4. Numerical Simulations of the Geodynamo

We now turn to the geodynamo and motion within the liquid core of the Earth (Figure 11). Here the motion is driven by buoyancy but dominated by the Coriolis force, and spatially segregated helicity (one sign in the north and another in the south) is thought to be crucial for the production of the magnetic field[11]. We start with the evidence of the numerical simulations.

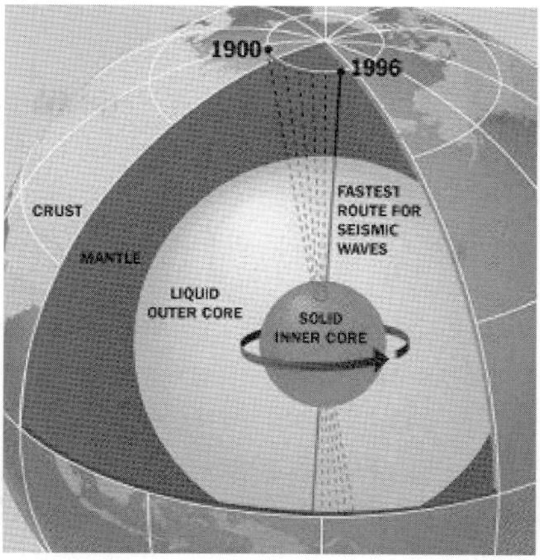

Fig. 11. Schematic of the interior of the Earth.(Figure courtesy of NASA/Science news).

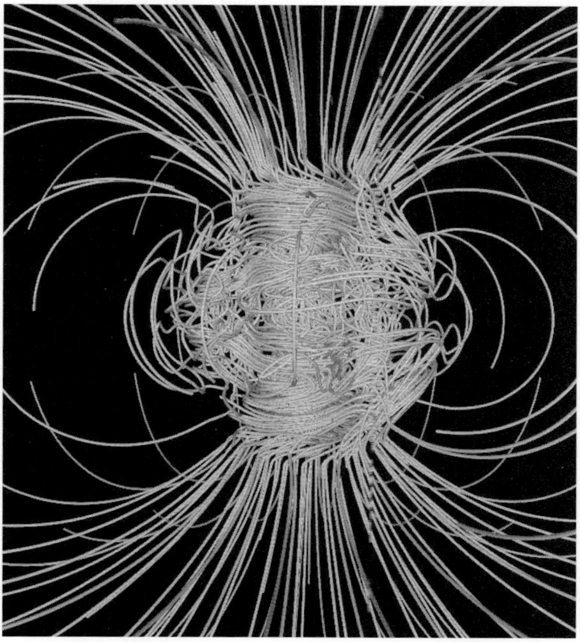

Fig. 12. Computer simulation of the terrestrial magnetic field generated by dynamo action in the liquid core of the Earth. (Figure courtesy of Gary Glatzmaier.)

Twenty years have elapsed since the pioneering dynamo simulations of Glatzmaier and Roberts (1995)[12], and it is striking that many of the current simulations manage to produce plausible Earth-like magnetic fields, i.e. dipolar and roughly aligned with the rotation axis (Figure 12). However, it is important to emphasise from the outset that the current numerical experiments still cannot get close to conditions in the Earth. In part, this is because the simulations cannot realise the extremely low values of the Ekman number in the core (which is around 10^{-15}), so that the computations are inevitably much too viscous, with a significant fraction of the dissipation arising from viscous stresses. By comparison, in the core of the Earth, the viscous stresses are completely negligible and play no significant dynamical role, the dissipation being almost entirely Ohmic. (The kinematic viscosity of liquid iron is smaller than that of air.) Moreover, the buoyant forcing in the numerical simulations is invariably much weaker than that in the core of the Earth, so that the resulting flow is often only mildly chaotic, rather than strongly turbulent. Finally, the time steps in the simulations are strictly limited by the Courant condition, which demands that each time step be of the order of the grid spacing divided by the inertial-wave group velocity. Consequently, in order to facilitate large (i.e. geophysical) time steps, the simulations invariably under-rotate. As a result, the Rossby numbers in the simulations are nearly always much too high. In summary, then, these simulations are overly viscous by a factor of around 10^9, as measured by the Ekman number, underpowered by a factor of at least 100, as measured by the Rayleigh number, and under rotate by a factor of around 10^3, as measured by the Rossby number[13–15].

There is another difference between the numerical simulations and conditions in the core of the Earth: i.e. the thermal boundary condition at the mantle. It turns out that the main driving force for convection in the core of the Earth is predominantly compositional buoyancy, resulting from the slow solidification of the solid inner core, although many of the simulations drive the flow through imposed temperature differences. In any event, the thermal resistance to heat escaping to the surface of the Earth is dominated by the poorly conducting mantle, and so the appropriate thermal boundary condition at the core-mantle boundary is one of fixed heat flux[16]. However, in the simulations it is usually more convenient to impose a fixed temperature difference across the core, which constitutes an additional source of uncertainty in the numerical dynamos.

Never-the-less, despite all of these uncertainties, many of the published numerical experiments exhibit plausible Earth-like features, such as a quasi-steady dipole aligned with the rotation axis and occasional field reversals. Evidently the simulations, while producing much which is misleading, seem to get something correct. An important but unresolved question is: what is it that they do get right?

When analysing the results of the simulations, it is useful to distinguish between events inside and outside the so-called tangent cylinder. This is an imaginary cylinder which is coaxial with the rotation axis, circumscribes the inner core, and spans the outer core from the south to the north. In the numerical dynamos most of the magnetic field generation usually takes place outside the tangent cylinder, where the flow is organised into long columnar convection cells, aligned with the rotation axis (Figure 8(c)). The sign of the rotation alternates from cell to cell, being cyclonic in one and anti-cyclonic in the next. However, as the rotation alternates in sign, so does the vertical velocity u_z, with u_z being anti-symmetric about the equator. Consequently, we find that the helicity density, $h = \boldsymbol{u} \cdot \boldsymbol{\omega}$, is uniformly of one sign in each hemisphere, being negative in the north and positive in the south. This helicity is thought to play a crucial role in dynamo action in the numerical simulations[17,18].

In those simulations which are weakly driven and highly viscous, the columnar vortices are rather static and span the core from north to south. The helicity in those cases comes mostly from Ekman pumping as the cyclones and anticyclones grind away on the mantle. In the moderately-driven, less viscous dynamos, the motion is more turbulent. The flow outside the tangent cylinder is not unlike that for a weakly-driven dynamo, consisting of long, thin columnar vortices, though the vortices are thinner, less regular and more dynamic. Unlike the weakly-forced simulations, the majority of these columnar vortices do not interact strongly with the mantle, so that Ekman pumping is much less effective at producing helicity when the forcing is moderate to strong. Never-the-less, the columnar vortices are observed to possess negative helicity in the north and positive helicity in the south, as in the weakly-forced cases[19]. The origin of this helicity, which is so crucial to dynamo action, is still a controversial and important question, to which we shall return below.

It is interesting to note that, in the moderately-forced dynamos, there tends to be a strong jet-like flow out along the equatorial plane that transports much of the heat and buoyant material from the inner core to the mantle, particularly when a fixed-heat-flux boundary condition is adopted at the mantle, as shown in Figure 13. When viewed in a cut through the equatorial plane this motion consists of random plumes which float out to the mantle. In the Earth, it takes around 300 years for light material to float out from the inner core to the mantle, which might be compared to the time taken by inertial waves to traverse the core, which is around one month. As suggested in the previous section, the flow of buoyant material out along the equatorial plane is likely to be a source of helical waves.

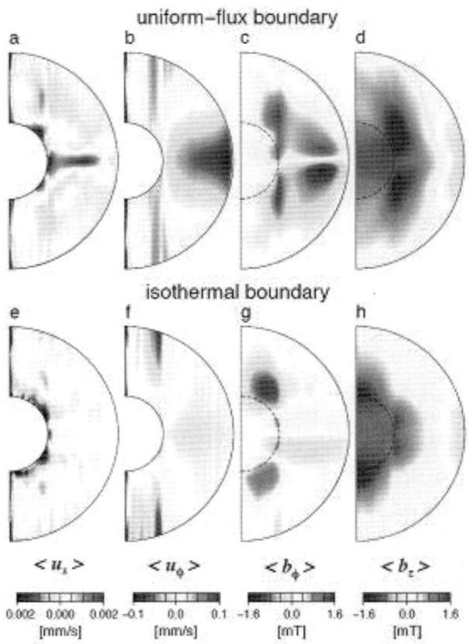

Fig. 13. Comparison of simulations with fixed-flux and fixed-temperature boundary conditions at the mantle. (Top row are fixed flux, bottom row fixed temperature.) From left to right: radial and azimuthal velocity, azimuthal and axial magnetic field[16].

Finally we note that the type of dynamo observed in most of the simulations is a classical α^2-dynamo operating outside the tangent cylinder, as we now discuss. In this respect, the key point to keep in mind is that the columnar vortices outside the tangent cylinder are observed to possess negative helicity in the north and positive helicity in the south. Moreover, classical dynamo theory tells us that helical columnar vortices will interact with an ambient magnetic field $\boldsymbol{B}_0$ to produce a local mean emf, $\langle \boldsymbol{u} \times \boldsymbol{b} \rangle$, which is parallel to $\boldsymbol{B}_0$ if the helicity is negative and antiparallel to $\boldsymbol{B}_0$ if the helicity is positive[11,20]. (At least this is the case when $\boldsymbol{B}_0$ is normal to the axis of the columnar vortices; no emf is generated if $\boldsymbol{B}_0$ is parallel to the columnar vortices.)

Let us adopt global cylindrical polar coordinates,(r, θ, z), centred on the Earth and suppose that the dipole field points to the north (Figure 14). Then the helical flow in the convection columns interacts with the radial

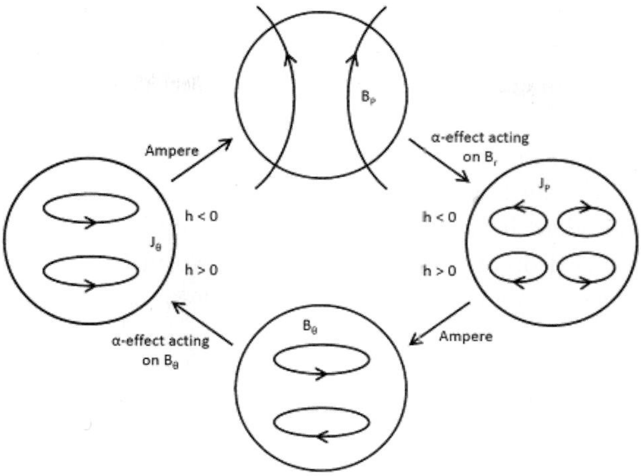

Fig. 14. A classical α^2-dynamo driven by helicity whose sign is antisymmetric about the equator. The dipole points to the north. When h is positive in the south and negative in the north we get B_θ to be positive in the north and negative in the south, which is indeed the distributions of h and B_θ seen outside the tangent cylinder in most of the simulations. (Figure from Davidson, 2014.[8])

field, B_r, to drive a radial emf $\langle \boldsymbol{u} \times \boldsymbol{b} \rangle_r$ which is positive in both the north and the south. (As B_r changes sign, so does the helicity, h.) This emf is weakest near the equatorial plane (where B_r is zero) and largest near the mantle, and so it drives a poloidal current, $\boldsymbol{J}_p$, which has a quadrupole structure, as shown in Figure 14. The associated east-west magnetic field, B_θ, is given by Ampère's law and this is readily seen to be anti-symmetric about the equatorial plane, being positive in the north and negative in the south. This is more or less the sign of B_θ observed in the numerical simulations outside the tangent cylinder. The helical convection rolls now operate on the azimuthal field, B_θ, to produce an azimuthal emf. Since h takes the opposite sign to B_θ in both the north and the south, the induced emf is positive across the core, driving a positive azimuthal current throughout the core. Finally we observe that this positive azimuthal current is exactly what we need to support the original dipole field. Thus we have a simple, self-consistent dynamo.

In this picture, everything rests on the observation that (at least in the simulations) the columnar vortices outside the tangent cylinder possess negative helicity in the north and positive helicity in the south. In the weakly-forced, highly-viscous simulations this asymmetric distribution in

h arises from the Ekman pumping on the mantle. However, the viscous stresses are completely negligible in the core of the Earth. Moreover, as the forcing in the numerical simulations gets stronger, fewer and fewer of the columnar vortices interact with the mantle. A key question, therefore, is what provides this asymmetric distribution of helicity (positive in the south, negative in the north) observed in the more strongly forced numerical dynamos, and what drives the helicity in the Earth's core. In this respect, perhaps a comparison of the equatorial jet shown in Figure 13(a) with the helical wave packets shown in Figure 10 provides a clue.

5. A Cartoon of the Geodynamo Based on Helical Waves

In Davidson (2014)[8] it is argued that the helicity in the more strongly forced simulations, and perhaps also in the Earth, comes from the buoyant plumes which float out along the equatorial plane. In particular, it is suggested that these plumes launch helical wave packets, just as in Figure 10, carrying negative helicity to the north and positive helicity to the south. Close to the equator, where the radial and azimuthal components of $\boldsymbol{B}$ are small, these waves are characterised as inertial waves, but away from the equator, where there is a significant magnetic field transverse to the rotation axis, they are likely to be so-called magnetostropic waves. However, magnetostrpohic waves transport helicity in much the same way as inertial waves; negative helicity to the north and positive helicity to the south.

Of course, in the core of the Earth, there is little direct evidence as to what the flow pattern looks like, but perhaps one important piece of evidence comes from Table 1[2]. Mercury, Earth, Jupiter and Saturn all have dipolar magnetic fields aligned, more or less, with the rotation axis. Moreover, the magnitude of the scaled axial magnetic field is very similar in all four cases. It seems likely, therefore, that the same type of dynamo is operating in all of these planets. However, the internal structure of these planets is very different, and in particular there is no mantle in the gas giants to produce Ekman pumping. The implication is that the all-important helicity in the planetary cores must be produced by a mechanism which is dynamically robust and also insensitive to mechanical boundary conditions. Helical wave generation by buoyant blobs and plumes in and around the equatorial plane satisfies both these conditions.

In Davidson & Ranjan (2015)[9] the magnitude of the emf generated by helical wave packets pushing through an ambient magnetic field is evaluated numerically. The waves are assumed to be launched near the equator

Table 1. Approximate properties of those terrestrial planets and gas giants which have dynamos, along with an estimate of the scaled magnetic field, $(\bar{B}_z/\sqrt{\rho\mu})/\Omega R_C$. Here ρ is the density, estimated as $\rho \sim 10^4 kg/m^3$ for the terrestrial planets and $\rho \sim 10^3 kg/m^3$ for the gas giants.

Planet	Rotation period (days)	Core radius $R_C(10^3 km)$	Dipole moment $\boldsymbol{m}(10^{22} Am^2)$	Mean axial core field $\bar{B}_z$ (Gauss)	Scaled axial field $(\bar{B}_z/\sqrt{\rho\mu})/\Omega R_C$
Mercury	58.6	1.8	0.004	0.014	5.6×10^{-6}
Earth	1	3.49	7.9	3.7	13×10^{-6}
Jupiter	0.413	55	150,000	18	5.2×10^{-6}
Saturn	0.444	29	4500	3.7	2.2×10^{-6}

and it is shown that, while such wave packets start out as inertial waves, they rapidly evolve into magnetostrophic waves as they punch through an ever-increasing east-west field. The resulting magnetostrophic wave packets are elongated along both the rotation axis and the ambient magnetic field (Figure 15). In any event, the size and distribution of the emf induced by these helical waves is shown in Davidson & Ranjan (2015)[9] to be sufficient to support a classical α^2-dynamo operating outside the tangent cylinder, of the type indicated schematically in Figure 14.

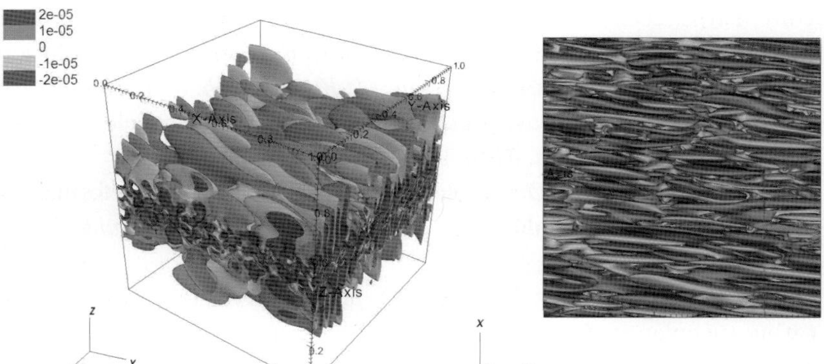

Fig. 15. Repeat of the calculation shown in Figure 9, but including the Lorentz force which elongates the wave packets along the magnetic field lines ($\Omega t = 10$). The magnetic field is in the y direction. (Figure from Davidson & Ranjan, 2015.[9])

6. Summary

We have shown evidence that linear inertial wave propagation is crucial to the dynamics of rapidly-rotating turbulence, spontaneously generating columnar vortices in both the laboratory and in numerical simulations. Indeed, these columnar vortices are often nothing more than linear inertial wave packets. We have also shown that inertial wave packets are particularly effective at generating and then spatially segregating regions of positive and negative helicity. The production and systematic segregation of helicity of opposite signs is central to the operation of planetary dynamos and we have argued that precisely this mechanism may be at work in the more planet-like dynamo simulations, and perhaps also in the core of the Earth, the helical wave packets being driven by buoyant equatorial plumes floating out towards the mantle.

References

1. P. A. Davidson, *Turbulence: an introduction for scientists and engineers* (Oxford Univ. Press, 2004).
2. P. A. Davidson, *Turbulence in rotating, stratified and electrically conducting fluids* (Cambridge Univ. Press, 2013).
3. P. A. Davidson, *An introduction to magnetohydrodynamics* (Cambridge Univ. Press, 2001).
4. P. A. Davidson, P. J. Staplehurst and S. B. Dalziel, On the evolution of eddies in a rapidly rotating system, *J. Fluid. Mech.* **557**, 135 (2006).
5. A. Ranjan and P. A. Davidson, Evolution of a turbulent cloud under rotation, *J. Fluid. Mech.* **756**, 488 (2014).
6. P. J. Staplehurst, P. A. Davidson and S. B. Dalziel, Structure formation in homogeneous freely decaying rotating turbulence, *J. Fluid. Mech.* **598**, 81 (2008).
7. Y. Bin Baqui and P. A. Davidson, A phenomenological theory of rotating turbulence, *Phy. Fluids* **27** (2015).
8. P. A. Davidson, The dynamics and scaling laws of planetary dynamos driven by inertial waves, *Geophys. J. Int.* **198**, 1832 (2014).
9. P. A. Davidson and A. Ranjan, Planetary dynamos driven by helical waves, part 2, *Geophys. J. Int.* **sub-judice** (2015).
10. B. Sreenivasan, Modelling the geodynamo: progress and challenges, *Perspectives Earth Sci.* **99**, 1739 (2010).
11. H. K. Moffatt, *Magnetic field generation in electrically conducting fluids* (Cambridge Univ. Press, 1978).
12. G. A. Glatzmaier and P. H. Roberts, A three-dimensional self-consistent computer simulation of a geomagnetic field reversal, *Nature* **377**, 203 (1995).

13. U. R. Christensen, Dynamo scaling laws and application to the planets, *Space Sci. Rev.* **152**, 565 (2010).
14. U. R. Christensen, Geodynamo models: Tools for understanding properties of earth's magnetic field, *Phys. Earth Plan. Int.* **152**, 565 (2011).
15. U. R. Christensen and J. Wicht, *Numerical dynamo simulations*, in *Treatise on Geophysics*, ed. P. Olson (Elsevier, 2007).
16. A. Sakuraba and P. H. Roberts, Generation of a strong magnetic field using uniform heat flux at the surface of the core, *Nature Geo.* **2**, 802 (2009).
17. C. A. Jones, Planetary magnetic fields and fluid dynamos, *Ann. Rev. Fluid Mech.* **43**, 583 (2011).
18. P. H. Roberts and E. M. King, On the genesis of the earth's magnetism, *Rep. Prog. Phys.* **76**, 1 (2013).
19. P. Olson, U. R. Christensen and G. A. Glatzmaier, Numerical modelling of the geodynamo: mechanisms of field generation and equilibration, *J. Geophys. Res.* **104** (1999).
20. E. N. Parker, Hydromagnetic dynamo models, *Astrophys. J.* **122**, 293 (1955).

Low-Order Models for Control of Fluids: Balanced Models and the Koopman Operator

Clarence W. Rowley

Department of Mechanical and Aerospace Engineering
Princeton University
Princeton, NJ 08544, USA
E-mail: cwrowley@princeton.edu

While the detailed dynamics of transitional and turbulent flows are complex and high-dimensional, many of the important characteristics can be captured by models of surprisingly low dimension. We review two approaches for constructing low-dimensional models from data: in particular, we discuss balanced proper orthogonal decomposition, which applies to linear systems and can give insight into transitional flows; and Koopman spectral analysis, which we use to extract coherent structures from snapshots of turbulent flows.

Keywords: Reduced-order models, balanced truncation, Koopman operator, dynamic mode decomposition

1. Balanced Models

We first describe a method for obtaining low-dimensional models for flows that may be described by linear equations. For instance, many of the important aspects of transitional flows are captured by the Navier-Stokes equations linearized about a laminar steady state. Here, we consider a general linear system

$$\begin{aligned} \dot{\boldsymbol{x}}(t) &= A\boldsymbol{x}(t) + B\boldsymbol{u}(t) \\ \boldsymbol{y}(t) &= C\boldsymbol{x}(t), \end{aligned} \tag{1}$$

where $\boldsymbol{x}(t) \in \mathbb{R}^n$ is the state vector (for instance, the velocity at a grid of points in space), $\boldsymbol{u}(t) \in \mathbb{R}^p$ is the vector of inputs, $\boldsymbol{y}(t) \in \mathbb{R}^q$ is the vector of outputs, and A, B, and C are matrices of appropriate dimension. Here, the inputs $\boldsymbol{u}$ could be unknown external disturbances, or in a flow control setting, actuators that manipulate the flow; the outputs $\boldsymbol{y}$ are typically sensor measurements, or quantities we are interested in predicting with our model. The system (1) may be obtained, for instance, by linearizing the Navier-Stokes equations about a particular flow state.

In principle, a linear system such as (1) can be used directly for control design. However, for fluids problems, the dimension of the system is typically too large for such an approach to be computationally tractable. For instance, for a simulation with 1 million grid points, and 3 flow variables, the state dimension is $n = 3\,000\,000$, and A is a $3\,000\,000 \times 3\,000\,000$ matrix. For this reason, we seek a smaller-dimensional system that captures approximately the same relationship between the inputs $\boldsymbol{u}$ and outputs $\boldsymbol{y}$.

1.1. *Balanced truncation for linear systems*

Balanced truncation is a common procedure for determining such a reduced-order model. The method was introduced by Moore[1], and the key is to balance the properties of *controllability* and *observability*. Controllability measures the degree to which a particular state can be excited by the *inputs* $\boldsymbol{u}(t)$: the idea is that, if a particular direction in state space is never excited by the inputs, then we can safely ignore it in a reduced-order model. This is, in fact, what is done by the standard Proper Orthogonal Decomposition (POD) procedure, as explained in Ref. 2 (§2.2.3). However, controllability is not the complete picture: we must also consider the dual concept of observability. Observability measures the degree to which a particular state excites future *outputs* $\boldsymbol{y}(t)$: from this perspective, if a particular state does not influence the outputs $\boldsymbol{y}(t)$, then we should be able to ignore it in a reduced-order model.

The difficulty is that in certain problems — for instance, non-normal systems as commonly arise in shear flows[3] — states that are weakly controllable may also be strongly observable, and thus they cannot be ignored. With standard POD models, this effect results in modes that have low energy (and are thus weakly controllable), yet are important for the dynamics (since they are strongly observable).

Fortunately, as long as a system is both controllable and observable, it is possible to change coordinates (with a linear transformation of the state $\boldsymbol{x}$) so that, in the new coordinates, the controllability and observability properties are identical. More precisely, these properties are characterized by matrices called the controllability and observability Gramians, and it is possible to find coordinates in which these matrices are equal and diagonal. The idea of balanced truncation is to switch to these coordinates, and then to truncate the states that are the least controllable and observable. The resulting procedure, while not optimal, is provably close to optimal, and has *a priori* error bounds — see Ref. 2 (§2.2.1) or standard texts such as Ref. 4 (§11.4) for details.

1.2. *Balanced POD*

The difficulty with applying balanced truncation directly to fluids problems is computational tractability. In order to determine the appropriate coordinate transformation, one first needs to solve two Lyapunov equations, linear matrix equations whose solutions are full (non-sparse) $n \times n$ matrices, and then find eigenvectors of a related $n \times n$ matrix. For fluids problems, n is usually too large for these computations to be numerically tractable (e.g., $n = 3\,000\,000$ in the example given above).

Balanced POD is a method that approximates balanced truncation, but with computational cost similar to POD. The method is based in spirit on the "method of snapshots" for POD modes, described in Ref. 5, and the notion of empirical Gramians presented in Ref. 6. For balanced POD, one requires data from simulations of the system (1) as well its corresponding adjoint system. In particular, one conducts a simulation for each of the p inputs: if $\boldsymbol{b}_j$ denotes the j-th column of the matrix B, one simulates the system

$$\dot{\boldsymbol{x}}(t) = A\boldsymbol{x}(t), \qquad \boldsymbol{x}(0) = \boldsymbol{b}_j$$

for each $j = 1, \ldots, p$, saving the (direct) snapshots $\boldsymbol{x}(t)$ at various times t. One then simulates the adjoint system for each of the q outputs: letting $\boldsymbol{c}_j$ denote the j-th column of C^T (where $(\cdot)^T$ denotes the matrix transpose), one simulates

$$\dot{\boldsymbol{z}}(t) = A^T\boldsymbol{x}(t), \qquad \boldsymbol{z}(0) = \boldsymbol{c}_j$$

for each $j = 1, \ldots, q$, saving the adjoint snapshots $\boldsymbol{z}(t)$ at various times t. One then takes inner products of all pairs of direct and adjoint snapshots, and finds a singular value decomposition of the resulting data matrix. A reduced-order model may then be found using the algorithm given in Ref. 2 (§3.3), and can be shown to approximate balanced truncation. The number of snapshots is typically much smaller than the number of states n (typically hundreds or thousands, instead of millions), so this procedure ends up being much more computationally tractable than full balanced truncation. For more details of the approach and its relation to balanced truncation, see Refs. 2 or 7 (Chap. 5).

1.3. *Example: linearized channel flow*

To illustrate the approach, we consider the flow through a plane channel, linearized about the (parabolic) laminar profile, as presented in Ref. 8. This flow is well understood: the linearized system is non-normal, and exhibits large transient growth[3], and thus it illustrates well the problems that can arise if the standard POD procedure is used for determining reduced-order models.

Here, we briefly summarize the resuits, but for the full details, see Ref. 8. We consider a plane channel that is periodic in the streamwise and spanwise directions, with a localized disturbance in the vertical velocity located at the center of the channel. An impulsive disturbance produces a large transient increase in energy, followed by an exponential decay back to the (stable) laminar profile. We attempt to model this behavior with POD and balanced POD.

The first 5 POD modes capture 99.72% of the energy, so one might expect a 5-mode POD model to work quite well at approximating the full-order linearized system. However, we find that the resulting 5-mode model performs extremely poorly, and completely fails to capture the transient growth. However, patient experimentation reveals that a 5-mode model consisting of POD modes 1, 2, 3, 10, and 17 performs quite well, and captures the transient growth accurately. Even though POD modes 10 and 17 have much less energy than modes 4 and 5, they turn out to be much more important to the dynamics.

When balanced POD is applied to this same system, even a 3-mode balanced POD model captures the transient energy growth well, and the accuracy of the low-order models increases if one retains more modes.

The reason for the surprisingly poor performance of the POD models is that, for this non-normal flow, some modes that are weakly controllable (POD modes with small energy) are strongly observable — modes 10 and 17 are examples of these — and thus cannot be neglected. Balanced POD, by contrast, captures both of these aspects, and at least for this example, produces models that are much more accurate and reliable.

2. Koopman Spectral Analysis

We now describe a different approach to extracting low-dimensional phenomena from complex systems, using spectral analysis of the Koopman operator. The goal here is not to extract a low-order model suitable for control design, as in the previous section, but rather to extract coherent structures from which we can better understand the dynamics of complex flows. Here, we briefly outline the theoretical foundations of the method; for examples of the method applied to turbulent flows, see, for instance, Refs. 9 or 10.

2.1. *Koopman operator*

Consider a discrete-time dynamical system in which a state $\boldsymbol{x} \in \mathcal{X}$ evolves according to $\boldsymbol{x} \mapsto \boldsymbol{f}(\boldsymbol{x})$. The Koopman operator acts on scalar-valued functions on $\mathcal{X}$, which we call *observables*. In particular, if $g : \mathcal{X} \to \mathbb{C}$ is a complex-valued function (an observable), then the Koopman operator U is given by $Ug = g \circ \boldsymbol{f}$: that is,

$$Ug(\boldsymbol{x}) = g(\boldsymbol{f}(\boldsymbol{x})). \tag{2}$$

It is easy to see that U is a linear operator (i.e., $U(\alpha g + \beta h) = \alpha Ug + \beta Uh$). As Koopman describes in Ref. 11, one can understand the behavior of the dynamical system by studying the eigenvalues and eigenfunctions of U.

Suppose U has eigenvalues λ_j and eigenfunctions φ_j, which satisfy

$$U\varphi_j(\boldsymbol{x}) = \lambda_j \varphi_j(\boldsymbol{x}), \qquad j = 1, 2, \ldots \tag{3}$$

and consider a vector-valued observable $\boldsymbol{g} : \mathcal{X} \to \mathbb{C}^q$. If each component of $\boldsymbol{g}$ may be written as a linear combination of eigenfunctions φ_j, then we may write (as in Ref. 12)

$$\boldsymbol{g}(\boldsymbol{x}) = \sum_{j=1}^{\infty} \varphi_j(\boldsymbol{x}) \boldsymbol{v}_j$$

and we call the vector coefficients $\boldsymbol{v}_j$ the *Koopman modes* associated with the observable $\boldsymbol{g}$. It is shown in Ref. 9 that, if the observable is the full state ($\boldsymbol{g}(\boldsymbol{x}) = \boldsymbol{x}$) and the dynamical system is linear, with $\boldsymbol{f}(\boldsymbol{x}) = A\boldsymbol{x}$, then the eigenvalues of A are also eigenvalues of the Koopman operator, and the eigenvectors of A are Koopman modes. However, the Koopman perspective applies to nonlinear systems as well.

2.2. *Dynamic mode decomposition*

Dynamic Mode Decomposition (DMD) is a method recently introduced to the fluid mechanics community[9,13] for extracting dynamically relevant structures from data, and can be used to approximate Koopman eigenvalues and modes. The description here follows that in Ref. 10, and differs from the original formulation in Refs. 9 and 13. Suppose we have two sets of data:

$$X = \begin{bmatrix} \boldsymbol{x}_1 \, \boldsymbol{x}_2 \cdots \boldsymbol{x}_m \end{bmatrix}, \qquad Y = \begin{bmatrix} \boldsymbol{y}_1 \, \boldsymbol{y}_2 \cdots \boldsymbol{y}_m \end{bmatrix}, \tag{4}$$

with $\boldsymbol{y}_j = \boldsymbol{f}(\boldsymbol{x}_j)$. The DMD modes are eigenvectors of the matrix

$$A = YX^+,$$

where X^+ denotes the Moore-Penrose pseudoinverse of X. It is shown in Ref. 10 (§4.1) that, under certain conditions, the DMD eigenvalues are Koopman eigenvalues, and the DMD modes are Koopman modes. See Theorem 4 of Ref. 10 for the precise conditions under which the DMD eigenvalues/modes correspond to Koopman eigenvalues/modes: in particular, the Koopman eigenfunctions must lie within the span of the observables.

In practice, however, one must be careful when applying DMD to a nonlinear system. Even for a simple system such as shown in Example 4.1.1 of Ref. 10, the Koopman eigenfunctions may not lie in the span of the observables, and this can cause DMD to fail utterly, for instance giving erroneous unstable eigenvalues for a nonlinear system with a single stable fixed point. In order to better understand the relationship between DMD and the Koopman operator, we consider an extended DMD procedure, as described below.

2.3. *Extended DMD*

We briefly summarize the Extended DMD approach described in Ref. 14. In this approach, we explicitly approximate the (infinite-dimensional) Koopman operator using a weighted residual method. We choose a set of observables $\{\psi_j\}_{j=1}^N$, and represent an observable φ and its image $U\varphi$ as linear combinations of the ψ_j:

$$\varphi(\boldsymbol{x}) = \sum_{j=1}^{N} a_j \psi_j(\boldsymbol{x}), \qquad U\varphi(\boldsymbol{x}) = \sum_{j=1}^{N} b_j \psi_j(\boldsymbol{x}).$$

The matrix representation of the Koopman operator U with respect to the basis $\{\psi_j\}$ is a matrix that maps $\boldsymbol{a} = (a_1, \dots, a_N)$ to $\boldsymbol{b} = (b_1, \dots, b_N)$. Of course, the subspace spanned by $\{\psi_j\}$ might not be invariant under U (i.e., $U\varphi$ might not lie in the subspace), and in a weighted residual method, one chooses $\boldsymbol{b}$ so that the error is orthogonal to a space spanned by a set of weighting functions $\{W_i\}_{i=1}^M$. Solving the resulting system gives $\boldsymbol{b} = \Psi_X^+ \Psi_Y \boldsymbol{a}$, with

$$\Psi_X = \begin{bmatrix} \langle W_1, \psi_1 \rangle & \cdots & \langle W_1, \psi_N \rangle \\ \vdots & & \vdots \\ \langle W_M, \psi_1 \rangle & \cdots & \langle W_M, \psi_N \rangle \end{bmatrix}, \quad \Psi_Y = \begin{bmatrix} \langle W_1, \psi_1 \circ \boldsymbol{f} \rangle & \cdots & \langle W_1, \psi_N \circ \boldsymbol{f} \rangle \\ \vdots & & \vdots \\ \langle W_M, \psi_1 \circ \boldsymbol{f} \rangle & \cdots & \langle W_M, \psi_N \circ \boldsymbol{f} \rangle \end{bmatrix},$$

where $\langle \cdot, \cdot \rangle$ denotes the inner product. In order to evaluate the inner products above, in general one needs knowledge of the map $\boldsymbol{f}$. However, if we are given only data, we may not have access to $\boldsymbol{f}$ explicitly. If we make the choice $W_i(\boldsymbol{x}) = \delta(\boldsymbol{x} - \boldsymbol{x}_i)$, where $\boldsymbol{x}_i$ are the data points in (4), then we have

$$\Psi_X = \begin{bmatrix} \psi_1(\boldsymbol{x}_1) & \cdots & \psi_N(\boldsymbol{x}_1) \\ \vdots & & \vdots \\ \psi_1(\boldsymbol{x}_M) & \cdots & \psi_N(\boldsymbol{x}_M) \end{bmatrix}, \qquad \Psi_Y = \begin{bmatrix} \psi_1(\boldsymbol{y}_1) & \cdots & \psi_N(\boldsymbol{y}_1) \\ \vdots & & \vdots \\ \psi_1(\boldsymbol{y}_M) & \cdots & \psi_N(\boldsymbol{y}_M) \end{bmatrix},$$

which may be evaluated directly from the available data points $\boldsymbol{x}_j, \boldsymbol{y}_j$. The approximation of the Koopman operator is then $K = \Psi_X^+ \Psi_Y$. Note the similarity between this formulation and the standard DMD approach described in §2.2: in particular, standard DMD is a (transposed) special case of Extended DMD, with the basis functions $\psi_j(\boldsymbol{x}) = [\boldsymbol{x}]_j$ (the j-th component of $\boldsymbol{x}$). By contrast, with Extended DMD, we explicitly choose the basis functions, and we may tailor these to the problem at hand. Many different choices are possible: orthogonal polynomials, spectral elements, and radial basis functions are some common choices considered in Ref. 14.

In practice, one may think of DMD as providing an approximation of Koopman eigenfunctions using basis functions that are *linear* in $\boldsymbol{x}$ (analogous to a first-order Taylor expansion). For problems for which the eigenfunctions can be approximated accurately using a linear approximation (e.g., in a small neighborhood of a fixed point), then one can expect DMD to produce accurate results. However, for nonlinear problems for which the Koopman eigenfunctions are sufficiently nonlinear (e.g., farther from a fixed point), Extended DMD can provide more accurate results, analogous to retaining more terms in the Taylor expansion.

Acknowledgments

This work was supported by the National Science Foundation and the Air Force Office of Scientific Research.

References

1. B. C. Moore, Principal component analysis in linear systems: Controllability, observability, and model reduction, *IEEE Trans. Automat. Contr.* **26**, 17 (February 1981).
2. C. W. Rowley, Model reduction for fluids using balanced proper orthogonal decomposition, *Int. J. Bifurcation Chaos* **15**, 997 (March 2005).
3. P. J. Schmid and D. S. Henningson, *Stability and Transition in Shear Flows*, Applied Mathematical Sciences, Vol. 142 (Springer-Verlag, 2001).

4. S. Skogestad and I. Postlethwaite, *Multivariable Feedback Control: Analysis and Design*, 2nd edn. (John Wiley and Sons, 2005).
5. L. Sirovich, Turbulence and the dynamics of coherent structures, parts I–III, *Q. Appl. Math.* **XLV**, 561 (October 1987).
6. S. Lall, J. E. Marsden and S. Glavaški, Empirical model reduction of controlled nonlinear systems, in *Proceedings of the IFAC World Congress*, July 1999.
7. P. J. Holmes, J. L. Lumley, G. Berkooz and C. W. Rowley, *Turbulence, Coherent Structures, Dynamical Systems and Symmetry*, 2nd edn. (Cambridge University Press, Cambridge, UK, 2011).
8. M. Ilak and C. W. Rowley, Modeling of transitional channel flow using balanced proper orthogonal decomposition, *Phys. Fluids* **20** (2008).
9. C. W. Rowley, I. Mezić, S. Bagheri, P. Schlatter and D. S. Henningson, Spectral analysis of nonlinear flows, *J. Fluid Mech.* **641**, 115 (December 2009).
10. J. H. Tu, C. W. Rowley, D. M. Luchtenburg, S. L. Brunton and J. N. Kutz, On dynamic mode decomposition: theory and applications, *J. Comput. Dyn.* **1**, 391 (December 2014).
11. B. O. Koopman, Hamiltonian systems and transformations in Hilbert space, *Proc. Nat. Acad. Sci.* **17**, 315 (1931).
12. I. Mezić, Spectral properties of dynamical systems, model reduction and decompositions, *Nonlin. Dyn.* **41**, 309 (2005).
13. P. J. Schmid, Dynamic mode decomposition of numerical and experimental data, *J. Fluid Mech.* **656**, 5 (2010).
14. M. O. Williams, I. G. Kevrekidis and C. W. Rowley, A data-driven approximation of the Koopman operator: extending dynamic mode decomposition, *J. Nonlin. Sci.* (June 2015).

Different Routes of Transition by Spatio-Temporal Wave-Front

Swagata Bhaumik*, Tapan K. Sengupta+ and Vidyadhar Mudkavi++

*Department of Mechanical and Aerospace Engineering, The Ohio State University, USA-43210

+Department of Aerospace Engineering, IIT Kanpur, Kanpur, India

++CSIR National Aerospace Laboratories, Bangalore, India

*E-mail: *swagatameister@gmail.com, +tksen@iitk.ac.in, ++vm@ctfd.cmmacs.ernet.in

Here, we demonstrate by DNS of 3D zero-pressure gradient boundary layer that both K- and H- or N-types of transition, as described in Kachanov (Ann. Rev. Fluid Mech., **26**, 1994) are consequences of low amplitude monochromatic deterministic excitation caused by the growth of spatio-temporal wave-front (STWF). In Bhaumik & Sengupta (Phys. Rev. E, **89**, 043018, 2014), STWF has been established as the precursor of 3D routes of transition. One of the main feature of present DNS is the extreme accuracy of the used compact schemes over a significantly longer computational domain, including the leading edge of the plate. The velocity-vorticity $(\vec{V}, \vec{\Omega})$-formulation used here helps achieving higher accuracy, maintain solenoidality of the vorticity vector and reduce aliasing error for solving Navier-Stokes equation (NSE). Results show that the H-type transition occurs for lower frequency of excitation, while K-type is seen to occur for higher frequency cases. This is in contrast to the theoretical view-point in the literature for H-type transition, which is claimed to occur via triad resonant interaction of spatial TS waves.

Keywords: Direct numerical simulation, Spatio-temporal wave front, Tollmien-Schlichting wave-packet, K- and H-type transition.

1. Introduction

Significant progress have been made for different aspects of early stages of 3D transition through experiments and by computations in recent times, yet the actual routes traversed by flows from laminar to eventual turbulent stage are not completely understood. To explain the unit processes and routes of 3D flow transition mechanisms, experimental efforts have been undertaken by various researchers by monochromatic deterministic excitation of wall bounded shear layers[1–4]. In Klebanoff *et al.*[2], a rectangular ribbon with spanwise spacers was vibrated monochromatically at 1489Hz and above (corresponding non-dimensional frequency $F_f = 2\pi\nu f/U_\infty^2 = 6.03 \times 10^{-4}$), near the surface of a flat plate. Such excitation caused evolution of an aligned pattern of Λ-vortices in the transitional zone, termed subsequently as K-type transition[5,6]. The Λ-vortices have the appearance of hairpin or horse-shoe shape whose central portion is lifted up with respect to the legs of the hairpin vortex. An alternate route of 3D transition by monochromatic excitation at a much lower frequency of 120Hz (corresponding non-dimensional frequency $F_f = 2\pi\nu f/U_\infty^2 = 1.37 \times 10^{-5}$) was reported[7] where Λ-vortices are found to be in staggered arrangement; classified later as H- or N-type transition[5,6].

Subsequent theoretical approaches attributed the H-type breakdown to a resonant mechanism, where a 2D disturbance wave interacts with two oblique 3D waves corresponding to half the frequency of the 2D wave, as theoretically predicted[8,9]. The computational efforts to induce H-type breakdown is attempted[10,11] by simultaneously exciting waves at fundamental frequency and its spanwise modulated sub-harmonic. None of these efforts register the role of STWF which has been highlighted recently[12] as the precursor of flow transition. This is because of the use of significantly shorter computational domain (10,11) and other similar efforts as compared to Ref. 12. In Refs. 13,14, it has been noted that STWF are the spatio-temporal eigenmodes of Navier-Stokes equation (NSE) as observed from the solution of OSE in spatio-temporal framework. It should be noted that the STWF originated due to the onset of excitation as a combination of spatio-temporal eigenmodes of the Orr-Sommerfeld equation (OSE), with weightage which strongly depends on the exciter location and excitation frequency[12] and therefore, properties of STWF are completely different from the pure spatial eigenmodes of OSE, which are termed in literature as Tollmien-Schlichting (TS) waves. Here, we show by the solution of full NSE by high accuracy dispersion relation preserving (DRP) method that transition of wall-bounded flows caused by the the growth of STWF follows both K- and H- or N-types of routes for low amplitude monochromatic deterministic wall-excitation. While the H-type transition is noted for lower frequency excitation cases, K-type is seen to occur for monochromatic excitations with higher frequency.

2. Governing Equations, Formulation and Numerical Methods

In Fig. 1, schematic of the computational domain represented by ABCDEFGH is shown which includes the sharp leading edge of the plate (along the z-axis). Retaining the sharp leading edge of the plate inside the com-

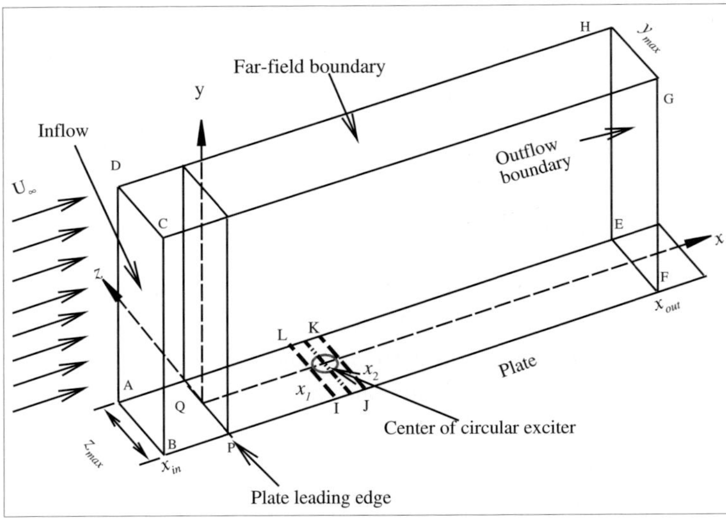

Fig. 1. Schematic diagram of a computational domain.

putational domain is done purposefully to capture disturbances which are created outside the shear layer, but near the leading edge. As shown earlier[15], such disturbances affect viscous instabilities created downstream of the leading edge. Presence of the sharp leading edge can also trigger inviscid instability as illustrated[16] for mixed convection flows. Due to these reasons, we have purposely included the leading edge in the computations domain, unlike the one suggested in Fig. 1 of Ref. 17 and subsequently followed in other similar efforts[10,11,18,19] to study disturbance evolution in a flat-plate boundary layer.

Uniform flow of velocity U_∞ is prescribed at the inflow ABCD. We perturb the flow deterministically by two different types of exciters placed on the wall: (i) Gaussian circular patch (GCP) indicated by the circle centered at the mid-point of IJKL, (ii) spanwise modulated (SM) exciter spanning IJKL which specifies a sinusoidal spanwise variation of excitation amplitude. The second excitation problem mimics the experiments reported[2], where spanwise spacers were used to introduce three dimensionality in the excited disturbances.

Rotational variant of the velocity-vorticity $(\vec{V}, \vec{\Omega})$-formulation of incompressible Navier-Stokes equation is used for simulations along with high accuracy optimized staggered compact schemes for spatial discretization and optimized three-stage Runge-Kutta scheme ORK_3 for time-integration[12]. To damp high-wavenumber upstream propagating spurious numerical oscillations[20] we have added fourth-order numerical diffusion along streamwise and wall-normal directions, while 6th-order implicit central filter[21] is used along periodic spanwise direction. Here, free stream speed, U_∞ is used as the velocity scale, L as the length scale and L/U_∞ as the time scale for nondimensionalizing all physical variables. The Reynolds number (Re_L) based on L is chosen as 10^5 for all the reported calculations. The computational domain extends from $-0.05L$ to $50L$ along the streamwise direction, 0 to L along the wall-normal direction and $-L/2$ to $L/2$ along the spanwise direction where periodic boundary conditions are applied. In terms of the local boundary layer displacement thickness δ^* at the exciter location, $L = 150.11\delta^*$.

The imposed wall-normal velocity component $v_w(x,z)$ for both GCP and SM exciters are given as

$$v_w(x,z) = \alpha A_m(x,z)\sin(\omega_0 t) \tag{1}$$

where α is the amplitude control parameter; $A_m(x,z)$ is the amplitude function with absolute value varying from zero to one and ω_0 is the nondimensional circular frequency given by $\omega_0 = F_f \times Re_L$. Here, F_f is already defined, which is related to the physical frequency f in Hz. The amplitude function $A_m(x,z)$ for the circular Gaussian type excitation[12,22] is given as

$$A_m(x,z) = \frac{1}{2}\left(1 + \cos\left(\frac{\pi r}{r_{max}}\right)\right)$$

for $r \leq r_{max}$ and $A_m(x,z) = 0$ for $r > r_{max}$. Here, $r = \sqrt{(x-x_0)^2 + (z-z_0)^2}$ is the radius of the circular patch, with x_0 and z_0 denoting the center of the circular patch. For the results reported here, $r_{max} = 0.09$, $x_0 = 1.5$ with corresponding $Re_{\delta^*} = 666$ and $z_0 = 0$ are used. For the SM exciter the amplitude function $A_m(x,z)$ is given as

$$A_m(x,z) = A_{FK}(x)\ \sin(8\pi z/z_{max})$$

where $A_{FK}(x)$ is the function corresponding to simultaneous blowing and suction, as described in Ref. 19. This exciter extends from $x_1 = 1.455$ to $x_2 = 1.545$, along the streamwise direction The spanwise modulated exciter is also centered around $x_0 = 1.5$ ($Re_{\delta^*} = 666$). Here, the discussed results are for the case when the amplitude control parameter is $\alpha = 0.01$. Hence, the maximum amplitude of the perturbation is 1% of the free-stream velocity.

We choose the $(\vec{V}, \vec{\Omega})$-formulation of incompressible NSE, over conventional primitive variable pressure-velocity formulation for its superior accuracy[22]. The chosen rotational variant of $(\vec{V}, \vec{\Omega})$-formulation automatically satisfies divergence-free condition on vorticity, while maintaining high accuracy of the computed solution at a lesser computational cost[22]. The attendant u- and w-components of velocity (streamwise and spanwise components of velocity) are obtained by solving corresponding velocity Poisson equations, whereas the v-component of the velocity (wall-normal component) is calculated by integrating the continuity equation from the wall, thereby satisfying the solenoidality of the velocity vector.

It is noted that unlike other effort[11], no random perturbations are added in the flow simulations. The streamwise extent of our computational domain is also considerably larger than similar numerical efforts reported[10,11], where a smaller domain from L to approx. $10L$ is used, which also includes the fringe buffer zone at outflow. Such a long computational domain as employed here is absolutely necessary to capture the complete dynamics and evolution of the STWF reported[12].

3. Brief Description of STWF

Instability theories based on spatial eigenvalue analysis predict only the asymptotic growing wave instead of a wave-packet, without revealing the details of the disturbance field near the exciter. However, when space-time dependent receptivity problems are solved, three elements of disturbance structures are noted[23–25]. The first component is the local solution which is obtained very close to the exciter. This is followed by the TS wave-packet corresponding to the frequency of excitation and its super-harmonics. The third component is the STWF composed of the spatio-temporal eigenfunctions of Orr-Sommerfeld operator[12]. The properties of STWF created due to monochromatic excitation is quite unique and completely different from the conventional Tollmien-Schlichting (TS) waves. In the spatio-temporal linearized parallel receptivity framework, the disturbance field corresponding to the imposed monochromatic wall excitation $v_d(x, 0, t) = \delta(x)\, e^{i\omega_0 t} H(t)$ is given by[23]

$$v_d(x, y, t) = \left(\frac{1}{2\pi}\right)^2 \int_{\alpha, Br} \int_{\omega, Br} \phi(y; \alpha, \omega) \exp\left[i(\alpha x - \omega t)\right] d\alpha\, d\omega \qquad (2)$$

Here, the integration is carried out along appropriately chosen Bromwich contours in α- and ω-planes[13]. Here, $H(t)$, $v_d(x, y, t)$ and ω_0 represent the

Heaviside function due to finite start-up of the excitation at $t = 0$; the disturbance wall-normal velocity component and the excitation frequency, respectively. In Eq. (2), α and ω are complex wavenumber and circular frequency, respectively. It is noted that Eq. (2) is obtained under linear assumption and thus, amplitude of excitation does not play any role for such analysis. The quantity $\phi(y)$ in Eq. (2) should satisfy the boundary conditions $\phi(0) = 0$ and $\phi'(0) = 1/(\omega - \omega_0)$ at the wall. The general expression for $\phi(y)$ subjected to these boundary conditions at the wall is given as[23]

$$\phi(y) = \left[\frac{i}{(\omega - \omega_0)}\right]\left[\frac{\phi_3(0)\phi_1(y) - \phi_1(0)\phi_3(y)}{\phi_1(0)\phi_3'(0) - \phi_3(0)\phi_1'(0)}\right] \tag{3}$$

where, $\phi_1(y)$ and $\phi_3(y)$ are the two fundamental solutions of OSE, which decay in the free stream ($y \to \infty$). The detailed derivation of Eq. (3) can be found in Ref. 23. One notes from the expression of $\phi(y)$ that all complex frequencies centered around the excitation frequency ω_0 are excited. The STWF created due to this time-harmonic excitation which is turned on at $t = 0$, is therefore a linear combination of all spatio-temporal eigenmodes of OSE with a weight proportional to $1/(\omega - \omega_0)$. In contrast, the excited TS waves are the spatial eigenfunctions of OSE at the excitation frequency ω_0. Therefore, the evolution and propagation of STWF exhibits completely different dynamics as opposed to the TS waves. The latter remains localized in space close to the exciter[23,26]. As the amplitude of the constituent eigenmodes of STWF are determined by its proximity to ω_0, the excitation frequency plays a vital role in its subsequent dynamics as it evolves and propagates downstream[12,26].

In Gaster & Grant[27], created wave-packets due to impulsive excitation of Blasius boundary layer is explained by numerically summing unstable TS modes obtained from spatial instability studies. In this context, it is pertinent to note that conventionally one identifies the TS waves as the eigenfunctions of OSE obtained from only spatial stability analysis. Therefore, there exists a definitive difference between STWF and TS waves, with former can not be classified as a part of the latter. Any type of switching of the exciter at any finite time-instant, however gradual it might be, would cause the generation of the STWF via the term $(\omega - \omega_0)(\phi_1(0)\phi_3'(0) - \phi_3(0)\phi_1'(0))$ in the denominator of Eq. (3). Although STWF contains a wide range of spatio-temporal modes with each mode having distinctive frequency and wavenumber content, the excited spatio-temporal modes and their relative amplitudes would strongly be influenced by the location and the frequency of excitation, as suggested by the the denominator of Eq. (3). A special feature of STWF is that unlike the impulsive excitation case considered[27], one can not express it purely as a summation of spatial TS-modes of equal or weighted amplitude. This aspect of spatio-temporal receptivity of boundary layer is generally overlooked in the literature. For example, even if while explaining the disturbance generated by a monochromatically vibrating ribbon[28], the problem is posed as a spatio-temporal problem (Eq. 2(a) of Ref. 28 which specifies the imposed wall perturbation as $v(0; x, t) = \cos(\omega_0 t)\delta(x)H(t))$, it is in essence treated as a signal problem, as the complicated resultant response of the linearized dynamical system can be nominally expressed as $\sum_i A_i(\omega_0; y)e^{(i\alpha(\omega_0)x-\omega_0 t)}$, where the summation is carried over discrete and continuous modes, as given by Eqs. (48) to (54) of Ref. 28. It is pertinent to note that a signal problem essentially suggests that the exciter is switched on at $t = -\infty$ and by the time one is looking at the solution, the STWF convects to $x = \infty$. Similar treatment of monochromatic excitation of shear-layers, where the exciter is switched on at $t = 0$, by corresponding signal problem is also reported in Hill[29] for wall-bounded flows and for free-shear layers in Huerre and Monkewitz[30].

The existence of STWF as the spatio-temporal eigenmodes of NSE as a distinct entity from the corresponding spatial eigen-modes are first reported from the solution of OSE in spatio-temporal framework in Sengupta, Rao & Venkatasubbaiah[13,14]. It is noted that STWF in the present context is significantly different in terms of the properties and characteristics of the constituent modes from the disturbance field reported[31–33] for the transient growth obtained for optimal initial perturbation. Similar distinction of STWF from different types of wave-packets described[34–36] in the context of bypass transition of wall-bounded flows can also be made. The detailed discussions in this regard corresponding to the monochromatic excitation are available[23].

It is pertinent to note that numerical simulations of boundary layer receptivity to deterministic time-periodic wall-excitation have also been reported[10,11,17–19,37–42]. However, unfortunately none of these references report the existence or dynamics of STWF, generated as a consequence of switching the exciter on at $t = 0$. This is due to the fact that in these references and the computational domain is short in the streamwise direction the STWF quickly convect out of the domain without displaying appreciable growth. In this context, We note the following comments[37] for the start-up of the exciter at $t = 0$: *"the disturbances are introduced for t > Q into a flow that was, prior to that, steady and undisturbed, the propagating, time-periodic Tollmien-Schlichting waves are preceded by a wave-front that is essentially a wave packet. Of course, the nonlinear behavior of the leading wave packet would be very different from that of pure Tollmien-Schlichting waves. The more complicated developments in this leading wave packet are not considered here, but will be the subject of another detailed investigation."*

Unlike the spatial TS wave-packet which remains localized in space, the STWF displays interesting dynamics as it propagates downstream[12,24]. It exhibits significant multi-modal interactions and also mode-switching as illustrated for 2D boundary layers. A special feature of the dynamics of STWF is self-regeneration during nonlinear stages of its evolution as explained[24,26] by which it keeps on inducing newer disturbance wave-packets upstream. These induced wave-packets in turn grows at a significantly higher rate, which considerably extends the highly-perturbed zone in both upstream and downstream directions. This self-regeneration mechanism sets STWF apart from any other transients noted in various other physical phenomena, thereby establishing it as the precursor of transition to turbulence in the context of 2D and 3D disturbances[12,24].

4. Results and Discussion

In Figs. 2(a) and 2(b) evolving disturbance streamwise velocity u_d at a height of $y = 0.00189$ (which is $0.283\delta^*_{ex}$ where δ^*_{ex} is the local boundary layer displacement thickness at the exciter location) are shown for time-harmonic GCP and SM exciters, respectively marking the various components of the disturbance field. Similar to 2D disturbance field[13,24], for 3D excitation cases also one notes the existence of three-component solution structure. The local solution is obtained very close to the exciter, and is followed by the TS wave-packet corresponding to the frequency of excitation and its super-harmonics. For the frequencies investigated, one also notices the presence of STWF for the displayed cases at all spanwise stations. The amplitude and phase of STWF shows significant variation as it propagates downstream. One notes from Figs. 2(a, b) that TS wave-packet does not grow and cause transition; instead STWF is the main precursor for it, as also conclusively established before[12]. The finding of STWF is indeed unique, as its existence and dynamics has not been reported in other simulations in the literature for ZPG boundary layer deterministically excited. The fact that the presented results of deterministic excitation for NSE could capture and explain the dynamics of the STWF, is related to

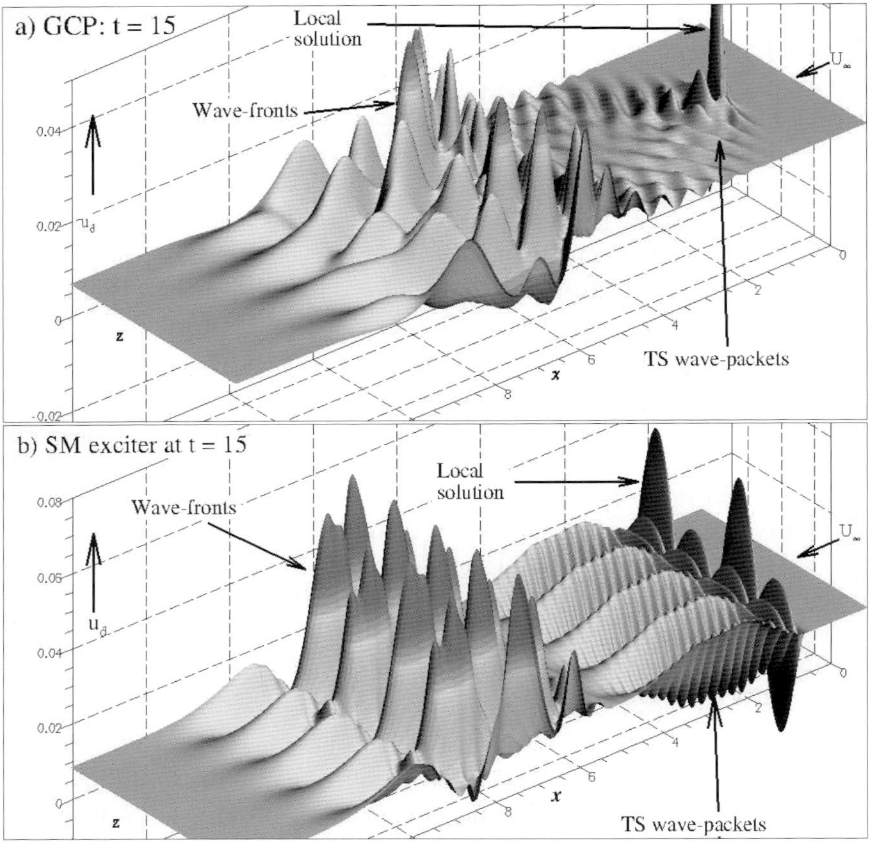

Fig. 2. Perspective plot of u_d at a fixed height shown for (a) GCP and (b) SM exciters. The frequency of excitation is $F = 2\pi\nu f/U_\infty^2 = 0.5 \times 10^{-4}$; ν is kinematic viscosity and f is frequency in Hz.

usage of DRP method in a longer domain.

In Fig. 3, we plot the time variation of maximum peak-to-peak amplitude of STWF at indicated spanwise stations for GCP excitation case of Fig. 2(a). Along $z = 0$, the amplitude of STWF initially decreases due to streamwise and spanwise dispersion caused by perturbation to spread over larger areas, which after $t = 3$ reaches a saturated state due to equilibrium between dispersion and linear amplification. For other spanwise locations also, the primary growth is followed by a saturation stage. This stage is followed by nonlinear secondary and steep tertiary growth. For $z = 0.37$ and 0.47, one does not note any secondary growth stage, unlike $z = 0$, due

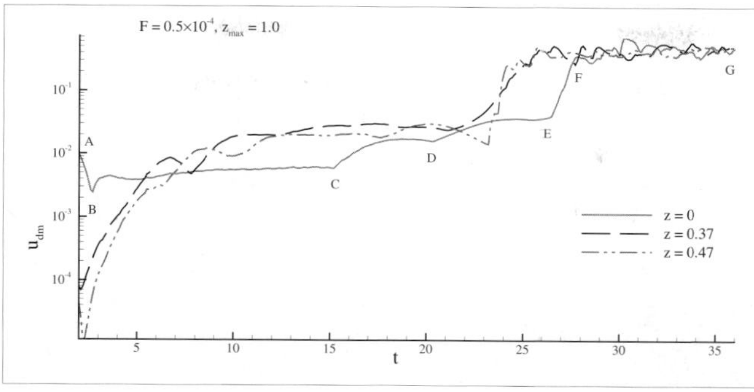

Fig. 3. Maximum peak-to-peak amplitude of STWF at a fixed height and indicated spanwise stations plotted as a function of time for GCP exciter cases with $F = 0.5 \times 10^{-4}$.

to the appearance of disturbances there after the linear stage is over. At these stations steep nonlinear growth commences earlier than at $z = 0$. At all the spanwise stations, secondary and tertiary growth lead to saturation of its amplitude at a level, which is almost 30% of U_∞.

4.1. *K- and H-type of transition*

For 3D flows it is challenging to characterize the disturbance evolution unlike for 2D flows, where one can visualize the flow structures from contour plots. Transition process in 3D flows are dominated by highly unsteady vortical structures with all three components equally important. Accurately representing these vortical structures unambiguously is quite difficult. Here, we capture the unsteady vortical structures by negative λ_2 iso-surfaces following the method proposed[43] to capture the unsteady vortical structures in later stages of 3D transition process. Here, λ_2 is the second eigenvalue of the symmetric matrix $S_{ik}S_{kj} + \Omega_{ik}\Omega_{kj}$ representing rate of strain tensor, where S_{ij} and Ω_{ij} are symmetric and anti-symmetric part of the velocity gradient tensor, respectively.

In Figs. 4(a) and 4(b) we have plotted the perspective of $\lambda_2 = -0.0025$ iso-surface (colored by streamwise velocity) for the GCP exciter cases with $F = 10^{-4}$ and $F = 0.5 \times 10^{-4}$, respectively. In Fig. 4, trailing part of the STWF is focused to show later stages of transition in the perspective plot. One notes the formation of Λ-type vortices in all cases with lifted Ω-like element of the hairpin vortices at the center. In Fig. 4(a), the darker spots denote the Ω-like element at the top of the Λ-vortices.

In Fig. 4(a), we note an aligned pattern of hairpin-vortices, while these

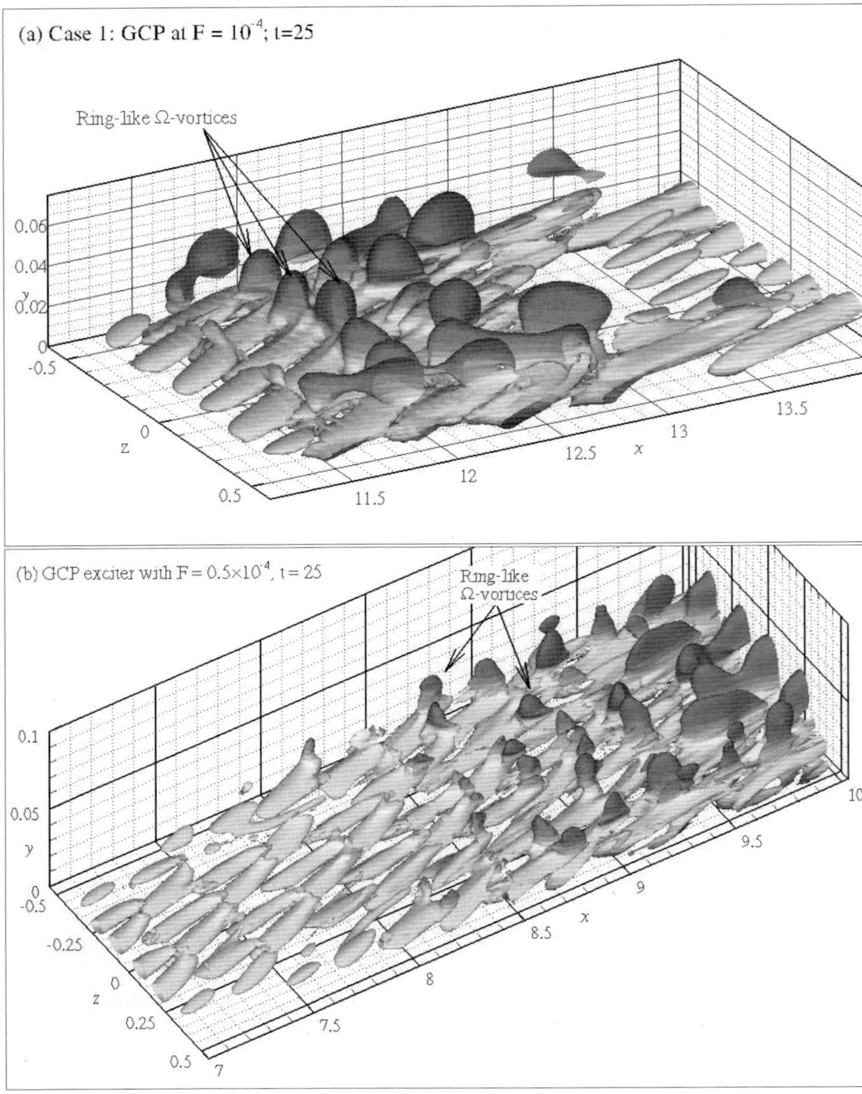

Fig. 4. Perspective view of $\lambda_2 = -0.0025$ iso-surface is shown in (x, z)-plane at $t = 25$ after the onset of excitation corresponding to GCP exciter case with (a) $F = 1.0 \times 10^{-4}$ and (b) $F = 0.5 \times 10^{-4}$. Flow is from left to right.

are arranged as λ vortices in Fig. 4(b), in a staggered arrangement. Therefore, the case shown in Fig. 4(b) has been conjectured to display subharmonic route of transition[5–7,11]. Here both the cases noted in Fig. 3 are due to monochromatic excitation, with the only difference in the excitation frequency. Thus, the present results clearly indicate that for moderate frequencies one notices K-type transition, as in Fig. 4(a), while H-type transition occurs at significantly lower frequencies.

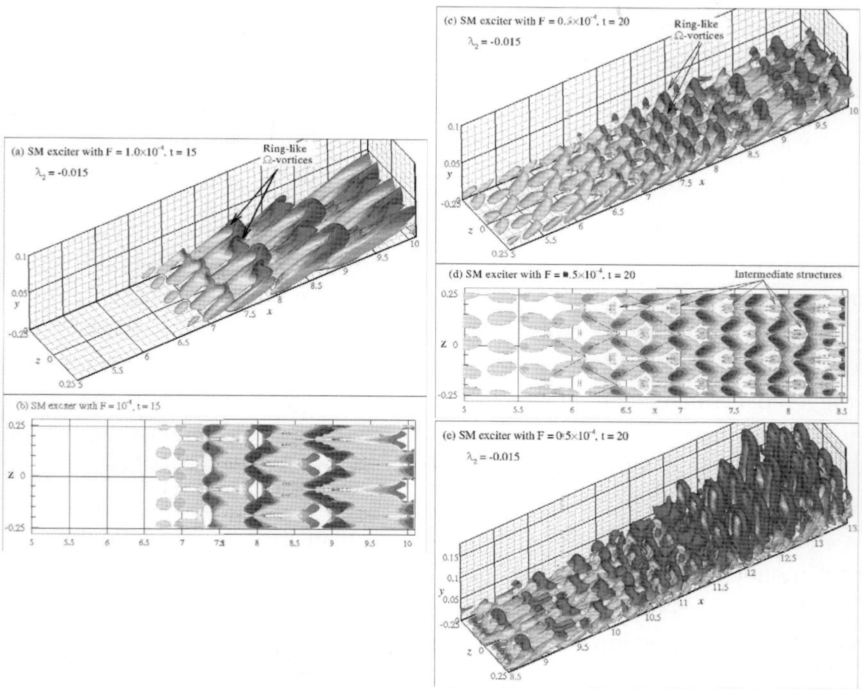

Fig. 5. Perspective and top view (in (x, z)-plane) of $\lambda_2 = -0.015$ iso-surface are shown for the SM exciter case with $F = 10^{-4}$ (a, b) and $F = 0.5 \times 10^{-4}$ (c, d, e). Flow is from left to right.

Similar differences in the arrangement of the vortices are also noted for the SM exciter cases in Figs. 5(a, b) and 5(c, d), for $F = 10^{-4}$ and $F = 0.5 \times 10^{-4}$, respectively. While a definitive staggered pattern of Λ-vortices are noted for $F = 0.5 \times 10^{-4}$ case, almost aligned arrangement of these vortices are seen for almost aligned for $F = 10^{-4}$ case. In both the cases, the existence of ring-like Ω-vortices connecting the legs of the vortices are noted. These are also known as hairpin vortices[18]. One notes from Figs. 5(a) and 5(c) that while for higher frequency, the hairpin vortices are aligned along the x-direction, these make an oblique angle for the lower frequency case. For $F = 10^{-4}$ case, Ω-vortices connect the two spanwise neighboring hairpin vortices (see Figs. 4(a) and 5(a)). For lower frequency case, Ω-vortices only connect the hairpin vortices, whose lifted front parts are close to each other. In Fig. 5(d), in between the dominant Λ-vortices, one notes the existence of some intermediate structures which appear due to the induction of pressure variation by the adjoining Λ-vortices[18]. At further downstream locations these structures become streamwise elongated, giving rise to the U-shaped vortices noted in Fig. 5(e), which penetrates the shear-layer thickness.

5. Summary and Conclusion

Present simulation results with non-dimensional frequency of $F = 10^{-4}$ shows existence of aligned Λ vortices at the early stages of flow transition by growing STWF, which are similar to K-type transition noted experimentally[2] for moderate to high frequency monochromatic excitation. For lower frequency of excitation of $F = 0.5 \times 10^{-4}$, H-type of transition with staggered arrangement of Λ vortices[7]. Thus, we demonstrate that both K- and H-type of transitions are triggered by monochromatic excitation via the growth of STWF, with the latter occurring for lower frequency of excitation.

It is noted that both K- and H-types of transition caused by deterministic monochromatic excitation are by the growth of STWF. All H-type transition routes demonstrated in experiments[7] and simulations[10,11] were thought to be via amplification of spatial TS waves, where a fundamental frequency along with its sub-harmonic is excited. This is the theoretical resonant triad interaction of a fundamental with its sub-harmonic frequencies[8,9]. Present results show the transition to occur for both K- and H-/N-type routes by monochromatic excitation solely by growth of STWF without requiring any subharmonic frequency excitation. These observations opens up new avenues of research for H-/N-type transition via detailed dynamics of the STWF and its role on eventual flow transition.

References

1. G. B. Schubauer and H. K. Skramstad, Laminar boundary layer oscillations and the stability of laminar flow, *J. Aero. Sci.* **14**, 69 (1947).
2. P. S. Klebanoff, K. D. Tidstrom and L. M. Sargent, The three-dimensional nature of boundary-layer instability, *J. Fluid Mech.* **12**, 1 (1962).

3. L. S. G. Kovasznay, H. Komoda and B. R. Vasudeva, *Detailed flow-field in transition, Proc. Heat Transfer and Fluid Mech. Inst.*, Stanford Univ. Press, Palo Alto, California, (1962).
4. F. R. Hama and J. Nutant, Detailed flow-field observations in the transition process in a thick boundary layer, *In Proc., Heat Transfer and Fluid Mech. Inst.* Stanford Univ. Press, **77** (1963).
5. Y. S. Kachanov, Physical mechanisms of laminar-boundary-layer transition, *Ann. Rev. Fluid Mech.* **26**, 411 (1994).
6. W. S. Saric and A. S. W. Thomas, *Experiments on the subharmonic route to turbulence in boundary layers. In Turbulence and Chaotic Phenomena in Fluids, Ed. T. Tatsumi*, North Holland: Elsevier, USA, (1984).
7. Y. S. Kachanov and V. Y. Levchenko, The resonant interaction of disturbances at laminar-turbulent transition in a boundary layer, *J. Fluid Mech.* **138**, 209 (1984).
8. A. D. D. Craik, Non-linear resonant instability in boundary layers, *J. Fluid Mech.* **50**, 393 (1971).
9. M. Zelman and I. Maslennikova, Tollmien-Schlichting-wave resonant mechanism for subharmonic-type transition, *J. Fluid Mech.* **252**, 449 (1993).
10. W. Würz, D. Sartorius, M. Kloker, V. I. Borodulin and Y. S. Kachanov, Detuned resonances of Tollmien-Schlichting waves in an airfoil boundary layer: Experiment, theory, and direct numerical simulation, *Phys. Fluids* **24**, 094103 (2012).
11. T. Sayadi, C. W. Hamman and P. Moin, Direct numerical simulation of complete H-type and K-type transitions with implications for the dynamics of turbulent boundary layers, *J. Fluid Mech.* **724**, 480 (2013).
12. S. Bhaumik and T. K. Sengupta, Precursor of transition to turbulence: Spatiotemporal wave front, *Phys. Rev. E* **89**, 043018 (2014).
13. T. K. Sengupta, A. K. Rao and K. Venkatasubbaiah, Spatio-temporal growing wave-fronts in spatially stable boundary layers, *Physical Review Letters* **96**, 224504(1) (2006a).
14. T. K. Sengupta, A. K. Rao and K. Venkatasubbaiah, Spatio-temporal growth of disturbances in a boundary layer and energy based receptivity analysis, *Phys. Fluids* **18**, (2006).
15. T. K. Sengupta, S. De and S. Sarkar, Vortex-induced instability of an incompressible wall-bounded shear layer, *J. Fluid Mech.* **493**, 277 (2003).
16. T. K. Sengupta, S. Bhaumik and R. Bose, Direct numerical simulation

of transitional mixed convection flows: Viscous and inviscid instability mechanisms, *Phys. Fluids* **25**, 094102 (2013).
17. L. Kleiser and T. A. Zang, Numerical simulation of transition in wall-bounded shear flows, *Ann. Rev. Fluid Mech.* **23**, 495 (1991).
18. B. A. Singer and R. D. Joslin, Metamorphosis of a hairpin vortex into a young turbulent spot, *Phys. Fluids* **6**, 3724 (1994).
19. H. Fasel and U. Konzelmann, Non-parallel stability of a flat-plate boundary layer using the complete Navier-Stokes equations, *J. Fluid Mech.* **221**, 311 (1990).
20. T. K. Sengupta, Y. G. Bhumkar, M. K. Rajpoot, V. Suman and S. Saurabh, Spurious waves in discrete computation of wave phenomena and flow problems, *Applied Math. Comput.* **218**, 9035 (2012).
21. M. R. Visbal and D. V. Gaitonde, On the use of higher-order finite-difference schemes on curvilinear and deforming meshes, *J. Comput. Phys.* **181**, 155 (2002).
22. S. Bhaumik and T. K. Sengupta, A new velocity–vorticity formulation for direct numerical simulation of 3D transitional and turbulent flows, *J. Comp. Phys.* **284**, 230 (2015).
23. T. K. Sengupta, *Instabilities of flows and transition to turbulence*, CRC Press, USA (2012).
24. T. K. Sengupta and S. Bhaumik, *Phys. Rev. Lett.* **107**, 154501 (2011).
25. S. Bhaumik, *Direct Numerical Simulation of Inhomogeneous Transitional and Turbulent Flows* Ph. D. Thesis, I. I. T. Kanpur, (2013).
26. T. K. Sengupta, S. Bhaumik and Y. G. Bhumkar, Direct numerical simulation of two-dimensional wall-bounded turbulent flows from receptivity stage, *Phys. Rev. E.* **85**, 026308 (2012).
27. M. Gaster and I. Grant, An experimental investigation of the formation and development of a wave packet in a laminar boundary layer, *Proc. Royal Soc. London. A.: Mathematical and Physical Sci.* **347**, 253 (1975).
28. D. E. Ashpis and E. Reshotko, The vibrating ribbon problem revisited, *J. Fluid Mech.* **213**, 531 (1990).
29. D. Hill, Adjoint systems and their role in the receptivity problem for boundary layers, *J. Fluid Mech.* **292**, 183 (1995).
30. P. Huerre and P. Monkewitz, Absolute and convective instabilities in free shear layers, *J. Fluid Mech.* **159**, 151 (1985).
31. K. S. Breuer and M. T. Landahl, The evolution of a localized disturbance in a laminar boundary layer. part 2. strong disturbances, *J. Fluid Mech.* **220**, 595 (1990).

32. L. Trefethen, A. Trefethen, S. Reddy, T. Driscoll *et al.*, Hydrodynamic stability without eigenvalues, *Science* **261**, 578 (1993).
33. D. S. Henningson, A. Lundbladh and A. V. Johansson, A mechanism for bypass transition from localized disturbances in wall-bounded shear flows, *J. Fluid Mech.* **250**, 169 (1993).
34. X. Wu, R. G. Jacobs, J. C. Hunt and P. A. Durbin, Simulation of boundary layer transition induced by periodically passing wakes, *J.Fluid Mech.* **398**, 109 (1999).
35. P. Schlatter and R. Örlü, Assessment of direct numerical simulation data of turbulent boundary layers, *J. Fluid Mech.* **659**, 116 (2010).
36. P. Schlatter and R. Örlü, Turbulent boundary layers at moderate Reynolds numbers: inflow length and tripping effects, *J. Fluid Mech.* **710**, 5 (2012).
37. H. Fasel, U. Rist and U. Konzelmann, Numerical investigation of the three-dimensional development in boundary-layer transition, *AIAA J.* **28**, 29 (1990).
38. P. R. Spalart and K.-S. Yang, Numerical study of ribbon-induced transition in blasius flow, *J. Fluid Mech.* **178**, 345 (1987).
39. R. Joslin, C. Streett and C.-L. Chang, Spatial direct numerical simulation of boundary-layer transition mechanisms: Validation of PSE theory, *Theo. Comput. Fluid Dyn.* **4**, 271 (1993).
40. R. Joslin and C. Streett, The role of stationary cross-flow vortices in boundary-layer transition on swept wings, *Phys. Fluids* **6**, 3442 (1994).
41. L. Kral and H. Fasel, Direct numerical simulation of passive control of three-dimensional phenomena in boundary-layer transition using wall heating, *J. Fluid Mech.* **264**, 213 (1994).
42. C. Stemmer, M. J. Kloker and S. Wagner, Navier-Stokes simulation of harmonic point disturbances in an airfoil boundary layer, *AIAA J.* **38**, 1369 (2000).
43. J. Jeong and F. Hussain, On the identification of a vortex, *J. Fluid Mech.* **285**, 69 (1995).

Bypass Transitional Flow Past an Aerofoil With and Without Surface Roughness Elements

Yogesh G. Bhumkar[1,2,*], Tony W. H. Sheu[1], Tapan K. Sengupta[3]

[1] *Center of Advanced Study in Theoretical Sciences, National Taiwan University, No. 1, Sec. 4, Roosevelt Road, Taipei, 10617, Taiwan, ROC.*

[2] *School of Mechanical Sciences, Indian Institute of Technology Bhubaneswar, Bhubaneswar, 751013, Odisha, India.*

[3] *High Performance Computing Lab., Aerospace Engineering Department, Indian Institute of Technology Kanpur, Kanpur, 208016, U.P., India.*

[*] *E-mail: bhumkar@iitbbs.ac.in*

Here, we have studied development of bypass transitional flow past a NACA 0012 aerofoil, with and without leading edge roughness elements. Calculations are performed for a Reynolds number of $Re = 1.25 \times 10^6$. We have used high accuracy, dispersion relation preserving (DRP) numerical schemes with appropriate numerical filters. We have discussed how bypass transition is triggered in an impulsively started flow past aerofoil. We have further analyzed the spectrum of temporal variation of vorticity at various locations on aerofoil surface along with variation of the RMS component of velocity fluctuations with the intermittency factor to characterize transitional flow.

Keywords: Vortex dynamics; Bypass transition; Boundary layer separation

1. Introduction

Flow past an aerofoil is an important research area for designing aircraft wings, turbine blades and windmills. Aerofoils are designed to achieve certain design parameters according to the requirements of the specific application. Study of fluid flow transition from laminar to turbulent stage has great practical importance and aerofoil section designers need to pay a close attention to various aerofoil characteristics which can be strongly affected by flow transition.

Transitional flows past different bodies have been studied experimentally as well as numerically in many previous studies. Initially researchers assumed that viscous action can only have stabilizing action on the flow development and focused their attention on understanding basic mechanisms for transition of inviscid flows[1]. Subsequently researchers focused their attention on the role of viscous action on flow transition via the solution of Orr-Sommerfeld equation[2,3]. These studies predicted the presence of Tollmien-Schlichting (TS) waves due to monochromatic excitation inside the shear layer, which presumably can undergo spatial growth resulting in flow transition depending on the nature of excitation parameters. This theoretical development predicted onset of flow criticality and provided properties of unstable waves. Schubauer and Skramstad[4] experimentally verified flow transition via the growth of TS waves. However, Morkovin[5] suggested that in many natural flows transition frequently occurs in the absence of TS waves. Classical flow transition route related to the growth of TS waves is bypassed in such flows and the flow transition process has been categorized as bypass transition. Any fluid dynamical system has a basic characteristic of receptivity and it is receptive to background as well as impressed deterministic disturbances. Depending on these disturbances, fluid dynamical system undergoes transition either by TS route or by bypass route. It has been observed that fluid dynamical system becomes very receptive to background disturbances in the case of flows experiencing strong adverse pressure gradient. In such situations, even a very small disturbance can amplify causing flow transition.

Strong adverse pressure gradient near trailing edge of an aerofoil is capable of amplifying minute background disturbances. Flow separation caused by surface roughness amplifies the boundary layer instability[6,7]. Large localized surface perturbations follow the bypass transition routes[5,8]. A detailed analysis of the boundary-layer development, as a result of large distributed roughness, which mimics the early ice accretion process near the leading-edge stagnation region is provided in[8]. Effects of roughness elements on flow transition have been identified as transition enhancement[9]. Aerodynamic characteristics such as lift, drag and pitching moments are adversely altered by the roughness induced effects[10,11]. Surface irregularities are caused due to the manufacturing and machining defects, icing accretion[10] and impact collisions with dirts and insects. In this respect, many studies have considered various severity of roughness induced perturbations[8,10,12,13]. Factors such as extent of roughness elements, their heights and respective locations from leading edge of the aerofoil have varying effects on the flow transition. Mechanism of premature transition induced by roughness elements was experimentally investigated earlier[8,9,14]. Small perturbations in the form of roughness elements can create eddies in the boundary layer causing reversal of the base flow. For an impulsively started flow, simulations were performed on a parabola with small perturbations[15].

Application of transient growth in giving rise to roughness induced transition has been reported[16,17].

In the present paper, we have investigated the details of transitional flow over an aerofoil, with and without roughness elements, by direct simulation of two-dimensional ($2D$) Navier-Stokes equation. We have focused attention on distinguishing how flow undergoes transition in the presence and absence of leading edge roughness elements by analyzing different characteristics from the laminar to the turbulent stage.

2. Governing 2D Navier-Stokes Equations and Formulation

Here, stream function (ψ) - vorticity (ω) formulation has been used to obtain numerical solution of the 2D Navier-Stokes equations. This formulation is very accurate as it ensures solenoidality of velocity and vorticity fields, simultaneously[18]. Orthogonal formulation of Navier-Stokes equation has fewer terms, as compared to the non-orthogonal formulation and causes lesser numerical error for incompressible flows. Consider a physical (x, y)-plane mapped to a computational (ξ, η)-plane, with ξ co-ordinate being along the azimuthal direction and the η co-ordinate in wall-normal direction. The governing Navier-Stokes equations can be expressed as[18,19],

$$\frac{\partial}{\partial \xi}\left[\frac{h_{22}}{h_{11}}\frac{\partial \psi}{\partial \xi}\right] + \frac{\partial}{\partial \eta}\left[\frac{h_{11}}{h_{22}}\frac{\partial \psi}{\partial \eta}\right] = -h_{11}h_{22}\omega \tag{1}$$

$$h_{11}h_{22}\frac{\partial \omega}{\partial t} + h_{22}u\frac{\partial \omega}{\partial \xi} + h_{11}v\frac{\partial \omega}{\partial \eta} = \frac{1}{Re}\left[\frac{\partial}{\partial \xi}\left(\frac{h_{22}}{h_{11}}\frac{\partial \omega}{\partial \xi}\right) + \frac{\partial}{\partial \eta}\left(\frac{h_{11}}{h_{22}}\frac{\partial \omega}{\partial \eta}\right)\right] \tag{2}$$

where h_{11} and h_{22} are the scale factors used in mapping. Equations (1) and (2) are solved using highly accurate, DRP schemes[18,19]. Potential flow past an aerofoil is the initial condition. On the aerofoil surface, no-slip conditions are used for prescription of the stream function and wall vorticity. Periodic boundary conditions are applied at a cut starting from the leading edge of the aerofoil to the outer boundary. At the outflow boundary, Sommerfeld boundary condition is used on the η-component of the velocity field to determine the stream function which is further used to evaluate the vorticity distribution at the outflow boundary. In addition, we have used numerical filters to control aliasing error and spurious q-waves which are responsible for unphysical early flow transition[18,20].

3. Results and Discussion

As a validation study, we compare numerical results for the flow past a NACA 0018 aerofoil at zero degree angle of attack (AOA) for $Re = 10^5$ with available experimental results[21]. In Fig. 1, we have compared the variation of mean velocity profiles over the aerofoil surface. Mean velocity profiles are obtained by time averaging data over a period of $0.05 \leq t \leq 30$, with 600 equi-spaced snapshots. Results show good match between the experimental[21] and present numerical results.

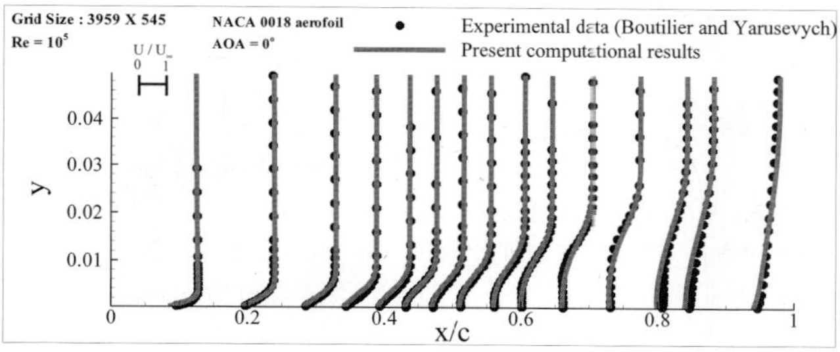

Fig. 1. Comparison of the mean velocity profiles obtained in the present simulation with the experimental[21] results.

Next, we report numerical results for flow past a smooth NACA 0012 aerofoil kept at zero AOA for $Re = 1.25 \times 10^6$. We have constructed an orthogonal O-topology grid around NACA 0012 aerofoil with 5060×545 grid points. Computations are performed with a time step of 5×10^{-6}. Figure 2(a) shows stream function contours at $t = 5$. Fluid flow up to 70% of aerofoil chord (c) is smooth and after this vortical disturbances appear as observed in the vorticity contours in Fig. 2(b). In the same frame different vortical disturbances present on the top surface of the aerofoil are identified as A, B, C, D, E. Instantaneous pressure distribution on the aerofoil surface at the same time is shown in Fig. 2(c). Near the leading edge, the pressure gradient is negative which causes flow to accelerate. The trend of the negative pressure gradient continues up to around 20% of the chord and subsequently adverse pressure gradient region starts. Flow remains attached approximately up to 70% of the chord ($x/c = 0.7$) and then separates forming small separation bubbles. These bubbles convect downstream with time. In addition, there are localized steep favorable and an adverse pressure gradient excitations imposed by the convecting vortical disturbances. Instabilities are progressively generated due to this excitation

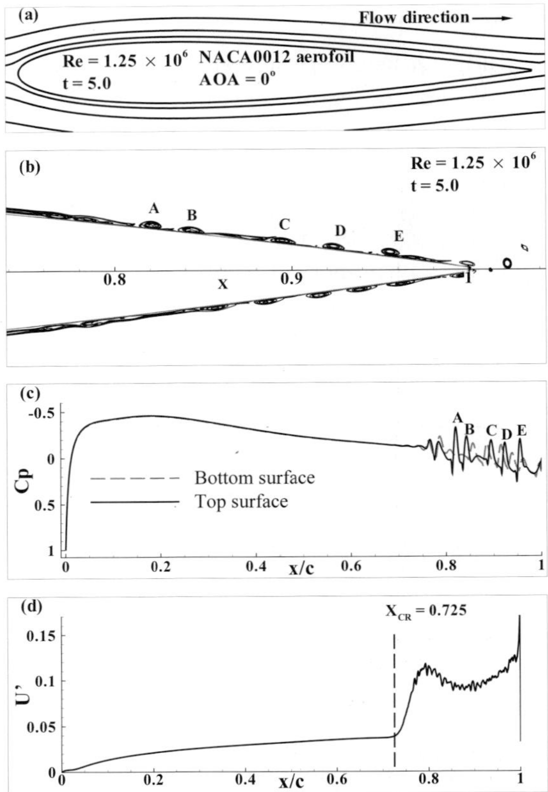

Fig. 2. Stream function and vorticity contours are shown in (a) and (b),respectively. Surface pressure variation and variation of RMS component of velocity fluctuation is shown in (c) and (d), respectively.

of the flow. The variation of RMS component of velocity fluctuations on the top surface of the aerofoil is shown in Fig. 2(d). In the laminar region there are no significant fluctuations. However, after the location of $x/c = 0.725$ there is a sudden growth in RMS value of velocity fluctuations indicating flow transition.

Vortical disturbance convecting over the surface has a distinctive intermittent characteristics. Intermittency factor gives the estimation of fraction of time during which the flow is turbulent. In earlier studies, laminar and turbulent parts of the signal have been differentiated using a second derivative of time series of flow quantities such as velocity or vorticity[22]. Differentiation operation amplifies the discontinuous portions in the sig-

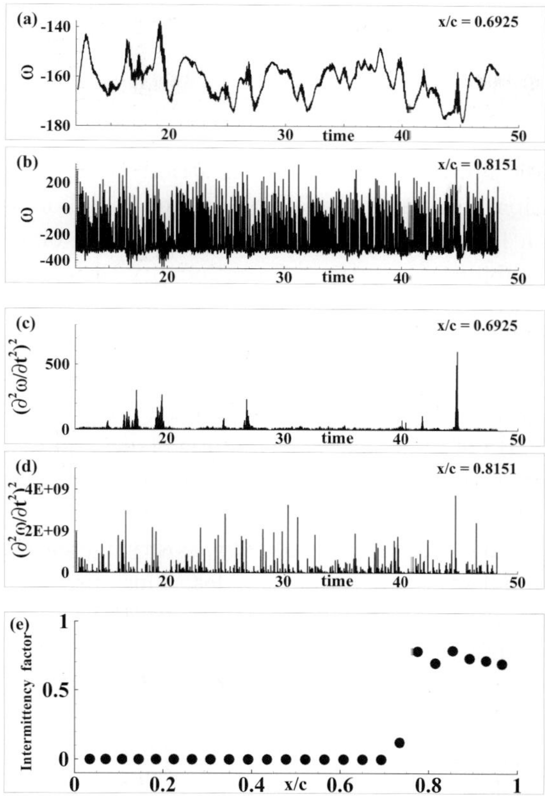

Fig. 3. Surface vorticity variation with time at $x/c = 0.6925$ and 0.8151 and square of its second derivative are shown in (a), (b) and (c), (d), respectively. Frame (e) shows intermittency factor variation for the smooth aerofoil case.

nal. We have shown the vorticity time series at $x/c = 0.6925$ and 0.8151 in Figs. 3(a) and (b), respectively. Corresponding square of the second derivative of vorticity are shown in Figs. 3(c) and (d), respectively. For our analysis, we have treated the flow as laminar if $(\frac{\partial^2\omega}{\partial t^2})^2 < 1000$ using information in Fig. 3(c). Corresponding intermittency factor is shown in Fig. 3(e). Intermittency picks up in the aft section of the aerofoil. Flow field over aerofoil is in transition phase and is not a fully developed turbulent flow as the maximum intermittency value is not close to one.

The energy expended by fluid flow in creating and sustaining rotationality is obtained by evaluating enstrophy property[23] and is a useful quantity to study transitional and turbulent flows. Enstrophy transport equation is given by[23],

$$\frac{D\Omega_1}{Dt} = \frac{2}{Re}\left[\frac{1}{2}\nabla^2\Omega_1 - (\nabla\omega)^2\right] \tag{3}$$

For a periodic flow, RHS of Eq. (3) is strictly negative due to role of viscosity in dissipating energy globally. However for inhomogeneous flows, the first term on RHS of Eq. (3) can be either positive or negative, which provides the possibility of creation of rotationality through diffusion term. The positive RHS quantity of Eq. (3) indicates the role of diffusion in triggering local instability and is a mechanism for creating rotationality at different scales by diffusion, in contrast to the concept of creation of smaller scales by vortex stretching in 3D flows[23]. We have used this here to evaluate the RHS of Eq. (3), as shown in Fig. 4 at $t = 50$. We have shown the positive part of the right hand side of enstrophy transport equation as black region while the negative part is shown by white region. The zoomed view near the trailing edge shows action of viscosity in creating rotationality in the form of small scale separation bubbles. This observation shows that although present calculations are 2D, still large bandwidth of scales get excited through diffusion in fluid flow.

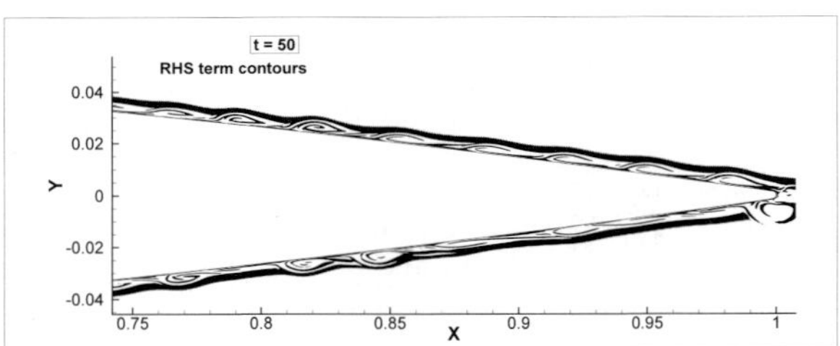

Fig. 4. Region where right hand side of enstrophy transport equation is positive has been shown as black region.

We have further considered bypass transitional flow past NACA 0012 aerofoil with leading edge roughness elements which are located near $x/c = 0.02$. Simulations are performed for a flow at zero AOA and $Re = 1.25 \times 10^6$. Average ratio of wavelength to amplitude $(\Delta S/\Delta H)$ of the roughness elements is close to 2 in all the cases, while ΔS is close to $0.015c$. We have again constructed orthogonal grids with 5060×545 points for the single, double and three roughness elements cases. Stream function contours for the flow past single roughness element case shown in Fig. 5(a), indicate multiple separation bubbles behind the roughness element. Once the vortical disturbances are detached from the roughness element, they convect downstream towards trailing edge. Amplitudes of these disturbances come down significantly as the flow accelerates near the leading edge of the aerofoil, as highlighted by the variation of intermittency factor in Fig. 5(b). Frequency spectrum corresponding to the vorticity variation at different streamwise locations for the single roughness element case have been shown in Figs. 5(c) to 5(e). Close to the roughness element, one observes the presence of a dominant primary frequency $R_{11} = 0.1067$. In addition, there are multiple higher harmonic frequencies denoted as R_{12}, R_{13}, R_{14}, R_{15} and R_{16}. As we move downstream, we further observe presence of dominant higher harmonic frequency events. However, as the flow is accelerating near the leading edge of the aerofoil, these disturbances decay further downstream. Near the trailing edge of the aerofoil ($x/c = 0.8151$), frequency spectrum completely fills up with a dominant frequency at $R_{18} = 0.03365$ as shown in Fig. 5(e).

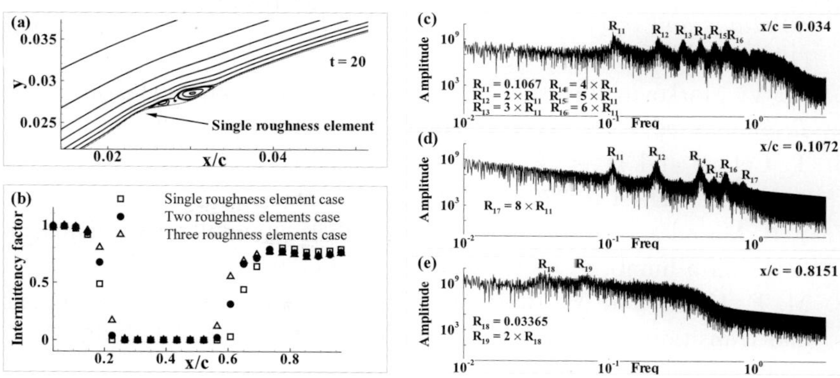

Fig. 5. (a) Stream function contours for the flow past a single roughness element case; (b) Intermittency factor variation for roughness elements case and (c)-(e) FFT of time variation of surface vorticity for the flow past single roughness element case.

4. Summary and Conclusions

Solutions of $2D$ incompressible Navier-Stokes equation for transitional flow over an aerofoil, with and without roughness elements, are obtained. Variation of different flow characteristics such as RMS component of velocity fluctuations, intermittency factors and frequency spectrum of surface vorticity signal are studied. For flow past a smooth aerofoil case, RMS component of velocity fluctuations becomes significant in the aft portion of the aerofoil. For flow past an aerofoil with leading edge roughness elements, flow undergoes transition just after the roughness elements. However, it undergoes re-laminarization due to favorable pressure gradient near the leading edge. Flow further undergoes transition in the aft portion of the aerofoil as indicated by variation of the intermittency factors.

References

1. T. K. Sengupta, Instabilities of Flows and Transition to Turbulence (CRC Press, USA, 2012.)
2. H. Schlichting, Zur entstehung der turbulenz bei der plattenströmung, *Nachr. Ges. Wiss. Göttingen, Math. -phys. Kl.* **42**, 181 (1933).
3. W. Tollmien, *Production of Turbulence*, Tech. Rep. 609, NACA Rep. (1931).
4. G. B. Schubauer and H. K. Skramstad, *Laminar boundary layer oscillations and transition on a flat plate*, Tech. Rep. 909, NACA Rep. (1947).
5. M. V. Morkovin, On the many faces of transition, in *Viscous Drag Reduction*, ed. C. S. Wells (Plenum Press, 1969) pp. 1–31.
6. T. Cebeci and D. A. Egan, Prediction of transition due to isolated roughness, *AIAA J.* **27**, 870 (1989).
7. J. A. Masad and V. Iyer, Transition prediction and control in subsonic flow over a hump, *Phys. Fluids* **6(1)**, 313 (1994).
8. M. F. Kerho and M. B. Bragg, Airfoil boundary-layer development and transition with large leading-edge roughness, *AIAA J.* **35(1)**, 75 (1997).
9. T. C. Corke, A. Bar-Sever and M. V. Morkovin, Experiments on transition enhancement by distributed roughness, *Phys. Fluids* **29(10)**, 3199 (1986).
10. W. W. Huebsch and A. P. Rothmayer, The effects of surface ice roughness on dynamic stall, *J. Aircraft* **39(6)**, 945 (2002).
11. W. W. Huebsch and A. P. Rothmayer, Numerical prediction of unsteady vortex shedding for large leading-edge roughness, *Comput. Fluids.* **33**, 405 (2004).
12. P. S. Klebanoff, K. D. Tidstrom and L. M. Sargent, On the evolution of a turbulent boundary layer induced by a three-dimensional roughness

element, *J. Fluid Mech.* **237**, 101 (1962).

13. Y. X. Wang and M. Gaster, Effect of surface steps on boundary layer transition, *Expts. Fluids* **39**, 679 (2005).
14. P. S. Klebanoff and K. D. Tidstrom, Mechanism by which two-dimensional roughness element induces boundary-layer transition, *Phys. Fluids.* **15(7)**, 1173 (1972).
15. R. Bhaskaran and A. P. Rothmayer, Separation and instabilities in the viscous flow past airfoil leading edges, *Comput. Fluids* **27(8)**, 903 (1998).
16. E. Reshotko and A. Tumin, Investigation of the role of transient growth in roughness-induced transition, *AIAA paper, AIAA 2002-2850* (2002).
17. E. B. White and F. G. Ergin, Receptivity and transient growth of roughness induced disturbances, *AIAA paper, AIAA 2003-4243* (2003).
18. T. K. Sengupta, High Accuracy Computing Methods: Fluid Flows and Wave Phenomena (Cambridge Univ. Press., New York, USA, 2013.)
19. T. K. Sengupta and Y. G. Bhumkar, Direct numerical simulation of transition over a natural laminar flow aerofoil, *Frontiers in Aerospace Engineering* **2**, 166 (2012).
20. T. K. Sengupta, S. Bhaumik and Y. G. Bhumkar, Direct numerical simulation of two-dimensional wall-bounded turbulent flows from receptivity stage, *Phys. Rev. E* **85(2)**, (2012).
21. M. S. H. Boutilier and S. Yarusevych, Separated shear layer transition over an airfoil at a low reynolds number, *Phys. Fluids* **24**, (2012).
22. T. K. Sengupta, S. De and K. Gupta, Effect of free-stream turbulence on flow over aerofoil section at high incidence, *J. Fluids Structures* **15(5)**, 671 (2001).
23. T. K. Sengupta, H. Singh, S. Bhaumik and R. R. Chowdhury, Diffusion in inhomogeneous flows: Unique equilibrium state in an internal flow, *Comput. Fluids* **88**, 440 (2013).

Global Stability and Transition to Intermittent Chaos in the Cubical Lid-Driven Cavity Flow Problem

J.-Ch. Loiseau* and J.-Ch. Robinet

DynFluid, Arts et Métiers ParisTech,
Paris, 75013, France
**E-mail: loiseau.jc@gmail.com*
www.dynfluid.eu

E. Leriche

Laboratoire de Mécanique de Lille, Université Lille-1,
Villeneuve d'Ascq, 59655, France

Transition from steady state to intermittent chaos in the cubical lid-driven cavity flow is investigated numerically. Fully 3D global stability analysis has revealed that the flow experiences a Hopf bifurcation at $Re_c = 1910$. As for the 2D-periodic LDC flow, the unstable mode is related to a centrifugal instability of the primary vortex core. Once unstable, DNS shows that the flow is driven toward a limit cycle before eventually exhibiting intermittent chaotic dynamics.

Keywords: Lid-driven cavity; Global stability; Transition to chaos

1. Introduction

The lid-driven cavity flow is an academic problem of fluid dynamics qualitatively presenting some of the key features responsible for transition. Depending on the Reynolds number and the spanwise wavenumber of the prescribed perturbation, the 2D-periodic LDC flow is unstable toward four different families of modes[1,2]. The four families are nonetheless all related to a centrifugal instability of the primary vortex core. The stability of the flow within a three-dimensional enclosure has received much less attention than its 2D-periodic counterpart. Because investigating the linear stability analysis of such three-dimensional flow still is computationally challenging, few references exist in the literature on the transition thresholds. To our knowledge, Feldman & Gelfgat[3] and Kuhlmann & Albensoeder[4] have only addressed the stability of the cubical lid-driven cavity flow by means of direct numerical simulations. As for the 2D-periodic global stability analysis, the exponentially growing perturbation in their DNS resemble the

2D-periodic ones and appears to be related to a centrifugal instability of the primary vortex core.

The aim of the present work is to investigate the transition from steady state to intermittent chaotic dynamics in the cubical lid-driven cavity flow. The linear stability of this three-dimensional flow and the underlying physical mechanisms are first investigated in sections 3.1 and 3.2, respectively. Section 3.3 provides some insights into the transition to intermittency in the non-linear regime while conclusions and perspectives are given in section 4.

2. Problem Formulation

The dynamics of the flow in a three-dimensional enclosure are governed by the incompressible Navier-Stokes (N-S) equations. Dimensionless variables are defined with respect to the depth/length/width L of the cavity and the constant velocity U_0 of the moving lid. Therefore, the Reynolds number is defined as $Re = U_0 L/\nu$, with ν being the kinematic viscosity. The origin of the axes system is assigned to be the geometrical center of the cavity such that the non-dimensional domain considered is: $\Omega = [-0.5, 0.5] \times [-0.5, 0.5] \times [-0.5, 0.5]$. The problem is discretized using the Legendre spectral element method implemented in the Nek 5000 solver. Ten elements, distributed according to a Gauss-Lobatto grid, are used in each direction along with Legendre polynomials of order 5 resulting in slightly more than half a million of degrees of freedom. Base flows $\mathbf{Q}_b = (\mathbf{U}_b, P_b)^T$ are computed using the selective frequency damping (SFD) approach. For further details, the reader is referred to Ref. 5. Once the base flow $\mathbf{Q}_b$ is computed, the N-S equations are linearised in its vicinity. Within the three-dimensional global stability framework employed in this work, seeking for the eigenvalues and eigenvectors of the Jacobian matrix is a hard computational task due to the large number of degrees of freedom involved in the computation. As a consequence, a time-stepper formulation of the eigenvalue problem is used. For further details on the time-stepper approach, the reader is referred to Ref. 6.

3. Results

3.1. *Global stability analysis*

Figure 1(a) depicts the eigenspectrum of the linearised Navier-Stokes operator for the cubical lid-driven cavity flow problem at $Re = 1930$. Two complex conjugate eigenvalues with circular frequency $\omega = \pm 0.58$ lie in the upper-half complex plane indicating that the steady base flow $\mathbf{Q}_b$ is unstable. Further stability calculations have provided a critical Reynolds number $Re_c = 1910$ in good agreements with the critical thresholds obtained in Ref. 3 ($Re_c = 1914$) and more recently in Ref. 4 ($Re_c = 1919$), both using DNS.

Figure 1(b) depicts the vertical velocity component of the leading unstable mode's real part while the motion it induces in the $y = -0.25$ horizontal plane is shown on figure 1(c). This mode is made of two structures: counter-rotating vortices along the upstream wall and high- and low-speed streaks along each vertical wall. Following the work in Ref. 1, this mode is decomposed into a component parallel to the flow's direction (i.e. $\mathbf{u}_{\parallel} = (\mathbf{u} \cdot \mathbf{U}_b)\mathbf{U}_b/\|\mathbf{U}_b\|^2$) and its complement (i.e. $\mathbf{u}_{\perp} = \mathbf{u} - \mathbf{u}_{\parallel}$). Such decomposition makes clear that, while the counter-rotating vortices are associated to a motion perpendicular to the flow's direction, the velocity streaks they induce are everywhere parallel to the direction of the flow.

3.2. *Instability mechanisms*

Investigating the kinetic energy transfer between the base flow $\mathbf{U}_b$ and the mode $\mathbf{u} = \mathbf{u}_{\parallel} + \mathbf{u}_{\perp}$, as performed in Ref. 1, has proven helpful in order to obtain a better understanding of the instability mechanisms at play. The transfer of kinetic energy is governed by the Reynolds-Orr equation:

$$\frac{\partial E}{\partial t} = -\int_{\mathrm{V}} \mathbf{u} \cdot (\mathbf{u} \cdot \nabla)\mathbf{U}_b \,\mathrm{dV} - \frac{1}{Re}\int_{\mathrm{V}} \nabla\mathbf{u} : \nabla\mathbf{u} \,\mathrm{dV} \tag{1}$$

where the first term on the right-hand side is the total production P, and the second one is the total dissipation D. Introducing the decomposition of the perturbation into the Reynolds-Orr equation (1), the production P can then be splitted into four different contributions, namely:

$$\begin{aligned} I_1 &= -\int_{\mathrm{V}} \mathbf{u}_{\perp} \cdot (\mathbf{u}_{\perp} \cdot \nabla)\mathbf{U}_b \,\mathrm{dV}, \; I_2 = -\int_{\mathrm{V}} \mathbf{u}_{\parallel} \cdot (\mathbf{u}_{\perp} \cdot \nabla)\mathbf{U}_b \,\mathrm{dV} \\ I_3 &= -\int_{\mathrm{V}} \mathbf{u}_{\perp} \cdot (\mathbf{u}_{\parallel} \cdot \nabla)\mathbf{U}_b \,\mathrm{dV}, \; I_4 = -\int_{\mathrm{V}} \mathbf{u}_{\parallel} \cdot (\mathbf{u}_{\parallel} \cdot \nabla)\mathbf{U}_b \,\mathrm{dV} \end{aligned} \tag{2}$$

To each of these four contributions is associated a different physical mechanism. The sign of the different integrals I_i then informs whether the associated mechanism acts as promoting (positive) or quenching (negative) the instability under consideration. Figure 2(a) provides the kinetic energy budget for the leading unstable mode at $Re = 1930$. As for the 2D-periodic lid-driven cavity flow (see Ref. 1), this energy budget is dominated by I_2.

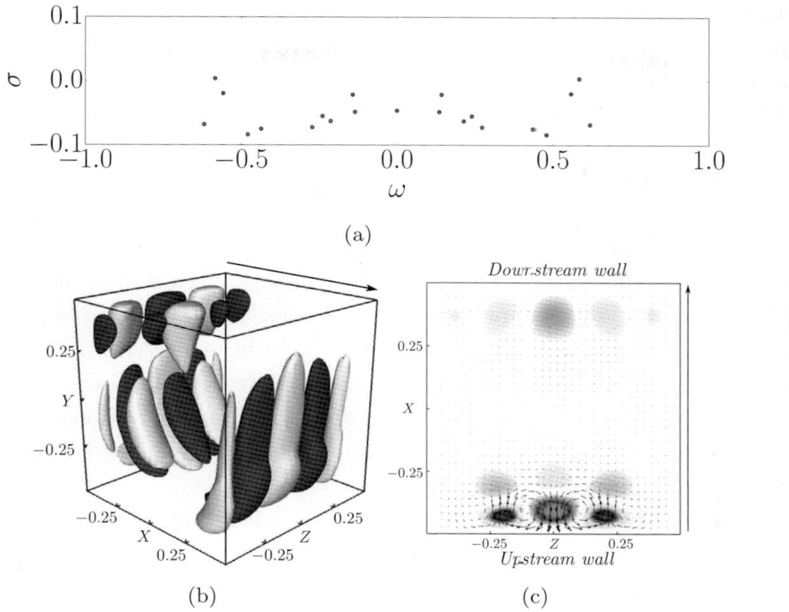

Fig. 1. (a) Eigenspectrum of the linearised Navier-Stokes operator for the cubical lid-driven cavity flow problem at $Re = 1930$. (b) Isosurfaces depicting $\hat{\mathbf{v}} = \pm 10\%$ of the maximum vertical velocity of the leading unstable mode's real part. (c) Motion it induces in the $y = -0.25$ horizontal plane. Shaded contours highlight the cross-plane velocity while vectors depict the in-plane motion. In both cases, the arrow shows the motion of the lid.

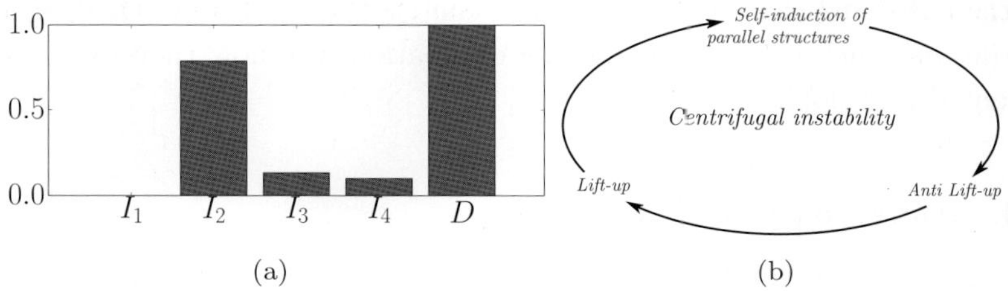

Fig. 2. (a) Kinetic energy budget for the cubical LDC at $Re = 1930$. (b) Re-interpretation of the centrifugal instability as a closed-loop mechanism.

According to Albensoder & Kuhlmann[1], the domination of I_2 indicates that the instability considered is of centrifugal type. Based on the definition of the production terms (2), another physical interpretation can nonetheless be given to the Taylor-Görtler instability taking place. Indeed, thanks to the decomposition of the production term $-\int_V \mathbf{u} \cdot (\mathbf{u} \cdot \mathbf{U}_b)\ \mathrm{dV}$ into four different contributions, as well as to the kinetic energy budget provided on figure 2(a), the centrifugal instability can be decomposed into a sum of three elementary mechanisms:

- I_2 characterises the transfer of kinetic energy from the vortical structure $\mathbf{u}_\perp$ to the parallel one $\mathbf{u}_\parallel$. The associated instability mechanism is thus closely related to the lift-up effect in boundary layer flows wherein streamwise velocity streaks (i.e. $\mathbf{u}_\parallel$) are promoted by streamwise-oriented counter-rotating vortices (i.e. $\mathbf{u}_\perp$).
- I_3 characterises the transfer of kinetic energy from the parallel structure $\mathbf{u}_\parallel$ to the vortical one $\mathbf{u}_\perp$. It can be shown that I_3 appears as the *dual* mechanism of the lift-up effect. Hence, it can be interpreted as the *anti lift-up* mechanism proposed by Antkowiak *et al.*[7].
- The mechanism associated to I_4 is a self-induction of the parallel structure occuring in the vicinity of the upper left corner flow of the cavity.

Based on these considerations, the centrifugal instability can be reinterpreted as a closed-loop instability as depicted on figure 2(b). It essentially relies on the possibility for the parallel structure $\mathbf{u}_\parallel$ generated by the lift-up effect (I_2) to reinforce the vortical structure $\mathbf{u}_\perp$, from which it has been generated, through the feedback provided by the anti lift-up effect (I_3) thanks to the curvature of the flow in the vicinity of the lower corner of the cubical enclosure. If the overall energy extracted from the base flow during one such cycle can overcome dissipation, the flow then experiences a centrifugal global instability.

3.3. *Transition to chaos*

The non-linear evolution of the flow is investigated by means of direct numerical simulation initialized using the appropriate unstable base flow. Figure 3(a) depicts the evolution in time of $E(t)$, *i.e.* the perturbation's kinetic energy, for a Reynolds number $Re = 1930$. Its time-evolution from $t = 0$ to $t \simeq 15\,000$ can be understood from the results of the linear stability analysis:

- From $t = 0$ to $t \simeq 7\,500$, the perturbation experiences a period of exponential growth due to the base flow being linearly unstable.

- From $t \simeq 7\ 500$ to $t \simeq 15\ 000$, non-linear effects kick in and the flow saturates on a limit cycle having a circular frequency ω very close to that predicted by linear stability analysis.

Though the dynamics of the flow appear to be relatively simple for $t \leq 15\ 000$, they become much more complicated for $t \geq 15\ 000$: irregular intermittent bursts of kinetic energy can be observed. Moreover, while the flow remains symmetric with respect to the mid-plane of the cavity for most of the time, a small asymmetry appears after each burst as illustrated on figure 3(b) by the spanwise velocity measured by a probe located in the $z = 0$ plane of the cavity. It is worthy to note finally that, after each burst, the perturbation comes back in the vicinity of the fixed point.

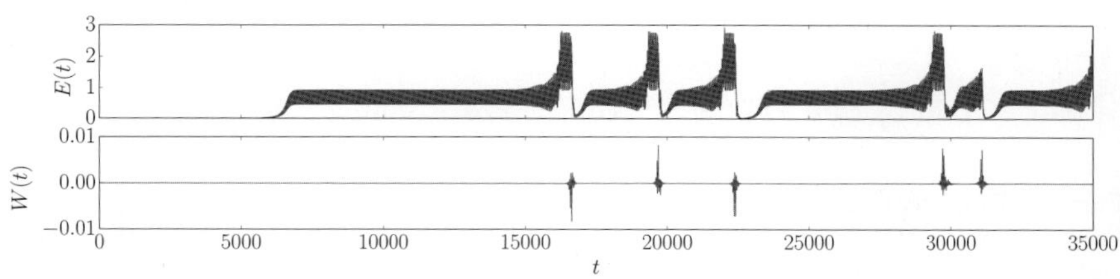

Fig. 3. Time evolution of the perturbation's total kinetic energy for $Re = 1930$ (top), and the spanwise velocity recorded by a probe located at $(x, y, z) = (-0.25, 0, 0)$ (bottom). The vertical axis of the top figure is scaled by a factor 10^4.

3.3.1. *Spectral analysis*

The signal $U(t)$, recorded by the same probe as before, is isolated over the period $14\ 000 \leq t \leq 18\ 500$ and analyzed using *Empirical Modes Decomposition*[8]. Results are displayed on figure 4. Despite its complexity, only one *Intrinsic Mode Funtion* (IMF) is necessary to represent most of the oscillatory behavior of $U(t)$. In order to have a better insight into the different frequencies involved in the dynamics, these IMFs are analyzed using *Hilbert Spectral Analysis*, resulting in what is known as the *Hilbert-Huang Transform*[8]. Figure 5 depicts the Hilbert spectrum of IMF_1. The frequency at which the system oscillates most of the time is the one predicted by linear stability analysis thus relating these dynamics to that of the unstable mode identified in section 3.1. Nonetheless, a sudden change of dynamics takes place during the bursts: the frequency switches from $\omega = 0.58$ to $\omega \simeq 0.14$. The underlying physical mechanism explaining this switch of frequency still is under investigation at the present time.

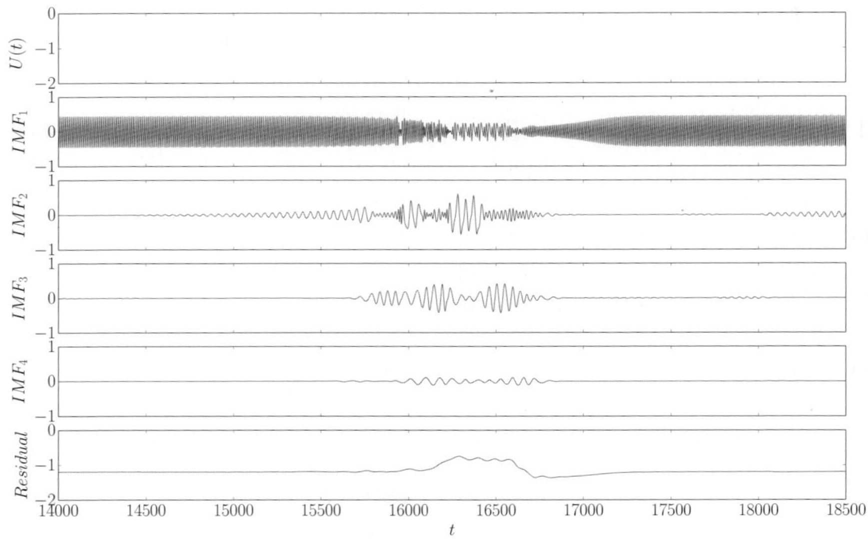

Fig. 4. Empirical modes decomposition of the perturbation's kinetic energy signal. From top to bottom: Original signal $U(t)$, first intrinsic mode function IMF_1, second intrinsic mode function IMF_2, third intrinsic mode function IMF_3, fourth intrinsic mode function IMF_4 and residual (*i.e.* $r = U(t) - \sum_{j=1}^{4} IMF_j$). The vertical axes are scaled by a factor 100.

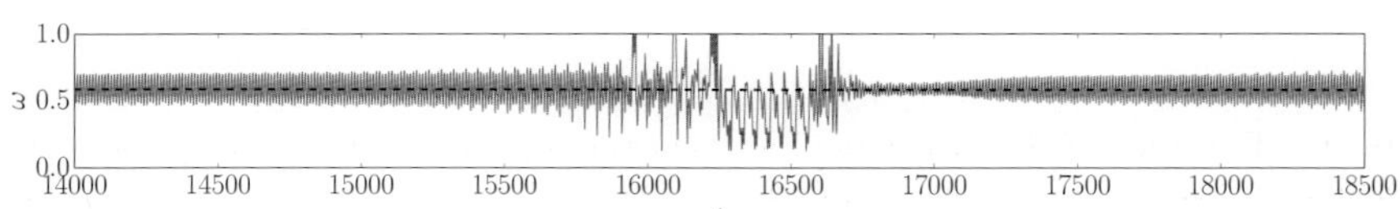

Fig. 5. Evolution of the instantaneous frequency ω obtained from the Hilbert Spectral Analysis. The black dashed line depicts the reference frequency predicted by linear stability analysis.

3.3.2. *Phase space representation*

Figure 6(a) depicts the complete dynamics of the system in a production-dissipation diagram. The traces of the bursts are clearly visible. Based on the symmetry considerations illustrated on figure 3(b), a symmetry boundary condition has been applied on the mid-plane of the cavity in order to stabilize the burst's dynamics. Such approach allows us to identify three different *exact* solutions:

- the fixed point located at $(0, 0)$ on such diagram,

- a low production-dissipation periodic cycle (represented by the blue loop) whose characteristics are well predicted by linear stability analysis,
- and a comparatively high production-dissipation periodic cycle (represented by the red loop).

Based on these considerations, the dynamics of the system observed on figure 3 can be interpreted in the phase space as a trajectory wandering around these three peculiar solutions. At the present time, only the linear stability of the fixed point has been investigated in details (see section 3.1). Preliminary results regarding the stability of the two periodic cycles provide some evidence explaining these dynamics as type II or type III intermittencies[9].

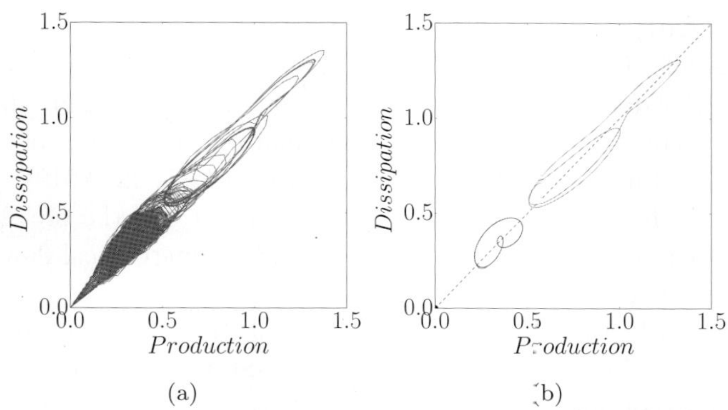

Fig. 6. Evolution of the dynamics in a production vs. dissipation diagram. (a) Complete dynamics. (b) The two limit cycles identified. Both horizontal and vertical axes are scaled by a factor 10^4.

4. Conclusion

The linear instability and subsequent transition to chaos of the cubical LDC flow has been investigated by means of global stability analyses and direct numerical simulations. The flow experiences a Hopf bifurcation at a critical Reynolds number $Re_c = 1910$ due to a centrifugal instability of the primary vortex core. Analysis of the different production terms highlights that the centrifugal instability can be re-interpreted as a closed-loop instability relying essentially on the lift-up and anti-lift-up mechanisms. Once unstable, the flow is then driven from the unstable steady equilibrium toward a periodic limit cycle. Direct numerical simulations have revealed that, though this periodic limit cycle appears to remain stable over quite a long period of time, bursts of kinetic energy are eventually observed. According to the classification by Pomeau & Manneville[9] and preliminary results about the stability of the limit cycles identified, such bursts of kinetic energy seem to be related to type II or type III intermittent chaos. The physical origin of these bursts still is unknown at the present time. Current work aims at getting a better understanding of these intermittent chaotic dynamics and aims at shedding some light on the underlying physical mechanisms.

References

1. S. Albensoeder, H. Kuhlmann and H. Rath, Three-dimensional centrifugal-flow instabilities in the lid-driven cavity problem, *Phys. Fluids* **12** (2001).
2. V. Theofilis, P. Duck and J. Owen, Viscous linear stability analysis of rectangular duct and cavity flows, *J. Fluid Mech.* **505**, 249 (2004).
3. Y. Feldman and A. Y. Gelgat, Oscillatory instability of a three-dimensional lid-driven flow in a cube, *Phys. Fluids* **22** (2010).
4. H. C. Kuhlmann and S. Albensoeder, Stability of the steady three-dimensional lid-driven flow in a cube and the supercritical flow dynamics, *Phys. Fluids* **26** (2014).
5. E. Akervik, L. Brandt, D. S. Henningson, J. Hoepffner, O. Marxen and P. Schlatter, Steady solutions of the Navier-Stokes equations by selective frequency damping, *Phys. Fluids* **18** (2006).
6. W. Edwards, L. Tuckerman, R. Friesner and D. Sorensen, Krylov methods for the incompressible Navier-Stokes equations, *J. Comp. Phys.* **110**, 82 (1994).
7. A. Antkowiak and P. Brancher, On vortex rings around vortices: an optimal mechanism, *J. Fluid Mech.* **578**, 295 (2007).
8. N. Huang, Z. Shen, S. Long, M. Wu, S. Shih, Q. Zheng, C. Tung and H. Liu, The empirical mode decomposition method and the Hilbert spectrum for non-stationary time series analysis, *Proc. Roy. Soc. London* **454**, 903 (1998).
9. Y. Pomeau and P. Manneville, Intermittent transition to turbulence in dissipative dynamical systems, *Commun. Math. Phys.* **74**, 189 (1980).

Spatio-Temporal Wave Front – Essential Element of Flow Transition for Low Amplitude Excitations

Akhil Mulloth, Prasoon Suchandra, T. K. Sengupta*

Department of Aerospace Engineering, IIT Kanpur,
Kanpur, U.P., 208 016, India
**E-mail: tksen@iitk.ac.in*

Various physical mechanisms of transition to turbulence in zero pressure gradient (ZPG) flow past a flat plate are reported with the help of theoretical and computational studies. Spatio-temporal wave-fronts (STWF) have been reported in [Phys. Rev. Lett., **107,** 154501 (2011); Phys. Rev. E., **89,** 043018 (2014)], from the solution of two- (2D) and three-dimensional (3D) Navier-Stokes equation (NSE), as the precursor of transition to turbulence. This causes a shift in our understanding of transition showing centrality of STWF, as opposed to the view-point that the growth of so-called Tollmien-Schlichting waves under favourable condition eventually leads to turbulence. Having shown the commonality of STWF's role in causing transition by 2D and 3D routes, here we investigate the role of STWF for different routes of transition from the accurate direct numerical simulation of 2D NSE. Present work explains the computational requirements for accurately capturing and studying STWF.

Keywords: Transition to turbulence; Receptivity; Direct Numerical Simulation; Navier-Stokes equation

1. Introduction

The problem of transition to turbulence in zero pressure gradient (ZPG) flow past a flat plate has intrigued researchers for a long time. However, it is nearing a solution via spatio-temporal simulation of linear and non-linear Navier-Stokes equation (NSE)[1–6]. Tollmien and Schlichting predicted theoretically Tollmien-Schlichting (TS) waves as precursor of transition, obtained with parallel flow approximation of the linearized NSE. The experiments in Schubauer & Skramstad[7] was an attempt to validate this instability theory for the first time. When similar arrangement was simulated as an excitation problem, for space-time dependent disturbance field by Bromwich contour integral method[2], three elements of disturbance field are noted with the following structures[1,3,4]: (i) a local solution followed by (ii) an asymptotic wave-packet and which is led by (iii) the growing spatio-temporal wave front (STWF), as shown in Fig. 1. The instability theory only predicts the asymptotic wave-packet in a qualitative manner, and is the TS wave-packet. Recent results[5,6] indicate some promising lead, which shows that STWF holds the key to understanding transition.

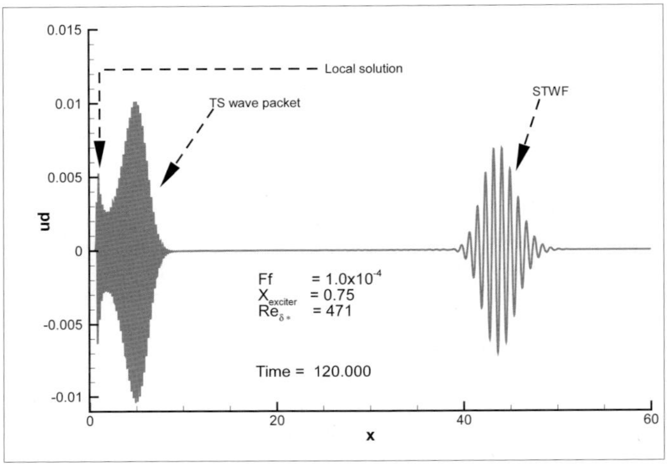

Fig. 1. Disturbance velocity at height $y = 0.00576$ inside the boundary layer (displacement thickness for unexited flow, $\delta^* = 0.059583$ at outflow, $x = 120$) plotted along the streamwise direction to show the three elements of disturbance field: (i) a local solution, (ii) TS wave-packet and (iii) spatio-temporal wave front (STWF). The non-dimensional frequency of excitation is $F_f = 1.0 \times 10^{-4}$, exciter's centre is located at $X_{exciter} = 0.75$, and Reynolds number based on displacement thickness at exciter's location is $Re_{\delta^*} = 471$.

Here, we have explained how STWFs are created over a ZPG Blasius boundary layer via the solution of 2D NSE for different excitation parameters. It is important to highlight that the genesis of STWF is independent of dimensionality of the problem, as it has been shown from the solution of 2D and 3D NSE[5,6]. It is equally important to note that STWF is also supported by linearized NSE with parallel flow approximation, as noted in the solution of Orr-Sommerfeld equation (OSE)[3,4]. In reporting the solution of NSE in the above references, high accuracy methods have been used, which are also known as dispersion relation preservation (DRP) methods[8]. Earlier solutions of NSE in the literature using second order implicit finite difference method[9] did not report any STWF. Even when the authors[10] increased the spatial accuracy to fourth order, STWF was not found. However, Fasel *et al.*[11] noted that *TS waves were preceded by a wave front that was essentially a wave-packet*, but they did not investigate it further. In the present study, we intend to explain why some of these references failed to report STWF and what are the basic requirements to capture STWF computationally. More importantly, we need to catalogue the properties and types of STWF possible for Blasius profile. Sengupta & Bhaumik[5] used DRP method with compact schemes, which has higher accuracy as compared to second and fourth order central difference (CD_2 and CD_4 respectively) methods along with implicit second order time integration. Use of an implicit method for time integration is inferior to explicit method for transitional flow, as has been shown in Sengupta *et al.*[12].

It is important to note that a relatively longer computational domain has been used to report the STWF[5], as compared to that have been taken earlier[9,10]. To reproduce the case of Fasel & Konzelmann (F&K)[10], we have also computed a case using CD_4 scheme and classical fourth order Runge-Kutta (RK_4) method for time integration. Exciter parameters are same as in F&K[10]: Exciter extends between $0.908 \leq x \leq 1.133$ with an amplitude of 0.00316 and non-dimensional frequency $F_f = \frac{2\pi\nu f}{U_\infty^2} = 1.4 \times 10^{-4}$ where ν is the coefficient of kinematic viscosity and f is the physical frequency of excitation in Hz. Here, spatial domain extends from -0.05 to 80 whereas spatial domain in F&K[10] extended only from 0.6839 to 4.062. This range is marked in Fig. 2(a). Another set of computations have been done using optimized compact scheme $OUCS3$ for spatial discretization and optimized third order Runge-Kutta (ORK_3) method for time integration[6,8]. Both these computations are done with and without ($0.6839 \leq x \leq 80$) the inclusion of the leading edge of the flat plate. The physical aspects of including the leading edge is highlighted in[13]. The Goldstein singularity at the leading edge creates a disturbance field which remains outside the shear layer due to shear sheltering effect, while moderating the growth of disturbances inside the shear layer. The amplitude of both TS wave-packet and STWF is lower in case of without the leading edge, as shown in Fig. 2. Also, the onset of non-linear growth of disturbance field is earlier when the leading edge is included.

It is noted that apart from the TS wave-packet, STWF forms at the exciter, whose spatio-temporal properties are different from the TS wave-packet and is noted to detach from the TS wave-packet in the downstream direction, for moderate to high frequencies of excitation. TS wave-packet is rooted in an almost fixed spatial location, as governed by the linearized stability analysis properties, while the STWF propagates downstream and grows via linear[3,4] and nonlinear[5] mechanisms. This observation was reported in Bhaumik[14] and Bhaumik & Sengupta[6] where the corresponding 3D transition routes have been investigated.

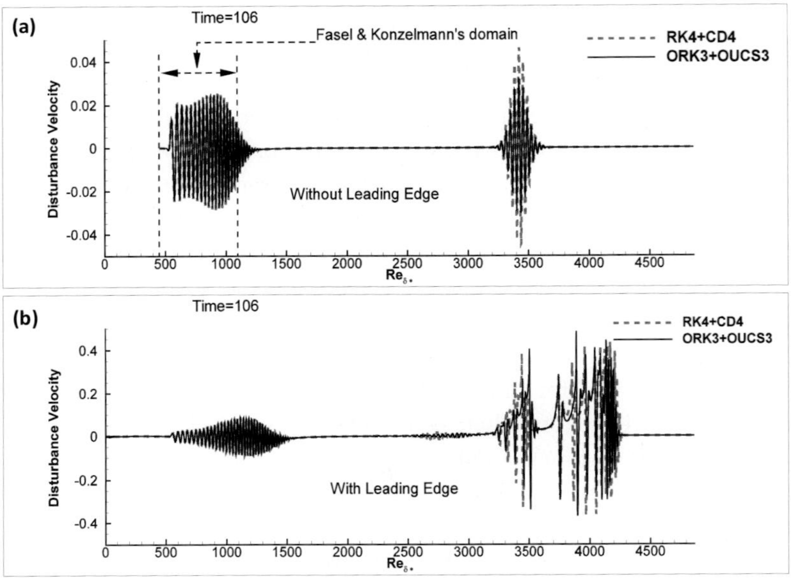

Fig. 2. Disturbance velocity plotted against $Re_{\delta*}$ (Reynolds number based on displacement thickness) to show the difference in disturbance field due to computation with different numerical methods. (a) shows results without leading edge and (b) shows results with leading edge. The extent of spatial domain used by Fasel & Konzelmann[10] is marked in (a) to emphasize the importance of taking longer computational domain.

2. Solution of Navier-Stokes Equation

The physical problem solved here is shown in Fig. 3, identifying the wall simultaneous blowing-suction (SBS) exciter positioned near the leading edge of the flat plate. To analyze non-linear flow receptivity with respect to the applied deterministic wall excitation through the SBS strip, disturbance quantities are evaluated by subtracting equilibrium solution from the instantaneous value. The SBS strip is placed between x_1 and x_2. For the simulations reported here, we have chosen a length scale (L) such that the Reynolds number based on L is 10^5 with U_∞ as the velocity scale. Details on governing equations, boundary conditions and SBS strip can be found in Sengupta 2012[1] (pp 155–159) and Sengupta & Bhaumik[5].

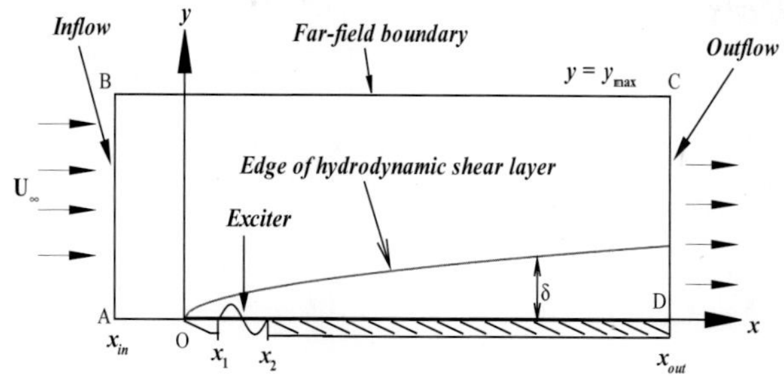

Fig. 3. Schematic of the flow and the computational domain including the simultaneous blowing and suction (SBS) strip.

A grid independence study is done to check the effect of spatial resolution on the solution of NSE. This is shown in Fig. 4, with number of grid points in streamwise direction mentioned at top-right corner of each frame. It's observed that high spatial resolution is required to accurately capture all the features of disturbance field. Results obtained with 8000 grid points onwards, not much of change in the solution is observed. Unstable flows computed with a neutrally stable algorithm are highly sensitive to background disturbances, which show up as a difference between the computations performed with 8000 and 12000 points. However the qualitative nature of the instabilities noted with these two grids are the same, with some difference noted to be present with the finer grid due to lower numerical diffusion. So, all our computations are reported here with 8000 grid points.

3. Classification of Spatio-Temporal Wave Fronts

As mentioned before, three elements of disturbance field are noted for shear flows[1,3,4]: (i) a local solution, (ii) an asymptotic wave-packet and (iii) the STWF. Depending upon the growth/decay of STWF, its interaction with asymptotic TS wave-packet and the dominant wavenumbers in solution spectrum, we propose the following classification of the observed STWFs. Apart from these four classes, a fifth type of STWF (class B) was observed when calculations were done with 4500 grid points in streamwise direction. This class showed initial growth followed by continuous decay. However, class B was not seen at all when computations were repeated with 8000 grid

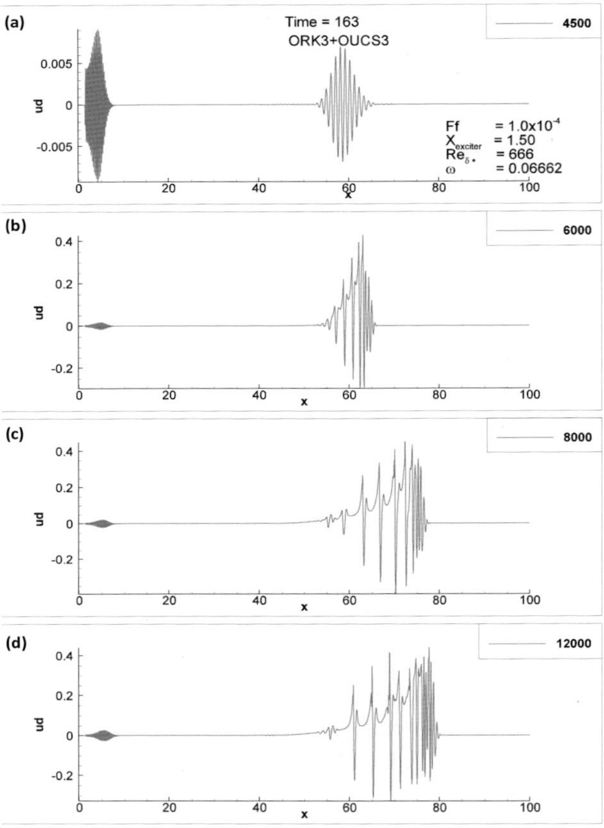

Fig. 4. Grid independence study – Effect of spatial resolution on disturbance field (at time = 163). Number of grid points in streamwise direction is mentioned at top-right corner of each frame. Exciter location $X_{exciter}$, Reynolds number $Re_{\delta*}$ at exciter's location, frequency of wall excitation F_f and circular frequency ω at exciter's location are mentioned in frame (a).

points. This was the motivation for the grid independence study and class B is excluded from further discussion. In all our following computations, the domain extends from $x = -0.05$ to $x = 120$ in streamwise direction with 8000 grid points, and from $y = 0$ to $y = 1.5$ in wall-normal direction with 400 grid points. We have used stretched grid in the range $-0.05 \le x \le 10$ and $0 \le y \le 1.5$. We have used $OUCS3$ for spatial discretization of the convective term and ORK_3 for time integration[5]. The disturbance velocity is plotted at a height $y = 0.00576$. The exciter width is 0.09 for all the cases but by varying the location of the exciter and frequency of excitation we observe the different classes reported here.

Class A : STWF of class A decays as it emerges from the local solution. Disturbance fields which show class A STWF may or may not even show distinct TS wave-packets. If TS wave-packet is created, it remains localized in space. Time evolution of class A disturbance field is shown in frames a, b and c of Fig. 5. In frame a, the STWF is developing and we don't see it clearly as a distinct packet in the plotted scale along y-axis. As the decay rate for class A is high, we see a highly damped STWF in frame b (a magnified view is presented therein). STWF is damped subsequently and is not visible in frame c. The inset of frame c shows the spectrum of the solution at time $t = 180$. Here, dominant wavenumbers correspond to the local solution and TS wave only.

Class C : STWF of this class shows initial attenuation of the packet due to dispersion[5,14] and but then many wavenumber components grow and the disturbance field becomes non-linear as shown in right frames of Fig. 5. A distinct TS wave-packet localized in space is observed for this case. Time evolution of class C disturbance field is shown in frames d, e and f of Fig. 5. Multiple STWFs are also common for this class. The spectrum of the solution at time $t = 180$ is shown in the inset of frame f. The TS wave-packet is very small compared to the non-linear STWF and hence, the TS wave-packet and the local solution don't have dominant wavenumbers visible in the spectrum.

Class D : Class D STWFs show strong interaction between STWF and TS wave-packet. In the solution spectrum, TS wave-packet has a larger peak with STWF always being a fraction of it. Time evolution of class D disturbance field is shown in frames a, b and c of Fig. 6. The disturbances grow to approximately 40% of the freestream velocity and saturate the flow field. The upstream part of the combined wave front is characterised by high wavenumbers which correspond to the TS wave-packet and the same is also observed from the solution spectrum with the wavenumbers in frame 6(c).

Class E : Class E STWFs also show strong interaction between STWF and TS wave-packet. In the solution spectrum, STWF peak and that of the TS wave-packet are of the same orders of magnitude, with STWF slightly dominating over the TS wave-packet. Time evolution of class E disturbance field is shown in frames d, e and f of Fig. 6. In Fig. 6, we can see the similarities and differences between class D and class E. Class E is similar to class D combined wave front which grows to about $0.4U_\infty$ and saturates. In contrast to Class D, the upstream part of the combined wave front is dominated by low wavenumbers for class E. The same is observed from the solution spectrum with lower wavenumbers (corresponding to STWF) having a higher peak shown in frame 6(f).

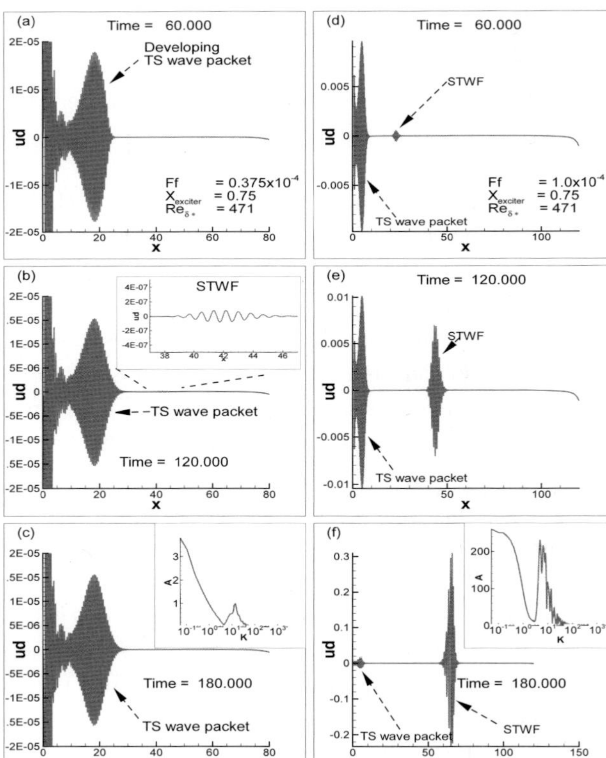

Fig. 5. STWF: Class A (shown in left frames a, b and c) and class C (shown in right frames d, e and f). Exciter location $X_{exciter}$, Reynolds number $Re_{\delta*}$ at the exciter's location and frequency of wall excitation F_f are mentioned for each class of disturbances. Frames c and f show the dominant wavenumbers in solution spectrum (amplitude A vs wavenumber K) for given time instant.

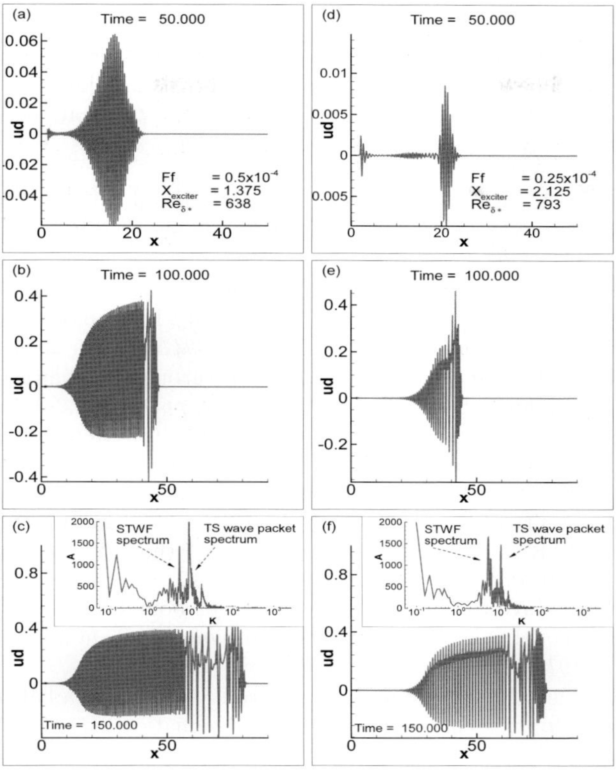

Fig. 6. STWF: Class D (shown in left frames a, b and c) and class E (shown in right frames d, e and f). Exciter location $X_{exciter}$, Reynolds number Re_{δ^*} at the exciter's location and frequency of wall excitation F_f are mentioned for each class of disturbances. Frames c and f show the dominant wavenumbers in solution spectrum (amplitude A vs wavenumber K) for given time instant.

4. Conclusion

Present work explains the computational requirements for accurately capturing and studying STWF. Based on the results obtained and analyses performed, we have classified the disturbances into four major classes. The computations are still going on with the aim of studying the genesis of these classes of disturbance and to investigate, if further sub-classifications can be made. We are also trying to map these classes on circular frequency – Reynolds number ($\omega - Re$) plane.

References

1. Sengupta T. K., Instabilities of Flows and Transition to Turbulence, *CRC Press, Taylor & Francis Group, Florida, USA*, 2012.
2. Sengupta T. K., Ballav M. and Nijhawan S., Generation of Tollmien-Schlichting waves by harmonic excitation, *Phys. Fluids A, Vol. 6(3), pp 1213–1222*, 1994.
3. Sengupta T. K., Kameswara Rao A. and Venkatasubbaiah K., Spatiotemporal growing wave-fronts in spatially stable boundary layers, *Phys. Rev. Lett. Vol. 96, pp 224504 (1–4)*, 2006.
4. Sengupta T. K., Kameswara Rao A. and Venkatasubbaiah K., Spatio-temporal growth of disturbances in a boundary layer and energy based receptivity analysis, *Phys. Fluids, Vol. 18, pp 094101(1–9)*, 2006.
5. Sengupta T. K. and Bhaumik S., Onset of turbulence from the receptivity stage of fluid flows, *Phys. Rev. Lett. Vol. 107, pp 154501*, 2011.
6. Bhaumik S. and Sengupta T. K., Precursor of transition to turbulence: Spatiotemporal wave front, *Phys. Rev. E. Vol. 89, pp 043018*, 2014.
7. Schubauer G. B. and Skramstad H. K., Laminar boundary layer oscillations and the stability of laminar flow, *J. Aero Sci., 14(2):69–78*, 1947.
8. Sengupta T. K., High Accuracy Computing Method: Fluid Flow and Wave Phenomena, *Cambridge Univ. Press, USA* 2013.
9. Fasel H. F., Investigation of the stability of boundary layers by a finite-difference model of the Navier-Stokes equations, *J. Fluid Mech., Vol. 78, pp 355–383*, 1976.
10. Fasel H. F. and Konzelmann U., Non-parallel stability of a flat-plate boundary layer using the complete Navier-Stokes equations, *J. Fluid Mech., Vol. 221, pp 311–347*, 1990.
11. Fasel H. F., Rist U. and Konzelmann U., Numerical investigation of the three-dimensional development in boundary layer transition, *AIAA J., Vol. 28, pp 29–37*, 1990.
12. Sengupta T. K., Sathyanarayanan V. K., Sriramkrishnan M. and Mulloth A., Role of Time Integration in Computing Transitional Flows Caused by Wall Excitation, *J. Sci Comput, DOI 10.1007/s10915-014-9967-1*, 2014.
13. Sengupta T. K., De S. and Sarkar S., Vortex-induced instability of incompressible wall-bounded shear layer, *J. Fluid Mech., Vol. 493, pp 277–286*, 2003.
14. Bhaumik S., Direct Numerical Simulation of Inhomogeneous Transitional and Turbulent Flows, *Ph. D. Thesis, I. I. T. Kanpur*, 2013.

Simulations Using Transition Models within the Framework of RANS

Y. C. Manu

Research Student,
SSIT, Tumkur, Karnataka 572107, India

A. Rajesh, M. B. Subrahmanya, D. S. Kulkarni and B. N. Rajani*

CSIR-National Aerospace Laboratories,
Bangalore, Karnataka 560017, India
**E-mail: rajani@ctfd.cmmacs.ernet.in*

The paper presents the results obtained using the two transport equation based transition models in the RANS framework, *viz.* $k_T - k_L - \omega$ model of Walters and Cokljat and $\gamma - Re_\theta$ SST model of Menter and Langtry. The relative performance of these two transition models for ERCOFTAC T3A flat plate test case at $Re = 6.12 \times 10^5$ and low Re SD7003 aerofoil at $Re = 6 \times 10^4$ are discussed in detail and compared with the available experimental/computational data. Finally, the two transitions models are applied to simulate three dimensional flow past a 6:1 prolate spheroid at $Re = 4 \times 10^4$.

Keywords: $k_T - k_L - \omega$ model; $\gamma - Re_\theta$ SST model; flat plate; low Re SD7003 aerofoil; prolate spheroid; pressure based finite volume method

1. Introduction

Transition is a complex phenomenon defining the process of the laminar flow changing to turbulent flow. The prediction and understanding of transitional flow plays a vital role in the field of aerospace, turbomachinery etc. Accurate prediction of laminar to turbulent transition is one of the major stumbling block in the development of a general purpose CFD tool. In CFD simulation, the transitional flow may be modelled using various methods ranging from the simplest empirical approach to DNS. It is well known that the RANS based approach is a viable tool for the designer as it gives required accuracy and lower computational cost. One of the simple approach to model flow transition is by using empirical correlations to estimate transition onset and/or the intermittency factor (γ) and coupling appropriately to turbulence models. The more recent approach is to use additional transport equations and/or model terms which include the effects of transition on the flow field prediction. In this direction, Walter and Leylek[1] proposed a single point transition model which was revised by Walter and Cokljat[2] and successfully applied to engineering problems. This model ($k_T - k_L - \omega$) addresses laminar, transition and fully turbulent flow within the framework of RANS. Furst *et al.*[3] have successfully used this single point transition model to simulate natural transition, bypass transition and transition due to incoming wake. Menter *et al.*[4] proposed a correlation based transition model using local variables with out disclosing the closure correlation functions. Menter and Langtry[5] later published these closure correlation functions. This model is based on Menter's SST model[6] and solves additional two transport equations, one for intermittency (γ) and another for momentum thickness based Reynolds number (Re_θ) for finding the transition onset. Simultaneously Malan *et al.*[7] also provided the closure correlation functions which were missing in Menter's paper[4] and successfully used this model for flow past 3D multi-element aerofoil. The present paper discusses the results obtained for the transitional flow past flat plate T3A test case, low *Re* SD7003 aerofoil and a 6:1 prolate spheroid using the $k_T - k_L - \omega$ model and $\gamma - Re_\theta$ SST model.

2. Mathematical Formulation

The time-averaged Navier Stokes equations for unsteady incompressible flow in the coordinate-free form is written as follows,

Mass conservation:

$$\bigtriangledown . \rho U = 0 \tag{1}$$

Momentum conservation:

$$D_t \rho U = - \bigtriangledown P + \bigtriangledown .((\mu + \mu_t)(\bigtriangledown U + \bigtriangledown^t U)) \tag{2}$$

where μ and ρ are fluid viscosity and density, p and U are the time-averaged pressure and velocity vector, respectively. The eddy viscosity μ_t is evaluated through turbulence/transition models. The transitional flows have been solved using $k_T - k_L - \omega$ model[2] and $\gamma - Re_\theta$ SST model[4,7]. The two transition models have been successfully implemented in the in-house flow code 3D-PURLES (Three Dimensional – Pressure based Unsteady Rans LES solver)[8] which is a pressure-based implicit finite volume algorithm. In the present implementation of $k_T - k_L - \omega$ model, the damping functions, correlations and model constants are same as that used by Furst *et al.*[3]. For the $\gamma - Re_\theta$ SST model the constants, damping function, *etc.* are used as given by Menter except for the two correlation functions *viz.* $Re_{\theta C}$ and F_{length} which are adopted from Malan *et al.*[7]. However, the $Re_{\theta C}$ correlation has been suitably modified as given below based on our validation study carried out for the flat plate T3A test case.

$$Re_{\theta C} = min\left(0.665\tilde{Re}_{\theta t} + 66.5, \tilde{Re}_{\theta t}\right) \quad (3)$$

3. Results and Discussion

3.1. *Flow over a flat plate*

The two transition models have been validated initially for the zero pressure gradient ERCOFTAC[9] T3A flat plate test case ($Re = 6.12 \times 10^5$, $Tu = 3.3\%$, $\mu_t/\mu = 13.3$). These simulations use a grid having 318×151 nodes with the near wall $y^+ < 1$ all along the plate. The relative performance of these models are shown in Fig. 1. The turbulent intensity (Fig. 1(a)) obtained by these two models are identical to the ERCOFTAC measurements. The variation of the skin friction coefficient (C_f) as shown in Fig. 1(b) clearly indicate the transition prediction capability of both the models with $\gamma - Re_\theta$ SST model having a closer agreement to the ERCOFTAC measurements[9]. Shape factor (Fig. 1(c)) predicted by both the transition models agree reasonably well with the measurement data in the laminar and transition region whereas the $\gamma - Re_\theta$ SST model is observed to deviate in turbulent region. This difference observed is due to the fact that $\gamma - Re_\theta$ SST model has predicted a thicker boundary layer in the turbulent region leading to a higher displacement thickness when compared to the $k_T - k_L - \omega$ model.

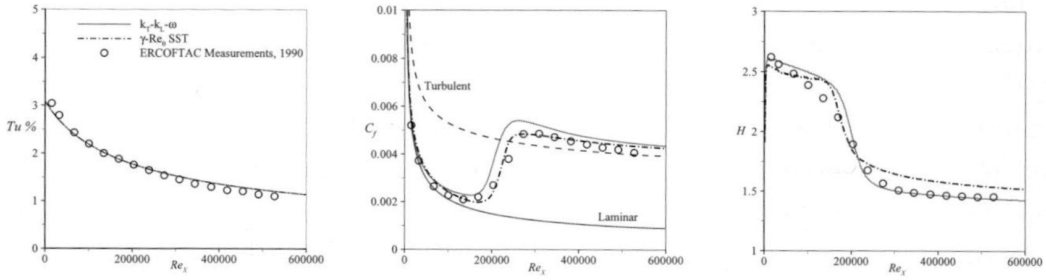

(a) Turbulent intensity (b) Skin friction coefficient (c) Shape factor

Fig. 1. Flat plate T3A test case ($Re = 6.12 \times 10^5$, $Tu = 3.3\%$, $\mu_t/\mu = 13.3$).

3.2. *Flow past SD7003 aerofoil section*

The relative performance of these two transition models have also been analysed for SD7003 aerofoil. This analysis uses C-topology having 512×101 grid points with the near wall y^+ kept to be less than one. The computations have been carried out using the QUICK scheme at $Re = 6 \times 10^4$ for different angles of attack. The performance of these transition models are presented in detail only for $\alpha = 4°$. The comparison of surface pressure (C_p) and surface skin friction coefficients (C_f) at $\alpha = 4°$ are shown in Fig. 2 which clearly indicate that $\gamma - Re_\theta$ SST model is in closer agreement to the Implicit LES data of Galbraith and Visbal[10]. Both these models could predict the laminar separation bubble (LSB) and onset of transition with $\gamma - Re_\theta$ SST model outperforming the $k_T - k_L - \omega$ model. The peak observed in the C_p curve indicates the onset of transition which is preceded by the LSB as indicated by the plateauing of the pressure distribution. The Reynolds shear stress contours (Figs. 3 and 4) and the change of sign observed in the C_f curve (Fig. 2(b)) also clearly indicate the formation of LSB. The $k_T - k_L - \omega$ model has predicted a larger laminar separation bubble compared to the $\gamma - Re_\theta$ SST model and measurement[11]. The location where the Reynolds shear stress (τ_{xz}/U_∞^2) exceeds more than 0.001 of the mean kinetic energy and exhibits a clear visible rise is another criteria to determine the transition onset location[11]. Reynolds shear stress contours and normal profiles clearly show that the transition onset location for the $\gamma - Re_\theta$ SST model ($x/c = 0.541$) is in closer agreement with the measurement[12] data ($x/c = 0.53$) and ILES[11] data ($x/c = 0.55$) compared to the $k_T - k_L - \omega$ model ($x/c = 0.477$). It is also observed that the $k_T - k_L - \omega$ model underpredicts the onset of transition based on the shear stress criteria, which is in contradiction to the pressure distribution plot where the transition onset is overpredicted ($x/c = 0.65$). This discrepancy may be attributed to the fact that the $k_T - k_L - \omega$ model is not designed to handle separated flow transition. The transition onset location, separation point and reattachment points on the suction surface of the aerofoil obtained by the transition models for different angles of attack are shown in Fig. 5 and it is observed that $\gamma - Re_\theta$ SST model is closer to ILES computations. As expected, it is observed that the transition onset location (Fig. 5(a)), separation (Fig. 5(b)) and reattachment points (Fig. 5(c)) predicted by the transition models shifts upstream as angle of attack increases. The transition onset is predicted in the wake region for $\alpha \leq 4°$ and $\alpha \leq 2°$ by the $k_T - k_L - \omega$ and $\gamma - Re_\theta$ SST model, respectively. In the $\gamma - Re_\theta$ SST model, the separated flow at $\alpha = 10°$ does not reattach on the aerofoil surface causing the aerofoil to stall which is underpredicted when compared to measurement data[12] ($\approx 12°$) and ILES data ($11° - 14°$). On the other hand, for the $k_T - k_L - \omega$ model, the separation bubble formed on the aerofoil is observed to grow in size till $\alpha = 16°$ leading to a continuous increase in the lift which may be due to the inability of this model to handle separated flow transition.

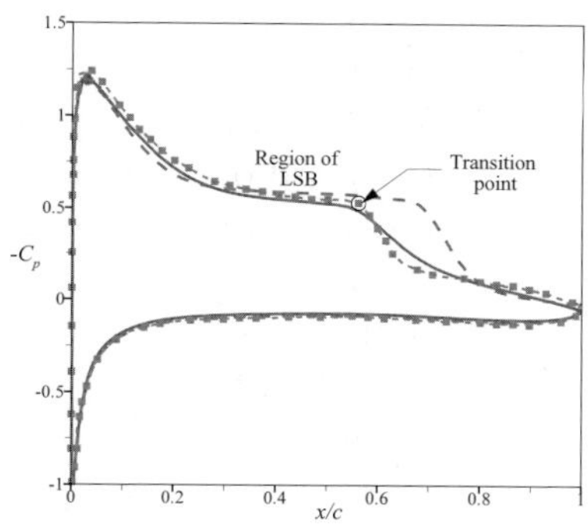

(a) Surface pressure coefficient

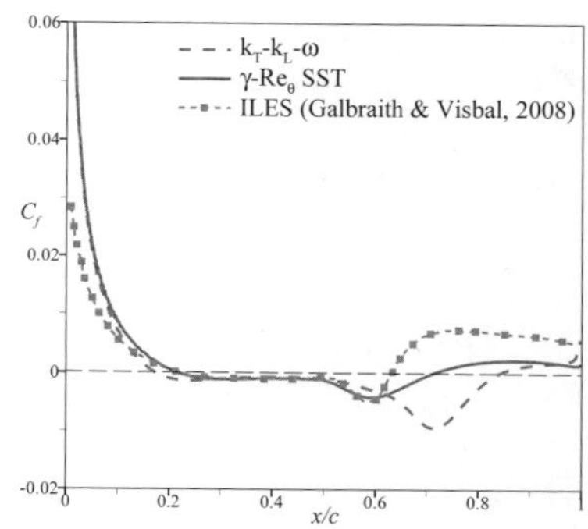

(b) Surface skin friction coefficient

Fig. 2. Flow past SD7003 aerofoil ($Re = 6 \times 10^4$, $\alpha = 4°$, $Tu = 1\%$, $\mu_t/\mu = 1$).

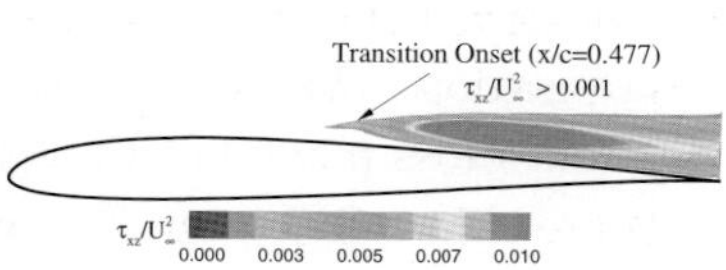

(a) Contours

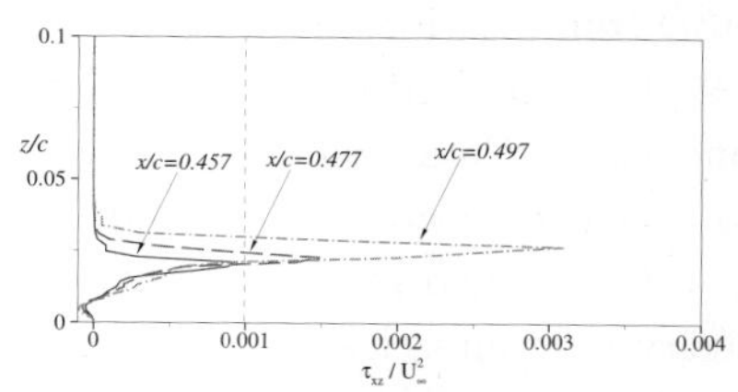

(b) Normal profiles

Fig. 3. Reynolds shear stress for SD7003 aerofoil using $k_T - k_L - \omega$ model ($Re = 6 \times 10^4$, $\alpha = 4°$).

3.3. *Flow past prolate spheroid*

The two transition models have been used to simulate flow past 6 : 1 prolate spheroid at $Re = 4 \times 10^4$ for different angles of attack, in order to understand

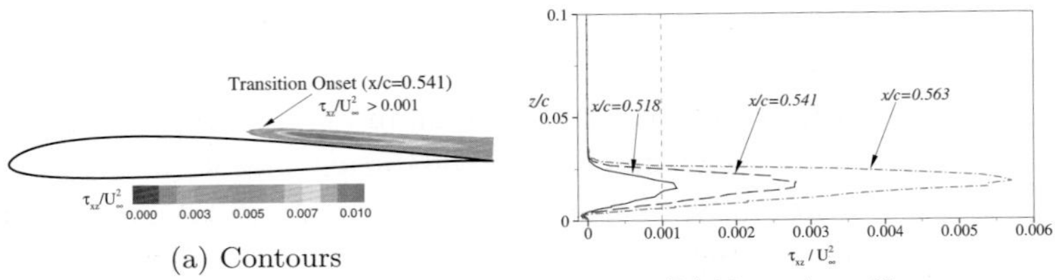

(a) Contours

(b) Normal profiles

Fig. 4. Reynolds shear stress for SD7003 aerofoil using $\gamma - Re_\theta$ SST model ($Re = 6 \times 10^4$, $\alpha = 4°$).

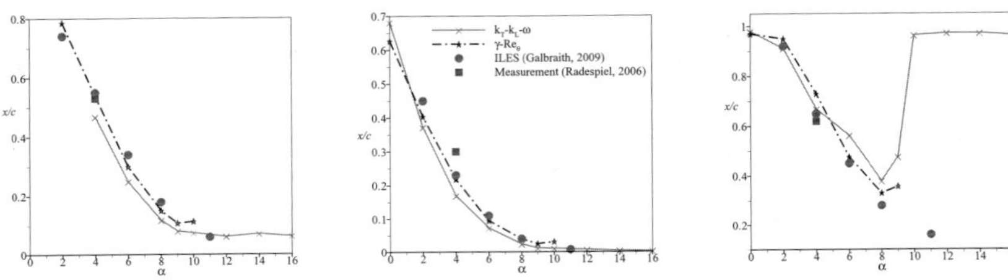

(a) Transition onset

(b) Separation point

(c) Reattachment point

Fig. 5. Variation of different parameters with α for SD7003 aerofoil.

its capability to predict three dimensional transitional and separated flows. The 3D computations have been carried out using an O-O topology having $111 \times 101 \times 312$ grid nodes (near wall resolution of $y^+ < 1$) coupled to second order CDS scheme for spatial discretisation. The eddy viscosity ratio (μ_t/μ) contours for 20° and 5° at different cross plane locations ($0 \leq x/L \leq 2$) obtained by the transition models are shown in Fig. 6. The eddy viscosity contours clearly indicate that the flow remains laminar ($\mu_t/\mu \approx 1$) on the spheroid with the transition occurring in the wake. As expected the transition location moves downstream with decrease in the angle of attack. It is further observed that the onset of transition predicted by the $k_T - k_L - \omega$ model ($x/L \geq 1.5$) is delayed for both the angles of attack when compared to the $\gamma - Re_\theta$ SST model ($x/L \geq 0.9$). This trend is also demonstrated in the turbulent kinetic energy (k_T) contours obtained by the $k_T - k_L - \omega$ model and intermittency (γ) contours obtained by the $\gamma - Re_\theta$ SST model, as shown in Fig. 7. The flow pattern obtained by the models at two angles of attack is observed to differ only near the aft region, as shown in Fig. 8, which may be attributed to the difference in the prediction of the transition phenomenon. The computed particle trace at $\alpha = 20°$ clearly show the separation and reattachment lines indicated by the converging and diverging streamlines along with saddle and nodal points near the aft region. At $\alpha = 5°$ the streamlines for both the models are smoother with only one nodal point and one saddle point in the aft region. The transition prediction capability of these two models cannot be affirmed with certainty due to lack of relevant measurement data.

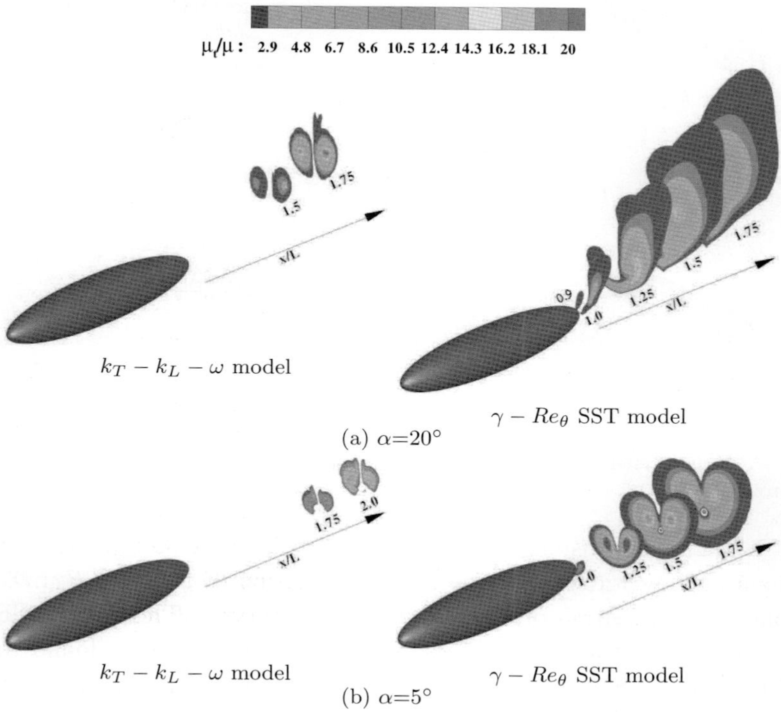

Fig. 6. Eddy viscosity ratio contours for prolate spheroid.

4. Concluding Remarks

The $k_T - k_L - \omega$ and $\gamma - Re_\theta$ transition models have been successfully implemented in the RANS framework in the in-house flow solution code. The transition onset and length predicted by the $\gamma - Re_\theta$ model for the ERCOFTAC T3A flat plate test case is found to be in a better agreement

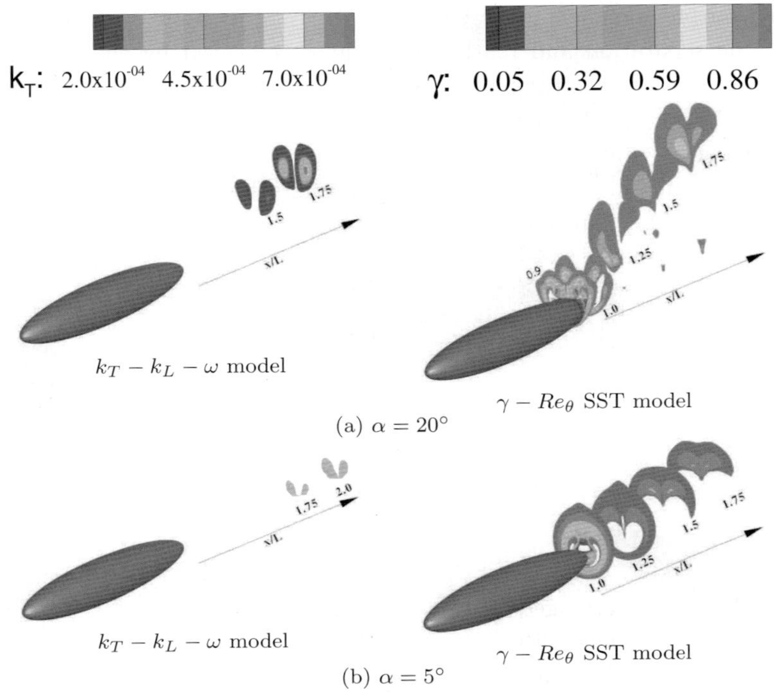

Fig. 7. Kinetic energy (left) and gamma (right) contours at $\alpha = 20°$ for prolate spheroid.

with measurement data than the $k_T - k_L - \omega$ model. For the low Re SD7003 aerofoil, the transition onset, separation point and the length of LSB at different angles of attack predicted by the $\gamma - Re_\theta$ model is found to be closer to the ILES/measurement data. The $k_T - k_L - \omega$ model for SD7003 aerofoil failed to predict the stall angle, which may be due to its inability to handle separated flow transition. The two models have demonstrated the ability to predict the effect of transition for 6:1 prolate spheroid. The relative performance of these transition models cannot be assessed due to the lack of measurement data.

References

1. D. K. Walters and J. H. Leylek, A New Model for Boundary Layer Transition using a Single-Point RANS Approach, *J. of Turbomachinery* **126**, 193 (2002).

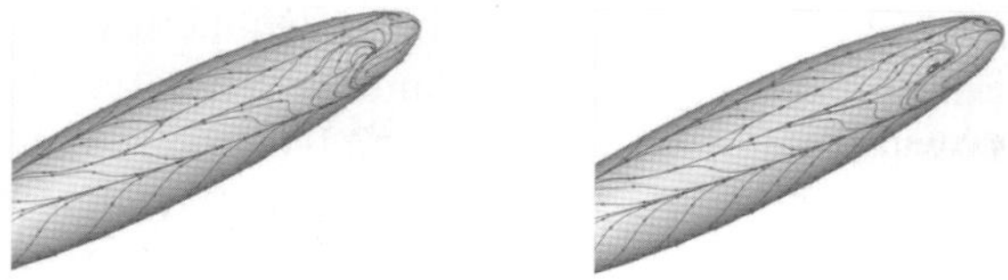

$k_T - k_L - \omega$ model $\gamma - Re_\theta$ SST model

(a) α=20°

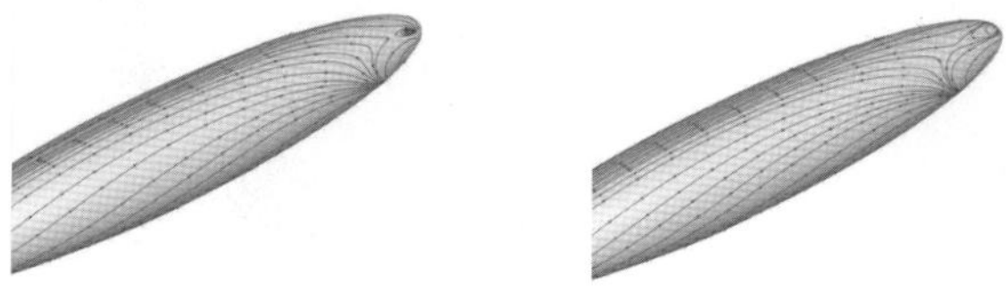

$k_T - k_L - \omega$ model $\gamma - Re_\theta$ SST model

(b) α=5°

Fig. 8. Computed particle trace for prolate spheroid.

2. D. K. Walters and D. Cokljat, A Three-Equation Eddy-Viscosity Model for Reynolds-Averaged Navier Stokes Simulations of Transitional Flow, *J. of Fluids Engg.* **130** (2008).
3. J. Furst, J. Prihoda and P. Straka, Numerical simulation of transitional flows, *Computing* **95**, 163 (2013).
4. F. R. Menter, R. Langtry, P. G. Huang and S. Volker, Transition modelling for general purpose CFD code, *Flow Turbulence and Combustion* **77**, 277 (2006).
5. F. R. Menter and R. Langtry, Correlation based transition modeling for unstuctured parallelized computational fluid dynamics code, *AIAA J.* **47** (2009).
6. F. Menter, Two-equation eddy-viscosity turbulence models for engineering applications, *AIAA J.* **32**, 269 (1994).
7. P. Malan, K. Suluksna and E. Juntasaro, Calibrating the $\gamma - Re_\theta$ Transition Model for Commercial CFD, *AIAA* (2009).
8. S. Pradeep, M. B. Subrahmanya, D. S. Kulkarni and B. N. Rajani, CFD simulation of flow past MAV wings, *International J. of Aerospace Innovations* .
9. J. Coupland, Flat plate transitional boundary layers, ERCOFTAC classic database, *http://cfd.mace.manchester.ac.uk/ercoftac* (1990).
10. M. Galbraith and M. Visbal, Implicit large-eddy simulation of low Reynolds number inflow past the SD7003 airfoil, *AIAA Sciences Meeting and Exhibit, Reno, NV* (2008).

11. R. Radespiel, J. Windte and U. Scholz, Numerical & experimental flow analysis of moving airfoils with laminar separation bubbles, *AIAA paper* **501** (2006).
12. M. Ol, B. McAuliffe, E. Hanff, U. Scholz and C. Kalher, Comparison of laminar separation bubble measurements on a low Reynolds number airfoil in three facilities, *AIAA paper 2005-5149* (2005).

DNS of Incompressible Square Duct Flow and Its Receptivity to Free Stream Turbulence

P. M. Bagade, N. Sawant, M. Sriramkrishnan and T. K. Sengupta*

Department of Aerospace Engineering, IIT Kanpur,
Kanpur, U.P., 208 016, India
**E-mail: tksen@iitk.ac.in*

Here DNS of incompressible flow through square duct and its receptivity to free stream turbulence (FST) is investigated for a sub-critical Reynolds number (Re) of 800 based on the side of the square duct and the oncoming velocity. An equilibrium flow is obtained first, whose receptivity to FST is studied, which is modelled by moving average time series model. This model corresponds to experimental results reported for channel flow.

Keywords: DNS, flow through square duct, effects of FST, receptivity, instability, transition.

1. Introduction

Flow in a channel with square cross section is a benchmark canonical problem for the study of instability and transition. For any instability problem it is paramount that one must first obtain an equilibrium flow, whose receptivity to different background disturbances are of interest [1,2]. Unfortunately, many internal and external flows do not allow computing the equilibrium state, as the numerical errors trigger flow instabilities. For example, inertia driven zero pressure gradient boundary layer flow over flat plate allows such a computation, while a wedge flow in the presence of adverse pressure gradient does not permit one to compute the equilibrium flow. However, pressure driven flow through square channels experiences favorable pressure gradient in the direction of the flow and one would expect to compute the equilibrium flow. In the present study, we first compute the equilibrium flow for a subcritical Reynolds number of $Re = 800$ and attempt to compute the flow for $Re = 1100$, where Re is defined in terms of the half width of the channel as the length scale and inflow velocity of the duct as the velocity scale. The equilibrium flow obtained for $Re = 800$ is further studied for effects of free stream turbulence (FST), on it.

Kim and Moin[3] solved the governing Navier-Stokes equations (NSE) in the primitive variable formulation using a spectral method, Fourier series in the streamwise and spanwise directions, and Chebychev polynomial expansion in the other direction is employed for the spatial derivatives and the time advancement was done using semi-implicit scheme. They treated the viscous diffusion terms implicitly.

In Bottaro *et al.*[4] a linear approach was followed to explain the coherent motion observed in turbulent square duct flows. The linear set of equations were accompanied by mixing length strategy in order to explain the structures. It was found that several coherent states are possible, in which some of there are perfectly symmetric as found in experimental studies. However it was noted[4] that the relation between secondary flows and exact coherent states is not yet understood. Interestingly in Uhlmann *et al.*[5], puffs have been identified in square duct flows within a Reynolds number regime of 800 to 1025. These have been identified as regions of localized turbulence, which sustains over time. They found the puffs to travel at approximately the bulk velocity of the flow, however the puffs were not considered as travelling waves[6]. The core of puffs resembled close to evanescent turbulent flows[7] which for square ducts was found to be $Re = 1100$. Uhlmann *et al.* have identified two strategies for generating puffs in square duct flows; one way is to take laminar base flow and apply forcing and the other way is by taking a fully developed turbulent flow and gradually reducing the Re until only puffs remain in the flow. Transient growth in channel flow has also been described in Schmidt *et al.*[8]

In many fluid dynamical applications, the oncoming flow has fluctuations with varying intensities of turbulence. The source of these background noises may be acoustic, vortical or entropic in origin. This calls for accurate modeling of FST to find its effects on flows. In the present work, effects of FST are studied in a square duct flow for $Re = 800$ based on centerline velocity and the side of the square as velocity and length scales. DNS is performed via parallel computing using 128 nodes with high accuracy numerical schemes[9].

2. Secondary Flows in Square Ducts

Flow through square duct is of importance, as it represents geometrical symmetry, which exhibits itself in flow structure also, even for turbulent flows. Flow through turbulent square channels is different from turbulent plane Poiseuille flows because of the secondary motions, first identified in

Göttingen[10,11] and are now known as Prandtl's secondary motion of second kind. Secondary motion magnitude is small of the order of 2 to 4 % of the main turbulent flow. Due to much smaller value of secondary flows in square channels, experimental measurement of this is difficult, even though some qualitative characteristics have been reported[12,13]. Numerical simulations are the preferred way to characterize these flows[14–16]. However, in none of the computational works reported so far, equilibrium flow has been obtained by solving the full NSE. Laminar steady flow obtained by accurate numerical methods[9,17] is the equilibrium flow which should be excited in studying its receptivity to vortical perturbations for incompressible flows, here.

3. Governing Equations and Numerical Methods

Rotational form of NSE equation in velocity-vorticity (V, ω) formulation are used[9,17,18]. The vorticity transport equation is given by

$$\frac{\partial \vec{\omega}}{\partial t} + \nabla \times (\vec{\omega} \times \vec{V}) + \nu \nabla \times (\nabla \times \vec{\omega}) = 0. \tag{1}$$

The velocity Poisson equation is given by

$$\nabla^2 \vec{V} = -\nabla \times \vec{\omega}. \tag{2}$$

Here a staggered grid is used[9,17] which is preferred in this formulation due to its higher accuracy and de-aliasing properties. In this grid, the velocity components are defined at the center of the planes on which the component is perpendicular, while the vorticity components are placed at mid-location on the edges of the cube, which are parallel to the direction of vorticity components. High accuracy dispersion relation preserving (DRP) optimized compact schemes[9,17] and optimized compact interpolation schemes are used for spatial discretization, whereas optimized third order Runge-Kutta scheme[17] is used for time integration. Also, in order to control numerical instabilities for high wavenumbers, both $1D$ and $2D$ filters[19] are used near the entrance section of the channel. A schematic diagram of the computational domain used here is shown in Fig. 1. A small upstream distance is provided to include the effects of the entrance section and here the non-dimensional upstream distance is chosen as -0.48. Including the entrance is necessary, as a similar exercise for flat plate has shown that the leading edge of the flat plate plays a significant role by providing site of perturbation. However, such disturbances stay outside the shear layer by "shear sheltering effect"[20].

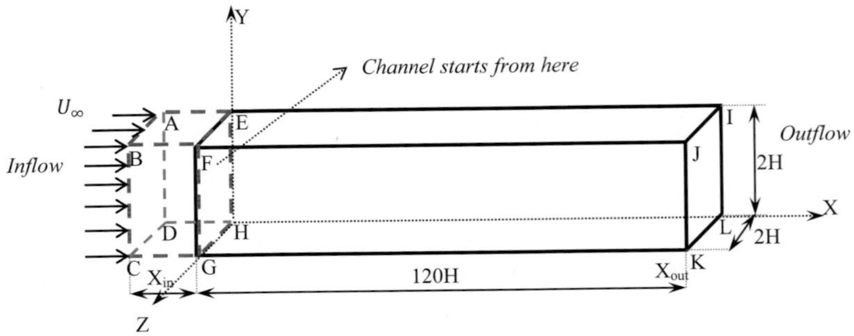

Fig. 1. Schematic diagram of the computational domain used for the present study.

4. Computing Equilibrium Flow at Subcritical Reynolds Number

Figure 2(a) shows the ω_z components at the spanwise mid-plane for $Re = 800$. It is found that initially a saddle point is formed as noted at $t = 25$, which degenerates into a wave-front moving to the right. A steady state solution is observed by $t = 200$, once the initial wave-front leaves the domain. Figure 2(b) shows the same for $Re = 1100$, a post-critical Re, for which the wave-fronts produced are strong enough to create vortices intermittently which convect downstream. A complete steady state is not reached for this Re, till $t = 400$.

Figure 3 shows the ω_x components at $x = 23.616, 71.808, 95.904$ for $Re = 800$ and 1100. For $Re = 800$ an eight-vortex pattern with two vortices associated with each corner are noted, whereas for $Re = 1100$, in addition to those two vortices at each corner are noted; there is an inner ring which is also associated with eight vortices. It has been observed experimentally[12] that the production of vorticity is always zero along a corner bisector and is of opposite sign on either side of it. Figure 4 verifies this for both Re 800 and 1100. The corresponding secondary flows are also noted. The flow is directed towards the corner. But the magnitude of secondary flows in this case is very small, of the order of $1/10^{th}$ of the magnitude of secondary flows associated with turbulent flows and hence they do not cause any distortion to the laminar parabolic velocity profile. Thus, laminar flows are associated with weak secondary motion which is captured accurately due to the use of high accuracy methods used in the present work.

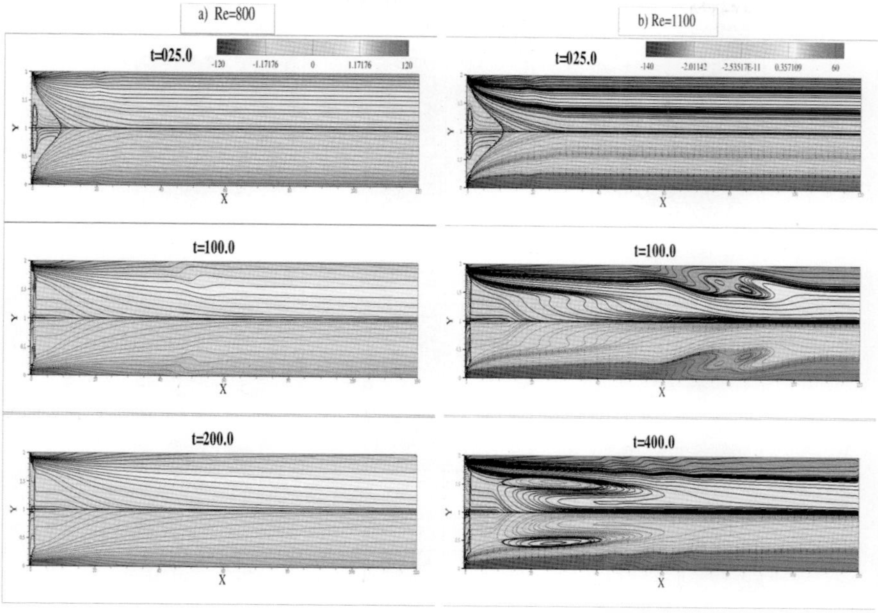

Fig. 2. a) ω_z contours at spanwise mid-plane for Re=800 case at indicated time instants. b) ω_z contours at spanwise mid-plane for Re=1100 case at indicated time instants.

Due of the cross-section of the duct being symmetric and the use of high accuracy DRP schemes, the symmetry of the flow field is maintained up to a very long time without imposing any such symmetry condition on the flow explicitly. However, for the flow at the supercritical Re, the symmetry is lost and regained intermittently. The supercritical flow is found to lose its 4− way symmtery at $t = 250$ and regain the symmetry at $t = 380$. The symmetry is lost again at $t = 400$, as can be seen from Fig. 3. Here, the flow is top-down symmetric and right-left symmetric but the symmetry about the diagonals is lost.

5. Free-Stream Turbulence Modeling

It is noted that solving NSE in the presence of FST, moving average model[21] is more realistic than solving stochastic differential equations for flows[22]. In a recent DNS, flow in a square duct has been studied with harmonic fluctuations with the oncoming flow to simulate the effects of FST[23].

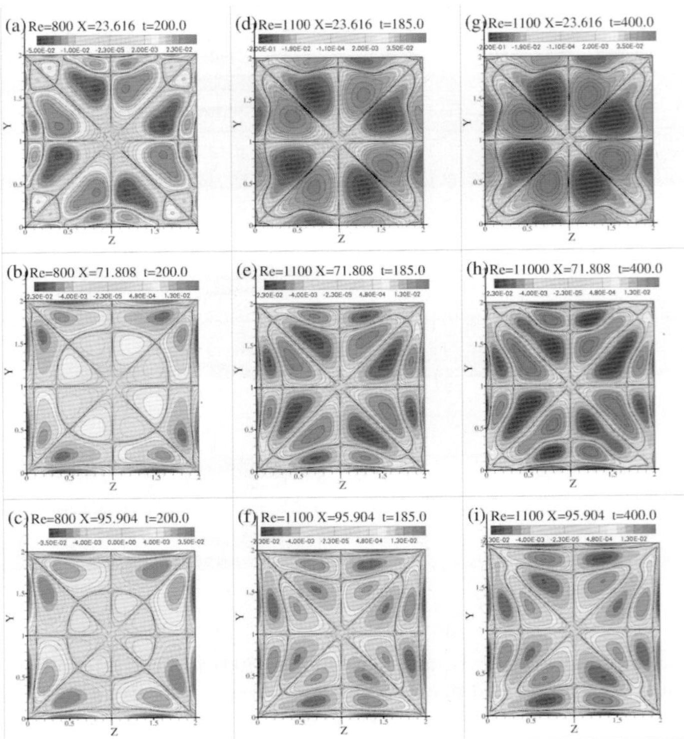

Fig. 3. ω_x contours at indicated time instants and streamwise locations for Re=800 and 1100.

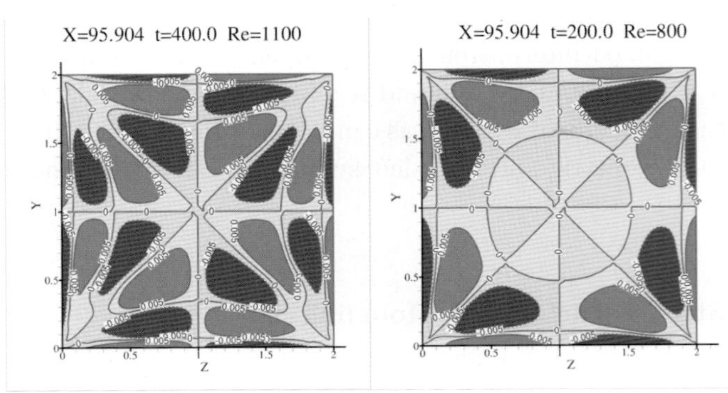

Fig. 4. ω_x contours at indicated time instants and streamwise location for Re=1100 and 800.

In the moving average model, from the histogram or probability distribution function (PDF) the central moments are obtained as

$$\mu_n = \frac{\sum [f(x_i) - \mu]^n}{\sum f(x_i)} \tag{3}$$

where, $f(x)$ represents the histogram and n indicates the corresponding moment about the mean, μ. A synthetic time series is generated at the inflow boundary using a standard first order moving average model, with the help of second and fourth moments[21]. Hence, in this model

$$x_t = e_t + \alpha_{fst} e_{t-1} \tag{4}$$

where, e_t is defined by a normal distribution with first central moment as zero and standard deviation defined as σ.

Subsequently, to include the effects of skewness (third moment) in the moving average model the low frequency component of FST was added, taken from experimental facility[24]. This is important since the large-scale anisotropy of the incoming flow is retained by this low wavenumber components. Thus, the net streamwise disturbance component becomes

$$u' = e_t + \alpha_{fst} e_{t-1} + \sum_{j=1}^{N} a_j e^{ik_j(x-ct)} \tag{5}$$

where first two terms are given by the Gaussian distribution and the last term shows the low frequency components of the disturbance that is facility- and speed-dependent[1]; c being the phase speed of propagation of these low wavenumber coherent structures. The Gaussian distribution is obtained from uniformly distributed random number generated, which is converted to normal distribution using box-Muller transformation[24]. The obtained values of σ and α_{fst} are used to obtain the synthetic signal and the low frequency part is added to model the FST.

This FST model is used to obtain the x-component of FST (u') and then y- and z- components are obtained by scaling for channel flow[25]. Figure 5 shows the x- and y-components of FST scaling contours at the inlet of the duct[25]. In the present study, we have taken $v' = w'$ due to geometric symmetry of the square duct.

A schematic diagram of the computational domain used is already shown in Fig. 1, with the inflow exactly at the entrance of the duct.

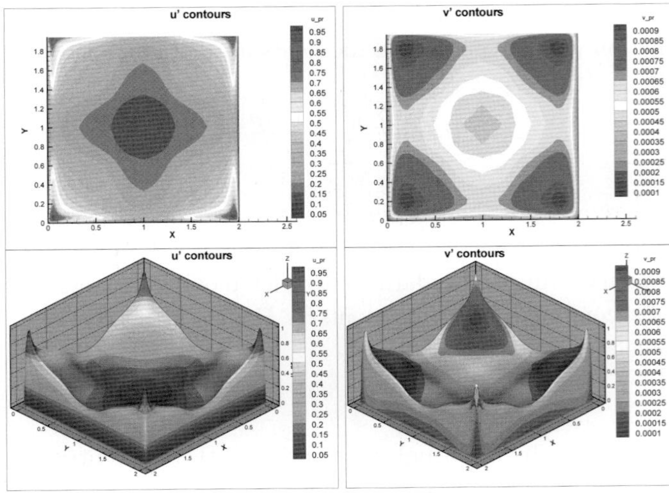

Fig. 5. Turbulence intensity contours at the inlet of the duct.

6. Results and Discussion

In Fig. 2, we noted a saddle point at $t = 25$ as was found for $2D$ channel flow in equilibrium state[9]. This disintegrates into a wave-front at later times and convects out of the domain. During early transient stages, this wave-front produces vortical structures seen at $t = 100$.

For $Re = 800$ with $Tu = 0.01\%$ started at $t = 204$, results are shown in Fig. 6. Figure 6(a) shows streamwise vorticity (ω_x) contours at the indicated times at the inlet ($x = 0$) on the top frames. Figure 6(b) shows the evolution of ω_x at $x = 24$ while the bottom frames in Fig. 6(c) show corresponding ω_x-contours at the same times for $x = 72$. In the absence of FST, the flow evolves into multi-fold anti-symmetries about the mid-point showing equiangular sectors in Fig. 3. Application of FST causes the multi-fold anti-symmetries to break down to pair about one of the diagonals. One sees the retention of such anti-symmetries for the total field even at later times.

The convecting nature of the disturbance is evident from the time, asymmetry is noted at each station. At all the stations, the effects of FST are noted by noting the break-up of perfect multi-fold symmetry of the vorticity component distribution to one of an asymmetric distribution into two cells demarcated by the diagonal line joining the bottom left to top right corner. The terminal frame shown here is in development stage.

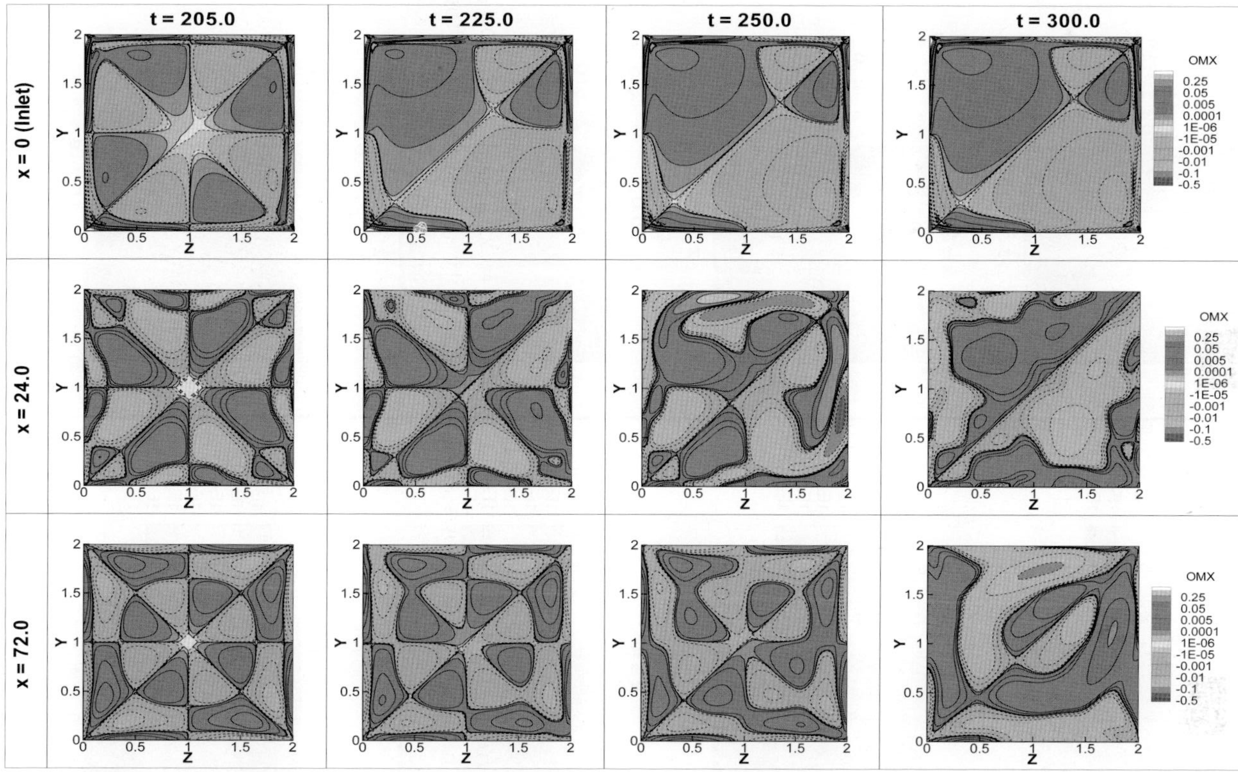

Fig. 6. ω_x contours at different sections at indicated times.

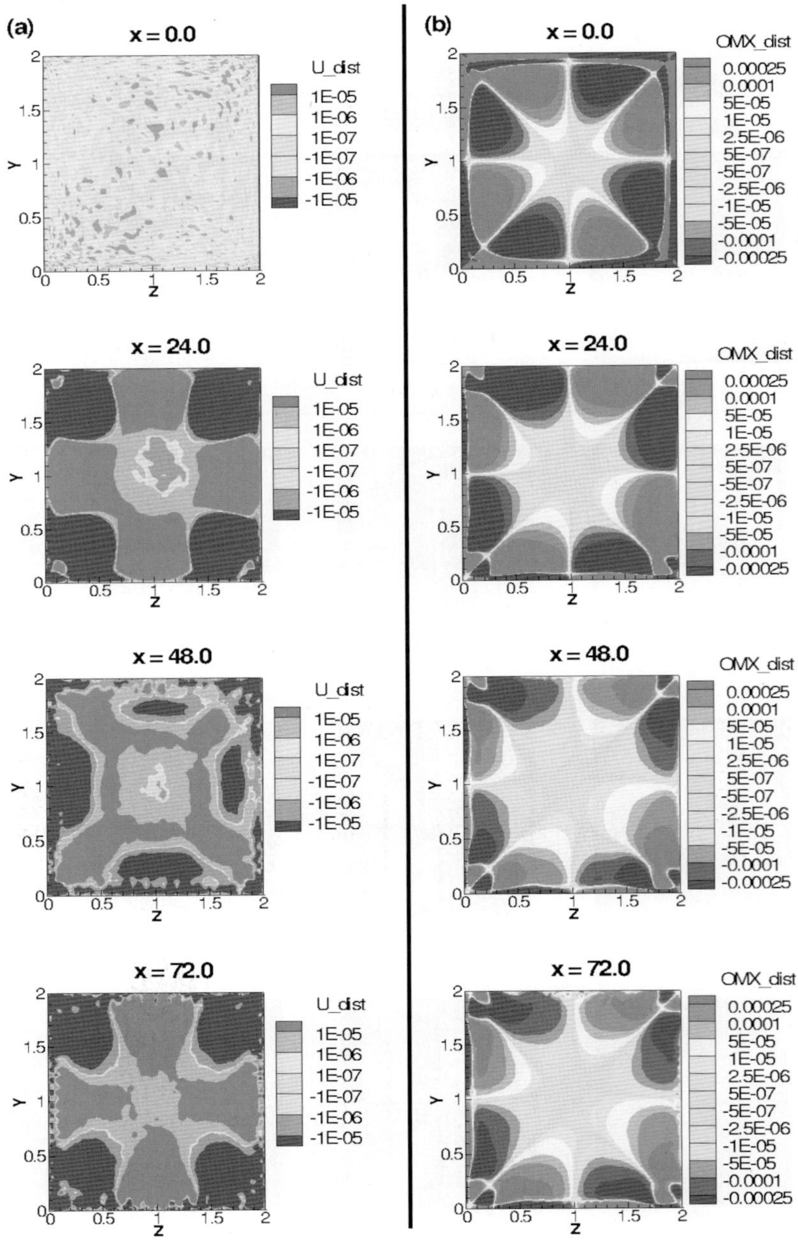

Fig. 7. (a) Disturbance velocity ($u^{'}$) contours at $t = 205$ (b) Disturbance vorticity ($\omega_x^{'}$) contours at same sections at $t = 205$.

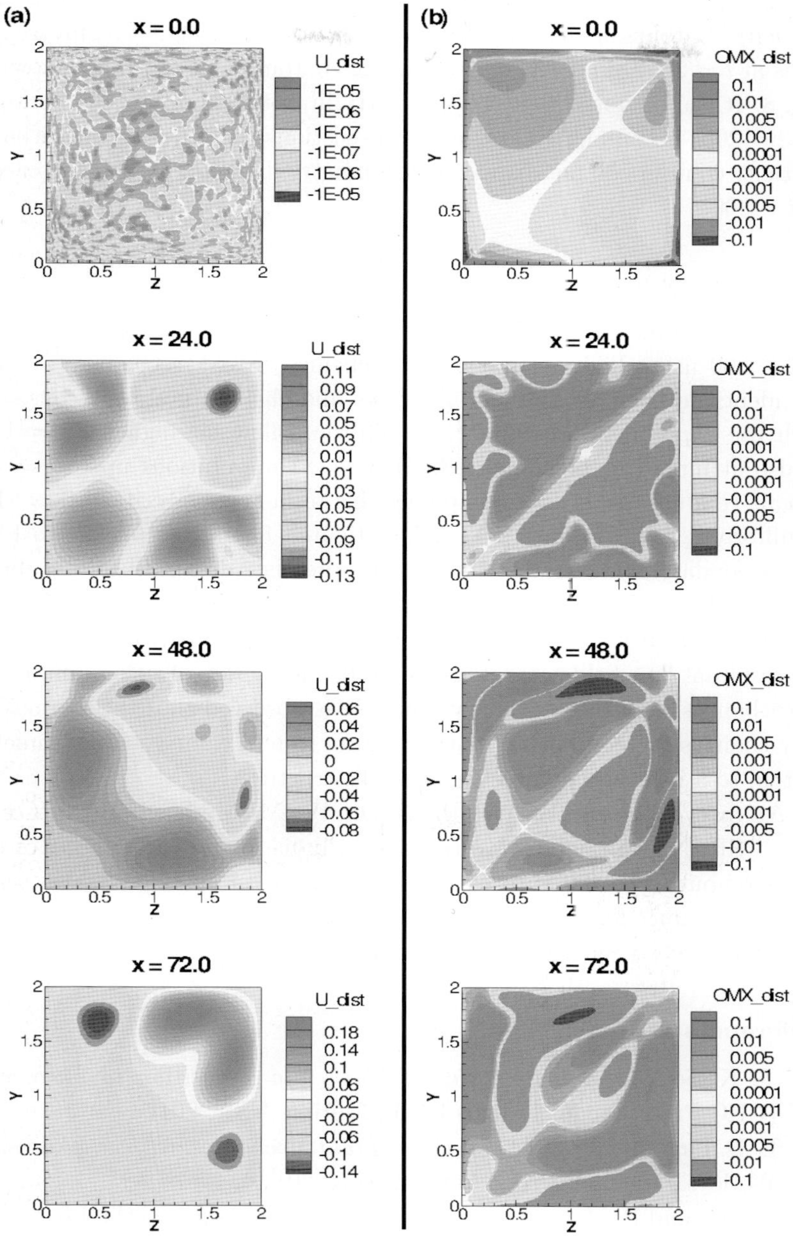

Fig. 8. (a) Disturbance velocity (u') contours at $t = 300$ (b) Disturbance vorticity (ω_x') contours at same sections at $t = 300$.

In Figs. 7(a) and (b), streamwise disturbance velocity and disturbance vorticity contours are shown respectively, for the indicated streamwise stations at $t = 205$. Similarly, corresponding contours at $t = 300$ are shown in Fig. 8. It can be noted that the symmetry in disturbance velocity contours (at $t = 205$) is completely lost at all the sections shown. Also, disturbance vorticity contours do not display any symmetry. The flow is inhomogeneous at $t = 300$.

7. Summary and Conclusions

The equilibrium flow is obtained for flow through square channels at $Re = 800$ using high accuracy DRP schemes. Weak secondary motions are identified even for the laminar flows and their structure is consistent with the experimental results[12]. Equilibrium flow for higher $Re = 1100$ is also attmpted to be computed. However, computations up to $t = 400$ reveals residual unsteadiness implying this to be a super-critical flow. The equillibrium flow for $Re = 800$ is investigated for its receptivity to FST. The secondary flows are ascribed to the presence of turbulence[14]. In an actual wind tunnel, the incoming flow possesses fluctuating components of velocity apart from the corner secondary vortices. It is essential to study the resultant flow field generated due to the interaction of such omnipresent disturbances with the secondary flows. It is observed that, FST considerably changes the equilibrium flow structure from a eight-vortex symmetric system to a near anti-symmetric two-cell structure.

Acknowledgements: We thank Dr. V. M. Mudkavi, Head, CTFD Div., CSIR-NAL Laboratories, Bangalore, India for providing access to its supercomputing facilities to carry out this research work. We also thank Dr. S. Bhaumik for making available a basic 3*D* code for flow over a flat plate which has been modified for the present work.

References

1. T. K. Sengupta, *Instabilities of Flows and Transition to Turbulence* (CRC Press, Taylor Francis Publ., Florida, USA, 2012).
2. L. D. Landau and E. M. Lifschitz, *Fluid Mechanics, 6th edi.* (Addision-Wesley, Pergamon Press, London, U.K., 1959).
3. J. Kim and P. Moin, Application of a fractional-step method to incompressible navier-stokes equations, *J. Comp. Phys.* **59**, 308 (1985).
4. A. Bottaro, H. Soueid and B. Galletti, Formation of secondary vortices in turbulent square duct flow, *AIAA Journal* **44**, 803 (2006).

5. M. Uhlmann, A. Pinelli, G. Kawahara and A. Sekimoto, Investigating puffs in square duct flow, *INI Workshop* (2008).
6. H. Wedin and A. Bottaro, Three-dimensional travelling waves in a square duct, *Phys. Review E.* **79**, 0653051 (2009).
7. M. Uhlmann, A. Pinelli, G. Kawahara and A. Sekimoto, Marginally turbulent flow in a square duct, *J. Fluid. Mech.* **588**, 243 (2007).
8. P. J. Schmid and D. S. Henningson, *Stability and Transition in Shear Flows* (Springer Science and Business Media, 2001).
9. T. K. Sengupta, *High Accuracy Computing Methods: Fluid Flows and Wave Phenomena* (Cambridge Univ. Press, New York, USA, 2013).
10. J. Nikuradse, Untersuchung *ü*ber die geschwindigkeitsverteilung in turbulenten strömungen, PhD thesis, University of Göttingen, Germany, (VDI Forsch, 1926).
11. L. Prandtl, *Über den reibungwiderstand strömenderluft* (Ergeb. Aerodyn. Versuch., Series III, University of Göttingen, Germany, 1927).
12. E. Brundrett and W. D. Baines, The production and diffusion of vorticity in duct flow, *J. Fluid. Mech.* **19**, 375 (1964).
13. B. E. Launder and W. M. Ying, Secondary flows in ducts of square cross-section, *J. Fluid. Mech.* **54**, 289 (1972).
14. B. Galletti and A. Bottaro, Large scale secondary structures in duct flow, *J. Fluid. Mech.* **512**, 85 (2004).
15. S. Gavrilakis, Numerical simulation of low-reynolds-number turbulent flow through a straight square duct, *J. Fluid. Mech.* **244**, 101 (1992).
16. A. Huser and E. Biringen, Direct numerical simulation of turbulent flow in a square duct, *J. Fluid. Mech.* **257**, 65 (1993).
17. S. Bhaumik and T. K. Sengupta, A new velocity-vorticity formulation for direct numerical simulation of 3*d* transitional and turbulent flows, *J. Comp. Phy.* **284**, 230 (2015).
18. S. Bhaumik and T. K. Sengupta, Precursor of transition to turbulence: Spatiotemporal wave front, *PRE* **89**, p. 043018 (2014).
19. T. K. Sengupta and Y. G. Bhumkar, New explicit two-dimensional higher order filters, *Comput. Fluids* **39**, 1848 (2010).
20. T. K. Sengupta, S. De and S. Sarkar, Vortex-induced instability of incompressible wall-bounded shear layer, *J. Fluid Mech.* **493**, 277 (2003).
21. W. A. Fuller, *Introduction to Statistical Time Series* (Cambridge Univ. Press, U. K., 1995).
22. D. Lucor and G. E. Karniadakis, Noisy inflows cause a shedding mode switching in flow past an oscillating cylinder, *Phys. Rev. Letters* **92**, 154501 (2004).

23. A. F. Shinn, S. P. Vanka and W. W. Hwu, Direct numerical simulation of turbulent flow in square duct using a gpu, *AIAA paper* **2010-5029** (2010).
24. T. K. Sengupta, D. Das, P. Mohanamuraly, V. K. Suman and A. Biswas, Modeling free-stream turbulence based on wind tunnel and flight data for instability studies, *Int. J. Emerging Multidisciplinary Fluid Sci.* **1 (3)**, 181 (2009).
25. H. Schlichting, *Boundary Layer Theory* (McGraw Hill Publ., 7th *edition*, 1978).

Evolution of RANS Modelling of High Speed Mixing Layers using LES

Arvind S. Iyer* and N. K. S. Rajan
CGPL,
IISc Bangalore
**E-Mail: arvind@cgpl.iisc.ernet.in*

Debasis Chakraborthy
Technology Director,
DRDL, Hyderabad, India

The current work evolves a model for shear stress for a compressible mixing layer, with M_c being a parameter. This model is evolved with the data obtained from LES. It is seen that the shear stress to strain rate relation is remarkably linear in the self similar region of the mixing layer, albeit the proportionality constant depends on the prevailing convective Mach number. The model constants obtained from the simulations match with those which can be derived from experiments. By using this model, the growth rate obtained falls well within the range of experiments even at high M_c.

Keywords: LES, U-RANS, Mixing Layer, Supersonic flow

Nomenclature

M_c	Convective Mach number
U	Stream velocity
ρ	Density
y_0	Center of the mixing layer
η	$\frac{y-y_0}{\delta(x)}$
r	$\equiv \frac{\rho_2}{\rho_1}$
s	$\equiv \frac{U_2}{U_1}$
$f(\eta)$	$\frac{\langle\rho\rangle(x,y)}{\rho_1}$
$g(\eta)$	$\frac{\{u\}(x,y)}{U_1}$
S	Strain rate
$_{1,2}$	Primary, Secondary Streams
$\{\cdot\}$	Favre average
$\langle\cdot\rangle$	time average
$\cdot'$	$\cdot - \langle\cdot\rangle$
$\cdot''$	$\cdot - \{\cdot\}$

1. Introduction

The turbulence modelling of high speed mixing layers, especially for the use in RANS has been a challenge since a long time. It has been shown in practice as well as reasoned out that the classical models of turbulence cannot capture the growth rate phenomenon accurately[1], which has been reported through experiments[2–4]. LES has been shown to be successful in predicting the growth rate of high speed mixing layers quite accurately[5–7]. The simulations also give a clear picture of the distribution of the shear stress, and its variation with that indicated by compressibility whose measure is provided by M_c.

It is seen from the analyses that the shear stress scales quite reasonably with the shear strain for most of the mixing layer, and that a simple mixing length approach is expected to yield the mixing layer reduction, albeit the length itself would be a function of M_c. This was estimated from the LES to be roughly an exponential function, and a series of experiments with RANS were tried and the functional form of the mixing length was established and coefficients determined. τ_{yy} at the same time was also related using a suitable algebraic relation to S_{xy}.

2. Terms Contributing to the Momentum Equation

To be able to model the stresses, it is necessary to know, which term is the most significant contribution to the respective momentum equations.

2.1. *Integral analysis*

Integral analysis connects the growth rate to the integrals of the mean flow and the shear stress yields

$$\delta'\left(\int_{-\infty}^{0} fg(g-r)d\eta + \int_{0}^{\infty} fg(1-g)d\eta\right) = 2T_{xy0};T_{xy} \equiv \frac{\left(\underline{\langle\tau_{xy}\rangle} - \langle\rho\rangle\{u''v''\}\right)}{\rho_1 U_1^2} \tag{1}$$

where $T_{xy0} = T_{xy}(y = 0)$. The derivation of the above expression, assumes absence of the pressure terms and the mixing layer not bending in any direction.

It was verified that the integrals on the LHS do not vary much, and are more or less dependent only on r and s in the self similar region. The integral analysis clearly leads to the fact that the most important term to the growth rate measurement is T_{xy} which is directly related to τ_{xy}. This is also supported by earlier findings of Sarkar *et al.*[8].

2.2. *The contributing terms in the stream-wise momentum equation*

First, looking at the x momentum equation, we have the relevant Reynolds stress terms as τ_{xx} and τ_{xy}. The actual source in the momentum equation

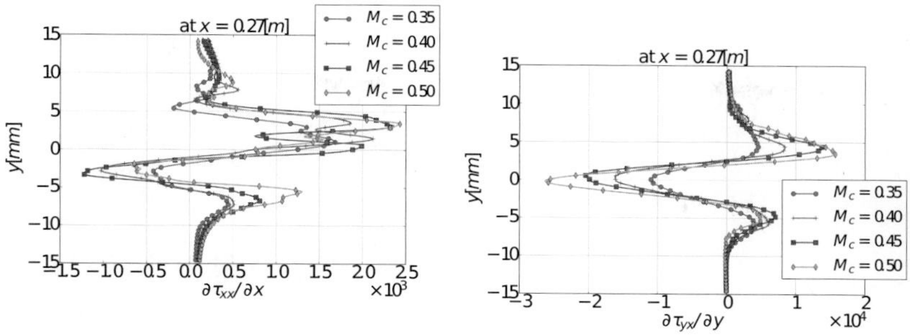

Fig. 1. Relative contribution of τ_{xx} and τ_{xy} to the x momentum equation.

is what is the respective contribution to the momentum, where it is the cross-wise gradient of τ_{xy} and the stream-wise gradient of τ_{xx} which are present. This term, which is plotted in Fig. 1.

It is clear from Fig. 1 that the contribution of τ_{xy} is an order greater than that of τ_{xx}. This is, of course, not unexpected that this represents the turbulent stresses which transport the most significant momentum in the direction perpendicular, and hence can be expected to be of a greater magnitude. Momentum mixing in a mixing layer, and the subsequent growth rate occurs mainly due to this component. It may hence be said that it is most important to model τ_{xy} accurately.

2.3. *The contributing terms in the cross direction momentum equation*

An examination of the y momentum equation indicates that the relevant Reynolds stress terms are τ_{yy} and τ_{yx}. As before, the contribution to the cross-wise momentum transport which appear as the cross-wise gradient of τ_{yy} and the stream-wise gradient of τ_{xy}. This is plotted in Fig. 2 from which it is clear that the contribution of τ_{yy} is an order greater than that of τ_{yx}. And for RANS simulations it is of a greater importance to model τ_{yy} as accurately as possible for the cross-wise momentum equation. However, this term does not directly influence the growth rate, and it is the transport of the non-dominant direction of momentum. Hence the importance of this term cannot be expected to be the same as that of τ_{xy}. It is thus clear from the above that the most important parameter for modelling is τ_{xy} and this is followed by τ_{yy}.

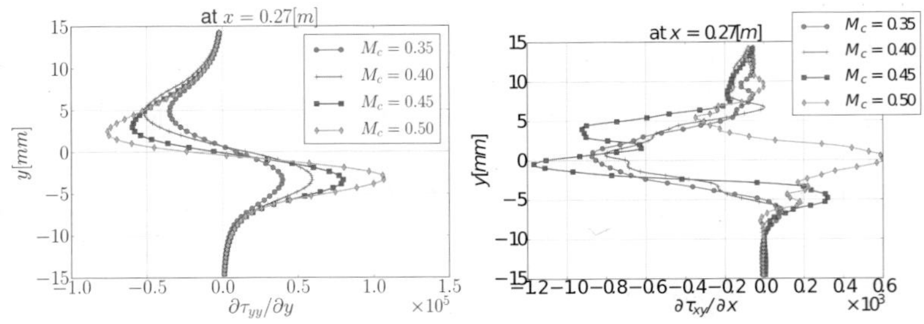

Fig. 2. Relative contribution of τ_{yy} and τ_{yx} to the y momentum equation.

3. Modelling of τ_{xy} and τ_{yy}

3.1. *Modelling of* τ_{xy}

The aim of this section is to express τ_{xy} as a function of some mean property. The logical choice of such a property is S_{xy}, taking inspiration from the Prandtl's mixing length hypothesis. This can be seen qualitatively in Fig. 3.

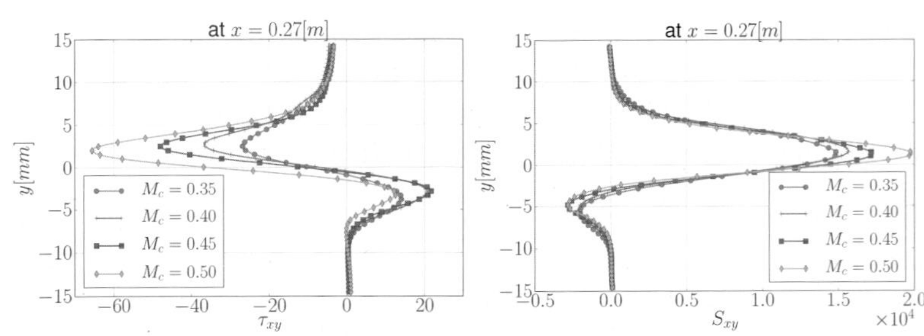

Fig. 3. τ_{xy} and S_{xy} at the same stream-wise location.

At once, we can express the relation in terms of non-dimensional parameters as

$$\frac{\tau_{xy}}{\bar{\rho}(\Delta U)^2} = L\frac{S_{xy}}{(\Delta U)/x} \text{ where } \Delta U \text{ is } U_1 - U_2 \text{ and } \bar{\rho} \text{ is } \frac{\rho_1 + \rho_2}{2}\,. \tag{2}$$

The above scaling is expected to yield simpler formulation because, it is known that under self similar conditions, τ_{xy} becomes almost a constant, whereas S_{xy} linearly decreases with x. Under such conditions, L is expected to be more or less independent of x, however that needs to be verified. Recognizing that S_{xy} can be made scale invariant with the mixing layer thickness and an appropriate velocity scale ΔU, non-dimensionalized quantity

$Sxy/(\Delta U)/x$ was chosen after noting that x is the independent variable on which δ depends strongly. This is shown in Fig. 4.

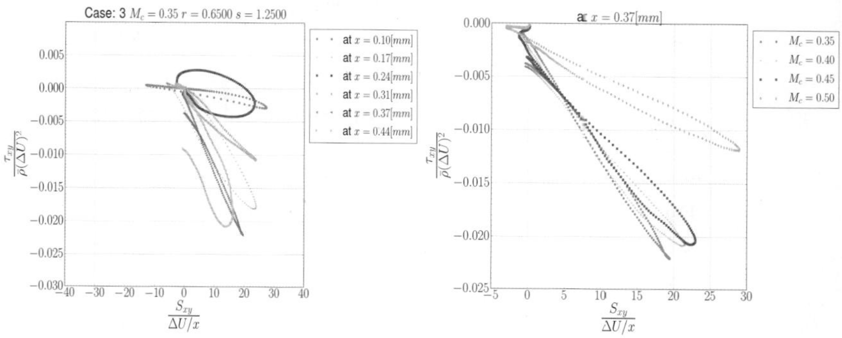

Fig. 4. τ_{xy} vs S_{xy} at various locations(left) and at different Mach numbers (right).

Similarly, with increasing M_c the behaviour of the τ_{xy} vs S_{xy} plot is shown in Fig. 4, which shows without doubt that *the shear stress in the region of self similarity at least, at all M_cs can be very well approximated by a straight line*. This implies the ratio of τ_{xy} to S_{xy}, while modelling, needs to incorporate the effect of M_c. That is $L = L(M_c)$ which needs to be obtained. From the LES computation, it is seen that L has the form $L = Ae^{-BM_c}$, where $A \approx -0.014$ and $B \approx 6.7$. This relation is used as a initial guess for the RANS computations which follow.

3.2. *Modelling of τ_{yy}*

The expected choice of the mean flow parameter to be used for modelling τ_{yy} is S_{yy} as shown in Fig. 3. It can be seen from Fig. 5 that though the variation of τ_{yy} is smooth, the variation of S_{yy} is absolutely not so. Furthermore, it is not appropriate to relate S_{yy} to τ_{yy} because of a basic reason that $\tau_{yy} \equiv \langle \rho v'' v'' \rangle \geq 0$ whereas no such requirement is present from S_{yy} which can and does assume both positive as well as negative values.

It is hence sought to relate τ_{yy} back to S_{xy}, noting that shear strain is the dominant strain in the mixing layer. The approximate equation thus obtained is

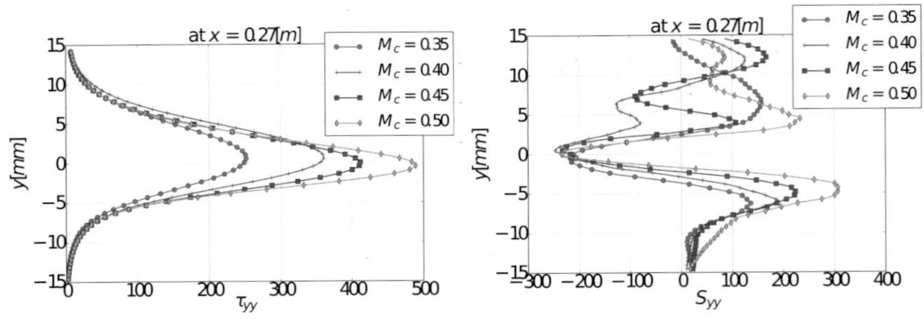

Fig. 5. τ_{yy} and S_{yy} at the same stream-wise location.

$$\frac{\tau_{yy}}{\rho \Delta U^2} = P + \frac{Q}{\left(\dfrac{M_c + 0.01}{0.25}\right)^2} \left(S^2(S+0.4)^2\right)^{0.125} \pm 0.01 \left(S^2 - S^4\right)^{0.125} \quad (3)$$

where $S = \dfrac{S_{xy}}{30(\Delta U/x)}$ and P and Q are constants evaluated to be nominally 0.02 and 0.03 respectively from the LES computations. In the above equation, it must be noticed that τ_{yy} is always positive while S can be both positive and negative.

4. Unsteady RANS Computations

Unsteady RANS was used for the computations for the compressible mixing layers, with the turbulence model for τ_{xy} as mentioned in the previous section. These simulations were carried out with the same fluid, air, in both the streams, and a constant $\mathrm{Pr_t}$ of 0.3 was used. These cases are defined in Table 1. With each of these test cases, an unsteady RANS simulation was performed, and averaging was performed over 10 sweeps, and the growth rate was measured. The growth rate showed an almost linear variation with L. This was used to estimate the value of L which would give the expected growth rate.

RANS Simulation Results The value of L which provides the expected growth rate is shown in Fig. 6. We proposed earlier an ansatz that the value of L is of the form $L \approx Ae^{-BM_c}$ Which is valid for low M_c to moderately large M_c. With this the best fit for the observed values of L is

$$L \approx -0.00294 e^{(-6.756 M_c)}. \quad (4)$$

Table 1. Specifications of RANS Simulations

Case	$U_{\text{avg}}[m/s]$	M_c	r	s	$U[m/s]$ Prm	$U[m/s]$ Sec	$T[K]$ Prm	$T[K]$ Sec	M Prm	M Sec	$p[Pa]$
1 - 7	500	0.27	0.68	1.25	597	402	358.1	286.49	1.58	1.19	7000
8 - 14	500	0.35	0.68	1.25	597	402	213.1	170.49	2.04	1.54	7000
15 - 21	500	0.45	0.68	1.25	597	402	128.9	103.14	2.62	1.98	7000
22 - 28	500	0.55	0.68	1.25	597	402	86.3	69.04	3.20	2.42	7000
29 - 35	500	0.65	0.68	1.25	597	402	61.7	49.43	3.79	2.86	7000
36 - 43	400	0.70	0.60	1.25	500	300	56.6	45.29	3.31	2.22	7000
44 - 49	400	0.75	0.55	1.25	516	283	66.5	53.20	3.16	1.94	7000
50 - 56	400	0.05	0.95	1.25	410	389	116.7	93.37	1.89	2.01	7000
57 - 63	400	0.10	0.90	1.25	421	378	122.9	98.35	1.89	1.91	7000

M_c	$L \times 10^4$
0.05	17.7
0.1	11.4
0.27	5.31
0.35	2.89
0.45	2.50
0.55	1.01
0.70	0.193
0.75	0.134

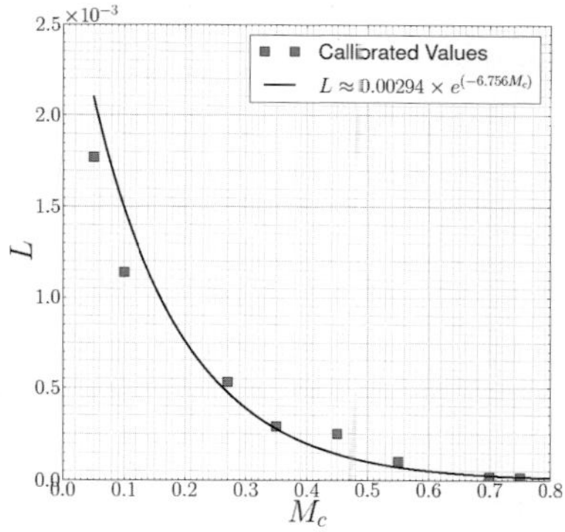

Fig. 6. Measured and interpolated variation of L with M_c.

Thus substituting in Eqn. 2 we get the model for τ_{xy} as

$$\frac{\tau_{xy}}{\bar{\rho}(\Delta U)^2} = \frac{S_{xy}}{(\Delta U)/x}\left(-0.00294e^{(-6.756M_c)}\right). \tag{5}$$

It can be seen that the exponent obtained is close to that obtained in LES. The scaling constant A cannot be compared because the value obtained from the LES is a time averaged value corresponding to all scales, whereas that obtained in URANS is for the instantaneous turbulent strains.

4.1. *Simulations with the model values*

The growth rate of the different cases were measured, and were compared with the experimental results. The plot of the growth is shown in Fig. 7. It can be seen that the growth rate of the mixing layer shows a clear decrease with increase in the M_c.

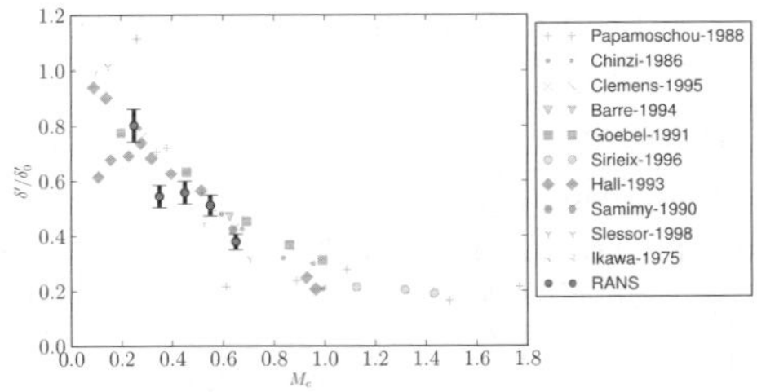

Fig. 7. Growth of the mixing layer with RANS at different M_cs.

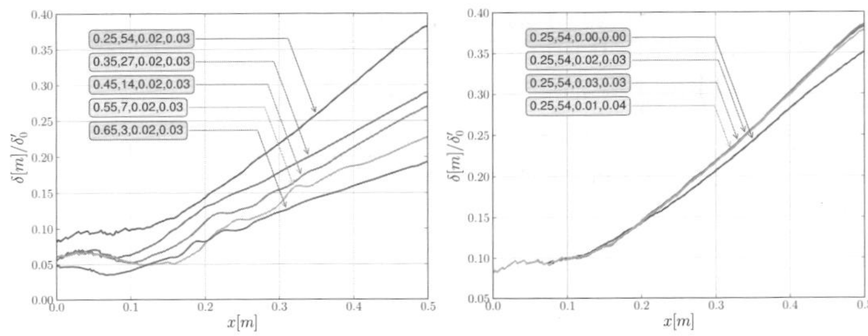

Fig. 8. Sensitivity of Growth rate with L and with A and B, the numbers indicate the M_c,$L \times 1e5$,A and B on left and M_c,$L \times 1e5$,P and Q.

Along with the model, it is also necessary to note the sensitivity of the growth rate to the coefficient in the model. This is plotted along with other data in Fig. 7 Furthermore, for the same L the effect of variation of A and B from Eqn. 3 was studied. This is shown in Fig. 8. It is seen clearly that whereas the growth rate is quite sensitive to the selection of the constants for τ_{xy}, it is hardly so in case of the coefficients related to τ_{yy}, unless for the case where the model for τ_{yy} does not make it identically zero.

5. Conclusions

The reason for the turbulent kinetic energy (k) based turbulence models are not suited for good prediction of the growth rates for mixing layers was identified from the LES simulations. τ_{xy} was determined to be the most important factor which contributes to the growth of the mixing layer. This was done using analysis of the data obtained from LES and integral analysis. It was seen from the analyses that the shear stress scales quite reasonably with the shear strain for most of the mixing layer, and that a simple mixing length approach is expected to yield the mixing layer reduction, albeit the length itself would be a function of M_c. This was estimated from the LES to be roughly an exponential function, and a series of experiments with RANS were tried and the functional form of the mixing length was established and coefficients determined. τ_{yy} at the same time was also related using a suitable algebraic relation to S_{xy}. The relation obtained after tuning was seen to predict the growth rates well within the band of experimental results. Further, it was shown that the growth rate was not sensitive to the modelling of τ_{yy}, indicating that a simpler model could have been used. The fact that the growth rate is not very sensitive to τ_{yy} (and of course τ_{xx}) also opens up the possibility of merging this with $k-\epsilon$ or $k-\omega$ model, where in the region of the mixing layer, the shear stress is predicted using the present zero equation model, and the normal stresses predicted by a two equation model. This will of course still be flawed in that the anisotropy would not be accounted for, even if the measure of the turbulent kinetic energy would roughly be correct. This particular aspect has not been dealt with in the present work.

References

1. J. Bardina, P. Huang and T. Coakley, Turbulence modeling validation, *AIAA paper*, 2121 (1997).
2. S. G. Goebel and J. C. Dutton, Experimental study of compressible turbulent mixing layers, *AIAA J.* **8**, 538 (1990).
3. D. Papamoschou and A. Roshko, The compressible turbulent shear layer: an experimental study, *J. Fluid Mech.* **197**, 453 (1988).
4. G. Elliott and M. Samimy, Compressibility effects in free shear layers, *Physics of Fluids A: Fluid Dynamics* **2**, 1231 (1990).
5. V. Sankaran and S. Menon, Large eddy simulation of scalar mixing in supersonic mixing layers, *Proceedings of the Combustion Institute* **30**, 2835 (2005).
6. A. S. Iyer and N. Rajan, Simulation of spatial high speed mixing layers using LES, *Comput. & Fluids* **109**, 113 (Mar 2015).
7. L. Doris, C. Tenaud and L. T. Phuoc, Les of spatially developing 3d compressible mixing layer, *Comptes Rendus de l'Académie des Sciences-Series IIB-Mechanics* **328**, 567 (2000).
8. C. Pantano and S. Sarkar, A study of compressibility effects in the high-speed turbulent shear layer using direct simulation, *J. Fluid Mech* **451**, 329 (2002).

Numerical Investigation of Centrifugal Instability Around a Circular Cylinder Rotated Impulsively

Akhilesh M. Prabhu, Ratnesh K. Shukla* and Jaywant H. Arakeri†

Department of Mechanical Engineering
Indian Institute of Science, Bangalore 560012, India
**E-mail: ratnesh@mecheng.iisc.ernet.in*
†E-mail: jaywant@mecheng.iisc.ernet.in

The inception and growth of centrifugally driven instabilities around an impulsively rotated cylinder is investigated through direct numerical simulations. Full three-dimensional simulation at a Reynolds number (Re) of 200, shows that the initial instability manifests as axisymmetric toroidal vortices stacked periodically along the cylinder axis. Based on this finding, two-dimensional axisymmetric simulations are performed for Reynolds numbers spanning seven orders of magnitude. A critical wavelength λ_c, defined as the wavelength of the first unstable mode is calculated over a range of Reynolds numbers based on the cylinder radius and wall velocity. A critical boundary layer thickness δ_c is defined based on vorticity and is found to scale as $\mathrm{Re}^{-2/3}$ at large Reynolds numbers. The critical wavelength λ_c exhibits a similar scaling, so that $\lambda_c \sim \delta_c$ for $\mathrm{Re} \gg 1$. A Taylor number based on the wavelength of the instability as the characteristic length scale is defined and its variation with the Reynolds number is discussed.

Keywords: Taylor-Couette flow; Centrifugal instability; Critical Taylor number; Direct numerical simulation.

1. Introduction

The internal flow created between rotating inner and outer cylindrical walls in the Taylor-Couette setup is often employed as an archetypal configuration for studying flows dominated by rotation. The rich fluid mechanics associated with this simple setup has instigated innumerable investigations encompassing a wide range of flow regimes from laminar, transitional, to fully developed turbulent states[1]. The stability of circular Couette flow was first investigated by G. I. Taylor[2] who identified a critical Reynolds number above which the steady Couette flow undergoes transition to a secondary flow with distinct poloidal circulation. The base flow in this linear stability analysis was assumed to be steady[3].

Our principal interest in this work is in the stability of the flow generated from the impulsive start of a rotating inner cylinder and a fixed outer cylinder. This special case of the general Taylor-Couette setup results in an unsteady base flow which is centrifugally unstable from the Rayleigh's stability criterion[4]. The instability manifests itself as axially periodic structures, similar to the Görtler vortices on curved boundaries. Previous experimental studies have shed light upon the development and growth of this primary instability, and in particular have reported on the critical wavelength and the onset time[5]. Bariclon *et al.*[6] and Wei *et al.*[7] explored the scaling of the Görtler vortex wavelength with the Taylor number for a statistically stationary flow. Numerical simulations though sparse in this regard, have captured the appearance of Taylor vortices and matched the corresponding critical wavelength with experimental findings[8]. These simulations were however, only carried out for a very small range of Reynolds numbers and the dependence of the critical wavelength on the Reynolds number was not explored. M.C. Kim *et al.*[9] studied the onset of the axisymmetric instability induced by a linear acceleration of the inner cylinder using propagation theory. Tan and Thorpe[10] proposed that the instability is analogous to the thermal instability in a deep pool of fluid heated from below. They defined a transient Taylor number and arrived at the critical time of onset of the instability and its wavelength which compared well with the experiments of Chen and Kirchner[5]. However, there has not been a systematic experimental or simulation based comprehensive investigation of this flow over a reasonably large range of Reynolds numbers. With this as our primary focus, we elucidate some of the salient features of the flow and discuss the scalings of important parameters which describe it.

Figures 1(a) and (b), show the schematic and the nature of the ensuing secondary flow, where δ_c denotes the boundary layer thickness at the onset of instability and λ_c the corresponding wavelength. The precise definitions of these quantities are given in the next section. The inner cylinder wall velocity $U_w = a\omega$, where a denotes the inner cylinder radius with ω as the constant rate of rotation about the vertical z axis. The Reynolds number (Re) is defined as aU_w/ν. The effect of flow rotation are characterized by the Taylor number (Ta $\sim$ Re^2). Analogous to the Rayleigh number based on plume spacing[11], we define a Taylor number (Ta_{λ_c}) using a length scale equal to the wavelength of the instability as $\mathrm{Re}^2\left({\lambda_c}^3/a^3\right)$. In the forthcoming sections we aim to establish the dependence of the Taylor number thus defined with the Reynolds number. Our initial investigations involving full three-dimensional simulations at $Re = 200$ reveal that the initial

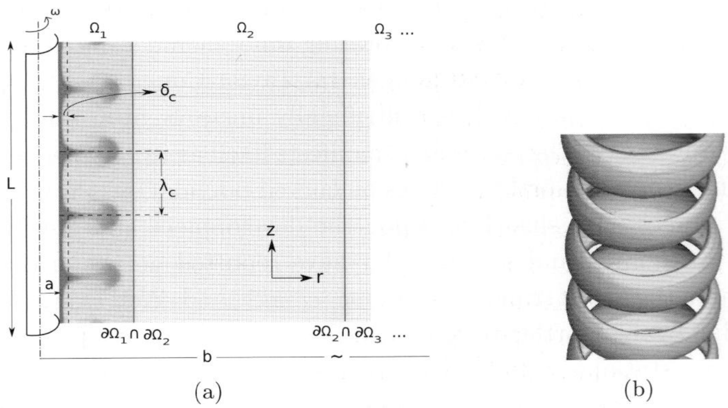

Fig. 1. (a) A schematic of the instability pattern illustrating the critical boundary layer thickness δ_c and the corresponding wavelength of the initial instability λ_c along with the computational domain employed (b) Contours of the axial vorticity for three-dimensional calculations at Re = 200.

instability is axisymmetric (see Fig. 1(b)). Guided by this observation, we perform axisymmetric simulations spanning seven orders of magnitude using a high-order accurate incompressible flow solver described next.

2. Simulation Methodology

We assume that all the flow quantities are independent of the circumferential coordinate and consider the primitive variable form of the incompressible Navier-Stokes equations in axisymmetric cylindrical coordinates (r, z) as shown in Fig. 1 (a). For spatial approximation we employ a tenth-order accurate non-uniform grid compact finite difference schemes[12,13] in the wall normal direction r along with Fourier spectral discretization[14] along the periodic axial direction z. A second-order accurate semi-implicit projection method is used for non-stiff temporal integration and to enforce incompressibility. The computational code has been extensively validated for both two and three-dimensional bluff body flows in previous works[15,16].

In addition we employ domain decomposition in the wall normal direction along with an influence matrix technique for efficient calculations on parallel architectures[17]. Figure 1(a) shows the distribution of the domains represented by Ω_1, Ω_2 and so on, and the domain boundaries by $\partial\Omega_1 \bigcap \partial\Omega_2$ etc. (marked by blue lines). For load balancing across processors, the number of mesh nodes is uniformly kept the same for each domain. The radial extent of the domains and the distribution of computational nodes within each domain is nonuniform and is such that a stable uniformly high-order accurate compact scheme is obtained in all the domains[12,13]. This distribution of nodes achieves minimum mesh spacing at the boundaries and hence allows for accurate resolution of the thin boundary layers at the two walls. No-slip boundary conditions on the inner and outer boundaries of the simulation domain correspond to the impulsive constant rotation of the inner cylinder and stationary outer cylinder. The length and the velocity scales are non-dimensionalized using the inner cylinder radius a and wall velocity U_w, respectively. Thus we set $a = 1$, $b = 12$ and $L = 120$. The numerical method accurately captures various flow instabilities in a Taylor-Couette cell including the formation of Taylor vortices and transition to the wavy vortex regime. Without any external disturbances the unsteady base flow is perturbed only by the growth of the discretization errors. Since the discretization methods used in the present investigation are highly accurate, the initial discretization error levels and their growth rates are excessively slow. This implies that without an external perturbation, the flow instabilities take relatively large simulation times to establish. To avoid computational expense we promote the growth of centrifugal instabilities at the inner cylinder by perturbing its otherwise constant rotation rate. The form of axial perturbation is such that it introduces disturbances of all possible wavelengths that can be accommodated by the computational mesh.

3. Results and Discussions

Before presenting the results from simulation runs, we consider an essential attribute of the wide-gap circular Couette flow that distinguishes it from the more commonly studied small-gap Taylor-Couette flow. The laminar axially homogeneous solution of the circular Couette flow with rotating inner cylinder and stationary outer cylinder can be realized only when the Reynolds number (based on the gap width and inner cylinder velocity) is below a critical value[2]. The initial unsteady Couette flow driven by viscous diffusion, is given by[18],

$$\frac{u_\theta(r,t)}{\Omega a} = \frac{a\left(b^2 - r^2\right)}{r\left(b^2 - a^2\right)} + \pi \sum_p \frac{J_1^2(pb)\left[J_1(pr)Y_1(pa) - Y_1(pr)J_1(pa)\right]}{J_1^2(pa) - J_1^2(pb)} e^{-p^2 t \mathrm{Re}} \tag{1}$$

where p are the roots of $J_1(pb)Y_1(pa) - Y_1(pb)J_1(pa) = 0$. The radial gradient of angular momentum is negative everywhere and therefore, from Rayleigh's criterion[4], the flow is centrifugally unstable in the inviscid limit. Hence, the ensuing instabilities develop over an unsteady base flow, which implies that the steady Couette flow may never be realized. It is then imperative to study the essential characteristic of this unsteady flow and search for universality if any.

We consider simulation runs for the flow configuration described above for a range of Reynolds numbers (Re) from 50 to 10^7. The secondary flow generated from the growth of the instability is illustrated in Fig. 2 through contours of angular momentum. The initial flow corresponds to a radial viscous diffusion of the angular momentum generated at the inner cylinder wall (with a profile similar to the one given by Eq. (1)). When the diffusion thickness (boundary layer thickness), defined in a specific sense later, reaches a critical value, the first unstable axisymmetric mode begins to appear and manifests as periodically stacked regions of radial inflow and outflow along the cylinder axis (see Fig. 2). The wavelength of this first unstable mode λ_c is extracted from the Fourier spectrum of the axial wall vorticity ω_z on the inner cylinder as shown in Fig. 3(a). It is observed that the amplitude spectrum consists of all the modes that can be accommodated by the computational grid in the axial direction. The mode with the largest amplitude corresponds to the one with the largest growth rate and therefore the critical wavelength λ_c. A boundary layer thickness can be defined using vorticity and can be thought of as a region that contains the total vorticity in the domain. The precise definition is made using the integral mean value theorem applied to the axial vorticity ω_z (see Fig. 3(c)). The critical boundary layer thickness δ_c is defined as the boundary layer thickness just before the onset of the secondary flow. Development of the boundary layer thickness with time is clearly evidenced in Fig. 3(b). We consider the maximum value of the boundary layer thickness to be equal to δ_c.

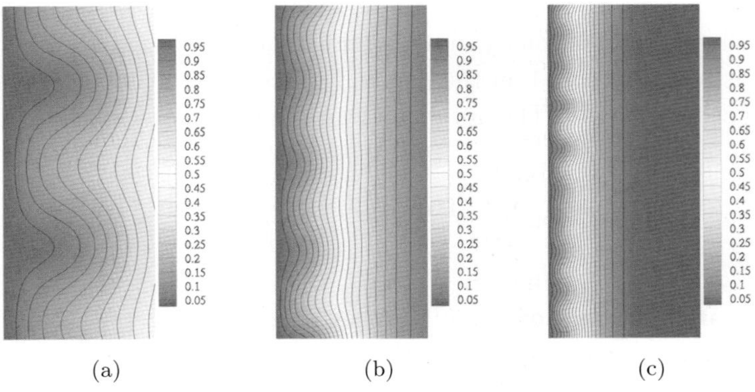

Fig. 2. Angular momentum contours for (a) Re = 50 (b) Re = 100 (c) Re = 200.

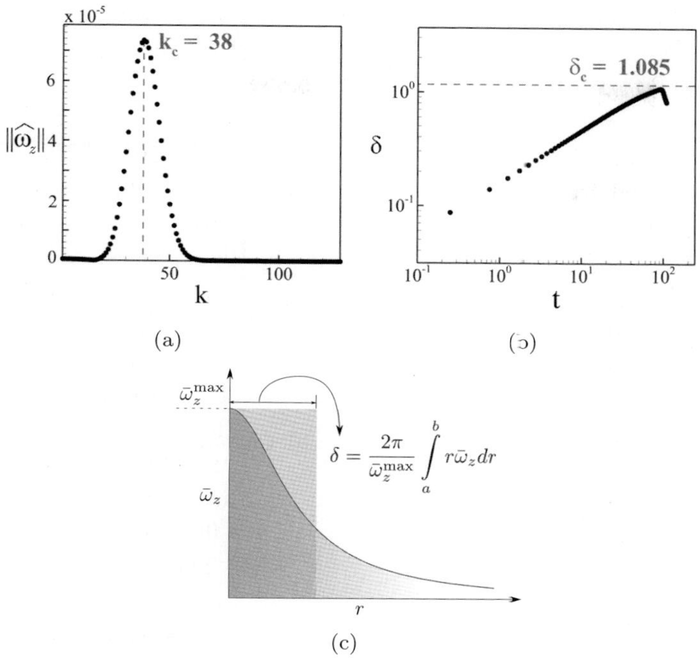

Fig. 3. (a) The magnitude of the Fourier transform of the axial wall vorticity on the inner cylinder $\hat{\omega}_z$ as a function of the Fourier wavenumber k. The critical wavelength is defined as $\lambda_c = L/k_c$. (b) Boundary layer thickness δ as a function of time and the critical boundary layer thickness δ_c for Re $= 150$. (c) Definition of the boundary layer thickness δ based on axial vorticity.

Figure 4(a) depicts the critical boundary layer thickness δ_c and wavelength λ_c as a function of the Reynolds numbers Re. Both δ_c and λ_c exhibit similar scaling of $\sim \mathrm{Re}^{-\frac{2}{3}}$ for Re $\gg 1$. Additional runs for different gap widths show no discernible change in the scaling behavior depicted in Fig. 4(a). Thus both δ_c and λ_c are independent of the gap between the inner and outer cylinders and depend only on Re. We remark here that this scaling holds only above a certain Reynolds number (for Re ≥ 400). As clearly evidenced in Fig. 4(a), for large Reynolds numbers both δ_c and λ_c are orders of magnitude lower than the inner cylinder radius ($\delta_c \sim 10^{-3}a$). It is also observed that the ratio λ_c/δ_c is independent of the Reynolds number over a significantly wide range of Reynolds numbers (Fig. 4(b)). This trend is expected and follows directly from the self similar behavior of the instability at moderate to high Reynolds numbers for which both λ_c and δ_c are significantly smaller than the radius of the inner cylinder a.

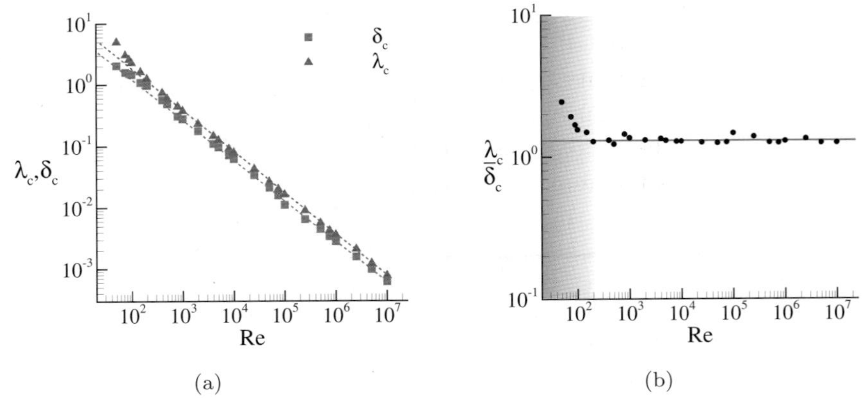

Fig. 4. (a) λ_c and δ_c as a function of Re. (b) The ratio $\frac{\lambda_c}{\delta_c}$ as a function of Re.

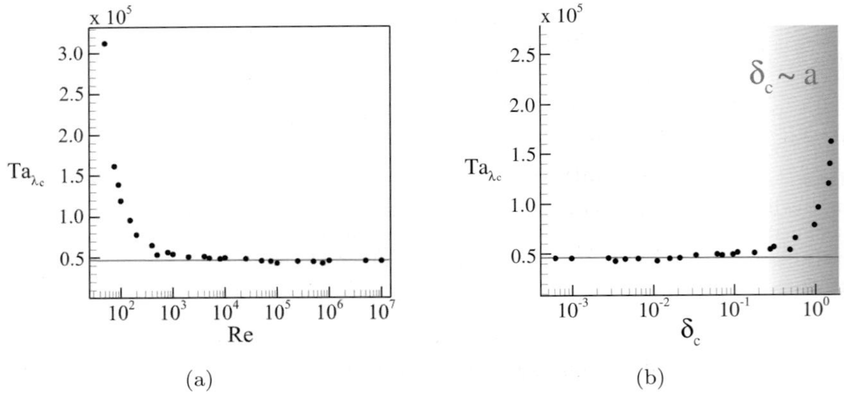

Fig. 5. Ta_{λ_c} as a function of (a) Re and (b) δ_c

From Fig. 5(a) we observe that Ta_{λ_c} is independent of Re for precisely the regimes in which both λ_c and δ_c scale as $\mathrm{Re}^{-\frac{2}{3}}$. This important result can also be derived by employing the definition of Ta_{λ_c}. As emphasized previously, for the regimes in which the scaling $\lambda_c, \delta_c \sim \mathrm{Re}^{-\frac{2}{3}}$ holds, both the length scales are substantially lower than the inner cylinder radius a. The corresponding flow can then be thought of as a small gap Taylor-Couette flow whose gap (boundary layer thickness) continues to increasing in time, till it reaches a critical threshold value (δ_c) above which the instability manifests. This line of reasoning along with the original definition of Taylor number leads directly to the scalings observed in Fig. 4(a). Figure

5(a) suggests that a critical Taylor-number does exist for the flow and its value is around 5×10^4. It is also interesting to note here that the above scalings and self similarity cease to exist in the shaded regions, as in this regime of Reynolds numbers, the boundary layer thickness is comparable to the inner cylinder radius. The variation of Ta_{λ_c} with δ_c as depicted in Fig. 5(b), confirms the above arguments.

4. Conclusion

Flow around an impulsively rotated cylinder is investigated through direct numerical simulations of circular Taylor-Couette setup for a large gap. Axisymmetric simulations were performed over a vast range of Reynolds numbers spanning seven orders of magnitude. The initial instability introduces a poloidal flow that exhibits a distinctive axial periodicity. The critical wavelength (λ_c) and the critical boundary layer thickness (δ_c) are found to scale as $\mathrm{Re}^{-2/3}$ for $\mathrm{Re} \gg 1$. A critical Taylor number defined based on either of the above two length scales, is found to be a constant whenever the above scaling holds. For the relatively low Reynolds number regime the critical Taylor number does have a Reynolds number dependence. This can be attributed to the fact that the thickness of the boundary layer at the onset of the instability is comparable to the inner cylinder radius. The self-similarity and the obtained scalings are therefore observed only when both λ_c and δ_c are a very small fraction of the radius of the inner cylinder.

References

1. R. Tagg, A guide to literature related to the Taylor–Couette problem, in *Ordered and turbulent patterns in Taylor-Couette flow*, (Springer, 1992) pp. 303–354.
2. G. I. Taylor, Stability of a viscous liquid contained between two rotating cylinders, *Philosophical Transactions of the Royal Society of London. Series A* , 289 (1923).
3. P. G. Drazin and W. H. Reid, *Hydrodynamic stability* (Cambridge University Press, 2004).
4. L. Rayleigh, On the dynamics of revolving fluids, *Proceedings of the Royal Society of London. Series A* **93**, 148 (1917).
5. C. Chen and D. K. Christensen, Stability of flow induced by an impulsively started rotating cylinder, *Physics of Fluids (1958-1988)* **10**, 1845 (2004).
6. A. Barcilon, J. Brindley, M. Lessen and F. Mobbs, Marginal instability

in Taylor–Couette flows at a very high Taylor number, *Journal of Fluid Mechanics* **94**, 453 (1979).
7. T. Wei, E. Kline, S.-K. Lee and S. Woodruff, Görtler vortex formation at the inner cylinder in Taylor–Couette flow, *Journal of Fluid Mechanics* **245**, 47 (1992).
8. D. Liu and C. Chen, Numerical experiments on time-dependent rotational Couette flow, *Journal of Fluid Mechanics* **59**, 77 (1973).
9. M. C. Kim, S. Kim and C. K. Choi, The convective stability of circular couette flow induced by a linearly accelerated inner cylinder, *European Journal of Mechanics-B/Fluids* **25**, 74 (2006).
10. K. K. Tan and R. B. Thorpe, The onset of convection caused by buoyancy during transient heat conduction in deep fluids, *Chemical Engineering Science* **51**, 4127 (1996).
11. S. A. Theerthan and J. H. Arakeri, A model for near-wall dynamics in turbulent Rayleigh-Bénard convection, *Journal of Fluid Mechanics* **373**, 221 (1998).
12. R. K. Shukla and X. Zhong, Derivation of high-order compact finite difference schemes for non-uniform grid using polynomial interpolation, *Journal of Computational Physics* **204**, 404 (2005).
13. R. K. Shukla, M. Tatineni and X. Zhong, Very high-order compact finite difference schemes on non-uniform grids for incompressible Navier–Stokes equations, *Journal of Computational Physics* **224**, 1064 (2007).
14. J. P. Boyd, *Chebyshev and Fourier spectral methods* (Dover Publications, 2001).
15. R. K. Shukla and J. H. Arakeri, Minimum power consumption for drag reduction on a circular cylinder by tangential surface motion, *Journal of Fluid Mechanics* **715**, 597 (2013).
16. J. H. Arakeri and R. K. Shukla, A unified view of energetic efficiency in active drag reduction, thrust generation and self-propulsion through a loss coefficient with some applications, *Journal of Fluids and Structures* **41**, 22 (2013).
17. R. Peyret, *Spectral methods for incompressible viscous flow* (Springer, 2002).
18. C. J. Tranter, *Integral transforms in mathematical physics* (Chapman and Hall London, 1971).

Direct Numerical Simulations of Riblets in a Fully-Developed Turbulent Channel Flow: Effects of Geometry

J. H. Ng, R. K. Jaiman* and T. T. Lim

Department of Mechanical Engineering, National University of Singapore
**E-mail: mperkj@nus.edu.sg*

Direct numerical simulations are performed to investigate the geometrical effects of riblets on the flow field in a fully-developed turbulent channel flow. The major focus of this paper is to compare the flow fields of the V-groove and the thin blade riblets. The objective is to determine the factors contributing to the more superior drag reduction performance of the thin blade riblets. In addition, the present work intends to elucidate flow phenomena discussed in the earlier studies, and to investigate the impact of riblet geometry on those phenomena.

Keywords: Riblets; Geometrical effects; Turbulence; Direct numerical simulations.

1. Introduction

Drag reduction in turbulent flows is currently an active area of research. A meager drag reduction could yield a substantial saving in operating expenditure crucial for sustaining the competitiveness of the aviation industry. It is customary to classify the aerodynamic drag on an aircraft into form drag, induced drag, compressibility drag and viscous drag[2]. The viscous drag is predominant during subsonic cruising and thus a considerable amount of efforts has been devoted to reduce it[8].

A successful control of turbulent flows demands a thorough understanding of the physics of turbulence. Unfortunately, such comprehension is still lacking even after a century of research[15,20]. Nonetheless, one of the milestones in turbulence research is the discovery of well-organized flow structures and the recognition of their importance in wall-layer dynamics. In particular, an ubiquitous feature of the inner layer in the form of low- and high-speed streaks have been identified. Studies have demonstrated that these streaks are responsible for the higher skin friction drag in turbulent boundary layers. Their formation has been traced to the near-wall streamwise vortices typically found in the buffer layer. Many endeavors have been attempted to manipulate these vortices for achieving viscous drag reduction[12].

Out of the several types of drag reduction techniques which have been explored historically, surface modification by means of tiny longitudinal striations, now commonly called riblets, give rise to an appreciable viscous drag reduction. The original concept underlying the development of riblets is to constrain the growth of low-speed streaks, because they are responsible for the bulk of turbulence production and fluctuation in a boundary layer[22]. Another view is that riblets restrain the motions of near-wall vortical structures and maintain their coherence. Stable coherent structures diminish the energy-cascading process to give lower skin friction drag[8,19].

Walsh[21] showed that the drag reduction performance of riblets depends on the non-dimensional height h^+ and spacing s^+, where the superscript $^+$ denotes a normalization by the viscous length scale ν/u_τ, while ν is the kinematic viscosity, and $u_\tau = \sqrt{\tau_w/\rho}$ is the friction velocity derived from the wall shear stress τ_w and density ρ. Walsh and Lindemann[23] concluded from their study on several different riblet shapes that the optimum triangular riblet size is $s^+ = 15$ and $h^+ = 13$, which produces up to 8% lower drag. Separately, Bechert *et al.*[1] have carried out an extensive experimental study of riblets in an oil channel. The study reported that the optimized thin blade riblets provide up to 9.9% viscous drag reduction.

Chu and Karniadakis[6] were the first to report turbulence statistics of flow over the V-groove riblets through their direct numerical simulations (DNS) in a fully-developed turbulent channel flow. Choi *et al.*[4] postulated that the interactions between the near-wall streamwise vortices and riblets have a strong influence on the amount of skin friction drag. Goldstein *et al.*[11] also confirmed that the size of riblets have to meet certain prerequisites in order to manipulate the near-wall flow field in a beneficial manner.

García-Mayoral and Jiménez[9] proposed a new length scale l_g^+ to reduce the geometrical dependence of drag reduction curves. l_g^+ is a measure of the groove cross-section area A_g in wall units: $l_g^+ = \sqrt{A_g}u_\tau/\nu$. Under this new scaling, the viscous slope of the drag reduction curves for various riblet geometries collapsed reasonably well. The authors reported that the optimum drag reduction occurred at $l_g^+ \approx 11$. Their subsequent DNS investigations on the thin blade riblets revealed the formation of two-dimensional spanwise rollers when l_g^+ is large. These Kelvin-Helmholtz-like instability were said to be the culprit of the higher viscous drag at large l_g^+. Whereas, the influence of the near-wall streamwise vortices was found to be minimal.

In spite of the new understandings on riblets brought about by both experiments and numerical simulations, there are still some vague areas that need further clarification. These include confirming the mechanism of riblets and determining a suitable dimensionless length scale to characterize the riblet performance. Furthermore, there is still a lack of comprehensive numerical data to illustrate the geometrical dependence of the flow field over riblets. In this vein, the present work aims to fill in this void in the literature.

A preceding work[16] has discussed the impacts of varying the size of V-groove riblets, defined in terms of the length scale l_g^+, on the near-wall flow field and the relevant turbulent statistics. A total of six V-groove riblet configurations have been simulated to investigate the variation trend of different flow quantities, and to explore whether there exists certain useful correlations between the resulting flow field and the length scale l_g^+. The primary focus of this ensuing study is to contrast the flow fields of the V-groove and the thin blade riblets at similar values of l_g^+ via direct numerical simulations. Such comparison could shed some light on how structured surfaces with size variation on the order of the thickness of the viscous sublayer can modify the near-wall flow field, and ultimately provide benign effects like viscous drag reduction.

2. Numerical Procedure

A fully-developed turbulent channel flow with a domain size of $L_x = 5\delta$, $L_y = 2\delta$ and $L_z = 2\delta$ is considered, where x, y, and z denote the streamwise, wall-normal and spanwise directions. The origin of the coordinate axes is situated at the center of the inlet plane. In this study, the chosen characteristic length and velocity scales are the channel half-width δ and the bulk streamwise velocity U_m, respectively. The riblets are mounted on the bottom wall to allow a direct comparison of flow properties between the smooth (top) wall and riblet (bottom) wall. Periodic boundary conditions are employed in the streamwise and spanwise directions. Whereas the no-slip condition is enforced in the wall-normal direction. In addition, a constant bulk velocity is also imposed in the channel domain. The initial condition is prescribed using the baseline turbulent channel flow solution computed based on a similar domain size and mesh resolution used by Kim *et al.*[13].

A spectral element solver[17] is employed for the current DNS investigations. The V-groove riblets have a spacing s and height h of 0.2δ. Whereas, the thin blade riblets have dimensions of $s = 0.2\delta$, $h/s = 0.1$ and $t_b/s = 0.1$, where t_b is the thickness of the blade. The spectral element

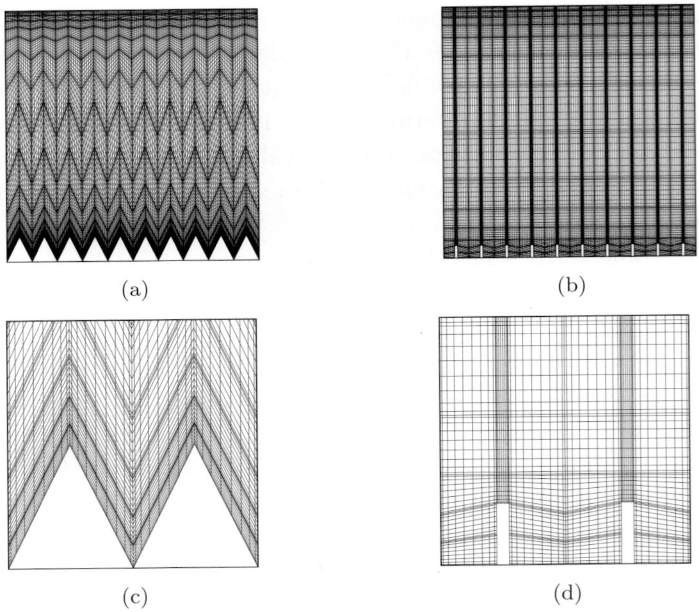

Fig. 1. Meshes of the current simulations: (a) V-groove riblets, (b) Thin blade riblets, (c) Close-up view near the V-groove riblets, and (d) Close-up view near the thin blade riblets.

meshes of these two configurations are shown in Fig. 1. Table 1 summarizes the mesh resolutions for both the V-groove (Case V) and the thin blade (Case B) configurations. The simulations are run at a bulk Reynolds number $Re_m = U_m\delta/\nu = 1842$. This translates to a friction Reynolds number $Re_\tau = u_\tau\delta/\nu$ of approximately 130. The dimensionless length scale l_g^+ proposed by García-Mayoral and Jiménez[9] is used to describe the size of the riblets, instead of the conventional length scales s^+ or h^+. The resulting values of l_g^+ is 12.15 and 11.99 for the V-groove and thin blade riblets, respectively. The current sizes of riblets in terms of l_g^+ are close to the optimum value of l_g^+ found by García-Mayoral and Jiménez[9].

Table 1. Simulation setup of the parametric study of V-groove riblets. The values in wall units are computed based on the surface-averaged friction velocity at the riblet wall.

Case	N	N_x	N_y	N_z	$\Delta x^+_{\min}$	$\Delta x^+_{\max}$	$\Delta y^+_{\min}$	$\Delta y^+_{\max}$	$\Delta z^+_{\min}$	$\Delta z^+_{\max}$
V	9	190	127	181	0.82	3.38	0.12	5.25	0.35	1.42
B		190	127	271	0.84	3.43	0.13	5.37	0.07	1.30

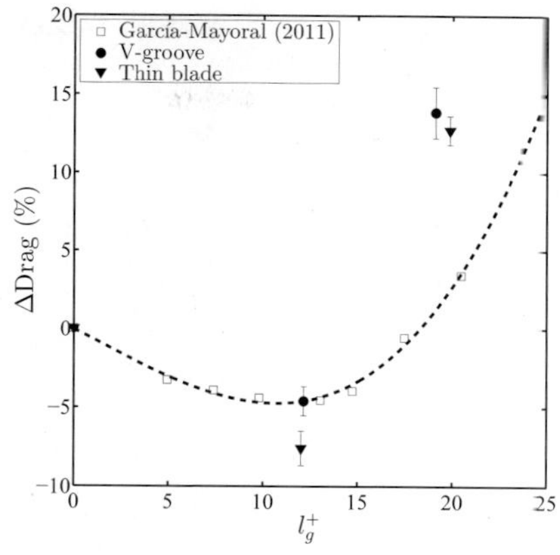

Fig. 2. The drag reduction curve of the present riblet configurations against l_g^+. The profile for the simulations of thin blade riblets performed by García-Mayoral[10] is included. Error bars represent a 95% confidence level. The values for two corresponding simulations at $Re_m = 2800$ are also included for illustration purpose. Error bars represent a 95% confidence level.

3. Results and Discussions

For the V-groove riblets, a skin friction drag reduction of 4.6% is computed. The thin blade riblets produced a higher reduction of 7.6%. Note that these values are subjected to approximately 1% uncertainty based on a 95% confidence level estimation obtained from the time history of drag evolution. Due to the streamwise homogeinity of the current computational domain, the pressure drag contribution is negligible. Figure 2 compares the present results with the drag reduction curve of the reference simulations of thin blade riblets with $t_b/s = 0.25$[10]. The present result of Case B is consistent with the findings of Bechert *et al.*[1] that a thinner blade at the same s^+ and h^+ will provide a greater drag reduction. Although not discussed here, the values for the two corresponding simulations at $Re_m = 2800$ are included in Fig. 2 to illustrate the trend and fitting at larger values of l_g^+.

The next part is to identify the factors which enable the thin blade riblets to perform better. This involves comparing the near-wall flow fields of the two riblet configurations. Preliminary findings show that there are more streamwise-aligned vortical structures spread over the thin blade riblets (see Fig. 4). In other words, the thin blade riblets are more effective in mitigating the spanwise interactions of the near-wall structures. Additionally, very few vortices are found to settle inside the grooves in both configurations.

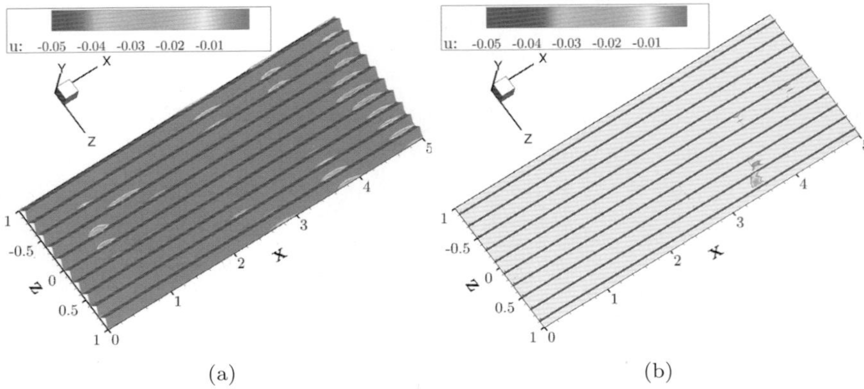

Fig. 3. Contoured regions of streamwise velocity u/U_m indicating the occurrence of instantaneous flow reversal near the riblet valleys at $Re_m = 1842$: (a) V-groove riblets, and (b) Thin blade riblets.

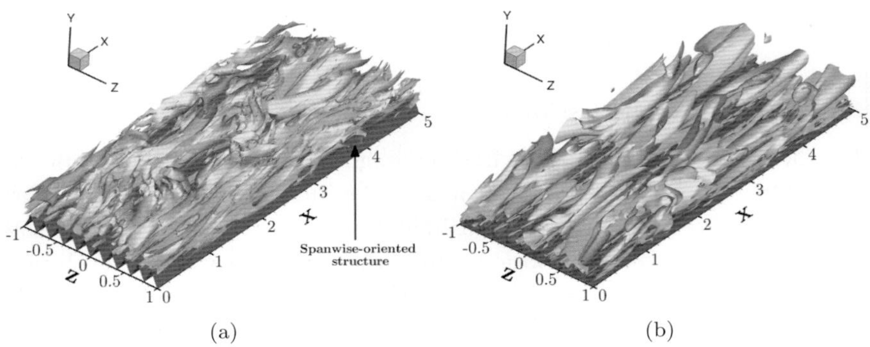

Fig. 4. Near-wall vortical structures deduced using the Q-criterion, where Q is measured in wall units: (a) V-groove riblets, and (b) Thin blade riblets. The arrow in Fig. 4(a) points out one of the spanwise-oriented structures responsible for causing the flow reversal at the corresponding location depicted in Fig. 3(a).

In addition, the present work intends to find out whether the flow phenomena reported in the preceding work[16] also exist in the thin blade riblet configuration. For example, instantaneous flow reversal near the valley is observed in the V-groove configuration (see Fig. 3(a)). This phenomenon is postulated to be caused by spanwise-oriented vortical structures (as indicated in Fig. 4(a)) forming near the tips. Some traces of this kind of structure are found on the thin blade riblets but their appearance is far less distinct than in the V-groove configuration. As a result, there are only some isolated instances that flow reversal is observed near the valley.

The occurrence of flow reversal in the vicinity of solid interface has been documented in the literature. Using the critical point theory[18], complex three-dimensional structures are shown to exist near critical points at the wall, and such structures are capable of causing flow reversal in the surrounding regions[5]. Recent researches have also identified strong spanwise vortices over regions with negative streamwise wall shear stress[3,14]. At this juncture, it is suffice to say that the surface geometry has an influence on the flow topology. A detailed investigation on this matter will be performed in the near future.

One observation when simulating the channel flow with riblets mounted on only one of the two walls is that the properties on the smooth wall are slightly modified. Such modification happens despite the simulations are performed at the same bulk Reynolds number. In particular, the friction velocity u_τ and in turn the friction Reynolds number Re_τ increase with the size of the riblets or l_g^+. It is uncertain at first whether such effect is caused solely by the change in the channel geometry, or it might be a ramification due to the interactions of the two walls. An attempt is thus made to quantify the changes of the friction velocity on the smooth wall when riblets of different sizes are incorporated. This approach stemmed from the classical Dean's correlation[7] which relates the skin friction coefficient C_f and the bulk Reynolds number Re of a fully-developed turbulent flow through a high aspect ratio rectangular duct: $C_f = 0.073/Re^p$, where $p = 0.25$, $Re = U_m H/\nu$, and $H = 2\delta$ is the channel height. Separately, Yakhot *et al.*[24] also demonstrated theoretically that $C_f \propto Re^{-\theta}$, where $\theta = 0.26 \sim 0.27$.

In order to apply the Dean's correlation to flow through a duct of other cross sections, a *hydraulic diameter* $d_h = 4A_c/c$, where A_c is the cross-sectional area of the duct and c is the wetted perimeter, can be employed to compute Re. It can be easily shown that $d_h = 2H = 4\delta$ in the case of the channel configuration with two planar walls[7]. Likewise, the Dean's correlation can be extended to the current channel configuration with different riblet geometry by computing an equivalent d_h. In this way, the values of C_f and Re_τ can be estimated for the different riblet configurations. The Dean's correlation is recast in the following form to aid the estimation:

$$C_f = 0.073 \left(\frac{U_m \delta'}{\nu} \right)^{-0.26} \tag{1}$$

where $\delta' = (d_h/d_0)\delta$ and $d_0 = 4\delta$. Figure 5 depicts a comparison of the C_f values obtained from the current set of simulations and the values estimated

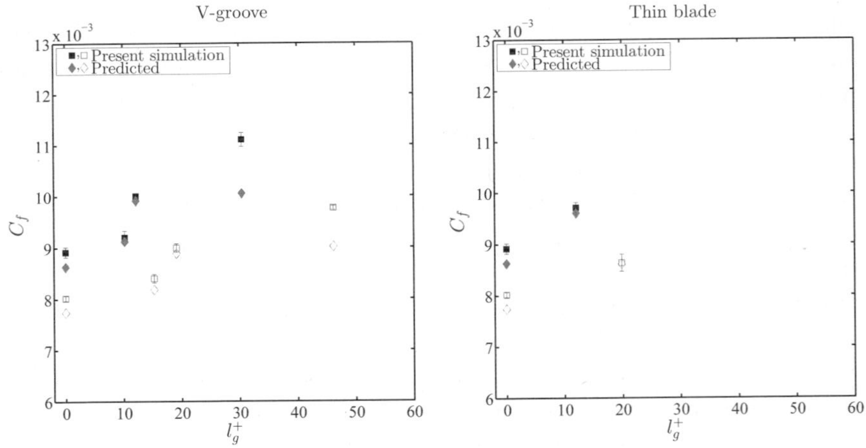

(a) (b)

Fig. 5. Comparison of skin friction coefficients C_f of the V-groove and thin blade riblet configurations at several values of l_g^+ with predicted value computed from : (a) V-groove riblets, and (b) Thin blade riblets. Error bars represent a 95% confidence level.

using Eq. (1). The agreement is within 3% for riblet configurations smaller than $l_g^+ \approx 20$ regardless of the shape. Hence, the properties modification on the smooth wall can be viewed as a consequence of the change in d_h. For larger l_g^+ configurations, the large discrepancy is most likely caused by the strong asymmetry in the flow in which the perturbations caused by the riblets severely interfere with the near-wall flow field on the top smooth wall. In this regard, the notion of a duct flow may not hold anymore.

Another useful information gained from this exercise is that the amount of drag reduction should be calculated with respect to a baseline channel flow at a *modified* bulk Reynolds number $U_m\delta'/\nu$ in order to have a fair comparison. Otherwise, a misleadingly greater amount of drag reduction would be computed. This can be achieved by either adjusting the channel height or the bulk flow velocity so as to maintain the same d_h. However, this approach is rather inconvenient as compared to computing the amount of drag reduction based on the smooth wall on one side of the channel.

4. Conclusions

A comparative DNS study has been conducted to determine the differences in a fully-developed turbulent channel flow over the V-groove and the thin blade riblets. A slight change in the riblet shape can have rather significant impact on the organization of the near-wall flow structures, and in turn their influence on the wall shear stress. The thin blade riblets give rise to a higher proportion of streamwise-aligned near-wall vortical structures. This helps in mitigating the motions and interactions of the near-wall structures in the spanwise direction, and thereby providing a higher viscous drag reduction than the V-groove riblets. The predicted skin friction drag reductions are 4.6% and 7.6% for the V-groove and the thin blade configurations, respectively. A switch from the V-groove to the thin blade shape also reduces the instances of flow reversal inside the riblet grooves. Additionally, this work also demonstrates that a small change of surface geometry on one wall of the channel can affect the properties on the opposite planar wall. Such effect on the skin friction coefficient, in particular, can be estimated using a *modified* Dean's correlation.

The authors wish to acknowledge supports from the National University of Singapore and the Ministry of Education, Singapore.

References

1. Bechert, D. W., Bruse, M., Hage, W., Van Der Hoeven, J. G. T. and Hoppe, G., Experiments on drag-reducing surfaces and their optimization with an adjustable geometry, *J. Fluid Mech.*, **338**, 59–87, (1997).
2. Bushnell, D. M., Aircraft drag reduction – a review, *Proc. Inst. Mech. Engineers*, Part G: J. Aerospace Engg., **217**, 1 1–18, (2003).
3. Cardesa, J. I., Monty, J. P., Soria, J. and Chong, M. S., Skin-friction critical points in wall-bounded flows, *J. Phys.: Conf. Series*, **506**, 1, (2014).
4. Choi, H., Moin, P. and Kim, J., Direct numerical simulation of turbulent flow over riblets, *J. Fluid Mech.*, 255, 503–539 (1993).
5. Chong, M. S., Monty, J. P., Chin, C. and Marusic, I., The topology of skin friction and surface vorticity fields in wall-bounded flows, *J. Turbulence*, **136**, 1–10 (2012).
6. Chu, D. C. and Karniadakis, G. E., A Direct numerical simulation of laminar and turbulent flow over riblet-mounted surfaces, *J. Fluid Mech.*, **250**, 1–42 (1993).
7. Dean, R. B., Reynolds number dependence of skin friction and other bulk flow variables in two-dimensional rectangular duct flow, *J. Fluids Engg., Trans. ASME*, 100, 215–223 (1978).
8. Gad-el-Hak, M., Drag Reduction, Flow Control: Passive, Active, and Reactive flow Management, (Cambridge Univ. Press, New York (2000)).

9. García-Mayoral, R. and Jimnez, J., Hydrodynamic stability and breakdown of the viscous regime over riblets, *J. Fluid Mech.*, **678**, 317–347 (2011).
10. García-Mayoral, R., The Interaction of Riblets with Wall-Bounded Turbulence, (Universidad Politécnica de Madrid, Madrid, Mayo, (2011)).
11. Goldstein, D. and Handler, R. and Sirovich, L., Direct numerical simulation of turbulent flow over a modelled riblet covered surface, *J. Fluid Mech.*, **302**, 333–376 (1995).
12. Kim, J., Control of turbulent boundary layers, *Phys. Fluids*, **15**, 5, 1093–1105 (2003).
13. Kim, J., Moin, P. and Moser, R., Turbulence statistics in fully developed channel flow at low Reynolds number, *J. Fluid Mech.*, **177**, 133–166 (1987).
14. Lenaers, P., Li, Q., Brethouwer, G., Schlatter, P. and ÖrlÖu, R., Rare backflow and extreme wall-normal velocity fluctuations in near-wall turbulence, *Phys. Fluids*, **24**, 3 (2012).
15. Lumley, J. L. and Yaglom, A. M., A century of turbulence, *Flow, Turbulence and Combustion*, **66**3, 241–286 (2001).
16. Ng, J. H., Jaiman, R. K. and Lim, T. T., Direct numerical simulation of geometric effects on turbulent flows over riblets, *7th AIAA Flow Control Conf.*, Atlanta, Georgia (2014).
17. Fischer, P. F., Lottes, J. W. and Kerkemeier, S. G., Nek5000 Web Page, (2008) "`http://nek5000.mcs.anl.gov` ".
18. Perry, A. E. and Fairlie, B. D., Critical points in flow patterns, *Adv. Geophysics*, **B18**, 299–315 (1974).
19. Sirovich, L. and Karlsson, S., Turbulent drag reduction by passive mechanisms, *Nature*, **388**, 6644, 753–755 (1997).
20. Spalart, P. R. and McLean, J. D., Drag reduction: Enticing turbulence, and then an industry, *Philo. Trans. Royal Soc. A: Mathematical, Physical and Engineering Sci.*, **369**, 1556–1569 (2011).
21. Walsh, M. J., Riblets as a viscous drag reduction technique, *AIAA J.*, **21**(4), 485–486 (1983).
22. Walsh, M. J., Riblets in, *Prog. Astro. and Aero. series* ***123**: Viscous drag reduction in boundary layers*, eds. Bushnell, D. M. and Hefner, J. N., AIAA, Washington, DC, 203–261 (1990).
23. Walsh, M. J. and Lindemann, A. M., Optimization and application of riblets for turbulent drag reduction, in *22nd AIAA Aerospace Sci. Meeting*, Reno, Nevada (1984).

24. Yakhot, V., Bailey, S. C. C. and Smits, A. J., Scaling of global properties of turbulence and skin friction in pipe and channel flows, *J. Fluid Mech.*, **652**, 65–73 (2010).

Computational Studies on Flow Separation Controls at Relatively Low Reynolds Number Regime*

K. Fujii†

Institute of Space and Astronautical Science, Japan Aerospace Exploration Agency
Yoshinodai, Chuo-ku, Sagamihara, 252-5210, Japan
†*E-mail: fujii@flab.isas.jaxa.jp*

Mechanism of flow separation control by the micro or small devices such as plasma actuators is discussed based on a series of computations of low speed flows over an NACA 0015 airfoil with devices attached near the leading edge. The DBD plasma actuator is shown to be very effective for controlling flow separation at a Reynolds number of 6.3×10^4. Flow attachment phenomenon induced by these devices is strongly related to the turbulent transition but the study reveals that it is not the only factor for the flow separation control. Analysis of the phase-average flow structures reveals that there exist remarkable structures in the flow near the airfoil surface, which characterizes the actuator-oriented flows to reduce flow separation.

Keywords: Flow Separation; Flow Control; Plasma Actuator; Airfoils; Micro Device

1. Introduction

Flow separation control has been one of the important research topics in fluid dynamics for many years. Among many types of flow control, strong interest lies on the control of flow separation over an airfoil at high angles of attack since such flow control would avoid aerodynamic stall and remarkably improve aerodynamic characteristics of aircraft. The technology is also applicable to fluid machines such as turbines/fans, and transportation systems on the ground as well.

Recent micro and small devices such as synthetic jets or DBD plasma actuators have been shown to be very effective for the control of flow separation over an airfoil (Fig. 1). Experimental studies proved that these

*This work is supported by the Strategic Programs for Innovative Research (SPIRE) of HPCI in Japan and both the "K" supercomputer at the AICS, Riken and the Fujitsu FX10 of University of Tokyo are used. The research was partly supported by the JSPS Grant-in-Aid for Scientific Research No.242461411 as well.

Fig. 1. Flow control with the DBD plasma actuator: Left: DBD off, Right: DBD on (from AIAA2009-531 by K. Asada *et al.* (Ref. 9)).

devices change separated flows to attached flows with very low power consumption[1,2]. The plasma actuator comprises two electrodes with a dielectric between them. When high alternating current (AC) voltage is applied on the electrodes, the plasma actuator induces weak flows.

Both the experimental and computational analysis showed that these devices are very effective for the Reynolds numbers less than 10^5.[3–9] One of the interesting facts of DBD plasma actuator is so-called "burst mode" of imposed voltage (schematically shown in Fig. 2) with duty cycles shows better separation flow control capability[3–6] with less energy consumption. Here, the burst mode actuation uses the modulation of base high frequency wave signals and switch-on and switch-off signals of such base frequency. Conventionally, non-dimensional burst wave frequency F^+ $(= f^+c/u_\infty)$, where f^+ is the burst frequency, u_∞ is the freestream velocity and c is the chord length. Many of the former studies showed that $F^+ = 1.0$ is effective, while both experimental and computational studies in the past showed that F^+ more than 2 and less than 20 is also effective[7–9]. Computational analysis of synthetic jets also showed that $F^+ = 6.0$ is effective[10]. Key features of DBD plasma actuators may be found if the reason why there are two efficient frequencies is clarified.

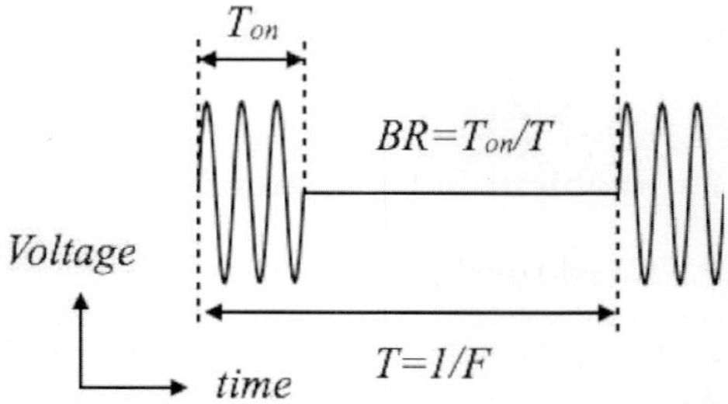

Fig. 2. Schematic picture of frequencies for the burst mode.

Computational Fluid Dynamics (CFD) became a strong tool for the researchers and engineers both in academia and industries[11,12]. In early days of CFD, clear distinction existed between practical applications to engineering problems using turbulence modeling and basic research for simple geometries toward better understanding of turbulence. Rapid progress of simulation methods and computer performance in the last 30 years is now removing the boundaries of these two types studies. Computational analysis of the flow fields now really contributes understanding fluid physics behind the mechanism of avoiding flow separation[13].

In this study, mechanism of flow separation control by DBD plasma actuators is mainly discussed based on a series of simulations of low speed flows over an NACA0015 airfoil with devices attached near the leading edge. Flow attachment phenomenon induced by these devices is strongly related to the turbulent transition but the study revealed that it is not the only factors. Analysis of the phase-average flow structure reveals that there exist remarkable flow structures in the flows near the airfoil surface, which may become a key for the flow control at higher Reynolds numbers.

Note that effects of these devices depends on the combination of velocities and model scales even at the same Reynolds number cannot be strictly discussed without specifying each scale because induced flows relative to the freestream velocity and the size of the plasma actuator relative to the chord length and other length scale would become different depending on the combination of these two scales. In this study, scales are defined based on the experiment in the authors laboratory[9].

2. Problem Setting and Numerical Approach

Three-dimensional compressible Navier-Stokes equations in body fitted coordinate system are employed as the governing equations in this study. The equations include body force terms that represent the effect of induced flows by the DBD plasma actuator in momentum and energy equations. The body force terms based on the one proposed by Suzen[14] is employed and the terms are extended to include unsteady features[8]. They consist of the electric force multiplied by the non-dimensional coefficient Dc, which translates the electromagnetic force to the magnitude of the body force in the flow field (see Ref. 14 for more details). The resultant push-push type sinusoidal body force changing in time is mapped onto the

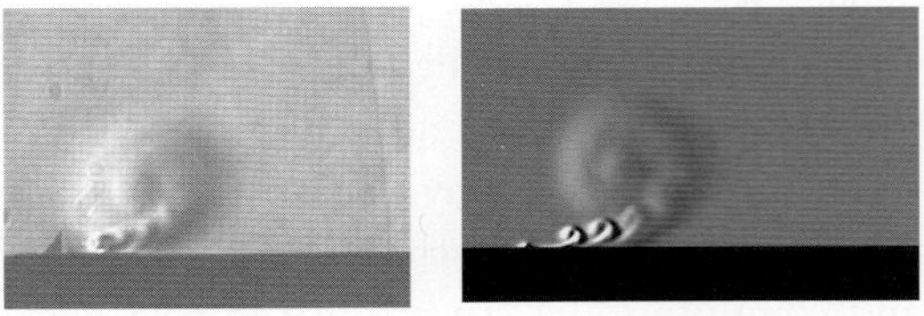

Fig. 3. Actuator induced flow field, left: experiment ($F^+ = 1, BR = 0.1, Vpp = 10$kv), Right: computation ($F^+ = 1, BR = 0.1, Dc = 0.0625$).

grid for the flow simulations. Figure 3 shows one of the validation results showing the comparison of the induced flows obtained in both the experiment and computation (see Ref. 15) for more details. Comparison of the velocities of induced flows over a flat plate with actuation both the computation and experiment identifies an appropriate Dc values for the condition of each velocity and model scale. Simulations with the more sophisticated plasma models by Koizumi and Nishida[16] confirmed reliability of the modified Suzen model used here.

The Reynolds number is set to be 6.3×10^4 based on the experiments[9] with the freestream velocity of 10 m/sec and the model length of 10 cm. This manuscript only handles the case of 14 degrees angle of attack in this study although simulations were conducted for many other angles of attack as well. Implicit LES (iLES) is adopted and high-resolution 6th-order compact difference scheme with 10th-order compact filtering is used[17]. Computational grids are properly distributed so that they are adequate for the iLES simulations. To enhance the grid resolution in the body force region near the leading edge, overset zonal method is employed[18]. The total number of grid points is roughly 20 million. Note that 6th order compact scheme has much higher spatial resolutions than conventional second order schemes. As a computational region, span length of 20% of the chord is prepared. A periodic boundary condition is imposed at the boundaries in the spanwise direction. Simulations are conducted for several types of airfoils but only the results for the NACA symmetric airfoils are presented. More details can be found in Refs. 19 and 20.

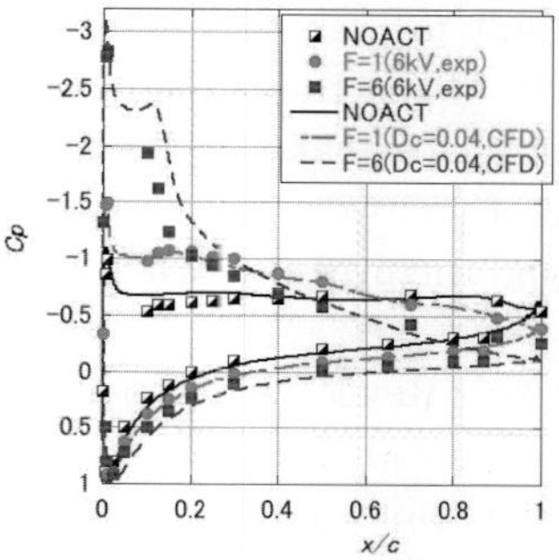

Fig. 4. Chordwise C_p distributions : $\alpha = 14°, 10\%$ bursting

3. Results and Discussions

Details of the plasma actuator setting to be simulated in this study can be found in Refs. 9 and 22, but the plasma actuator of the length of 30 mm and the thickness of 220 μm (including both the exposed and ground copper electrodes and polyimide film for dielectric) is located near the leading edge. Computed surface pressure distributions over the airfoil surface at $\alpha = 14°$ are shown in Fig. 4 together with the experimental results, which typically demonstrates clear effect of the burst frequencies. Solid line is the result for the actuator off, alternating long and short dashed line is the result for $F^+ = 1.0$ and dashed line is the result for $F^+ = 6.0$. Distinct symbols are for the experimental results. The simulation results agree well with the corresponding experiments. Similar agreement is also obtained for simulations with $Dc = 0.01$ and the experiment with 5kV. Note that Dc values are semi-empirically chosen for the comparison and difference exists in the maximum induced velocities computed in the still-air condition and those obtained in the experiments in still air. Although not shown here, the result for the continuous actuation had no effect and showed almost same result as that of DBD actuator off (NOACT in the figure), which identifies clear effect of burst modes. The result for $F^+ = 6.0$, in contrast to the other two cases, shows that suction peak appears near the leading edge and the pressure distributions indicate attached flows over the surface.

At this Reynolds number, there appears so-called leading-edge separation bubble at lower angles of attack. The flat C_p region just after the suction peak for $F^+ = 6.0$ corresponds to this phenomenon. The results along with Figs. 3 shows that the present approach with iLES simulations can be used for the discussion of the mechanism of DBD plasma actuators for flow separation control.

Figure 5 shows the lift-to-drag ratios for three different NACA symmetric airfoils having different thicknesses. The angle of attack is 14 degrees.

Note that this angle of attack is a few degrees higher than the stall angle as massive separation from the leading edge occurs at an angle of attack slightly after 11 degrees both in the experiments and computations. The numbers with "%"

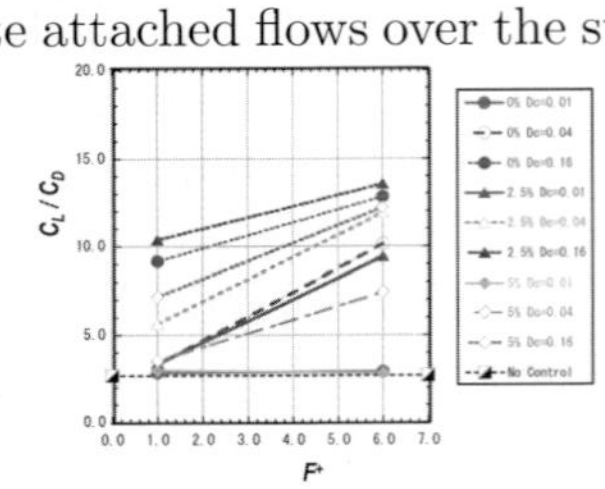

(a) NACA0015

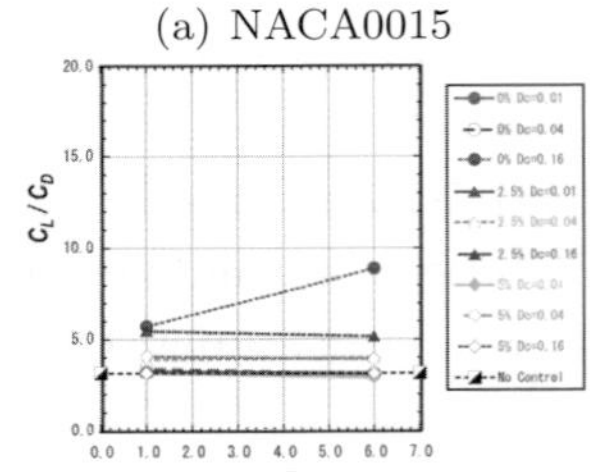

(b) NACA0012

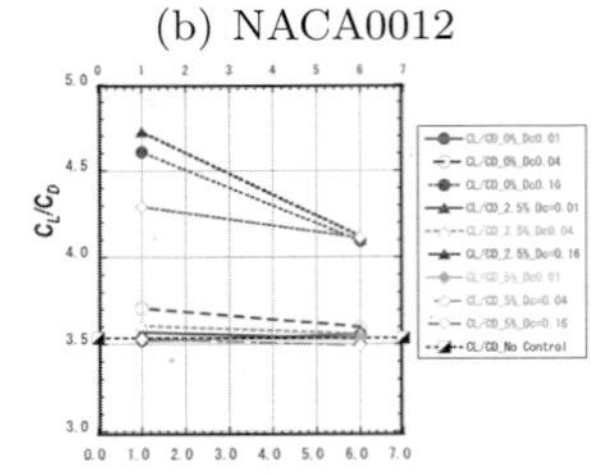

(c) NACA0006

Fig. 5. L/D ratio .vs. burst frequency F^+: $\alpha = 14°, 10\%$ bursting.

marks in the captions show the location of the actuator and Dc values correspond to the strength of the momentum induced by the plasma actuator (roughly speaking, which corresponds to the voltage in the experiment).

Higher burst frequency has better aerodynamic characteristics in terms of the L/D for the two thicker airfoils. This is true for any actuator locations. On the other hand, lower burst frequency has better performance for the thin airfoil although the L/D increase is not as large as the results for thicker airfoils. The plots in Fig. 5 show that it becomes more difficult to control massive flow separation for thin airfoils. This may be due to the fact that separated shear layer stays away from the thin airfoil, whereas separated shear layer stays closer to the airfoil surface for thicker airfoils because of the difference of leading edge radius. Our former studies[5,9] showed that flow separation control with the plasma actuator authority for high burst frequencies (such as 5-10 in non dimensional units) relies on the laminar-to-turbulent transition. It is imagined that $F^+ = 6.0$ does not work for NACA0006 because disturbances introduced by the plasma actuator to promote turbulent transition effect is limited to the region near the airfoil surface and is not sufficient for the separated shear layer located away from the surface. Both experiments and computations identified the existence of two-dimensional spanwise vortex structures induced by the bursting of plasma actuators. Such two dimensional vortex structures become larger and stronger when lower frequencies are applied and that may promote momentum exchange in wider flow region for the case of NACA0006.

Lift coefficients for NACA0012 at same 14 degrees are plotted in Fig. 6. The L/D ratio in the middle plot of Fig. 5 corresponds to this Fig. 6. Lift coefficients are larger for $F^+ = 1.0$ than the result for $F^+ = 6.0$ for this airfoil although L/D has opposite trend or almost no effects for most of the cases. It indicates that strong drag reduction occurs for $F^+ = 6.0$. It depends how we use the benefits of plasma actuators, but it is important to measure both C_L and C_D for the evaluation of plasma actuators authority of flow separation control.

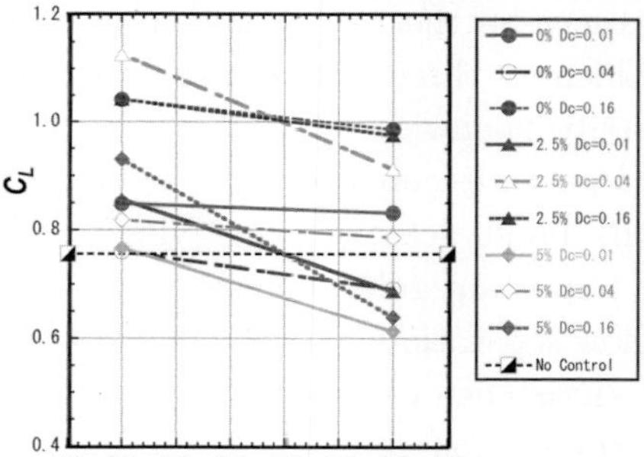

Fig. 6. Lift coefficient .vs. burst frequency F^+: NACA0012, $\alpha = 14°$, 10% bursting.

Separation points at this angle of attack (14 degrees) is 2.5% of the chord for NACA0015 airfoil, 1.5% for NACA0012 airfoil and almost 0% for NACA0006 airfoil. Figure 5 also indicates that plasma actuators work better when located near the separation points of the flows without actuation as pointed out by many of the former studies.

More than 200 iLES simulations were conducted[20]. The result identified that the DBD plasma actuators show good control performance for wide range of flow conditions. Large numbers of the simulations above also showed that rapid growth of the Reynolds stress has strong relation with the flow separation control effect. Instantaneous flow fields, phase-averaged flow fields, frequency spectrum, and linear stability were used for the analysis of the computed results. A few different representative flow structures are identified[19,20] by such effort.

Firstly, turbulent transition seems to be the key for the flow separation control at this Reynolds number as has been mentioned. To further identify this fact and find good burst parameters, the linear unstable modes in the separation shear layer profile and spatial growth rates are investigated[19]. The result shows that the most unstable frequency is approximately $St = 15$, where St is the frequency defined based on the chord length and freestream velocity. A few two-dimensional vortices appear at the beginning and after the end of every bursting. They merge together and a strong two-dimensional vortex with the same frequency as F^+ is eventually created as shown below. Namely, the initial vortices have the Strouhal number similar to unstable frequency. This may be the reason that actuation is most effective for F^+ around 5, which is three times lower than the unstable frequency, $St = 15$. Secondly, spanwise vortex structures created by the burst actuation play important role for the flow separation control. Existence of such structures is found from the animation of the phase and spanwise averaged flow fields.

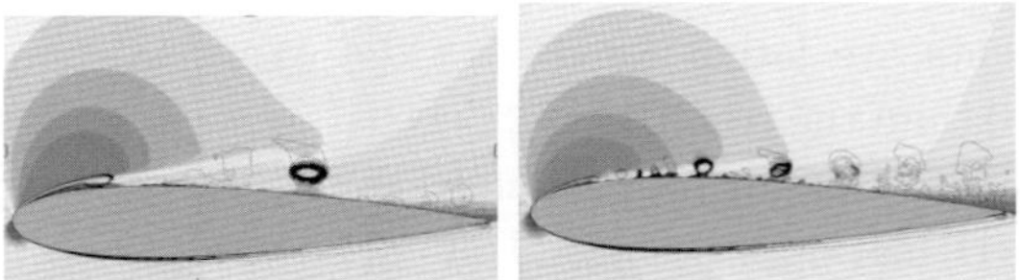

(a) $F^+ = 1.0$ (b) $F^+ = 6.0$

Fig. 7. Ensemble and moving averaged flow fields, 10% bursting, background contours: chordwise velocity, Black contour lines: second invariant of the velocity gradient tensors (picked up from Ref. 22).

Figure 7 shows one image shot of the ensemble and moving averaged flow fields. There appears a sequence of vortices, the number of which corresponds to the burst frequency F^+. Same structures are observed in the phase-lock averaged flow fields in the experiment as well[21]. These two-dimensional spanwise vortices promote large-scale momentum exchange, which influences to the flow control authority of plasma actuators. Although it is speculative and further study is necessary, these vortices would work when the shear layer stays away from the airfoil surface, while small disturbance induced by the plasma actuators work when the shear layer stays rather close to the airfoil surface. That may be why $F^+ = 1.0$ having larger and stronger two-dimensional vortices work better for NACA0006 airfoil.

DBD plasma actuators especially in the burst mode promote transition and the boundary layer becomes attached. The boundary layer, however, keeps the periodicity of the burst mode as observed in Fig. 6. It is true that the time-averaged boundary layer profile is an attached one, but the time history of such analysis indicates the role of the DBD plasma actuator is not to attach the flows but to prevent vortices to move away from the airfoil surface and keep them close to the airfoil surface. It should be noted that very similar structure is observed for synthetic jets when periodic components from the phase and spanwise averaged Reynolds stress are plotted[10].

Thirdly, streamwise momentum introduced by the plasma actuator is important for flow control for all the cases. The two flow structures above depend on the parameters of burst mode but the third one always exists in the induced flows by the plasma actuator. The induced momentum becomes largest in the continuous mode and becomes smaller for lower burst ratios.

All the three flow mechanism stay together in the flow field created by the plasma actuator. Some of them become dominant depending on the flow and geometric conditions. This is why plasma actuators work more efficiently than any other flow control devices.

4. Importance of HPC Use

As has been mentioned, discussion here is the result of numbers of the large-scale flow simulations. Important phenomena occur in the boundary layers over the airfoil surface and it is necessary to capture small disturbance induced by the plasma actuators inside the boundary layer. Such phenomena would not be simulated with the RANS model in general even though some features of plasma actuators may be simulated. Therefore, iLES computations with high-resolution spectral-like scheme are considered and simulations are mostly conducted on "K" supercomputer in Japan having 10 PFLOS. Fluid dynamics is strongly nonlinear and, as a result, such numbers of large-scale simulations are very useful for understanding of the flow structures even for the preliminary or initial phase of the design process.

5. Summary

Mechanism of flow separation control by DBD plasma actuators is discussed based on a series of simulations of low speed flows over NACA airfoils with the actuator devices attached near the leading edge. Three-dimensional compressible Navier-Stokes equations with the body force terms representing the induced flows by the DBD plasma actuator are solved. Simulations are validated by the comparison with the experiments conducted in the authors laboratory at the ISAS/JAXA.

Large-scale computations revealed key structures of flow separation control mechanism especially for the burst mode actuations. There seem to be three mechanisms that are the keys for flow control authority of plasma actuators; (1) direct addition of momentum, (2) certain-scale two dimensional vortices that exchange large-scale momentum and pull down the separated shear layer toward the airfoil surface, (3) laminar-to-turbulent transition. It is important that these three mechanisms stay together and some of them become dominant factors, depending on the actuator parameters and flow conditions. This explains why plasma actuators work better and more efficient than other flow control devices.

Acknowledgments

T. Nonomura, H. Aono, M. Sato, and A. Yakeno at ISAS/JAXA and Dr. H. Nishida at Tokyo Noko University, members of the HPC SPIRE program as well as former and current graduate students at the ISAS are the contributors of the present study.

References

1. T. C. Corke, M. L. Post, and D. M. Orlov, Single dielectric barrier discharge plasma enhanced aerodynamics: Physics, Modeling and Applications, *Expts. Fluids*, **46** (2009).
2. A. Glezer and M. Amitay, Synthetic Jets, *Ann. Rev. Fluid Mech.*, **34** (2002).
3. D. P. Rizzetta and M. R. Visbal, Large eddy simulation of plasma-based control strategies for bluff body flow, *AIAA J.*, **47** (2009).

4. I. Kaneda, *et al.*, An effective three-dimensional layout of actuation body force for separation control, *Int. J. Aerospace Engineering*, **2012** (2012).
5. K. Asada and K. Fujii, Computational analysis of unsteady flow-field induced by plasma actuator in burst mode, AIAA-2010-5090 (2010).
6. N. Benard, J. Joliboris and E. Moreau, Lift and drag performances of an axisymmetric airfoil controlled by plasma actuator, *J. Electrostatics*, **67** (2009).
7. A. A. Sidorenko, *et al.*, Pulsed discharge actuators for rectangular wing separation control, AIAA-2007-941 (2007).
8. M. Visbal, D. V. Gaitonde, S. Roy, Control of transitional and turbulent flows using plasma-based actuators, AIAA-2006-3230 (2006).
9. K. Asada, Y. Ninomiya, A. Oyama and K. Fujii, Airfoil flow experiment on the duty cycle of DBD plasma actuator, AIAA-2009-531 (2009).
10. Y. Abe, K. Okada, T. Nonomura and K. Fujii, The effects of actuation frequency on the separation control over an airfoil using a synthetic jet, 5th European Conf. Aero. and Space Sci., **FP10-1** (2013).
11. J. S. Shang, A glance back and outlook of computational fluid dynamics, ASME FED SM 200345420 (2003).
12. K. Fujii, Progress and future prospects of CFD in aerospace-wind tunnel and beyond, Prog. Aerospace Sci., **41** (2005).
13. K. Fujii, High-performance computing based exploration of flow control with micro devices, Philo. Trans. A, The Royal Soc., **372**, 1471–2962 (2014).
14. Y. B. Suzen, and P. G. Huang, Simulations of flow separation control using plasma actuator, AIAA-2006-877 (2006).
15. H. Aono, *et al.*, Flow structures induced by a DBD plasma actuator in static air, Ann. Meeting Japan Soc. Fluid Dyn., (in Japanese) (2014) submitted to *Mech. Engg. J. JSME.*
16. T. Koizumi and H. Nishida, Coupling of discharged plasma and induced flows by the DBD plasma actuators, Ann. Meeting JSASS (in Japanese) 2013.
17. S. K. Lele. Compact finite difference schemes with spectral-like resolution, *J. Comput. Phys.*, **103**(1) (1992).
18. K. Fujii, Unified zonal method based on the fortified solution algorithm, *J. Comput. Phys.*, **118** (1995).
19. T. Nonomura, *et al.*, Control mechanism of plasma actuator for separated flow around NACA0015 at Reynolds number 63,000 - separation bubble related mechanisms, AIAA-2013-0853 (2013).

20. M. Sato, *et al.*, Massive parametric study by LES on separated-flow control around airfoil using DBD plasma actuator at Reynolds number 63,000, AIAA-2013-2750 (2013) to appear in *AIAA J.* (2015).
21. S. Sekimoto, Flow control mechanism of a DBD plasma actuator for airfoils in low speed free-streams, Ph.D. thesis, Univ. of Tokyo (2015).
22. K. Asada, *et al.*, LES of transient flows controlled by DBD plasma actuator over a stalled airfoil, to appear in *Int. J. Comput. Fluid Dyn* (2015).

Frequency Dependent Capacitance SDBD Plasma Model for Flow Control

P. M. Bagade, T. K. Sengupta*

Department of Aerospace Engineering, IIT Kanpur,
Kanpur, U.P., 208 016, India
**E-mail: tksen@iitk.ac.in*

S. Sengupta

Department of Mechanical Engineering, Ohio State University,
Columbus, Ohio, USA
E-mail: soumyoss08@gmail.com

H. D. Vo

Department of Mechanical Engineering,Ecole Polytechnique,
Montreal, Quebec, Canada
E-mail: huu-duc.vo@polymtl.ca

Here, a new capacitance based plasma model is developed using with frequency and presssure dependence incorporated. This model shows a good match for the induced body force with experimental results. High accuracy numerical schemes are used to solve time-accurate Navier-Stokes equation embeded with the plasma model.

1. Introduction

A new frequency-dependent capacitance model has been proposed for direct numerical simulation (DNS) to calculate the spatio-temporal distribution of surface charge density to capture the effects of dielectric barrier discharge (SDBD) plasma actuator. Figure 1 shows a schematic of a SDBD actuator. When a radio frequency AC voltage signal of low power and high amplitude of the order of kV is applied across the electrodes, air is weakly ionized over the dielectric above the covered electrode with little or no heating. Fujji[8] has investigated flow separation controls at low Reynolds number (Re) using LES methodology. The author has noted that for different velocity and model scales, the computational results need to be multiplied by non-dimensional coefficient(s) to match with the experimental results. Direct numerical simulation of 2D Navier-Stokes equation (NSE) is performed using this time accurate model. Instantaneous velocity profiles for the wall jet at three downstream locations are plotted which shows a good match with the experimental results of Versailles *et al.*[4]. This frequency-dependent model shows better match with the body force with experimental data in Versailles *et al.*[4] compared to the previous frequency independent models due to Orlov[1], Suzen *et al.*[2] and Lemire-Vo[6]. Further validation of this DNS procedure is made with respect to experimental results of Whalley and Choi[7].

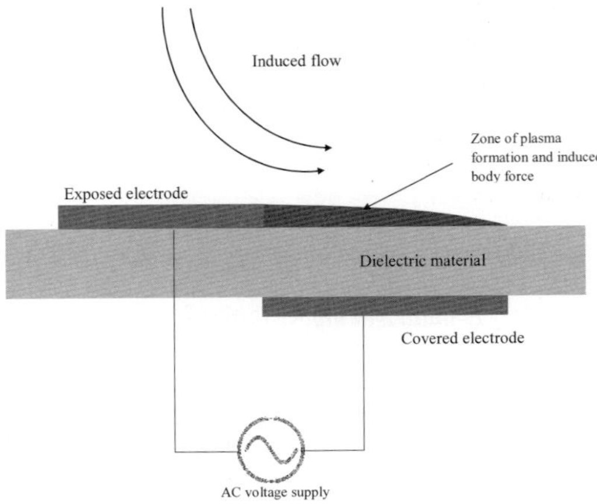

Fig. 1. Schematic of plasma actuator configuration.

2. Governing Equations and Plasma Model

Maxwell's equations are used as governing equations to model plasma dynamics as described in Orlov[1], Sengupta *et al.*[5], Lemire-Vo[6]. If magnetic forces are neglected, then the electrohydrodynamic force is expressed as[1]

$$\vec{f_B} = \rho_c \vec{E} \tag{1}$$

where $\vec{f_B}$ is the body force per unit volume; ρ_c is the net charge density and $\vec{E}$ is the electric field. According to Orlov[1], the time scale associated with the flow response is six and two orders of magnitude lower than those are associated with the ionization process and the AC input cycle, respectively. Since plasma formation is virtually instantaneous in convection scale of the flow, equations of electrostatics are adequate in describing discharges. The electric field can thus be derived from the gradient of a scalar potential as[2]

$$\vec{E} = -\nabla \Phi \tag{2}$$

From Maxwell's equation for charge density (ρ_c), one obtains

$$\nabla \cdot (\epsilon \vec{E}) = \rho_c \tag{3}$$

Using Eq. (2) in (3), one obtains

$$\nabla \cdot (\epsilon \nabla \Phi) = -\rho_c \tag{4}$$

where ϵ is the permittivity of the medium, which can be expressed as $\epsilon = \epsilon_r \epsilon_0$, with ϵ_r being the relative permittivity of the medium and ϵ_0 being the permittivity of free space.

Suzen *et al.*[2] proposed a model in which the total electric potential Φ is the resultant of the potential due to the external electric field (ϕ) and that due to the plasma charge density (φ). The characteristic length for electrostatic shielding in plasma is denoted by the Debye thickness (λ_d). Assuming that λ_d is small and that the surface charge is not large, separate equations for the two potentials are obtained as

$$\nabla \cdot (\epsilon \nabla \phi) = 0 \tag{5}$$

$$\nabla \cdot (\epsilon_r \nabla \varphi) = -\rho_c/\epsilon_0 \tag{6}$$

At any point within the plasma, one can relate ρ_c and φ using the definition of Debye length (λ_d) as[1]

$$\varphi = -\rho_c \lambda_d^2/\epsilon_0 \tag{7}$$

Using Eqs. (7) and (6), one obtains the equation for ρ_c as[2]

$$\nabla \cdot (\epsilon_r \nabla \rho_c) = \rho_c/\lambda_d^2 \tag{8}$$

Here, we propose a new frequency dependent plasma actuation model to provide the boundary conditions for the charge density over the dielectric surface as given below

$$\rho_{c,w}(t)|_n = \frac{C_{nd} \times V_n(t) \times (\frac{f}{f_0})^k \times (\frac{p}{p_0})^{-m}}{Vol_{ln}} \tag{9}$$

where, $\rho_{c,w}$ is the surface charge density; C_{nd} is dielectric capacitance; V_n is voltage on the dielectric surface of the nth sub-circuit for spatio-temporal lumped-element circuit model[5]; f is frequency of AC voltage source and Vol_{ln} the volume of the nth sub-circuit as shown in Fig. 2(a). The charge on the dielectric surface at a particular time depends only on the current and applied voltage of the AC source. The plasma sub-circuit was solved and the values of the body force were compared with the experimental data in Versailles *et al.*[4] to choose a good value of k. The best value of k was found to be $k = 0.1016$ for this data set. In the previous frequency-independent model, this value is $k = 0$. Here, f_0 is the reference frequency taken as f_0 = 1kHz and p_0 is the reference pressure (atmospheric pressure = 100kPa).

Figure 2(a) shows a schematic where air and dielectric are represented as a parallel network of resistors and capacitors. Each parallel circuit consists of an air capacitor (Cna), a dielectric capacitor (Cnd) and an air resistive element (Rnf or Rnb). The plasma and the discharge characteristics are modeled by the resistor. Zener diodes (Dnf or Dnb) sets the plasma initiation, which in turn, set the voltage threshold levels for the ionization to occur. Two Zener diodes are necessary to represent the ionization process with each active during one part of the applied AC voltage cycle. The circuit model with its elements are shown in Fig. 2(b).

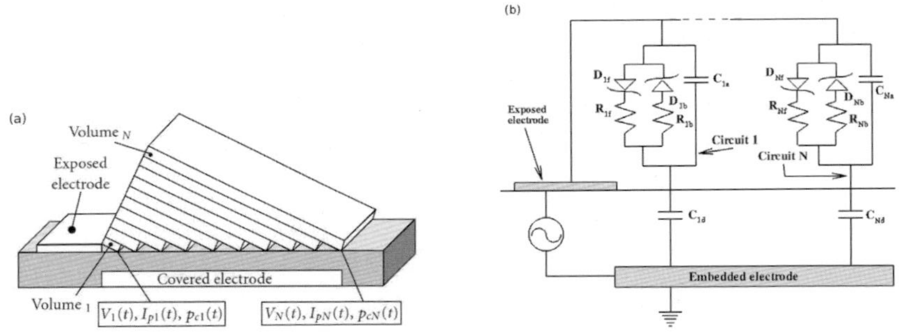

Fig. 2. (a) Plasma formation region divided into a number of volumes, and (b) Electric circuits used to calculate charge density and potential on dielectric surface.

Solving Eqs. (5), (6), (8) and using Eq. (2) in Eq. (1), provides the body forces. Present model (Eq. (9)) is used for solving Eq. (8). The body force obtained is incorporated into the NSE for simulating the effects of plasma on the flow. Numerical solution of NSE are obtained using stream function (ψ) - vorticity (ω) formulation for the incompressible flow. The efficacy of this formulation in obtaining correct results in a faster way is explained in Sengupta[3]. The governing NSE in the (ψ, ω)-formulation expressed in an orthogonal grid[5] are expressed by the dimensional equations

$$h_1 h_2 \frac{\partial \omega}{\partial t} + h_2 u \frac{\partial \omega}{\partial \xi} + h_1 v \frac{\partial \omega}{\partial \eta} = \nu \left[\frac{\partial}{\partial \xi}\left(\frac{h_2}{h_1}\frac{\partial \omega}{\partial \xi}\right) + \frac{\partial}{\partial \eta}\left(\frac{h_1}{h_2}\frac{\partial \omega}{\partial \eta}\right)\right] + \frac{\partial (h_2 F_2)}{\partial \xi} - \frac{\partial (h_1 F_1)}{\partial \eta} \tag{10}$$

$$\frac{\partial}{\partial \xi}\left(\frac{h_2}{h_1}\frac{\partial \psi}{\partial \xi}\right) + \frac{\partial}{\partial \eta}\left(\frac{h_1}{h_2}\frac{\partial \psi}{\partial \eta}\right) = -h_1 h_2 \omega \tag{11}$$

where h_1 and h_2 are the scale factors used in mapping physical (x, y)-plane to the computational (ξ, η)-plane[3,5], where ξ is in streamwise direction and η is in the wall-normal direction. The scale factors can be defined by $h_1 = \sqrt{x_\xi^2 + y_\xi^2}$ and $h_2 = \sqrt{x_\eta^2 + y_\eta^2}$. In the above, (F_1, F_2) are the curvilinear components of the body force $\vec{f}_B$ generated by plasma actuation. In Eq. (10), ν is the kinematic viscosity of air whose value here is taken to be $1.5 \times 10^{-5}\ m^2/s$.

3. Results

Figure 3 shows the time averaged total body force plotted as a function of peak-to-peak applied voltage for indicated frequencies. It can be noted that compared to the frequency-independent plasma model, new proposed frequency-dependent model shows a better match with the body force measured in Versailles *et al.*[4]. Figure 4 shows the evolution of velocity field

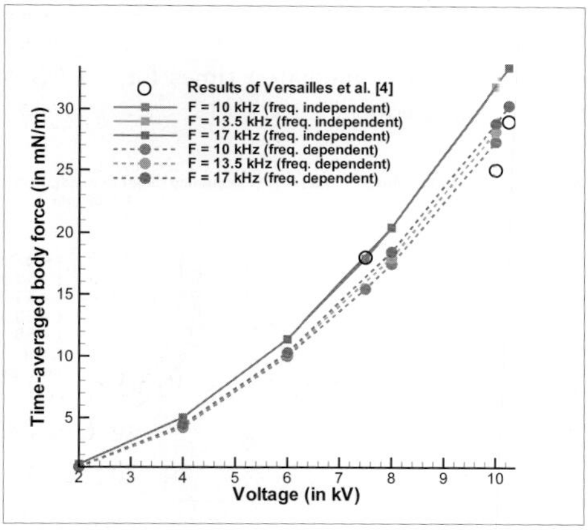

Fig. 3. Time-averaged total body force plotted as a function of peak-to-peak applied voltage for indicated frequencies.

at indicated times for the solution of 2D NSE with the new model, with $V_{app} = 7.5kV$ and frequency $10kHz$. It can be noted that the effects of plasma actuation are similar to effects of wall jet. Figure 5 shows the near-wall velocities at indicated locations downstream of the actuator. Experimental time-averaged velocity data obtained by PIV measurements[4] are compared with the computed solution. It can be observed that the time-averaged computational data is in good agreement with the experimental data of Versailles *et al.*[4]. Also, the difference in the experimental

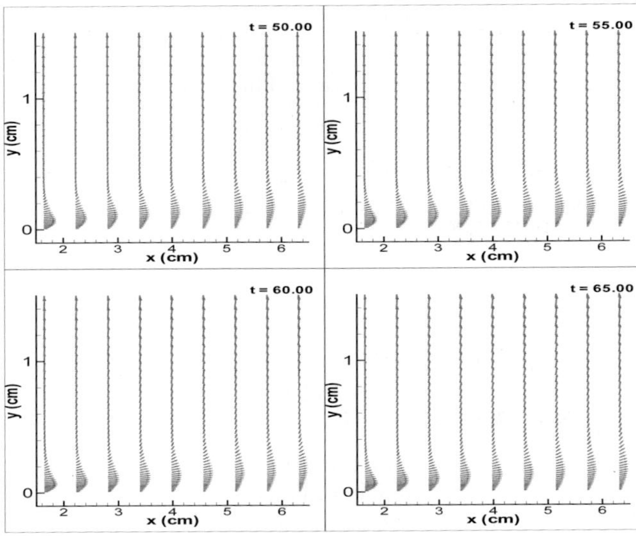

Fig. 4. Evolution of velocity field at indicated times for 2D NS equation with capacitance model ($V_{app} = 7.5kV$ and frequency $10kHz$).

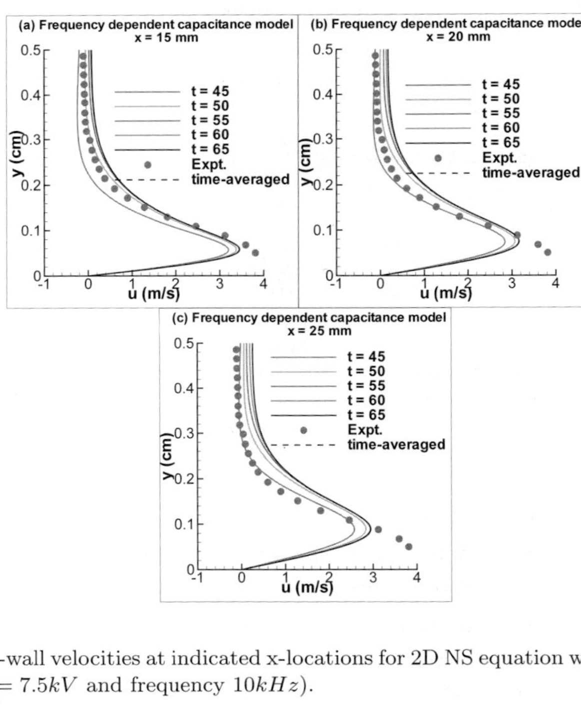

Fig. 5. Near-wall velocities at indicated x-locations for 2D NS equation with capacitance model ($V_{app} = 7.5kV$ and frequency $10kHz$).

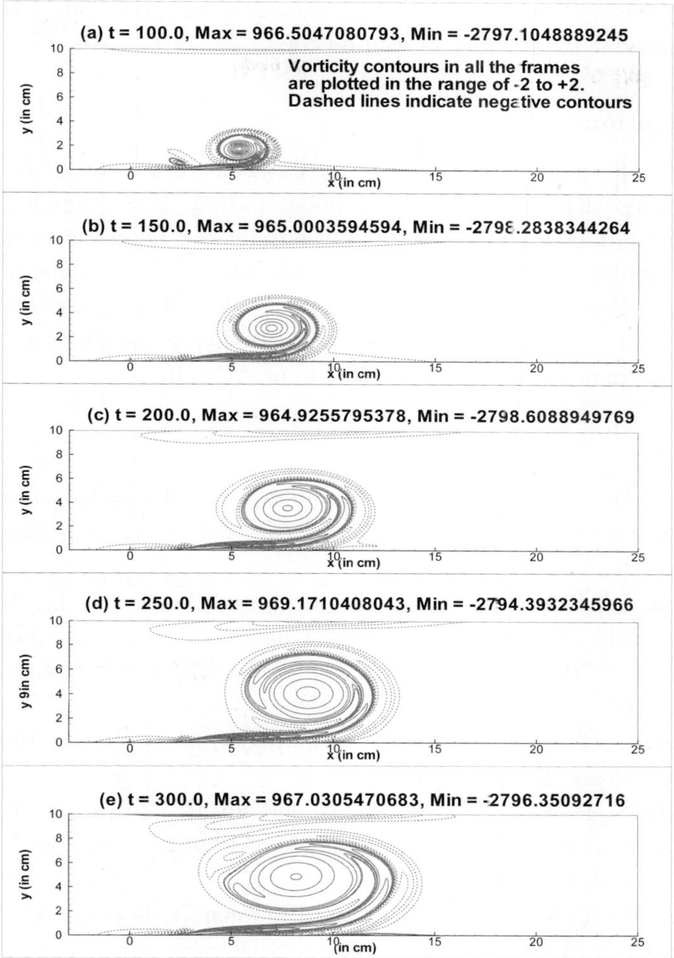

Fig. 6. Time evolution of vorticity with time when 2D NSE is solved along with capacitance model for plasma actuation. Voltage = 5.5kV, frequency = 20kHz corresponding to case-A of Whalley and Choi[7].

and computational profiles is much lesser for the location nearest to the actuator ($x = 15mm$) than the other two locations; with better match observed in the outer part of the boundary layer. Figure 6 shows evolution of vorticity contours at indicated times for sinusoidal voltage input of 5.5kV and frequency 20kHz, corresponding to case-A of Whalley and Choi[7]. The computations are done in a box and it is observed that the evolved vorticity structures move downstream and upwards. However, these are stretched in the domain due to presence of wall on top.

4. Conclusions

A new plasma model based on the capacitance is proposed to calculate the surface charge density for SDBD plasma actuation. Total body force due to plasma actuation calculated based on the frequency-dependent capacitance model provides a good qualitative match with the experimental results reported in Versailles *et al.*[4], as in Fig. 3. This model provides a realistic and physical representation for the surface charge distribution and also shows a frequency dependence on the input AC source.

References

1. Orlov, D. M., Modeling and Simulation of Single Dielectric Barrier Discharge Plasma Actuators, *Ph.D. Thesis, University of Notre Dame, Notre Dame, Indiana*, 2006.
2. Suzen, Y., Huang, P. G., Jacob, J. D. and Ashpis, D. E., Numerical Simulation of Plasma Based Flow Control Applications, *AIAA-2005-4633, In 35th Fluid Dynamics Conference and Exhibit, Toronto, Ontario*, 2005.
3. Sengupta, T. K., *High Accuracy Computing Methods*, Cambridge University Press, New York, USA, 2013.
4. Versailles, P., Gingras-Gosselin, V. and Vo, H. D., Impact of Pressure and Temperature on the Performance of Plasma Actuators, *AIAA J.*, **48(4)**, 859–863, 2010.
5. Sengupta, T. K., Suman, V. K. and Bhumkar, Y. G., Control of Bypass Transitional Flow past an Aerofoil by Plasma Actuation, *I.J.E.M.F.S*, Vol. 3(2+3), 2012.
6. S. Lemire and H. D. Vo, Reduction of Fan and Compressor Wake Defect using Plasma Actuation for Tonal Noise Reduction, *ASME Paper GT 2008-50821*, Germany, 2008.
7. R. Whalley and K. S. Choi, The starting vortex in quiescent air induced by dielectric-barrier-discharge plasma, *J. Fluid Mech.*, **(703)**, 192–203, 2013.
8. K. Fujji, Computational Studies on Flow Separation Controls at Relatively Low Reynolds number Regime, *Proceedings of IUTAM Symp. on Advances in Computation, Modeling and Control of Transitional and Turbulent Flows, India*, Dec. 2014.

Effects of Uniform Blowing or Suction on the Amplitude Modulation in Spatially Developing Turbulent Boundary Layers

Y. Kametani*, R. Örlü, and P. Schlatter
Linné FLOW Centre, KTH Mechanics, Osquars backe 18, 10044 Stockholm, Sweden
*E-mail: yukinori.kametani@mech.kth.se

K. Fukagata
Keio University, Hiyoshi 3-14-1, Kohoku-ku, Yokohama, 223-8522, Japan

Large-eddy simulations of spatially evolving turbulent boundary layers with uniform blowing or suction were performed at a moderate Reynolds number (based on free-stream velocity U_∞ and momentum thickness θ) of up to $Re_\theta \approx 2500$ in the uncontrolled case. Aiming at the control of external flows, the influence on the skin friction drag and turbulent statistics is investigated with the control amplitude of 0.1% of U_∞. While uniform blowing reduces the skin-friction coefficient and enhances the amplitude of all components of the turbulence intensities, uniform suction reverses the effect on the skin friction and scales turbulent intensities with wall units throughout the inner layer. A net-energy saving rate S of around 13% could be achieved by blowing over the controlled surface. Since the locally defined S develops in the streamwise direction, the global S increases as the streamwise length of the controlled region is expanded. The fact that the control gain has similar trend in the streamwise direction indicates that the control efficiency increases the streamwise length of the uniform blowing region. The 2D amplitude modulation map shows that the correlation between the energy of the small scale turbulence and the large scale modulation velocity is remarkably increased (decreased) around the diagonal and off-diagonal peaks by blowing (suction). These results are directly linked to the observed enhancement of turbulence intensities in the inner layer for the blowing and obtained inner-scaling for the suction case.

Keywords: turbulent boundary layers; flow control, amplitude modulation

1. Introduction

In our modern society, commercial transport such as bullet trains or aircrafts play an important role. The skin fiction drag due to viscosity, however, has a huge impact on the global environment due to the consumption of mostly fossil fuels. Numerical simulations are now a powerful method to investigate turbulent flows close to solid walls since Kim *et al.*[1] performed their pioneering direct numerical simulation (DNS) of a turbulent channel flow. Since then, various drag reducing methods applied on the wall, such as riblets (Choi *et al.*[2]), active opposition control (Choi *et al.*[3]), and wall deformation (Tomiyama & Fukagata[4]), have been examined numerically in internal flows. However, the examination in external flow still needs further investigations.

Modification of the shear flow by uniform blowing or suction is used for engineering applications such as film cooling on turbine blades or slotted wings. Kametani & Fukagata[5] performed a DNS of a spatially evolving zero-pressure-gradient turbulent boundary layer with uniform blowing or suction from the wall. As a result, with the wall blowing fixed at 0.1% of the free-stream velocity U_∞, around 10% drag reduction was achieved despite the simple constant input from the wall. On the other hand, the decrease of the Reynolds shear stress and turbulent vortices were confirmed in the case of wall suction. In engineering applications, however, the Reynolds number (Re) is much higher than that in their DNS. Therefore the control effects on the turbulent structures at high Re such as large-scale structures (see e.g. Marusic *et al.*[6]) needs to be investigated. Recently, the turbulent structures in turbulent boundary layers at relatively high Re have been investigated by means of large-eddy simulations (LES) (Eitel-Amor *et al.*[7]).

In this study, the effect of uniform blowing and suction with a finite streamwise length of the blowing and suction region at a moderate Reynolds number and the control efficiency are discussed. The effect of the control on the outer layer structures is investigated by considering the amplitude modulation and its influence on the turbulence intensity.

2. Numerical Simulation

A series of large-eddy simulations (LES) have been performed. The present LES uses ADM-RT model[8]. The governing equations are the spatially filtered Navier–Stokes equations with an active relaxation term and continuity equations (see Kametani *et al.*[9]). The streamwise, wall-normal and spanwise lengths of the computational domain are $(L_x, L_y, L_z) = (3000\delta_0^d, 100\delta_0^d, 960\delta_0^d)$, respectively, where δ_0^d denotes the inlet displacement thickness. The corresponding numbers of grid points are $(N_x, N_y, N_z) = (2048, 257, 1536)$, while the grid resolution is $(\Delta x^+, \Delta y^+_{max}, \Delta z^+) = (20.9, 13.3, 8.9)$. The boundary layer starts from a Blasius profile at the inlet of the computational domain and is tripped to a turbulent boundary layer via

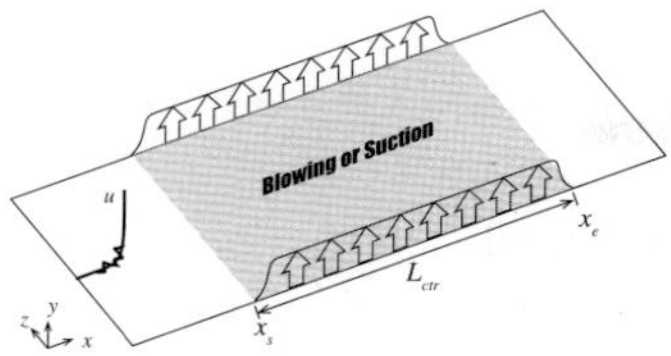

Fig. 1. Schematic of the controlled region as well as definition of the coordinate system and relevant parameters.

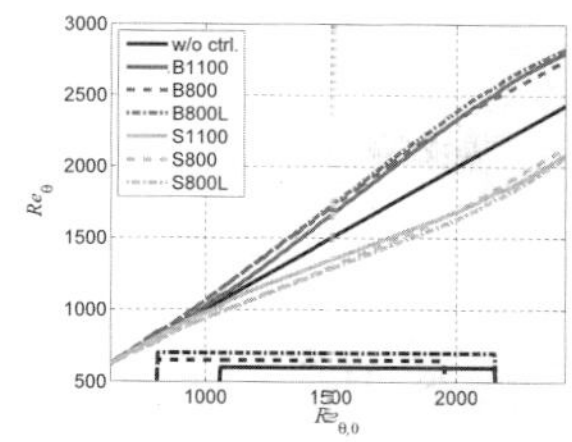

Fig. 2. Reynolds number relation. Lines at the bottom indicate the control region for each case.

a trip force (detailed in Eitel-Amor *et al.*[7]). Fully turbulent condition is achieved for the range of $Re_\theta \geq 700$. Blowing/suction is achieved by the constant wall-normal velocity imposed on the wall, $V_w/U_\infty = \pm 0.001$, as shown in Fig. 1. In order to consider the area-dependency of the control, blowing or suction is applied over three different areas: Control at $800 \leq Re_{\theta,0} \leq 1800$ is referred to as B/S800, $1100 \leq Re_{\theta,0} \leq 2100$ as B/S1100 and $800 \leq Re_{\theta,0} \leq 2100$ as B/S800L, respectively. The subscript 0 refers to the uncontrolled (reference) case. The averaging time in wall units is $T^+ \approx 2000$ for the uncontrolled case and B/S1100, and $T^+ \approx 1000$ for the other cases.

3. Statistics

The Reynolds number based on the momentum thickness in each case is plotted as a function of the streamwise location in Fig. 2. The lines at the bottom visualize the control region in each case. The figure shows that the spatial growth of the boundary layer is promoted by blowing and delayed by suction. Although the amplitude of blowing or suction is small, the thickness of the boundary layer in the blowing case is approximately 13% larger than that of the uncontrolled flow at $Re_\theta = 2100$, while it is approximately 18% lower in the suction case. Instantaneous streamwise vorticity of B/S800L is shown in Fig. 3 and compared to the uncontrolled case. It can be confirmed that the development is promoted by blowing and delayed by suction.

The statistics at some streamwise locations in the B/S1100 cases are considered next. Figure 4 shows the mean streamwise velocity scaled by local wall units. Gray lines show the linear and logarithmic ($U^+ = 1/0.41 \ln y^+ + 5$) velocity profile. While the linear law can be found in

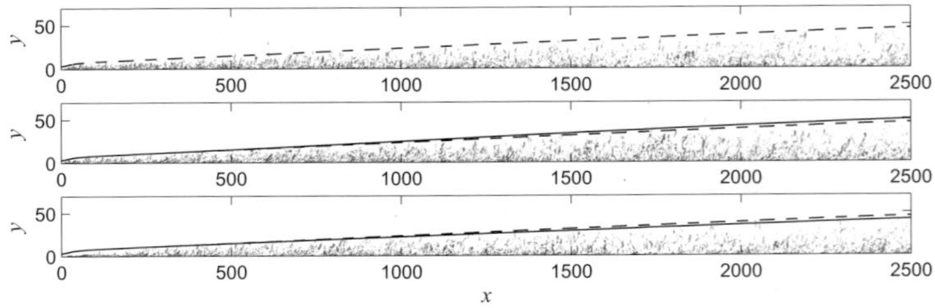

Fig. 3. Instantaneous streamwise vorticity, $-0.1 < \omega_x < 0.1$, in $x-y$ plane. The representation is stretched by a factor of 3 in the wall-normal direction. Top, uncontrolled; middle, blowing (B800L); bottom, suction (S800L). Solid line, 99% boundary layer thickness, δ; dashed line, δ of the uncontrolled case. The x-extent corresponds to $7 \times 10^4 < Re_x < 1.2 \times 10^6$.

all cases, the logarithmic profile is not appropriate for the blowing and suction cases. The root-mean-square (rms) and the turbulent intensity of streamwise velocity are plotted in Fig. 5, where the wall-normal location is scaled by local wall units. Due to the reduced (enhanced) skin friction coefficient as will be shown later, turbulent intensities seem to be increased (decreased) by blowing (suction). The lines in the uncontrolled case collapse in the buffer region, i.e. $4 < y^+ < 40$, while the extent of the collapse shrinks in the case of blowing and expands for suction. For the suction case, inner-scaling can be observed throughout the inner layer. The profile is supposed to approach that of the asymptotic suction turbulent boundary layer as the suction amplitude increases, cf. Schlatter and Örlü.[10] It can be concluded that the intensity of the outer-layer structures are effectively reduced by the weak amount of suction. Because of the small amplitude of suction compared to Kametani & Fukagata[5], the suppression of turbulence is rather small in the present simulation.

In order to consider the effect of the control on the turbulence without being biased by the changed skin friction or the thinned or thickened boundary layer, the diagnostic plot á la Alfredsson *et al.*[11] is considered next. Hereby, the streamwise turbulence intensity (u^{rms}/U) is plotted against the velocity (U/U_∞) as shown in Fig. 6. In such a plot, the turbulence intensity in the outer layer is found to scale linearly with the velocity independent of the Reynolds number for the case of a zero-pressure-gradient turbulent boundary layer. Also shown as a black straight line is the linear relation (Alfredsson *et al.*[11])

$$\frac{u^{rms}}{U} = a + b\frac{U}{U_\infty}, \tag{1}$$

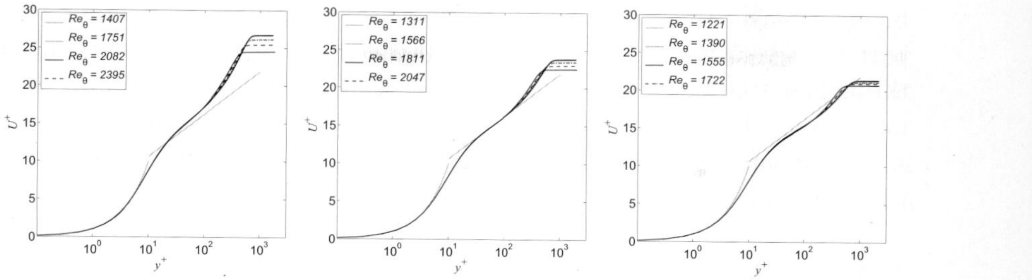

Fig. 4. Mean streamwise velocity. Left, blowing; center, uncontrolled; right, suction.

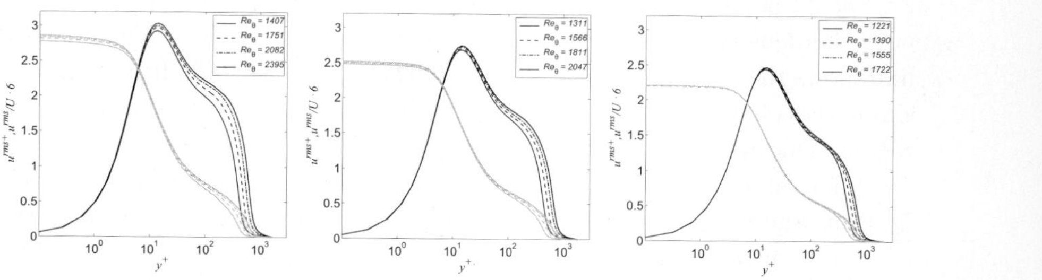

Fig. 5. Root-mean-square (black) and turbulent intensity (gray) of streamwise velocity. Left, blowing; center, uncontrolled; right, suction.

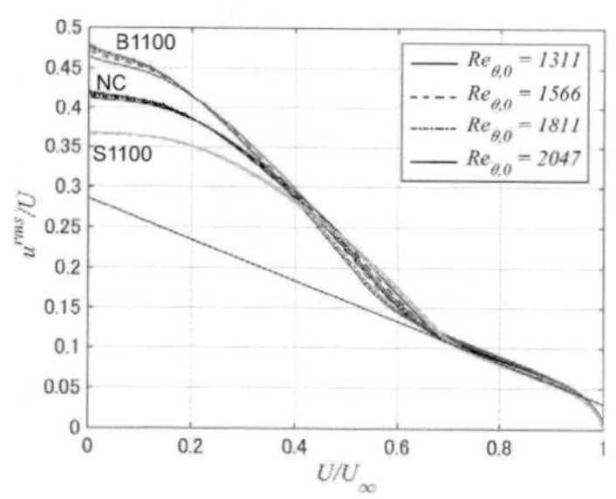

Fig. 6. Diagnostic profile á la Alfredsson *et al.*[11]. Straight line represents Eq. 1.

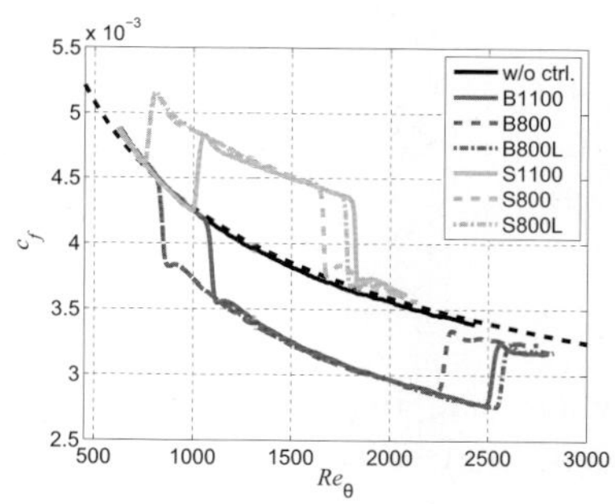

Fig. 7. Friction coefficient as a function of Re_θ. The black dashed line, $c_f = 0.024 Re_\theta^{-1/4}$.

with the empirical constants $a = 0.286$ and $b = -0.255$. It can be appreciated, that despite the apparent changes in the inner-scaled mean and rms profiles (Figs. 4–5), the influence of the control is to act mainly on the wall (thereby setting the friction velocity, i.e. skin friction) and to alter the boundary-layer thickness, but the turbulence intensity corresponding to a particular velocity within the boundary layer remains largely unchanged throughout the outer region. For the suction case, the rms profiles were found to scale in inner-scaling throughout the inner layer, however, this scaling expands throughout the entire boundary layer when considered in the diagnostic plot.

4. Skin Friction Drag Reduction

The effect of blowing or suction on the skin friction coefficient, $c_f = 2\tau_w/\rho U_\infty^2$, is plotted as a function of the Re in Fig. 7, where ρ and τ_w denote the density and the wall-shear stress, respectively. The black dashed line shows the power-law relation $c_f = 0.024 Re_\theta^{-1/4}$, which indicates that all simulations have reached a developed turbulent state prior to the controlled region. Due to the variation of the boundary-layer thickness, the range of Re differs among the cases. It can be inferred from the figure that the skin friction drag is reduced by blowing and enhanced by suction. Although the range of control is varied, it appears that the reduced or enhanced profiles of c_f approach and collapse on a single line for the controlled case.

The local drag reduction rate, R^L, the net-energy saving rate, S^L, and the control gain, G^L, are defined as

$$R^L(x) = \frac{c_{f,0}(x) - c_f(x)}{c_{f,0}(x)}, \tag{2}$$

$$S^L(x) = \frac{c_{f,0}(x) - (c_f(x) + w_{in}(x))}{c_{f,0}(x)}, \tag{3}$$

$$G^L(x) = \frac{c_{f,0}(x) - c_f(x)}{w_{in}(x)}, \tag{4}$$

where w_{in} denotes the local input power from the wall flux calculated as $w_{in}(x) = 0.5V_w(x)^3/U_\infty^3$. Due to the quite weak blowing amplitude, i.e. $w_{in}(x) \sim \mathcal{O}(10^{-9})$, the net-energy saving rate S^L is nearly equivalent to the drag reduction rate R^L. The mean energy saving rate S can be defined as $S = \frac{1}{x_e - x_s}\int_{x_s}^{x_e} S^L \mathrm{d}x$, where x_s and x_e denote the starting and ending edge location of the blowing region, respectively. Similarly, the mean control gain, G, can be defined. From the present simulation, approximately 13% net-energy saving is achieved by blowing. The B1100, B800 and B800L cases achieve $S = 17.6\%$, 17.2% and 18.0%, respectively. Since the locally defined S^L increases in the streamwise direction, the mean value, S, is expected to increase as the blowing region is extended in the streamwise direction. It should be noticed that the measures of the mean control efficiency, G and S, increase as the streamwise length of the controlled region is elongated because the control gain G^L increases as the streamwise length increases as well as the net-energy saving rate, S^L. The result of B800 and B1100 further indicates that the blowing starting position seems to affect the energy saving rate even though the area of the controlled region is equal; starting at higher Re_θ achieves slightly higher net-energy saving.

5. Amplitude Modulation

The interaction between the scales of the turbulent structures is discussed in this section. Mathis *et al.*[12] decomposed the streamwise velocity fluctuation into the modulating large-scale and modulated small-scale fluctuations as $u' = u'_L + u'_S$ in order to discuss the interaction between them by investigating the amplitude modulation ($\mathrm{AM} = \langle u_L(y_1)' u_S(y_2)'^2 \rangle$). The Reynolds number dependency of the peak locations in the 2D AM map as function of Re was investigated by Eitel-Amor *et al.*[7] Their results indicate that the distribution around the diagonal peak, $(y_1^+, y_2^+) = (10, 10)$, is dominant at low Reynolds number; however, the off-diagonal peak, $(y_1^+, y_2^+) = (3.9Re_\tau^{0.5}, 10)$, grows as the Reynolds number increases. Figure 8 depicts the AM values affected by the blowing or suction at $Re_{\theta,0} = 1800$. The cutoff values of the temporal period and the spanwise wavelength are set to be $(\lambda_t^+, \lambda_z^+) = (400, 400)$, which is supposed to separate the inner-peak and the outer peak, found from the spectral analysis. The distribution of the AM value around the diagonal and off-diagonal peak is remarkably increased by blowing, while it decreases by suction. The increase of the AM value in the inner region is supposed to come from the increased turbulent generation by blowing. On the other hand, the one in the outer region located distant from the wall is supposed to be caused by the turbulence shifted away from the wall by blowing. The increase in the off-diagonal peak region by blowing does not seem to be similar to the one due to an increase of the Reynolds number as discussed in Eitel-Amor *et al.*[7] because the increase of the distribution occurs around peaks, rather than in the intermediate region between the peaks. The increase of the friction Reynolds number, Re_τ, might cause a slight shift of the off-diagonal peak location towards larger y_1 values. For the suction case, the value in the outer region is drastically decreased and it is difficult to discuss the off-diagonal peak. The shifts of the peaks are more apparent in the one-dimensional AM plot as a function of y_1^+ at $y_2^+ = 10$ shown in Fig. 8(d). The increase in

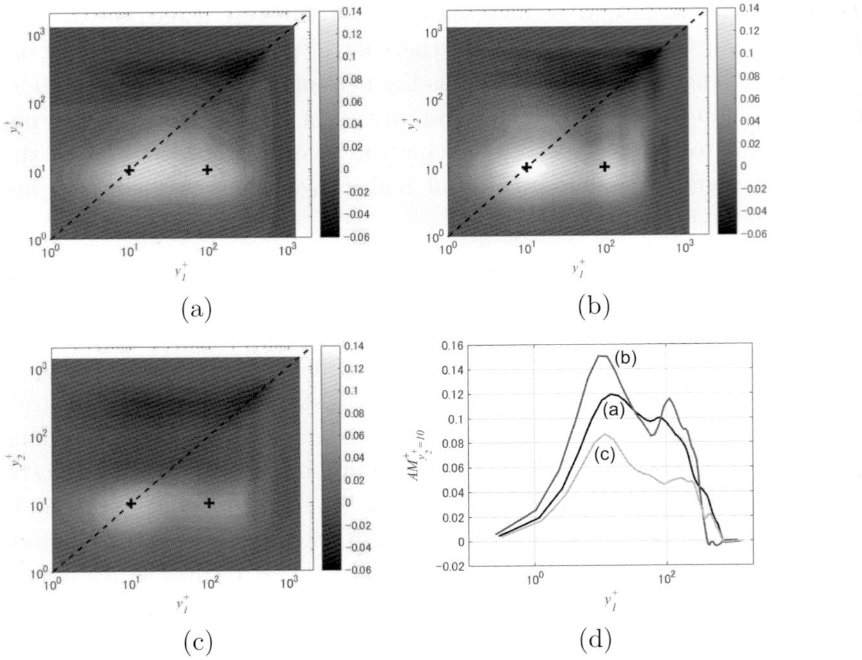

Fig. 8. 2D Amplitude modulation map, $\langle u_L(y_1)' u_S(y_2)'^2 \rangle / u_\tau^3$, at $Re_{\theta,0} = 1800$. (a) uncontrolled case, (b) B1100, (c) S1100, (d) one-dimensional plot as a function of y_1^+ at $y_2^+ = 10$. Black crosses mark the locations of diagonal peak and off-diagonal peak in the uncontrolled case, respectively.

the off-diagonal peak in case of blowing indicates that there is an increased modulation of the near-wall small-scale activity, which is responsible for the increased turbulence intensities (cf. Fig. 5), while the nearly vanished off-diagonal peak in case of suction explains why turbulence quantities become Reynolds number independent.

6. Conclusions

Large-eddy simulations of spatially evolving turbulent boundary layers with uniform blowing or suction were performed at a moderate Reynolds number. Friction reduction was achieved by blowing, while suction leads to turbulence suppression. Over 10% of mean friction drag reduction could be achieved despite the low blowing amplitude of only 0.1%. It was found that the control efficiency, i.e. the net-energy saving rate and the control gain of blowing, increases as the controlled area is extended in the streamwise direction. Consideration of the 2D amplitude modulation map, $\langle u_L(y_1)' u_S(y_2)'^2 \rangle$, revealed that the amplitude modulation effect is increased (decreased) in the case of blowing (suction). These observations are directly linked to the behaviour of the turbulence intensities, which are enhanced in case of blowing and nearly independent of the Reynolds number in the case of suction. The latter is therefore a result of the inhibited amplitude modulation of the near-wall small-scale activity by the large-scale outer-layer structures.

Acknowledgments

This work was partly supported through Grant-in-Aid for Scientific Research (C) (No. 25420129) and Grant-in-Aid for JSPS Fellows (No. 24-3450) by Japan Society for the Promotion of Science (JSPS). Financial support by the Knut and Alice Wallenberg Foundation is also gratefully acknowledged. Computer time is provided by Swedish National Infrastructure for Computing (SNIC).

References

1. J. Kim, P. Moin and R. Moser, Turbulence statistics in fully developed channel flow at low Reynolds number, *J. Fluid Mech.* **177**, 133 (1987).
2. H. Choi, P. Moin and J. Kim, Direct numerical simulation of turbulent flow over riblets, *J. Fluid Mech.* **255**, 503 (1993).
3. H. Choi, P. Moin and J. Kim, Active turbulence control for drag reduction in wall-bounded flows, *J. Fluid Mech.* **262**, 75 (1994).
4. N. Tomiyama and K. Fukagata, Direct numerical simulation of drag reduction in a turbulent channel flow using spanwise traveling wave-like wall deformation, *Phys. Fluids* **25**, 105115 (2013).
5. Y. Kametani and K. Fukagata, Direct numerical simulation of spatially developing turbulent boundary layers with uniform blowing suction, *J. Fluid Mech.* **681**, 154 (2011).
6. I. Marusic, R. Mathis and N. Hutchins, Predictive model for wall-bounded turbulent flow, *Science* **329**, 193 (2010).
7. G. Eitel-Amor, R. Örlü and P. Schlatter, Simulation and validation of a spatially evolving turbulent boundary layer up to $Re_\theta = 8300$, *Int. J. Heat Fluid Flow* **47**, 57 (2014).
8. P. Schlatter, P. Stolz and L. Kleiser, LES of transitional flows using approximate deconvolution model, *Int. J. Heat Fluid Flow* **23**-3, 549 (2004).

9. Y. Kametani, R. Örlü, P. Schlatter and K. Fukagata, Drag reduction in turbulent boudary layers: Effect of uniform blowing and suction, in *Proc. Int. Symp. on Engineering Turbulence Modelling and Measurements (ETMM10)*, (Marbella, Spain, 2014).
10. P. Schlatter and R. Örlü, Turbulent asymptotic suction boundary layers studied by simulation, *J. Phys.: Conf. Ser.* **318**, 022020 (2011).
11. P. Alfredsson, R. Örlü and A. Segalini, A new formulation for the streamwise turbulence intensity distribution in wall-bounded turbulent flows, *Eur. J. Mech. B Fluids* **36**, 167 (2012).
12. R. Mathis, I. Marusic, N. Hutchins and K. Sreenivasan, Relationship between the velocity skewness and the amplitude modulation of the small scale by the large scale in turbulent boundary layers, *Phys. Fluids* **23** 121702 (2011).

Turbulent Drag Reduction in Channel Flow Using Weak-Pressure Forcing

B. A. Khan and M. F. Baig*

Department of Mechanical Engineering,
Aligarh Muslim University, Aligarh (India)
**E-mail: mfbaig.me@amu.ac.in,*
basheerahmad61@gmail.com

In the present study, an attempt is made to employ a novel technique to reduce the friction drag significantly while spending less actuating power. Regarding this, we have introduced periodically actuated weak sources of monopoles/dipoles in the shear layer region of turbulent channel flow at $Re_\tau = 180$. The control is achieved by pressure perturbations using monopoles/dipoles as sources for generating weak pressure waves in shear layer region in both spanwise as well as in streamwise directions and this proposed scheme is practically achievable. A total of four simulation cases have been performed with different orientation of monopoles, dipoles and quadrupoles placed at the bottom wall of the turbulent channel. Results have been analysed based on spatio-temporal response, modification of turbulent structures and statistical analysis. Increased mass-flow rate (for fixed mean pressure-gradient) in the spatio-temporal response accompanied by reduction in magnitude of high-speed streaks as well as spatial density of near-wall coherent structures is a clear indication of reduction in turbulent drag. It is observed that upto 2-5% skin-friction turbulent drag-reduction is achieved with the current technique.

Keywords: monopoles, dipoles and quadrupoles, wall shear-stress, near-wall streaks, power spectrums

1. Introduction

There is no shortage of ideas to reduce the skin-friction drag by perturbing the flow-field by both active and passive devices (Gad-el Hak, 1989), but it is required to apply the effective control technique in real-life with positive saving of power. Most control strategies for turbulent drag-reduction have primarily concentrated on effective modification of regeneration-cycle of near-wall turbulence by focussing either on the transient growth of turbulent-streaks or by weakening the streamwise longitudinal vortices present in the near-wall region or by suppressing the essential mechanisms that are prime suppliers of dominant instabilities which aid in formation of near-wall streaks.

We aim to employ a novel technique of introduction of periodically actuated weak sources of monopoles/dipoles in the shear layer region so as to affect the perturbation flow-field significantly without disturbing the mean flow. The effect of control in a fully developed channel flow is studied by means of a direct numerical simulation at a fixed friction Reynolds number of $Re_\tau = 180$.

2. Mathematical Formulation and Numerical Technique

The incompressible non-dimensional form of governing equations having constant density ρ and constant kinematic viscosity ν and are expressed in tensorial form as :-

$$\frac{\partial u_i}{\partial x_i} = 0 \tag{1}$$

$$\frac{\partial u_i}{\partial t} + u_j \frac{\partial u_i}{\partial x_j} = -\frac{\partial p}{\partial x_i} + \frac{1}{Re_\tau}\Big(\frac{\partial^2 u_i}{\partial x_j \partial x_j}\Big) + \delta_{i1}\Big(-\frac{\partial \overline{p}}{\partial x_i}\Big). \tag{2}$$

Lengths are scaled with δ (half-channel height), velocities are scaled with u_τ (friction-velocity), the time is scaled by $\frac{\delta}{u_\tau}$, and the pressure by ρu_τ^2, respectively. $\frac{dP}{dx_i} = -1$ is the mean-pressure gradient driving the channel flow. A schematic representation of the physical domain with four spanwise monopoles is shown in figure 1.

The initial condition is provided from a previous simulation of fully developed turbulent channel flow while the physical boundary conditions mainly consist of no-slip and no-penetration at both the walls. Periodic boundary conditions are used in streamwise as well as in spanwise directions.

The lengths of channel in streamwise, spanwise and in wall-normal direction are correspondingly $2\pi\delta$, $\pi\delta$ and 2δ having $130 \times 130 \times 106$ grid points, respectively. The numerical solutions are obtained using semi-implicit, pressure correction type scheme proposed by Cheng and Armfield (1995) on a non-staggered structured grid. The details of the numerical scheme can be seen in Ahmad *et al.* (2014).

3. Control Strategy and Numerical Experiments

The mathematical expressions for the spherical pressure disturbance waves generated using monopoles, dipoles or quadrupoles in non-dimensionalized

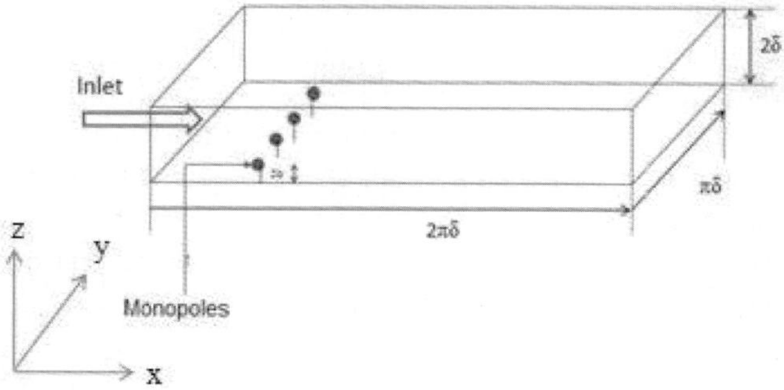

Fig. 1. Geometry of the physical domain under study alongwith spanwise monopoles.

form are given as (Russel *et al.* 1998):
For monopoles,

$$p(\theta, t, r) = \frac{\kappa Q c}{4\pi r} \sin(\omega t - \kappa r) \tag{3}$$

For dipoles,

$$p(\theta, t, r) = \frac{\kappa^2 Q c d}{4\pi r} \cos(\theta) \sin(\omega t - \kappa r) \tag{4}$$

For quadrupoles,

$$p(\theta, t, r) = \frac{\kappa^3 Q c d D}{\pi r} \cos^2(\theta) \sin(\omega t - \kappa r) \tag{5}$$

where parameters Q, c, d, D, ω and κ correspond to source strength, phase speed of wave, distance between two identical monopoles, distance between two identical dipoles, wave frequency and wave number. The term 'r' represents the distance from source to any point in the flow domain where induced pressure is to be ascertained and θ is angle from the vertical and $c = \frac{\omega}{\kappa}$ is characteristic phase speed of the pressure disturbance and the disturbance pressure which is impulsively applied is felt everywhere in the flow-field at the same time.

The periodicity in streamwise and spanwise direction is employed to confine the disturbance pressure within the control volume while the top and bottom walls reflect the disturbance pressure in the flow-field. These pressure disturbance sources are actuated periodically in the shear layer region so as to modify the near-wall turbulence in such a way that it gets attenuated. The disturbance pressures are added explicitly to the pressure flow-field as the magnitude of disturbance pressures is less than 2% of the flow pressures.

Four different numerical experiments Case 1 to Case 4 have been conducted to evaluate the optimum actuating parameters needed to achieve near to maximum control performance.

Case 1 pertains to two monopoles in spanwise orientation placed at a wall-normal height $z^+ = 5$ with $\kappa = 30$, Q $= 0.5$, $\omega = 10$ and the actuation periodicity is $\tau^+ = 22.2$.

Case 2 pertains to four monopoles in streamwise orientation at a wall-normal height $z^+ = 10$ with $\kappa = 30$, Q $= 0.5$, $\omega = 10$ and the actuation periodic interval is $\tau^+ = 0.011$.

Case 3 pertains to four dipoles in spanwise orientation at a wall-normal height $z^+ = 5$ with $\kappa = 30$, Q $= 0.5$, $\omega = 10$ and the forcing actuation interval is $\tau^+ = 2.7$.

Case 4 pertains to eight quadrupoles in spanwise orientation at a wall-normal height $z^+ = 10$ with $\kappa = 20$, Q $= 0.5$, $\omega = 10$ and the forcing actuation interval is $\tau^+ = 22.2$.

Case	Orientation of forcing	κ	ω	Q	z^+	Time period, τ^+	$No.of sources$	$\triangle S_f\%$
1	spanwise monopoles	30	10	0.50	5	22.2	2	5.0
2	streamwise monopoles	30	10	0.50	10	0.011	4	4.1
3	spanwise dipoles	30	10	0.50	5	2.7	4	1.9
4	spanwise quadrupoles	20	10	0.50	10	22.2	8	1.2

4. Spatio-Temporal Response

A precise measurement of turbulent-drag reduction can be obtained by analysing the spatio-temporal response. The instantaneous percent-change in turbulent skin-friction is expressed as:

$$\triangle S_f\% = \left(\frac{S_{fc} - S_{fuc}}{S_{fuc}}\right) \times 100\% = \left[1 - \frac{U_{b_{uc}}^2}{U_{b_c}^2}\right] \times 100\% \tag{6}$$

The bulk velocity of uncontrolled flow $U_{b_{uc}}$ is almost equal to 15.7, whereas for controlled cases the bulk velocity is higher for all the cases as shown in Figures 2(a) to (d). Case 1 shows a statistically-stationary value of bulk velocity U_{bc} nearly 16.05, thereby implying an effective decrease of skin-friction drag. For the Case 2, the mass-flow rate also increases and its value at stationarity is in around 16 while for the cases 3 and 4 correspond to actuation of the dipoles and quadrupoles show a considerably smaller increase in mass-flow rate compared to monopole cases, thereby implying a smaller reduction of turbulent drag vis-a-vis monopole actuation sources.

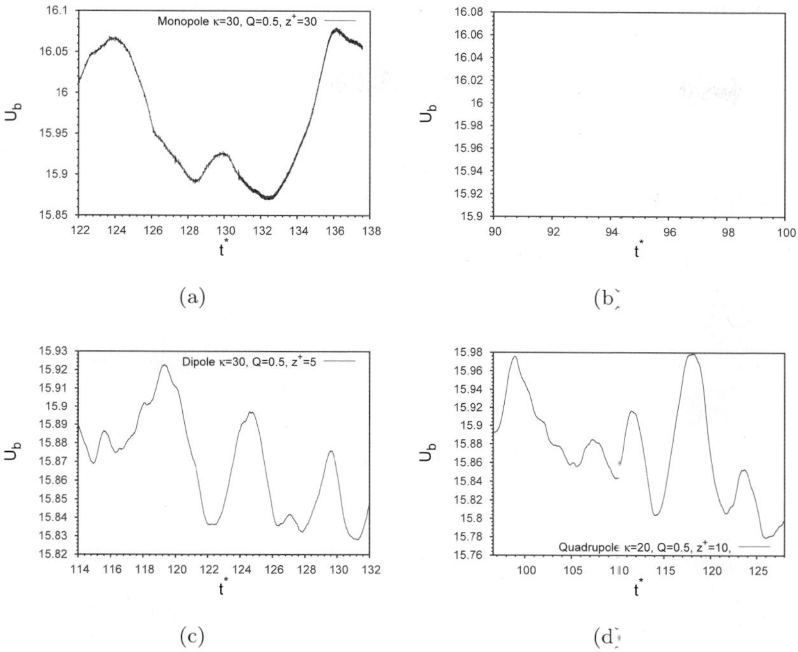

Fig. 2. Figures showing bulk-flow rate for all controlled cases (a) Spanwise monopoles: Case 1 (b) Spanwise monopoles: Case 2 (c) Spanwise Dipoles: Case 3 and (d) Spanwise Quadrupoles: Case 4, respectively.

5. Statistical Analysis of Flow

The statistical-computation of mean and second-order moments of fluctuating velocities ($\overline{u'}$,$\overline{v'}$,$\overline{w'}$) are obtained by spatial averaging over homogeneous directions and then averaging the results in time. The profiles obtained by statistical analysis of the control cases are compared with that of uncontrolled case to elucidate the variation of flow properties.

The streamwise velocity fluctuating component u_{rms} is unaffected by the actuation applied. The fluctuations of r.m.s. velocity in the spanwise direction v_{rms}, show a reduction in magnitude especially in the zone $10 < z^+ < 50$. The peak of w_{rms} at both the lower and upper walls are slightly decreased in the zone $10 < z^+ < 50$ as compared to uncontrolled case. A small amount of reduction in Reynolds stress $\overline{u'w'}$ is achieved especially in the buffer & inner log-law region for the control cases thereby implying weakening of near-wall turbulence.

5.1. *Near-wall streaks*

On visualizing Figures 3(a) to (e), It is observed that there is increased spatial density of the low-speed streaks for the controlled cases (except case 4) and this is accompanied by reduction in magnitude of high-speed streaks. This implies that for the controlled cases (except case 4), there is slight reduction of near-wall turbulent kinetic energy.

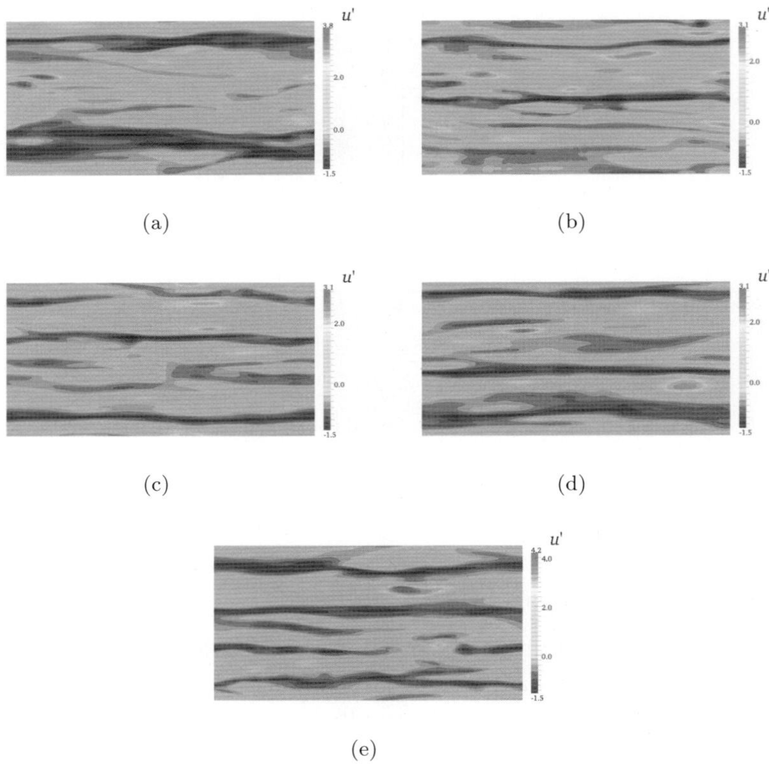

Fig. 3. Comparison of near-wall turbulent streaks at $z^+ = 20$ from bottom wall for uncontrolled and controlled flow (a) Uncontrolled case (b) Spanwise Monopoles: Case 1 (c) Spanwise Monopoles: Case 2 (d) Spanwise Dipoles: case 3 and (e) Spanwise Quadrupoles: case 4, respectively.

5.2. *Turbulent vortical-structures*

Iso-surfaces of Q invariant are visualised in the bottom-half of the channel for threshold magnitude of 120 as shown in the Figures 4(a) to (e). For the controlled cases (except case 4), we see marked reduction of coherent-structures as compared to the uncontrolled-case which indicates that the small-scale vorticity gets reduced leading to a weaker turbulence. This reduction in near-wall turbulence is less marked for cases involving dipoles & quadrupoles compared to monopoles.

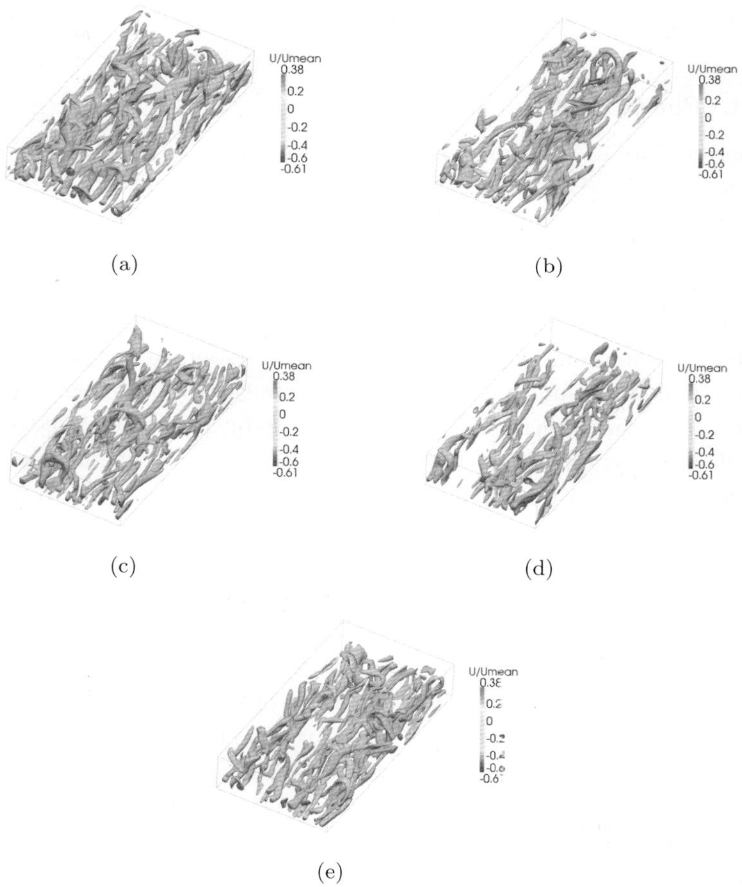

Fig. 4. Comparison of vortical structures in the lower-half of the channel for uncontrolled and controlled flow (a) Uncontrolled case, (b) Spanwise Monopoles: Case 1, (c) Spanwise Monopoles: Case 2, (d) Spanwise Dipoles: Case 3, and (e) Spanwise Quadrupoles: Case 4, respectively.

5.3. *Turbulent burst-events computation using VITA technique*

The bursting phenomenon comprises sweep and ejection events where ($u^{'} > 0$, $v^{'} < 0$) are termed 'sweeps' and ($u^{'} < 0$, $v^{'} > 0$)are termed 'ejections'. These burst-events are obtained using VITA conditional sampling technique (Blackwelder and Kaplan (1983)). The number of turbulent burst-events for controlled cases 1 & 2 are reduced by 15-25% in comparison to uncontrolled-case, while for Case 4, the reduction is only by 10%.

5.4. *Power spectrums*

Figures 5(a) to (e) shows the power spectrums of a turbulent signal of $u(at\ z^{+} = 5)$. For the uncontrolled case, the power is concentrated in the lower frequencies while the higher frequencies contain small power almost of the order of 10^{-7}. For the case of monopole actuators, small scales corresponding to higher frequencies contain comparatively much higher power of the order of 10^{-2} to 10^{-4} as can be seen from Figures 5(b) and (c). Thus there is a thousand fold or higher increase in power content of smaller scales compared to the uncontrolled case. For the case of dipole and quadrupole actuators, the power increase at higher frequencies compared to uncontrolled case is only 10 to 10^{2}. Hence, the power spectrums suggest that periodic pressure forcing for a range of wavenumbers leads to inverse energy cascade with a consequent rise in energy content of inertial scales and thereby lowers dissipation of turbulent kinetic energy.

6. Conclusions

In the present study, we performed DNS of the turbulent channel flow in which application of weak-pressure waves in shear layer region have been employed using monopoles, dipoles and quadrupoles, respectively. The results obtained show a 2-5% reduction in skin-friction turbulent drag. On collectively analysing the results for spatio-temporal, statistical and instantaneous quantities, it is observed that the mass-flow rate has been enhanced slightly for all the control cases (except case 4). Moreover, higher spatial density of low-speed streaks, lower rotationality of near-wall vortical structures and less number of bursting events for the controlled cases (except case 4) imply that near-wall turbulence has been reduced leading to a weaker turbulent kinetic energy and thereby reduction in skin-friction drag. The suggested mechanism for the present control scheme implies that

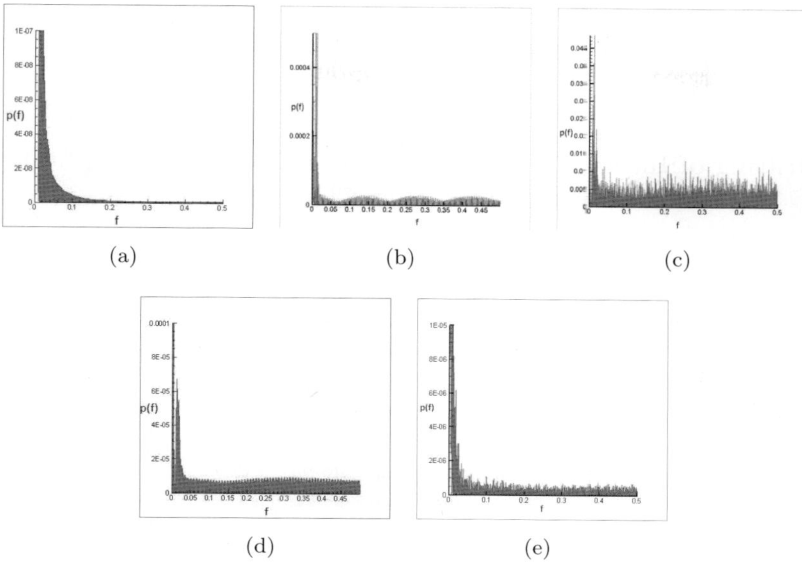

Fig. 5. Comparison of power spectrum for uncontrolled and controlled cases (a) Uncontrolled case, (b) Spanwise Monopoles: Case 1, (c) Spanwise Monopoles: Case 2, (d) Spanwise Dipoles: Case 3, and (e) Spanwise Quadrupoles: case 4, respectively.

weak-pressure forcing leads to an inverse energy cascade which weakens the near-wall turbulence resulting in skin-friction drag reduction.

References

1. Gad-el-Hak, M., Flow Control, *Appl. Mech. Rev.* **42**, 261–293 (1989).
2. Cheng, L. and Armfield, S., A simplified marker and cell method for unsteady flows on non-staggered grids. *Intl. J. for Num. Meth. in Fluids* **21**, 15–34 (1995).
3. Ahmad H., Baig M. F. and Fuaad P. A., Numerical investigation of turbulent-drag reduction induced by active control of streamwise travelling waves of wall-normal velocity. *European J. Mech. - B/Fluids* **49**, 250–263 (2015).
4. Russell D. A., Titlow J. P., and Bemmen Y., Acoustic monopoles, dipoles, and quadrupoles: *AJP J.* (1999).
5. Blackwelder, R.F., Haritonidis, J.H., The bursting frequency in turbulent boundary layers. *J. Fluid Mech.* **132**, 87–103 (1983).

Drifting of Internal Gravity Wave in a Non-Boussinesq Stably Stratified Turbulent Channel Flow

S. M. Yahya* and S. Sanghi

Applied Mechanics Department, Indian Institute of Technology Delhi,
New Delhi,110016, India
**E-mail: yahya_syed@rediffmail.com*

S. F. Anwer

Mechanical Engineering Department, Aligarh Muslim University,
Aligarh, 202002, India
E-mail: sfahadanwer@zhcet.ac.in
www.amu.ac.in

In a stably stratified turbulent flow, computation based on Boussinesq assumption (uniform fluid properties) for strongly temperature dependent fluid properties like viscosity and thermal conductivity results in an inaccurate estimation of momentum and heat transfer coefficients. Role of thermo-fluid properties adversely affect the turbulent strucures and shows evidence of relaminarization [1] at the hot wall of the channel. In this article the role of internal gravity wave, its behavior in Bousinesq and non-Boussinesq situation is elaborated via turbulent structure and energy spectra for turbulent channel flow of air with varying viscosity and thermal conductivity. Due to temperature dependent properties internal gravity wave shifted towards hot wall from the core of the channel.

Keywords: Channel flow; non-Boussinesq; Internal gravity wave; turbulence.

1. Introduction

The flow generated by the combined effect of buoyancy and pressure gradient in a confined channel under high temperature is of great importance in research [2,3]. In such stratified turbulent flows controlling of momentum and heat transfer efficiency has found wide range of application in industries. These flows also have utility in geophysical and environmental applications. In literature, we have only stably stratified flows analysed using Oberbeck-Boussinesq (hereafter referred as OB) approximation (i.e, the density varies only through gravitational body force linearly with temperature) except Lessani & Zainali [3]. However this approximation is not quite good practically when fluid is subjected to high temperature gradient (Ahlers *et al.*[4–6]). Many researchers[7–9] try to emphasize the stable stratification in a channel but they all uses constant property or quasi-variable assumption. However, literature still lacks in an insight physics of stably stratified flow of air under high temperature gradient. The aim of this work is to contribute to the physical understanding the effects of non-uniform physical properties on the flow structure (specially internal gravity wave) in a turbulent channel flow. In this study, a $\Delta T_{walls} = T_h - T_c = 300K$ ($T_h = 600k$ and $T_c = 300k$), $Re_\tau = 150$ and Grashof number $Gr = 1.12 \times 10^7$. The flow is driven by a constant pressure gradient along the stream wise 'x' direction. The reference geometry consists of two horizontal (infinite) flat parallel walls (shown in Fig. 1) and the x-, y- and z-axes point in the stream-wise, span-wise and wall-normal directions. The typical channel dimensions considered are $4\pi\delta \times \frac{4\pi\delta}{3} \times 2\delta$ with $\delta = 1$.

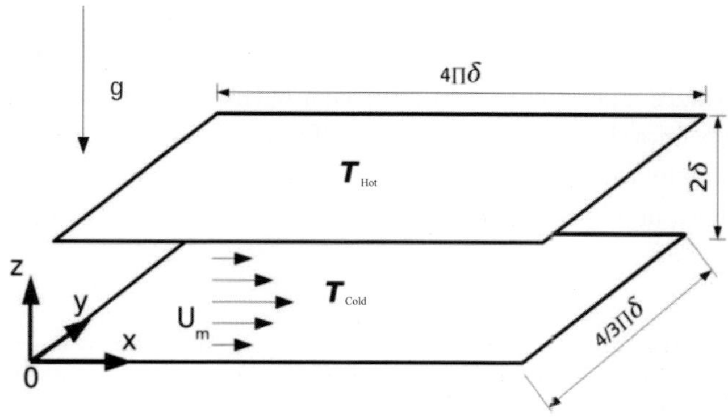

Fig. 1. Schematic of channel flow

2. Governing Equations and Numerical Modelling

The governing equations for stably stratified channel flow after Favre averaging and implicit filtering can be written as[10]

$$\frac{\partial \bar{\rho}}{\partial t} + \frac{\partial \bar{\rho}\tilde{u}_j}{\partial x_j} = 0 \tag{1}$$

$$\frac{\partial (\bar{\rho}\tilde{u}_i)}{\partial t} + \frac{\partial (\bar{\rho}\tilde{u}_i\tilde{u}_j)}{\partial x_j} = -\frac{\partial \bar{P}^{(1)}}{\partial x_i} + \frac{1}{Re_\tau}\frac{\partial \tilde{\tau}_{ij}}{\partial x_j} + F\delta_{i1} - \bar{\rho}Ri_\tau\delta_{i3} \tag{2}$$

$$C_p\left[\frac{\partial\bar{\rho}T}{\partial t}+\frac{\partial\bar{\rho}\tilde{u}_j\tilde{T}}{\partial x_j}\right]=\frac{1}{RePr}\frac{\partial}{\partial x_j}\left((\kappa+\kappa_t)\frac{\partial\tilde{T}}{\partial x_j}\right)+\frac{(\gamma-1)}{\gamma}\frac{d\bar{P}^{(0)}}{dt} \tag{3}$$

$$\bar{P}^{(0)}=\bar{\rho}\tilde{T} \tag{4}$$

$$\frac{\partial\bar{P}^{(0)}}{\partial x_i}=0 \tag{5}$$

where $\tilde{\tau}_{ij} = 2(\mu+\mu_t)\left(2\tilde{S}_{ij}-\frac{2}{3}\tilde{S}_{kk}\delta_{ij}\right)$ and $Ri_\tau = \frac{Gr}{Re_\tau^2}$ is the friction Richardson number with Gr being Grashof number. where symbols have there usual meaning (see these papers[3,10] for more detail). In the present work, For modeling sub-grid stresses, we have used Wall Adaptive Layer Equation (WALE) model suggested by Nicoud and Ducros[11] while for thermal part, we have used a dynamic Smagorinsky model. he employed scheme is an extension to pressure correction method proposed for variable density flows under low Mach number approximation, Anwer *et al.*[12] and Yahya *et al.*[13]. It utilizes a staggered grid approach with second order time stepping of Adam Bashforth with only central differencing is utilized for evaluating convective terms while viscous terms are implicitly calculated. Pressure correction is obtained by solving a Pressure Poisson equation using a multi-grid solver.

3. Results and Discussion

3.1. *Turbulent structure*

The dynamics of centre of the channel is entirely different for stably stratified flow from non-buoyant flow. The most unstable region of the flow lies close to the boundaries where large temperature gradient exist and dominant features are large eddies originate from there. Consider the upper hot wall: there are intermittent eruptions of hot fluid away from the wall. Colder fluid tries to mix with the fluid in an eruptions (hot fluid) but it is forced back to its original position by the action of gravitational force. This makes the situation oscillating about mean position and form the thick region associated with wave near the core which blocks the transport of momentum and heat transfer across the channel. Temperature contour is the the most common way for identifying the the internal gravity waves(IGW) at the center: temperature advection through velocity replicates the effects and physical structure of the flow field prudently. Comparison were made with buoyant flow having uniform fluid properties to that of varying viscosity and thermal conductivity. In Fig. 2(a), the case with constant property assumption is depicted via the temperature contours in a 2D plane on $Ri_\tau = 498$. It is clear that IGW and turbulence appear simultaneously in the core of the channel, whereas bursting process related with turbulence marks the near wall region. As observed from Fig. 2(a) internal waves and thermocline (a thick interface where large temperature gradient exist) coincided in the same region, confirm the occurrence of IGW in a thermocline[14]. Thermocline reduces the mixing efficiency by hampering the fluid lumps as it does not allow to pass through by this thick interface. In buoyancy affected shear flow, high temperature fluid lumps are not covered by low temperature fluid as in the case of non-buoyant flow and keep the hot plumes away from the interface resulting in a reduced mixing. Non-Boussinesq case is shown in Fig. 2(b), where a strong asymmetry and a clear demarcation of relaminarized flow is observed with the eviction of thermocline (IGW) towards the hot wall. These results are also analytically explained with scaling arguments. It is obvious that viscosity increases at the hot wall results in an decrease of Reynolds number, which makes a large surge in the local Richardson number ($Ri_\tau = \frac{Gr}{Re_\tau^2}$) at the hot wall and depletion in its value at cold wall. This makes the situation strongly stratified at the hot wall (locally relaminarized) and mildly stratified at the cold wall. Also thermal conductivity plays the same role it increases the the temperature drop across the thermal boundary layer on the hot side, eventually increases the Grashof number locally beacuse it depends on ΔT_{BL} (temperature accross thermal boundary layer). Again it assisting in local relaminarization at the hot wall because of strong stratification ($Ri \propto Gr$) there, however at the cold wall turbulence prevails. At higher Reynolds number the effects of viscosity fades out and the relaminarization is solely due to the thermal conductivity variation. Since both μ and κ have positive dependence, so the observed effects (local laminarization at hot wall) are same and thermal conductivity pays the toll for μ at high values of Reynold number.

3.2. *Energy spectra*

To speculate the effect of buoyancy and temperature dependent fluid properties on the structure of the flow, we have evaluated the streamwise spectra of wall normal velocity fluctuations[8,15]. For this we compute the streamwise energy spectra of w'^2 at the hot wall, cold wall and the centerline of the

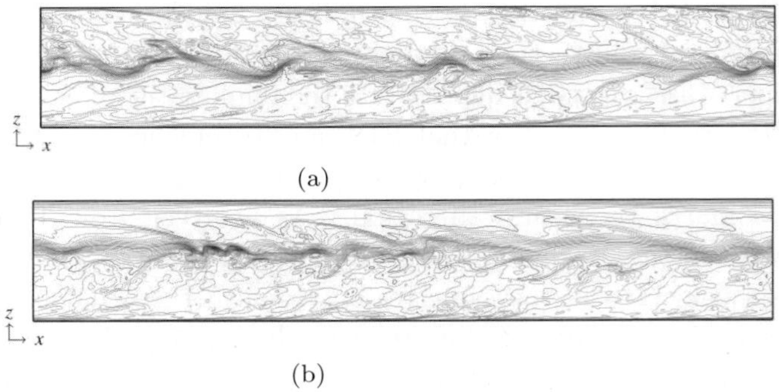

Fig. 2. The contour of temperature field (a) constant fluid properties, (b) varying fluid properties at $Ri_\tau = 498$ and $Re_\tau = 150$.

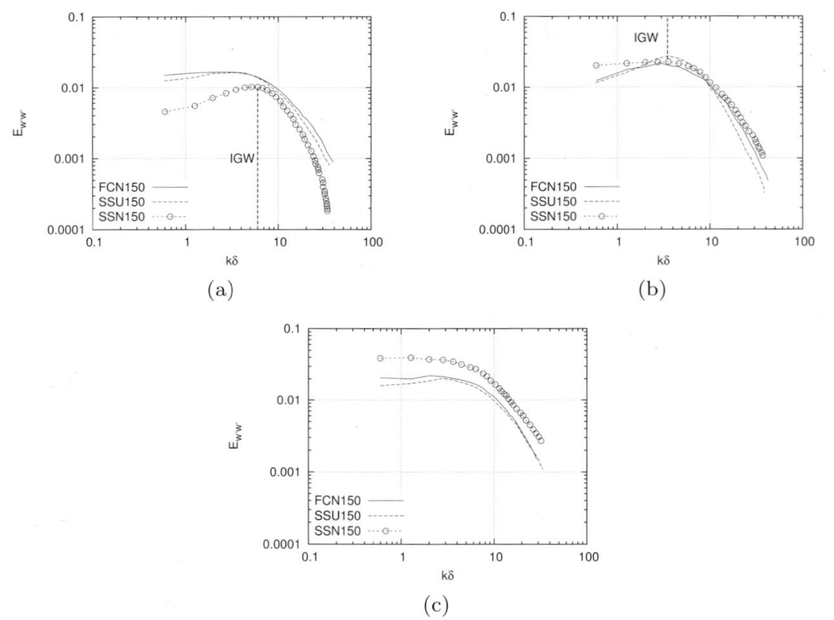

Fig. 3. Energy spectra of wall normal velocity fluctuations at $Ri_\tau = 498$ for three different position (a) $z^+ = 200$ (hot wall), (b) $z^+ = 150$ (centreline), (c) $z^+ = 100$ (cold wall) at $Re_\tau = 150$.

channel to enriched the physical understanding of stable stratification. In the Fig. 3, we clearly mentioned the three situation of the flow namely, non-buoyant (FCN), Boussinesq (SSU) and non-Boussinesq (SSN) and discussed their effects on spectra. At first we comapre the case of constant property stratified flow with forced convective turbulence, and deduced that due to the influence of buoyancy the energy spectra decreases outside the central region (Fig. 3(a&c)) at low wavenumber (k), emphasizing the sole effect of buoyancy in selective modulation of turbulent scales with more impact on slower (low wavenumber) and larger scales of the flow. However at the center (see Fig. 3(b)) their is a increase in the value of energy spectrum for a range of $1 < k\delta < 4$ with peak occuring at $k\delta \simeq 2.6$ reflecting the impression of IGW. Considering the case with non-uniform fluid properties, larger effects are observed outside the central region of the domain especially on the hot side where energy spectrum first increase upto $k\delta \simeq 3.7$ and then decrease sharply. This non-monotonic behavior is due to the presence of IGW on the hot side of the channel.

References

1. R. Narasimha and K. R. Sreenivasan, Relaminarization of fluid flows, *Adv. App. Mech.* **19**, 221 (1979).
2. J. S. Turner, *Buoyancy Effects in Fluids*, 1 edn. Cambridge univ. press, United Kingdom (1973).
3. B. Lessani and A. Zainali, Numerical investigation of stably stratified turbulent channel flow under non-boussinesq conditions, *Phys. fluids.* **10**, 8 (2009).
4. G. Ahlers, E. Brown, F. F. Araujo, D. Funfschilling, S. Grossmann and D. Lohse, Non-oberbeckboussinesq effects in strongly turbulent rayleighbenard convection, *J. Fluid Mech.* **569**, 409 (2006).
5. A. Sameen, R. Verzicco and K. R. Sreenivasan, Non-boussinesq convection at moderate rayleigh numbers in low temperature gaseous helium, *Phys. Scr.* **T132**, 014053 (2008).
6. P. Urban, P. Hanzelka, V. Musilova, T. Kralik, M. L. Mantia, A. Srnka and L. Skrbek, Heat transfer in cryogenic helium gas by turbulent rayleighbnard convection in a cylindrical cell of aspect ratio 1, *New J. Phys.* **16**, 053042 (2014).
7. V. Armenio and S. Sarkar, An investigation of stably stratified turbulent channel flow using large-eddy simulation, *J. Fluid Mech.* **459**, 1 (2002).
8. O. Iida, N. Kasagi and Y. Nagano, Direct numerical simulation of turbulent channel flow under stable density stratification, *Int. J. Heat Mass Transfer* **45**, 1693 (2002).

9. R. P. Garg, J. H. Ferziger, S. G. Monismith and J. R. Koseff, Stably stratified turbulent channel flows. i. stratification regimes and turbulence suppression mechanism, *Phys. Fluids* **12**, 2569 (2000).
10. S. M. Yahya, S. F. Anwer and S. Sanghi, Phenomenological and statistical analyses of turbulence in forced convection with temperature-dependent viscosity under non-boussinesq condition, *Eur. Phys. J. E* **36**, 120 (2013).
11. F. Nicoud and F. Ducros, Subgrid-scale stress modeling based on the square of velocity gradient tensor, *Flow, Turb. and Combust.* **62**, 183 (1999).
12. S. F. Anwer, H. N. Khan, S. Sanghi, A. Ahmad and S. M. Yahya, Extension of smac scheme for variable density flows under strong temperature gradient, *AIP Conf. Proc.* **1440**, 683 (2012).
13. S. M. Yahya, S. F. Anwer and S. Sanghi, A conservative pressure correction method on collocated grid for low mach number flows, *World J. Mech.* **5**, 253 (2012).
14. J. H. Ferziger, J. R. Koseff and S. G. Monismith, Numerical simulation of geophysical turbulence, *Comput. Fluids* **31**, 557 (2002).
15. K. Yeo, B. G. Kim and C. Lee, Eulerian and lagrangian statistics in stably stratified turbulent channel flows, *J. Turbulence* **10**, 1 (2009).

Numerical Study of Sink Flow Turbulent Boundary Layers

Saurabh S. Patwardhan and O. N. Ramesh
Department of Aerospace Engineering,
Indian Institute of Science, Bangalore, India

Direct numerical simulations of sink flow turbulent boundary layers are performed. These simulations have been performed at higher Reynolds numbers than the fairly recent experimental sink flow studies of Dixit and Ramesh (2008). The mean velocity profiles are found to be logarithmic, albeit with non-universal constants. The structure of the outer layer in a sink flow shows important differences with respect to a corresponding zero pressure gradient turbulent boundary layer.

1. Introduction

Research in the area of turbulent boundary layers has traditionally focussed on zero-pressure-gradient (ZPG) turbulent boundary layer (TBL) where there is no external pressure gradient influencing it. ZPG TBL is referred to as a canonical boundary layer given the experimental as well as computational ease with which it can be simulated. An excellent summary of the physical mechanisms at work in an ZPG TBL is given by Sreenivasan[1]. However, boundary layer flows with non-zero streamwise pressure gradients have not received as much attention as ZPG TBLs in the literature. In particular, we consider a sub class of favourable pressure gradient boundary layers which are known to be the only smooth-wall boundary layer in equilibrium – the so-called sink flow boundary layer. The sink flow boundary layer is a turbulent counterpart of Falkner-Skan boundary layer flows.

Sink flow boundary layers fall in a category of boundary layers with "exact equilibrium"[2] which differentiates it from the boundary layers with approximate equilibrium. The differentiation is due to Reynolds number effects or the constancy of Reynolds number in sink flow boundary layers which ensures that the flow is self-preserving even in the region close to the wall[3]. Townsend[2] and Rotta[4] showed that the only condition under which the exact equilibrium is possible in a turbulent boundary layer is for the converging flow between two plates. Sink flow boundary layer has many interesting properties apart from self preservation or self-similarity of the mean velocity profile, Reynolds stress and other turbulence statistics. The Reynolds number defined by any local velocity and length scale is invariant along the streamwise flow. The skin friction coefficient (C_f) and pressure gradient parameter K (defined as) are also constants along the stream.

Dixit and Ramesh[5] observed that the logarithmic law in the mean velocity profile has a non-universal behaviour in sink flows. The log-law constants showed a systematic departure from their standard values. The self-similarity of velocity profiles was observed not only in the conventional inner and outer scaling but also in the laminar-like scaling. A higher order asymptotic matching revealed the pressure-gradient dependence of log-law constants of the velocity profile. Dixit and Ramesh[6] studied large-scale structures in sink flow TBLs. The observations about large-scale structures were based on the wall shear stress-streamwise velocity correlations in the flow. They observed that the structure angle is a function of pressure gradient parameter and decreases with increasing intensity of pressure gradient. The structure also gets elongated in the streamwise direction with increasing pressure gradient. An analysis of near-wall turbulent motions showed that the sweep events are less in sink-flow boundary layers compared to ZPG TBL of the same Reynolds number. The sink flow boundary layer also shows difference in the diffusion of turbulent kinetic energy (TKE) towards the edge of the boundary layer. In the case of ZPG TBL, the TKE diffusion suddenly becomes large[7,8]. The spectra of these sink flow boundary layers also showed the existence of k_1^{-1} region[10] which is observed in very high Reynolds number ZPG flows[11].

The pioneering work on simulations of sink flow TBLs was done by Spalart[8]. Recently, Yuan and Piomelli[9] have studied sink flow boundary layers numerically. However, their work was focussed on rough wall boundary layers. In this paper, two different sink flow boundary layers are studied and analyzed using direct numerical simulations. The K values studied in this work are 7.71×10^{-6} and 2.71×10^{-7} in the increasing order of Reynolds number. This first corresponds to PG1 case of Dixit[7] and the latter is the highest Reynolds number case studied by Jones *et al.*[12]. This Reynolds number is the highest simulated till now for the sink flow boundary layers. The aim of the present study is to generate a comprehensive database of the sink flow boundary layers. We specifically look at the non-universal nature of logarithmic laws in the mean velocity profile of sink flow TBL. This non-universal nature was earlier observed in experimental study of Dixit and Ramesh[5] at lower Reynolds numbers. In the present study, we investigate a higher Reynolds number sink flow to see whether non-universality exists at lower pressure gradients. Further, the outer layer structure of sink flow TBL is examined. The paper is organized as follows – Section 2 describes basic characterization of simulated sink flow TBLs and numerical method. Section 3 presents results of simulation and section 4 presents the main conclusions of the paper.

2. Basic Characterization and Validation of Sink Flow TBL

The sink flow turbulent boundary layer is self preserving and has self-similar mean velocity profile along streamwise direction. The Reynolds number based on any length scale of the flow also remains constant in the flow. The other boundary layer parameters like shape factor, skin friction coefficient are also constant in these flow. Different integrals parameters for these flows are shown in Table 1. The mesh used for these two simulations consisted of 1536 points in streamwise direction, 344 points in wall-normal direction and 512 points in the spanwise directions. These numbers were arrived at after grid-independence exercise. The domain extended 210 l_0, 60 l_0 and 70 l_0 in the streamwise, wall-normal and spanwise directions respectively where l_0 is the reference length scale. The reference length scale is about 0.65 θ_{in} where θ_{in} is the normalized inlet momentum thickness which is roughly same in both cases. The reference length scale in flow S1 is 2.3 times the reference scale for flow S1. The resolution in terms of inner co-ordinates and other details are shown in Table 2. The resolution is found to be satisfactory and is as fine as the one used for simulations[13].

Table 1: Details of integral parameters for the sink flow turbulent boundary layers.

Sink flow	K	C_f	R_θ	δ_+	H
S1	7.71×10^{-7}	0.004121	1231	788.5	1.34
S2	2.71×10^{-7}	0.003337	2960.9	1726.4	1.286

Table 2: Details of computational mesh used for direct numerical simulations of sink flow turbulent boundary layers. N_x,N_y,N_z denote the number of mesh points in each direction. L_x,L_y,L_z are the domain lengths in respective directions. Re is the Reynolds number for based on reference length and velocity scale.

Sink flow	(N_x,N_y,N_z)	$(L_x,L_y,L_z)/\ \theta_{in}$	$Re=U_0 l_0/\nu$	$(\Delta x_+, \Delta z_+)$	Δy_{+min}
S1	1536,344,512	(323,92.3,107.7)	800	(5.25,5.25)	0.2
S2	1536,344,512	(323,92.3,107.7)	1846	(12.11,12.11)	0.46

Sink flow boundary layers exhibit a single-layer scaling like laminar boundary layers[15,5]. Hence, the streamwise momentum equation can be converted to an ordinary differential equation (ODE). The derivation of the ODE is shown in the next section where the equilibrium conditions for more generalised equilibrium flows with wall transpiration are derived. The equation can be written as,

$$f'' + 1 - f'^2 - g' = 0 \tag{2.1}$$

where the notations have standard meaning from laminar boundary layer, here f is a non dimensional mean stream function and primes refer to derivatives in the wall-normal direction. Mean velocity is given by and $f = U / U_\infty$ the Reynolds stress gradient is represented by g' where $g = \frac{\overline{u'v'}}{U_\infty^2 \sqrt{K}}$. Rearranging and integrating Eq. 2.1,

$$-\frac{\overline{u'v'}}{U_\infty^2} = \sqrt{K}\left(f'_w - f' + \int_0^\eta (1-f^2)\,d\eta \right) \qquad (2.2)$$

where f'_w is velocity gradient at the wall. From the above equation, a relation between Reynolds shear stress and mean velocity can be obtained (see Eq. 2.2). A similar expression was derived by Jones & Launder[15] where the limits of integral were from η to ∞ instead of 0 to η in the present case. Figures 1(a) and 1(b) show a comparison of the Reynolds shear stress calculated using the above expression and the mean velocity profile obtained from the DNS. The comparison of the two in simulated flows is excellent depicting the accuracy of the calculation and ensuring that the grid resolution is adequate. Interestingly, Eq. 2.2 is a relation between the mean velocity and Reynolds shear stress. A relation between mean velocity and Reynolds shear stress lies at the heart of modeling turbulent flows using an eddy-viscosity approach.

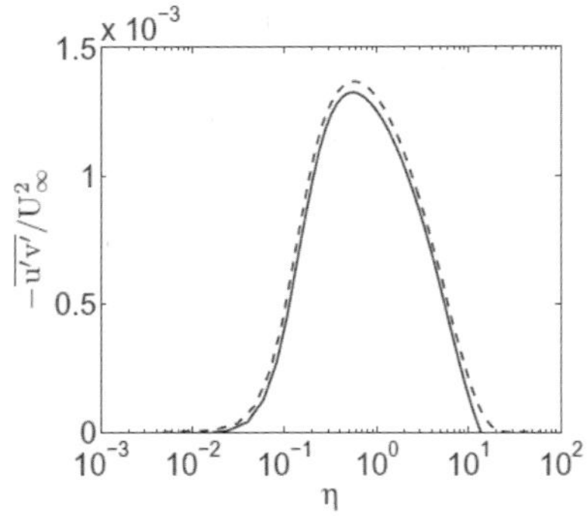

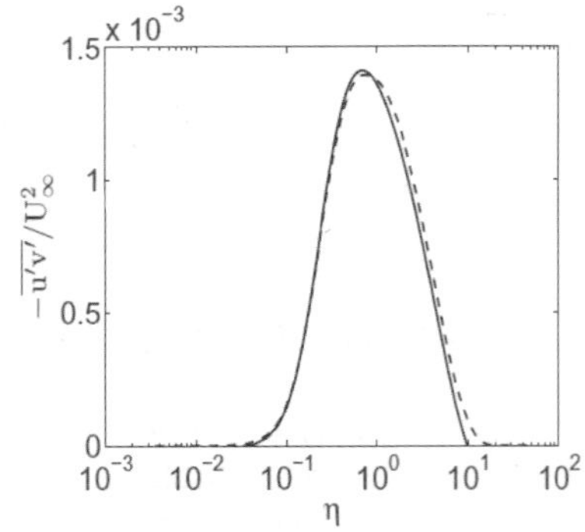

Figure 1: Comparison of Reynolds shear stress in sink flow TBL (a) K = 7.71 × 10^{-7} (b) K = 2.71 × 10^{-7}. Dotted line shows DNS data while continuous line shows Reynolds stress calculated from Eq. 2.2.

3. Results and Discussion

Figure 2(a) shows the variation of Reynolds number with pressure gradient parameter K. The solid line corresponds to the equilibrium curve[12] for the sink flow boundary layers. The data points of experimental results[5] are also plotted along with the present DNS results. The agreement is fairly good. The log-law in the mean velocity profile was assumed to have universal constants while deriving the equation for the equilibrium curve[12]. The small deviation from the equilibrium curve can be attributed to non-universality of log-laws[5] which exists even at higher Reynolds numbers which we will see next. Figure 2(b) shows the self similarity of mean velocity profiles for S1 case. The profiles collapse well with each other demonstrating the self-similar nature of the sink flow turbulent boundary layers.

Self-similar profiles were also observed for S2 case but are not shown here. The mean velocity profiles are compared with the experimental data[5]. The experimental and numerical results compare well with each other as seen from figure 2(b).

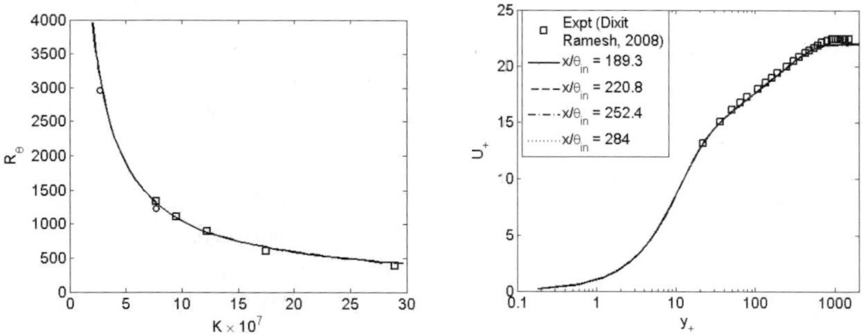

(a) (b)

Figure 2: (a) Variation of Reynolds number based on momentum thickness with pressure gradient parameter K solid line: Equilibrium curve[12]; □ : Experiments[5]; o : present DNS results. (b) Self similarity of mean velocity profiles and comparison with experimental data[5].

3.1. *Non-universality of logarithmic laws*

The mean velocity profile in sink flow TBLs show a logarithmic region which extends well into the outer layer of the flow as shown in figure 3. These logarithmic law constants are different from the so-called "universal constants" of ZPG boundary layers (Dixit and Ramesh, 2008). They also showed a systematic variation of these constants with pressure gradient parameter K. Figure 3 shows the mean velocity profiles for the cases S1 and S2.

It can be observed that the mean velocity profiles of these boundary layers are different from the "universal logarithmic law" both in terms of slope and the additive constant. Further, it can be observed that the slope of the logarithmic portion of these profiles as well as additive constants differ from each other as expected. The higher Reynolds number boundary layer profile has a logarithmic law with slope smaller than the low Reynolds number profile. The so-called Karman constants (inverse of slope) are 0.415 and 0.38 in low Reynolds number and high Reynolds number cases respectively and the additive constants are 6.5 and 5.5 respectively.

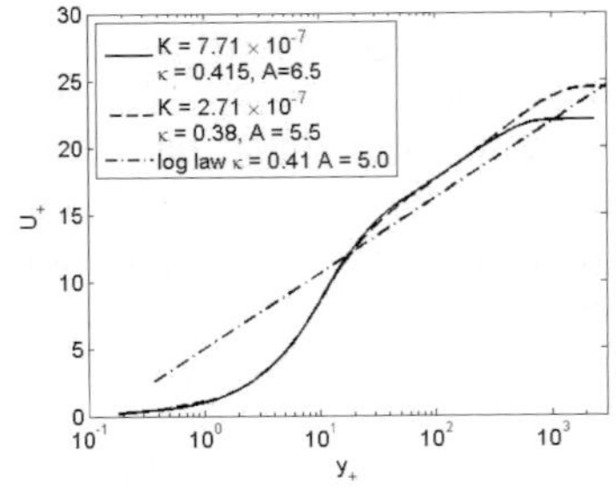

Figure 3: Comparison of mean velocity profiles for the flows S1 ($K = 7.71 \times 10^{-7}$) and S2 ($K = 2.71 \times 10^{-7}$) along with the "universal logarithmic profile".

3.2. *Outer layer structure in sink flow turbulent boundary layers*

Figure 4 shows a comparison of turbulent intensity profile in a sink flow TBL and a zero-pressure-gradient (ZPG) turbulent boundary layer. The details of DNS of a ZPG TBL can be found elsewhere (PhD thesis, Saurabh Patwardhan). The Reynolds numbers based on momentum thickness for the two flows are similar. Re_θ for ZPG is 1432 while for sink flow TBL it is 1231. Two important observations can be made by looking at these turbulent intensity profiles. While the peak of inner-scaled turbulent intensity is roughly of same magnitude, it is shifted towards the wall in the case of sink flow. This is due to the effect of favourable pressure gradient. The second important observation is the absence of outer energy site that occurs between $y_+ = 100$ to 200 in this case of ZPG TBL. For more details on the outer energy site, one can refer to Marusic *et al.*, (2010). Also, the overall turbulent energy levels in the outer layer of the sink flow TBL are lower than the corresponding ZPG TBL which is also presumably due to the pressure gradient. This is also consistent with the observation of Spalart (1986) that the energy budget of ZPG and sink flow TBL are different in the outer layer towards the edge of boundary layer. Turbulent energy transport by turbulent diffusion in ZPG TBL was observed to be 20 times higher compared to that in sink flow TBL (see figure 10 of Spalart (1986)).

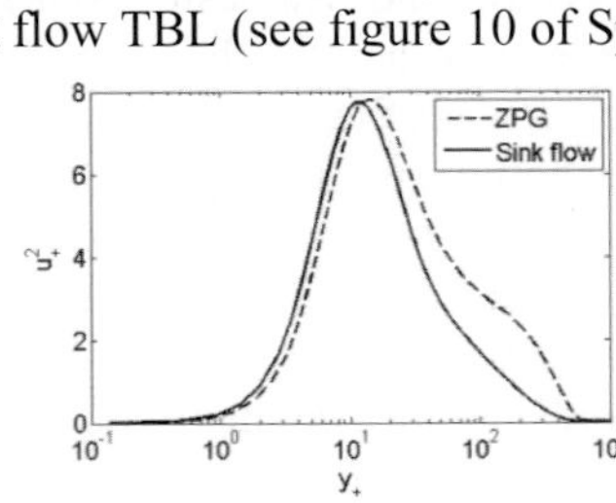

Figure 4: Comparison of turbulent intensity profile in zero-pressure-gradient (ZPG) and sink flow TBL. ($Re_\theta = 1432$ for ZPG and $Re_\theta = 1231$ for sink flow $K = 7.71 \times 10^{-7}$)

4. Conclusions

Properties of sink flow turbulent boundary layer are studied using direct numerical simulations. Sink flows with two different pressure gradient parameter are studied. The sink flow S2 is the highest Reynolds number sink flow simulated till now. The variation of Reynolds number based on momentum thickness shows a very good agreement with the equilibrium curve calculated by Jones *et al.*, (2001). The mean velocity profiles as well as second order statistics show self similarity. The mean velocity profiles for S1 flow show a very good agreement with the experimental data of Dixit and Ramesh (2008). The mean velocity profiles exhibit a non-universal logarithmic law along the expected lines. The Karman constant (inverse of logarithmic profile slope) decreases with decreasing K. The additive constant also decreases with K.
Outer layer structure of sink flow TBL is investigated further. Compared to ZPG boundary layer, turbulence energy levels in sink flow are lower. The outer energy site is also absent in the case of sink flow. The peak values of streamwise turbulent intensity are close to each other in ZPG and sink flow but the location is closer to the wall in sink flows. These effects are presumably due to favourable pressure gradient.

References

1. Sreenivasan, K. R., The turbulent boundary layer. In *Frontiers in Experimental Fluid Mechanics* (ed. M. Gad-El-Hak). Springer, Berlin (1989).
2. Townsend, A. A., Equilibrium layers and wall turbulence. *J. Fluid Mech.* **11**, 97–120 (1961).
3. Nickels, T. B., Equilibrium turbulent boundary layers with lateral streamline convergence or divergence. *J. Fluid Mech.* **623**, 273–282 (2010).
4. Rotta, J. C., Turbulent boundary layers in incompressible flow. *Prog. Aeronaut. Sci.* **2**, 1–220 (1962).
5. Dixit, S. A. and Ramesh, O. N., Pressure-gradient-dependent logarithmic laws in sink flow turbulent boundary layers. *J. Fluid Mech.* **615**, 445–475 (2008).
6. Dixit, S. A. and Ramesh, O. N., Large-scale structures in turbulent and reverse-transitional sink flow boundary layers. *J. Fluid Mech.* **649**, 233–273 (2010).
7. Dixit S. A., Structure of Sink Flow Boundary Layers, PhD Thesis, Indian Institute of Science (2010).
8. Spalart, P. R., Numerical study of sink flow boundary layers. *J. Fluid Mech.* **172**, 307–328 (1986).
9. Yuan J. and Piomelli U., Numerical simulations of sink-flow boundary layers over rough surfaces. *Phys. Fluids*, **26,** 015113 (2014).
10. Dixit, S. A. and Ramesh, O. N., On the k_1^{-1} scaling in sink-flow turbulent boundary layers.*J. Fluid Mech.* **737**, 329–348 (2013).
11. Nickels, T. B., Marusic, I., Hafez, S. & Chong, M. S., Evidence of the k_1^{-1} law in a high-reynolds-number turbulent boundary layer. *Phy. Rev. Lett. 95, 074501*, 1–4 (2005).

12. Jones, M. B., Marusic, I. & Perry, A. E., Evolution and structure of sink-flow turbulent boundary layers. *J. Fluid Mech.* **428**, 1–27 (2001).
13. Wu, X. and Moin, P., Direct numerical simulation of turbulence in a nominally zero-pressure-gradient flat-plate boundary layer. *J. Fluid Mech.* **630**, 5–41 (2009).
14. Marusic, I., McLeon, B. J., Monkewitz, P. A., Smits, A. & Sreenivasan, K. R., Wall-bounded turbulent flows at high Reynolds numbers: Recent advances and key issues. *Phys. Fluids* **22**, 0651031–23 (2010).
15. Jones, W. P. & Launder, B. E., Some properties of sink-flow turbulent boundary layers. *J. Fluid Mech.* **56**, 337–831 (1972).
16. Schlichting, H. & Gersten, K. , Boundary Layer Theory, McGraw Hill, USA (2000).

Coherent Structure in Oil Body Embedded in Compund Vortex

T. O. Chaplina and Yu. D. Chashechkin
A.Yu. Ishlinsky Institute for Problems in Mechanics of the RAS, Moscow, Russia, 119526
E-mail: tanya75.06@mail.ru

Transport of miscible or immiscible admixtures in a compound vortex generated by a uniformly rotating disk located at the bottom of the cylindrical container is investigated. All admixtures form inside and on the surface of vortex flow well outlined coherent structures. Parameters of the structures depend on flow components and conditions of the experiment. All registered coherent structures are rather stable and reproduced. Their patterns change monotonously with variations of the flow parameters.

Keywords: Vortex; Coherent structure; Flows.

Introduction

With the growth and development of industry, more and more chemically active substances and compounds, including environmental hazards, are discharged into the environment (such as dioxin, Bhopal, India, 1984, Seveso, Italy, 1976, discharges into the Amur River, China, 2005, and the accidents with the reprocessed nuclear fuel, e.g. UK Sellafield, 2005). To control the level of pollution and environmental safety, or even to plan for people evacuation the mass transfer from a compact source should be evaluated as applied to prevailing meteorological conditions. The transport of matter is affected by flows, vortices and waves in the atmosphere and hydrosphere.

Theoretical studies of vortex flows are based on the theory of an ideal quiescent[1,2] and rotating fluid, using a priori assumption about the dynamics of the phenomenon. Many works are devoted to study of the passive admixture transfer in the vortex flows. At the same time in some experiments, the admixtures were found in compact areas that have typical forms of "dye walls"[3] and helical filaments. The patterns of vortex structures are widely represented in the well-known albums and books[4,5]. The goal of the investigation is to register the coherent structure of admixture in a compound vortex.

1. Experimental Technique and Parameters of the Flow

The transport of finite volumes of different admixtures introduced on the free surface of water drawn into the vortex motion in the vertical cylindrical container has experimentally been studied. The photo of the experimental setup is shown in Fig. 1(*a*). The vortex flow was induced by a disk located on the bottom of open vertical cylindrical container *1* with diameter 29.4 cm. To decrease optical distortions of the recorded flow pattern, cylindrical container is put into the rectangular container *2* with plane walls. The activator of vortex motion, the flat disk of 2 mm thick *3* is located in the center of the container bottom. To flatten the bottom of the cylindrical container on the level of the upper edge of the disk, the false bottom *4* is mounted in some experiments. The photo- and video-recordings of the flow pattern are performed simultaneously from the top 5 and side 6 views. The control of the experiment and the data recording has been carried out by the PC 7. The water volume with the admixture in the container is illuminated by white-light source with dispersive sheet 8.

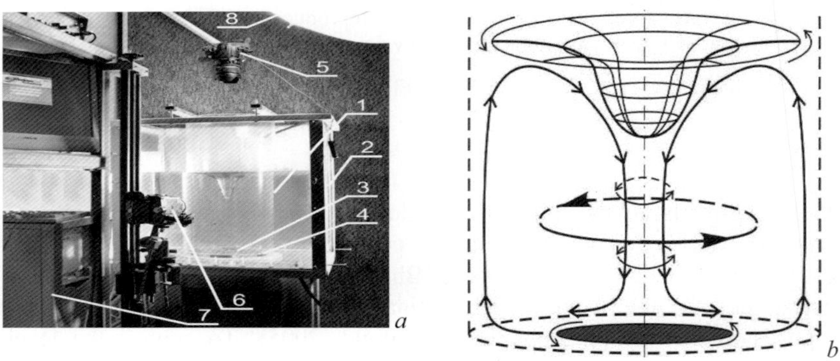

Fig 1. Photo of the experimental set-up (*a*) and the scheme of the compound-vortex, (*b*) induced by the rotating disk in the cylindrical container.

The complex flow in the container, the schematic pattern of which is shown in Fig. 1(*b*), contains both the vortex and wave components. The observed flow pattern is schematically reduced to a combination of two vortices, one of which is vertical and cylindrical with the angular velocity ω_c, and the second vortex is toroidal ω_t, with a circular axis encircling the central vertical axis. As a result of their coupled action, a compound vortex is formed; within it, liquid particles move along spiral and helical trajectories with the general characteristic frequency $\omega = \omega_c + \omega_t$. The ratios of the parameters of the experiment form the traditional set of dimensionless parameters, including Reynolds number $\mathrm{Re} = (R^2\Omega)/\nu$, Froude number $Fr = (R^2\Omega^2)/gH$, Weber number

$We = \rho L^3 \Omega^2 / \sigma$, and for two-layer or two-component medium (water, oil) – and the Bond number $Bo = gH^2(\rho_1 - \rho_2)/\sigma$ and Atwood number $A = \rho_1 - \rho_2 / \rho_1 + \rho_2$, where ρ_1 and ρ_2 – density of fluid components. The flows studied in these experiments are characterized by Reynolds numbers in range 5000-50000, Froude number – 50-1800, Atwood number –0.009-0.20, Bond number – 1.0-4.5. The variety of specific dimensional and dimensionless parameters of the problem highlights the complexity and transience of the studied flow that is taken into account in designing the experimental setup and methods.

2. The Compound Vortex in Clean Water

In these series of experiments, the radii of the container R_o and disc-activators R, the depth of the layer of fluid H were constant. The size and shape of the surface of the cavity depend on the values of the radial and tangential component of fluid velocity at the edge of the disk. At relatively low frequencies of rotation of the disk, the activator forms a smooth surface in the cavity (Fig. 2(*a*)). The depth of the crater increases monotonically toward its center, the lower part has a parabolic shape, typical of rigid-body rotation of the liquid. With increasing frequency of rotation of the disk, shape of the free surface gets complicated, and ceases to be smooth on the surface. Large-scale perturbations with large amplitude are clearly visible Fig. 2(*b*). In the central part of the cavity, shape of the liquid surface is distorted and spatially periodic perturbations with sharp crests and troughs start appearing.

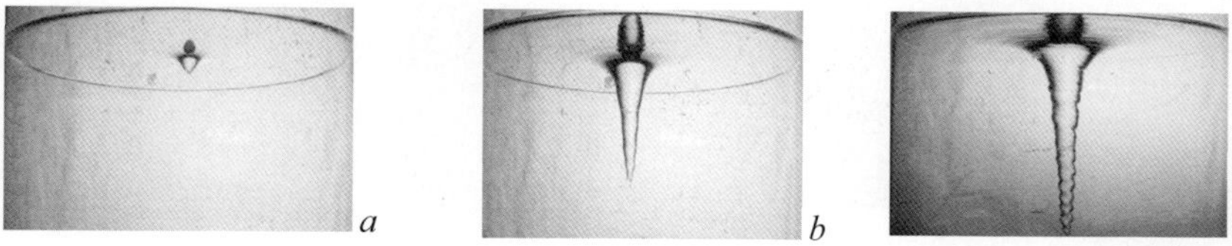

Fig. 2. The characteristic shape of the surface of a fluid over a rotating disk (H=40 cm, R=2.5cm): (a-c), Ω = 430, 980, 1530 RPM.

The third type of surface liquid in a cavity with surface perturbations of two types – large and small-scale inertial spiral waves – there is a further increase in the frequency of rotation of the disk-activator (Fig. 2(*c*)).

3. Evolution of Coloured Miscible Fluid Structure Inside and on the Surface of Compound Vortex

It is well known that a drop of a dye placed on the liquid surface is transformed into cascade of sinking dye rings connected by arcs[6]. The existence of two spirals extended from two different spots of dye is illustrated on the Fig. 3. Drop of blue ink was put on the free surface 2sec later than purple ink drop. The number of spiral revolutions grows with time. The distances from the center are 1 and 3 cm, respectively. The inner spiral grows faster and in 7sec duration has 1.5 revolutions; the outer spiral has no complete revolutions.

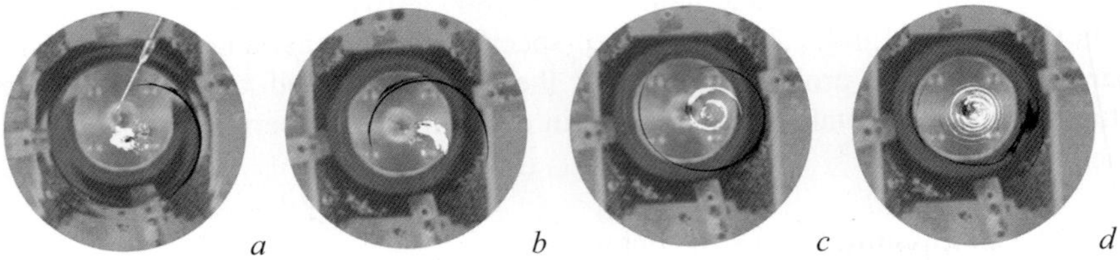

Fig. 3. Changing the region occupied by the spiral arms on the surface of the compound vortex: (a-d) t= 2, 4, 7, 23 sec (H=40cm, Ω = 180 RPM, R=5cm).

The results of the other series of experiments show the pattern of dye penetration into the liquid body (Fig. 4). The dye drop is introduced on the free surface in 4.0 cm distance from the centre of rotation and immediately formed "dye wall" (Fig. 4(*a*)). By the time of surface spiral structure formation, the dye forms two cylindrical envelopes in the liquid layer depth (Fig. 4(*b*)). Separate filaments of each color fluid exist for a long time.

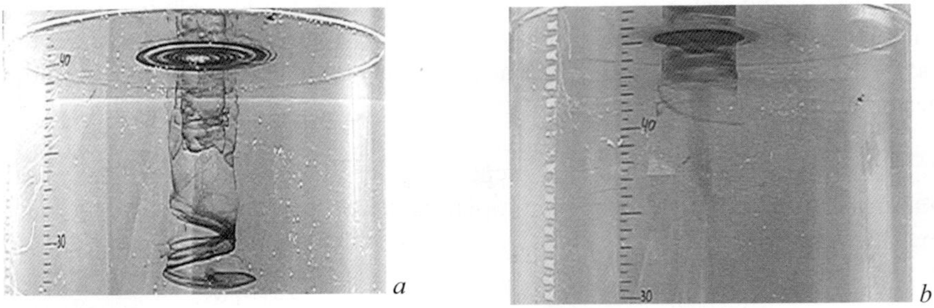

Fig. 4. Separate dye structures (H=40 cm, R=5 cm): (a-b) Ω = 200, 800 RPM.

All the experiments with different types of miscible admixtures considered here showed stability of the dye distribution patterns, their structure and dynamics on the free surface and in the liquid depth.

4. Transport of Immiscible Admixture

In compound vortex the flow oil patch with smooth boundary placed onto free surface is also transformed into a set of spiral arms and separate[7]. Inside the fluid the oil is aggregated into smooth body of revolution, shape of which is inquired in[8].

As shown in the experiments, dynamics and the pattern of distribution of oil on the surface depend on the experimental conditions and the properties of the oil. The spot of lightweight oil (150 ml) partially remains on the surface, partially assembled in the central oil body at moderate angular speeds (Fig. 5(*a*)). The dynamic state of the bulk liquid free surface structure and shape of individual elements are changing with increasing speed to 1170 RPM (Fig. 5(*b*)). We can observe as oil spiral arms, separated by thin strips of water and oil extended body in the flow pattern over a disk of smaller radius (5 cm).

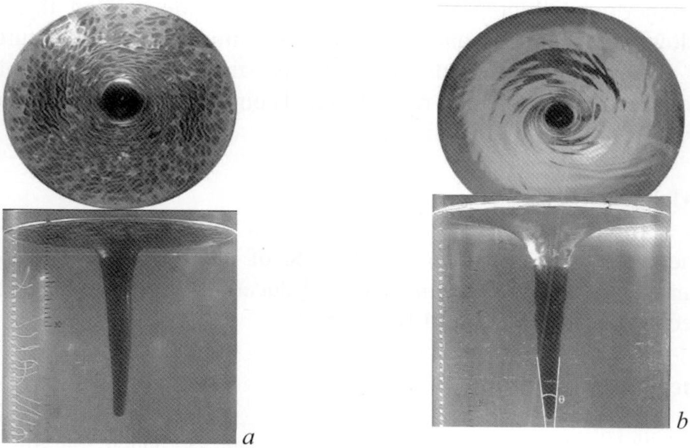

Fig. 5. Compound vortex with a portion of sunflower oil Vo = 150ml, Hw = 40 cm: a) R=5.0cm, Ω=320 RPM; b) R=7.5 cm, Ω=1170 RPM (side and top view).

Comparison oil body sizes depending on the number and physical impurities immiscible with different initial values of the depth of the water layer indicates that an increase in viscosity of the detection agent leads to a decrease in vertical resolution oil body. In pure water a cavity is formed, the cross section of the central part of which is close to the classical parabolic Rankine vortex, the origin is placed on the lower pole of the cavity (Fig. 6(*a*)). The spiral waves are observed on the liquid surface. Addition a small amount of oil will increase the depth of the cavity and a change of its shape (Fig. 6(*b*)).

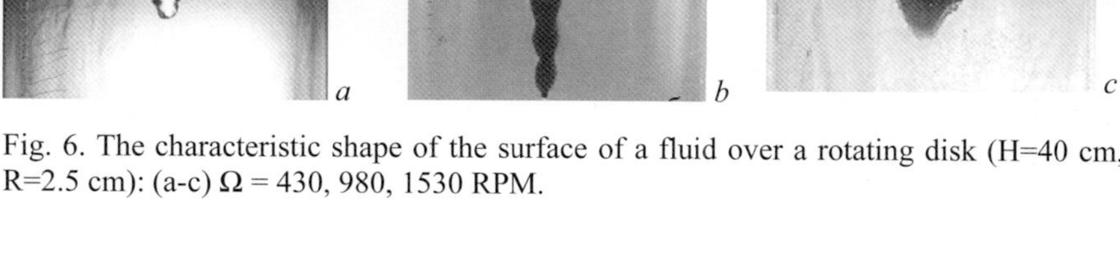

Fig. 6. The characteristic shape of the surface of a fluid over a rotating disk (H=40 cm, R=2.5 cm): (a-c) Ω = 430, 980, 1530 RPM.

The portion of the light impurities is drawn into the water, where oil forms a rotating body at promotion in two-layer fluid (Fig. 6(*c*)). Contact surface oil - air remains smooth at low angular speeds disk. Inertial and capillary waves appear on the surface with increasing of rotation speed.

Conclusions

The transport of miscible and immiscible fluids in a compound vortex was studied experimentally. All admixtures formed inside and on the surface of vortex flow well outlined coherent structures. Patterns of the structures depend on flow components and conditions of the experiment. All registered coherent structures are rather stable and reproduced. Their patterns change monotonously with variations of the flow parameters.

Acknowledgments

Experiments are performed on Stand VTK of unique facility «GFK IPMech RUS» under support of RF Ministry on Education and Science. The study is supported by RFBR Grant № 14-01-00015a.

References

1. L. M. Milne-Thomson, *Theoretical hydrodynamics*, London, McMillan and Co. LTD, 4-th edn. 632 (1960).
2. K. V. Koshel, S.V. Prantz, *Chaotic advection in the Ocean*, Moscow-Izhevsk, 360 (2008).
3. Taylor G. I. *Experiments with rotating fluids*, Roy. Soc. Proc. A. 114-12, (1921).
4. M. Van Dyke, *An Album of Fluid Motion*, Parabolic Press, 176 (1982).
5. G. K. Batchelor, *An Introduction to Fluid Dynamics*, Cambridge: CUP, 61 (1967).
6. Yu. D. Chashechkin, T. O. Chaplina, E. V. Stepanova, *Transport of admixture in a stationary vortex flow,* ICTAM, Book of abstracts, 312–313 (2012).
7. E. V. Stepanova, T. O. Chaplina, Yu. D. Chashechkin, *Transport of oil in a compound vortex*, Fluid Dyn., **46**, (2), 214–224 (2011).
8. T. O. Chaplina, *Transport of Oil in a Compound Vortex* // Wave in Fluids: effects of non-linearity, rotation, stratification and dissipation. IUTAM symposium 12-3, 45–48 (2012).

Quantitative Characterization of Single Orifice Hydraulic Flat Spray Nozzle

D. M. Sharma†
TSI Instruments India Pvt. Ltd,
Bangalore, Karnataka 560037, INDIA
†E-mail: deepak.sharma@tsi.com
www.tsi.com

W. T. Lai
TSI Inc., 500 Cardigan Road
Shoreview, MN 55126, USA
E-mail: wlai@tsi.com

The single orifice hydraulic flat spray nozzle was evaluated with two global imaging techniques to characterize various aspects of the resulting spray. The two techniques were high resolution flow visualization and Particle Image Velocimetry (PIV). A CCD camera with 29 million pixels was used to capture shadowgraph images to realize ligament formation and collapse as well as droplet interaction. Quantitative analysis was performed to give the sizing information of the droplets and ligaments. This camera was then applied with a PIV system to evaluate the overall velocity field of the spray, from nozzle exit to droplet discharge. PIV images were further post-processed to determine the inclusion angle of the spray. The results from those investigations provided significant quantitative understanding of the spray structure. Based on the quantitative results, detailed understanding of the spray behavior was achieved which can be useful of computing validation. Further a comparative table determining the usage various non-intrusive techniques as per the development of the flows at various region of interest were concluded.This helps in selection of the most feasible technique required to be used in order to seek the detailed information of the certain flow characteristics.

Keywords: Single Orifice Flat Spray Nozzle; Shadowgraphy; PIV, Spray characterization.

1. Introduction

The characteristics associated with spray formation are of great importance to many in the spray community. Sprays are applied in numerous circumstances, for example, in the case of internal combustion engines, pharmaceutical processes, paint application, agricultural fertilizers and pesticides, fire suppression, and a multitude of other applications. Spray performance has significant impacts on daily life. Subsequently, continual research of spray diagnostics has been carried out for the past several decades to explore methods to achieve a complete understanding of spray characteristics.

This investigation offers to further advancement in the spray analysis methodology[1,2]. Spray measurements of a single orifice flat spray nozzle were collected through a variety of techniques to fully characterize the attributes of the spray. As such, this case-study offers an approach to spray characterization, progressing from qualitative flow visualization, to highly accurate quantitative results. In this way, the spray attributes from the nozzle exit, and down-stream, were fully realized.

Initially the imaging-based analysis of the spray was performed with high resolution CCD camera with 29 million pixel camera and a continuous LED illumination module. The next system employed was the particle image velocimetry (PIV) system using the same camera and a pulsed laser. The imaging system was setup in a shadowgraph configuration with an illuminated diffusion surface located directly opposite of the camera and behind the spray. The illumination was re-arranged using a Nd:YAG pulsed laser into a traditional orthogonal lightsheet configuration to achieve velocity field measurements of the spray. These spray images were further post-processed to derive the spray angle as well.

It was realized that the employment of the two analytical techniques, shadowgraphy and high resolution PIV, provided the insightful approach to spray characterization. Conducting measurements in such a sequence offered an evolving understanding of the spray, which aided in isolating the critical regions in the spray, and best suited method of interrogation. With the quantitative measurements performed in those critical regions, sufficient information was obtained to provide reproducible characterization of the spray behavior.

2. Experimental Setup

The spray nozzle used in the investigation was a single orifice hydraulic flat spray nozzle from Spraying Systems Company, model TPU650050-TC. The nozzle operated under a liquid pressure of 4 bar, with an estimated flow rate of 225 mL/min. Liquid water at room temperature was sprayed from the nozzle under atmospheric conditions.

Although configurable, the two systems were not used simultaneously. The experiment operated under steady state conditions. Thus each measurement technique was applied separately. Systems were arranged together so that the measurements were taken with minimal disturbance. This experimental configuration helped to maintain the steady state nature of the spray. The shadowgraph flow visualization allowed the shadow of the spray structure to be capture. Such arrangement was beneficial because the distribution of the illumination could be very uniform with the arrangement of a "diffuser" plate in the path of the illumination. Uniform illumination was important such that the light scattering from the different structures (wave fronts, ligament, large and small droplets) in the spray could be captured. In addition, the forward scattering arrangement by the camera provided the highest sensitivity of the capture.

The 29 million pixel CCD camera (at 4 fps) was used for the shadowgraph and PIV systems for the spray so that the detailed structure of the spray could be captured. Image capture was synchronized with a LED illuminator for the flow visualization. Due to the high pixel resolution of 29 million pixels, the images of the ligament and droplets in spray were very quantitative, meaning that the size of those objects could easily be measured by counting the number of pixels representing the objects. The Nd:YAG laser source was positioned to illuminate a plane of light for PIV image capture and velocity analysis. The velocity profile for the entire flat spray was measured, including the velocity distribution at the exit of the nozzle.

Results

Before analyzing the high resolution results, flow visualization on the nozzle spray was carried out using a high speed camera with 256x 256 pixel resolution and at 25000 fps. The technique is schematically demonstrated in the Figure 1.

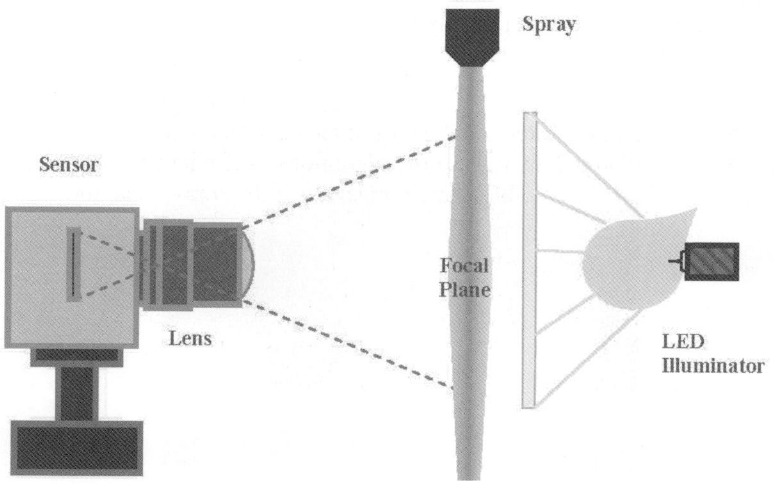

Figure 1: High speed imaging technique for flow visualization.

Figure 2 demonstrates the results of flow visualization using high speed camera imaging at 25000 fps. Typical frames are sorted out to observe the development of liquid sheet, growth of ligaments and generation of liquid droplets.

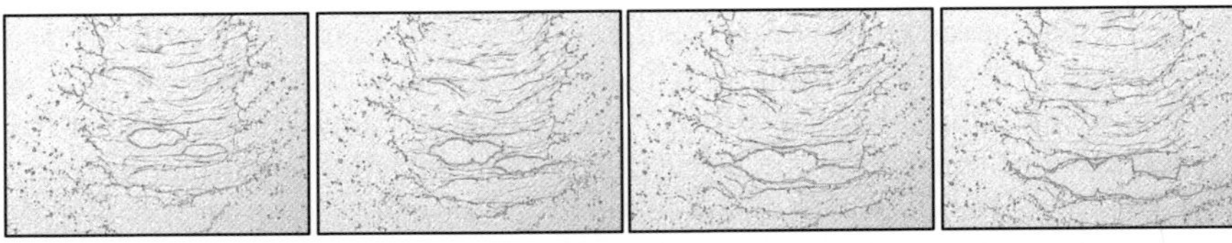

Figure 2: Acquired High speed flow visualization images at 25000 fps.

The data was analyzed in a way that utilized strengths of each of the measurement techniques in providing information about specific aspects of the spray. As such, discussion of results progresses from a qualitative global to quantitative point analysis.

High resolution shadowgraph images were analyzed to characterize general spray attributes. Although the image was captured with back illumination, the grey-scale values of the images were inverted to better highlight the attributes of the spray. Specifically, the sheet breakup length was determined to be 16.55mm below the nozzle exit, by inspecting the instantaneous image in Figure 3. The middle portion of the image where the breakup occurred was magnified and shown at the lower right corner of the figure where both a ligament formation and collapse were observed. Additionally, the high resolution images depicted larger droplets at the edges of the spray and smaller toward the center. It was also clear that droplet interaction was more prevalent in the center portion of the spray than the outside edges. It was also observed that droplets captured in the shadowgraph images appeared mostly spherical toward the bottom of the image, although there were some irregularities.

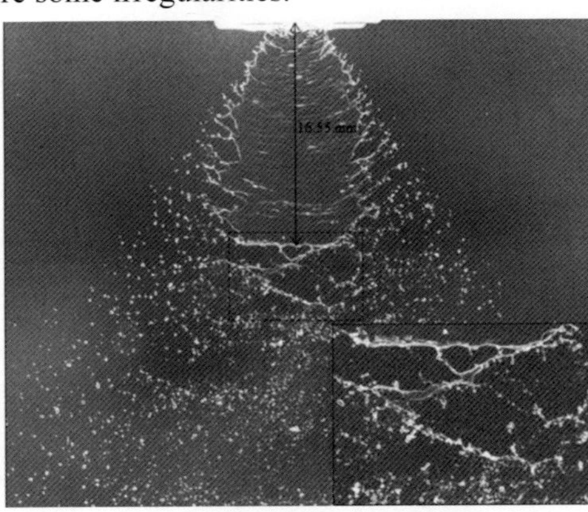

Figure 3: High resolution shadowgraph image depicting sheet break-up length, ligament characteristics, and droplet interaction.

Due to the high pixel resolution of the 29 million pixel camera, the image of the spray has high pixel resolution to represent the structure (droplets and ligaments). Subsequently it is possible to get the quantitative size information of the individual droplets from the image. Figure 4 shows the sizing result of some of the droplets. The orange circle represents the identification of the droplet and the number shows the size of the droplet in pixels. The analysis was performed with the Insight 4G software module called Size-Shape-Analysis. The module provides an algorithm to detect the edge of the object and then calculates the number of the pixels within the object.

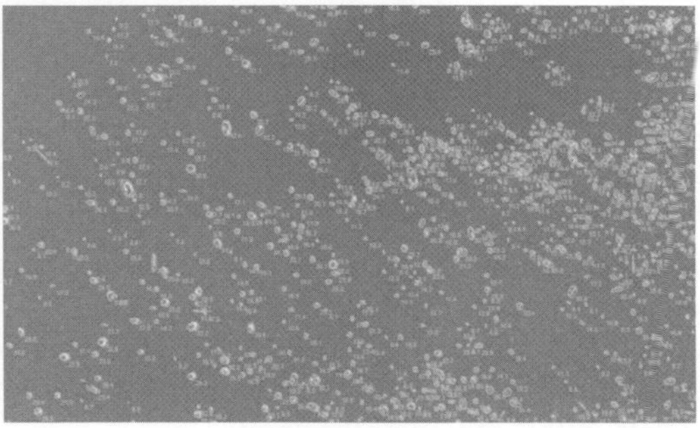

Figure 4: Quantitative sizing of the droplets

The spray was further analyzed using particle image Velocimetry (PIV) in order to determine the average velocity on a plane located along the centerline. The 29 megapixel CCD camera was used to collect 500 image pairs. The total field of view was approximately 90 x 70mm, and the resultant vector pitch was 0.55mm. As shown in Figure 5, ensemble correlation was used to determine the average velocity field. Ensemble correlation is a technique by which the correlation maps at a given location and across multiple realizations are added to determine the average displacement at a given location.

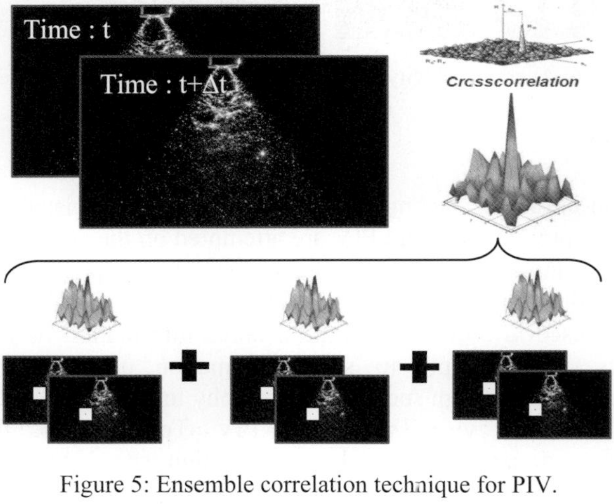

Figure 5: Ensemble correlation technique for PIV.

The advantage of this technique is that high SNR data is achievable in regions of relatively low seeding density, such as very close to the nozzle exit, where the spray is still breaking up into droplets. The ensemble-averaged vector field can be seen in Figure 6. The contour quantity shown is the vertical velocity, with darker structures indicating greater negative velocity and lighter shades indicating lesser negative velocity. The velocity field reveals a dual-lobed structure which exhibits a very slight asymmetry in the spray plume, as the lobe of higher velocity extends further downstream on the left side than on the right (see, for example, the yellow contour level). This asymmetry is minimal however, and the overall spray structure is quite uniform in terms of velocity. The velocity very near the nozzle exit is greater than -25 m/s and decreases to a velocity of -20 m/s at 50mm downstream, and -18 m/s at 70mm downstream.

A zoomed-in, single image of the near-nozzle measurement region can also be seen in Figure 6, with the ensemble vector field overlaid. The liquid sheet is visible very close to the nozzle exit, as well as the ligaments before breakup in the downstream portion of the spray. The ensemble correlation is beneficial in achieving reliable data in this region of the spray.

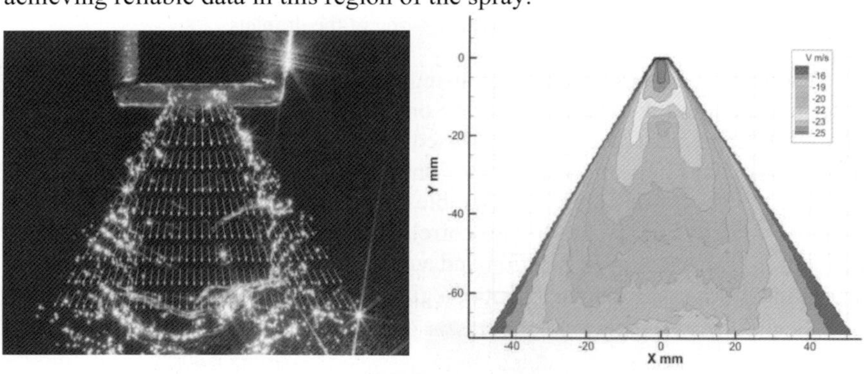

Figure 6: Average velocity field obtained from 500 PIV vector fields.

3. Conclusion

All the non-intrusive techniques inclusive offlow visualization, high speedshadowgraphy imaging and PIV are attempted on the spray. Usefulness of these techniques in characterizing the various regimes in spray development are summarized in the Table 1. This helps in selection of the most feasible technique required to be used in order to seek the detailed information of a certain flow characteristics. For example to seek information on droplet interaction, sphericity and diameter, high speed shadowgraphy imaging is a better technique as compared to flow visualization and PIV. To determine the ligament characteristics and secondary breakup, both high speed imaging and PIV technique provide better information as compared to flow visualization. But shadowgraphy won't be useful at all if we are intending to determine the liquid sheet breakup length, sheet instabilities and velocity profiles.

Table 1: Derived comparisons of various non-intrusive methods applied to sprays.

	Flow Visualization	Shadowgraph Imaging	Particle Image Velocimetry	
Liquid Sheet	√		√	Sheet breakup length
	√			Sheet instabilities
			√	Velocity Profile
Ligaments		√	√	Ligament Characteristics
	√	√	√	Secondary Breakup
Droplets			√	Droplet velocity
		√		Droplet Interaction
		√		Droplet Sphericity
		√		Droplet Diameter

References

1. W. D. Bachalo, "Spray Diagnostics for the Twenty-first century", Atomization and Sprays, **10**, 439–474 (2000).
2. K. M. Bade, R. J. Schick, "Post-Processing of Phase Doppler Interferometry Data for Planar SprayCharacteristics", Spray Analysis and Research Services, Spraying Systems Company, IL, USA, ILASS-Americas 25th Annual Conference on Liquid Atomization and Spray Systems, Pittsburgh, PA, (2013).

Shell Model for Buoyancy-Driven Turbulent Flows

Abhishek Kumar* and Mahendra K. Verma+

Department of Physics, Indian Institute of Technology Kanpur, Kanpur, 208016, India

**E-mail: abhkr@iitk.ac.in* +*E-mail: mkv@iitk.ac.in*

In this paper we construct a comprehensive shell model for the buoyancy-driven turbulence, which is applicable to convective turbulence and stably-stratified turbulence. We simulate these models in the turbulent regime and show that the stably-stratified turbulence exhibit Bolgiano-Obukhbov scaling ($E(k) \sim k^{-11/5}$), and the convective turbulence shows Kolmogorov spectrum ($E(k) \sim k^{-5/3}$).

Keywords: Shell model; Buoyancy-driven turbulence; Rayleigh Bénard convection.

1. Introduction

Turbulent fluid flows exhibit complex behaviour. Strong nonlinear interactions among large number of modes of a turbulent flow makes the theoretical analysis highly intractable. Shell models are simplified sets of the equations of fluid mechanics in wavenumber space. They are low-dimensional models which are successful in explaining some features of turbulence[1,2], e.g., the models for fluid turbulence reproduce the Kolmogorov's theoretical predictions. In a shell model, the wavenumber space is logarithmically divided into N shells, where N is the total number of shells. The energy exchange takes place between three nearest neighbouring shells, e.g. $(n+1, n, n-1)$, and each shell contains only one mode. A major advantage of shell model over direct numerical simulation (DNS) is its ability to solve flow simulations with very high Reynolds number (Re). For instance, a DNS of a turbulent flow with Re $\approx 10^9$ would require $\mathrm{Re}^{9/4} \approx 10^{81/4} \approx 2 \times 10^{20}$ grid points and significantly large computational time, which are impossible even in most sophisticated supercomputers of today, but such large-Re flows can be approximately solved using a shell model with $N = 76$. We remark that in the present paper we have performed shell model simulation for Re $\approx 10^9$ (see Table I). A point to note is that the shell model assume spherical isotropy in wavenumber space.

There have been several advances in understanding of fluid turbulence using shell model. However, till date, no convergence has been reached on the shell model for buoyancy driven turbulence[3–5], which is very important in the flows of geophysics, astrophysics, atmospheric and solar physics, as well as in engineering. Buoyancy-driven flows exhibit variety of phenomena—waves, patterns, weak turbulence, and strong turbulence (see the companion paper by Verma *et al.*[6] in this volume). Turbulence in buoyancy-induced flows come in two categories: (a) convective turbulence in which the hotter and lighter fluid at the bottom rises, while the colder and heavier fluid at the top comes down. These flows are unstable; (b) Stably-stratified turbulence in which the lighter fluid is above the heavier fluid.

The shell models describe the turbulent flows in the isotropic regime, hence we limit ourselves to the isotropic turbulent flows. For the isotropic stably-stratified flows, the buoyancy should be smaller than the nonlinear term $\mathbf{u} \cdot \nabla \mathbf{u}$ that corresponds to Richardson number (the ratio of buoyancy and nonlinear term) less than unity. Strong buoyancy leads to quasi two-dimensionality[7,8], whose analysis is beyond the scope of the shell model. For the convective turbulence, the flow is approximately isotropic[9,10], hence it can be analysed using a shell model.

For stably stratified flows, Bolgiano[11] and Obukhov[12] first proposed a phenomenology, according to which the kinetic energy (KE) spectrum $E_u(k) \sim k^{-11/5}$, the entropy spectrum $E_\theta(k) \sim k^{-7/5}$, KE energy flux $\Pi_u(k) \sim k^{-4/5}$, and entropy flux $\Pi_\theta(k) = \text{constant}$. This scaling is referred to as Bolgiano-Obukhov (*BO*) phenomenology[6,13,14]. Here $\mathbf{u}$ and θ are the velocity and temperature fluctuations.

Since buoyancy drives the convective turbulence, Procaccia and Zeitak[15], and L'vov[16] proposed the *BO* scaling for convective turbulence as well. However Kumar *et al.*[13] showed that the energy flux for the convective turbulence does not decrease with wavenumber since buoyancy feeds the kinetic energy (contrary to *BO* scaling). Kumar *et al.*[13] analyzed the energy spectrum and the energy feed due to buoyancy for the Rayleigh Bénard convection, and showed that for a set of parameters, the kinetic energy feed due to the buoyancy was compensated by the dissipation rate in a region of wavenumbers, for which they observed Kolmogorov's spectrum.

In the present paper we provide a comprehensive shell model for the buoyancy-induced turbulence, and show that the kinetic energy spectrum in Rayleigh Bénard convection is close to Kolmogorov's theory of fluid turbulence, while that in stably-stratified flows is proportional to $k^{-11/5}$. We also show the existence of ultimate regime[17] in shell model for convective turbulence. For convective turbulence, we have computed the Richardson number and compare it with the Péclet number scaling, similar to the analysis for Verma *et al.*[6]. Our shell model results are consistent with the numerical findings of Kumar *et al.*'s[13].

2. Shell Model for Buoyancy-Induced Turbulence

A Sabra shell model[18,19] of buoyancy-driven turbulence under Boussinesq approximations is

$$\frac{du_n}{dt} = N_n[u,u] + \alpha g \theta_n - \nu k_n^2 u_n + f_n, \tag{1}$$

$$\frac{d\theta_n}{dt} = N_n[u,\theta] - \frac{d\bar{T}}{dz} u_n - \kappa k_n^2 \theta_n, \tag{2}$$

where u_n is the velocity shell variable, θ_n is the shell variable for the temperature fluctuations, f_n is the external forcing, g is acceleration due to gravity, α, ν, κ are the thermal expansion coefficient, the kinematic viscosity, and the thermal diffusivity respectively, and $k_n = k_0 \lambda^n$ is the wavenumber of the shell. Since the shell models are applicable to a periodic box without walls, the above can be considered as a part of an unbound system with a vertical temperature gradient of $d\bar{T}/dz$.

In convective turbulence, the buoyancy compensates the dissipative losses, hence the shell model for convection does not require any external forcing to maintain a steady state. However, the shell model for the stably stratified turbulence needs to be forced for obtaining a steady state. In our shell-model simulations of stably stratified turbulence, we force n_f small wavenumber shells randomly so as to feed a constant energy supply rate ε to the system. The energy supply rate ε is divided equally to the n_f shells, hence

$$f_n = \sqrt{\frac{\varepsilon}{n_f \Delta t}} \exp(i\phi_n) \tag{3}$$

where ϕ_n is the random phase of the nth shell chosen from the uniform distribution in $[0, 2\pi]$. In our simulation we force the shells $n = 3$ and 4, hence $n_f = 2$.

The nonlinear terms $N_n[u,u]$ and $N_n[u,\theta]$ of the above equations are

$$N_n[u,u] = -i(a_1 k_n u^*_{n+1} u_{n+2} + a_2 k_{n-1} u^*_{n-1} u_{n+1} - a_3 k_{n-2} u_{n-1} u_{n-2}) \quad (4)$$

$$\begin{aligned} N_n[u,\theta] = & -i[k_n(d_1 u^*_{n+1}\theta_{n+2} + d_3\theta^*_{n+1}u_{n+2}) \\ & + k_{n-1}(d_2 u^*_{n-1}\theta_{n+1} - d_3\theta^*_{n-1}u_{n+1}) \\ & - k_{n-2}(-d_1 u_{n-1}\theta_{n-2} - d_2\theta_{n-1}u_{n-2})] \end{aligned} \quad (5)$$

where $a_1 = d_1 = 1$, $a_2 = d_2 = \lambda - 2$, and $a_3 = d_3 = 1 - \lambda$, where λ is chosen as golden mean whose value is $(\sqrt{5}+1)/2$[1]. The above model conserves kinetic energy $\sum_n u_n^2/2$ when the viscous and the buoyancy terms are turned off. $N[u,\theta]$ is chosen as above so that

$$\Re\left(\sum_n \theta^*_n N_n[u,\theta]\right) = 0, \quad (6)$$

where $\Re$ is the real part of the argument.

It is convenient to work with the nondimensionalized equations, which is achieved by using the characteristic length d as the length scale, $\sqrt{\alpha g|d\bar{T}/dz|d^2}$ as the velocity scale, and $|d\bar{T}/dz|d$ as the temperature scale. Therefore, $u_n = u'_n\sqrt{\alpha g|d\bar{T}/dz|d^2}$, $\theta_n = \theta'_n|d\bar{T}/dz|d$, $k_n = k'_n/d$, and $t = t'(d/\sqrt{\alpha g|d\bar{T}/dz|d^2})$. In terms of nondimensonalized variables, the equations are

$$\frac{du'_n}{dt'} = N'_n[u',u'] + \theta'_n - \sqrt{\frac{\mathrm{Pr}}{\mathrm{Ra}}}k'^2_n u'_n + f'_n, \quad (7)$$

$$\frac{d\theta'_n}{dt'} = N'_n[u',\theta'] - Su'_n - \frac{1}{\sqrt{\mathrm{RaPr}}}k'^2_n\theta'_n, \quad (8)$$

where $S = (d\bar{T}/dz)/(\Delta/d)$, where Δ is the absolute temperature differences in height d. $S = 1$ for positive $d\bar{T}/dz$ (stably stratified flows), and $S = -1$ for negative $d\bar{T}/dz$ (convective flows).

The two nondimensional control parameter in the above equations are Rayleigh number Ra, which is a ratio of the buoyancy force and dissipative force, and the Prandtl number Pr, which is a ratio of the kinematic viscosity and the thermal diffusivity. We also remark that our shell model differs from that of Brandenburg[3].

A stably-stratified flows are stable, hence the fluctuations would vanish asymptotically. Hence, to generate turbulence in stably-stratified flow, we force the large scale flows. We apply forcing f_n to shells with small n that represents large length scales. Note that for $f_n = 0$, we obtain $u_n \to 0$ and $\theta_n \to 0$ asymptotically.

3. Numerical Methods

We simulate the shell models by solving Eqs. (7) and (8). For stably-stratified turbulence, $S = 1$, but for convective turbulence, $S = -1$. For time stepping we use fourth-order Runge-Kutta (RK4) method. We took 76 shells for the convective turbulence simulations, and 36 shells for the stratified turbulence simulations. For both convective and stably stratified turbulence we take $\mathrm{Pr} = 1$.

For the stably-stratified turbulence, we performed two sets of simulations, one with $\mathrm{Ra} = 10^5$ and the other with $\mathrm{Ra} = 10^{10}$. To generate turbulence, we employ random forcing to the shells 2 to 4. The Richardson number, which is a ratio of the buoyancy term and the nonlinear term $\mathbf{u} \cdot \nabla \mathbf{u}$, is an important parameter for stably stratified turbulence. We compute the spectra and fluxes of the kinetic energy and entropy $(\theta^2/2)$ by averaging over many time frames ($\sim 10^8$) of the steady-state flow.

Table I. Parameters of our simulations of convective turbulence: Rayleigh number Ra, Nusselt number Nu, Reynolds number Re, kinetic energy dissipation rate ϵ_u, entropy dissipation rate ϵ_θ, and Kolmogorov's constant K_{Ko}. We choose number of shells $N = 76$ and Prandtl number $\mathrm{Pr} = 1$ for all our runs.

Ra	Nu	Re	ϵ_u	ϵ_θ	K_{Ko}
10^{10}	6.0×10^6	8.6×10^5	36.5	48.8	1.2
10^{11}	1.9×10^7	2.7×10^6	76.5	61.8	0.9
10^{12}	6.4×10^7	8.7×10^6	51.0	54.3	1.0
10^{13}	1.9×10^8	2.7×10^7	61.8	80.6	0.9
10^{14}	6.4×10^8	8.7×10^7	36.4	48.5	1.2
10^{15}	2.2×10^9	2.9×10^8	94.1	95.2	0.9
10^{16}	6.7×10^9	8.9×10^8	42.1	108.1	1.1

For convective turbulence, we perform simulations for various Rayleigh numbers as listed in Table I and compute various quantities listed in Table I using the steady-state data. We also compute the Nusselt number for different Ra's and derive a scaling relationship between Nu with Ra. Note that the Nusselt number is the ratio of the total (convective plus conductive) heat flux to the conducive heat flux, and it is expressed as[20]

$$\mathrm{Nu} = \frac{-\kappa\frac{d\bar{T}}{dz} + \langle u_n\theta_n\rangle}{-\kappa\frac{d\bar{T}}{dz}} = 1 - \left\langle \frac{u_n}{\kappa}\frac{\theta_n}{\frac{d\bar{T}}{dz}} \right\rangle = 1 + \sqrt{\mathrm{RaPr}}\langle u'_n\theta'_n\rangle. \tag{9}$$

Here the nondimensionalized shell velocity and temperature field are $u'_n = u_n/\sqrt{\alpha g d^2 d\bar{T}/dz}$ and $\theta' = (\theta/d)/(d\bar{T}/dz)$ respectively.

4. Results and Discussions

4.1. *Shell model for stably-stratified turbulence*

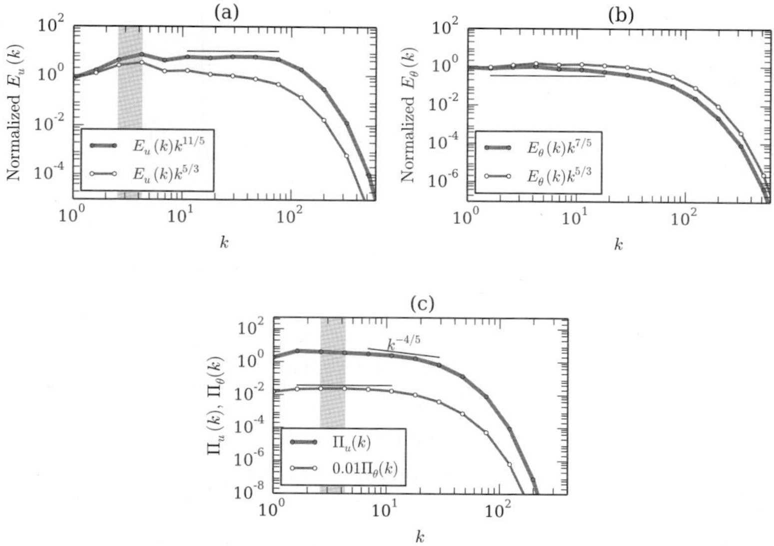

Fig. 1. Stably stratified turbulence with Pr = 1, Ra = 10^5, and Ri = 0.10: plots of (a) normalized KE spectra, (b) normalized entropy spectra and (c) KE flux $\Pi_u(k)$ and entropy flux $\Pi_\theta(k)$. The plots shows that *BO* scaling fits with the data better than the *KO* scaling.

For stably stratified flow, we take Pr = 1, Ra = 10^5. The energy supply rate was chosen to be $\varepsilon = 50$. Under steady state, we achieved Ri = 0.10, Reynolds number Re = 10^3, kinetic energy dissipation rate $\epsilon_u = 3.9$, and entropy dissipation rate $\epsilon_\theta = 3.1$. In Fig. 1(a) we plot the normalized KE spectra, $E_u(k)k^{11/5}$ for the *BO* scaling, and $E_u(k)k^{5/3}$ for the *KO* scaling. Shadow region in the figure is the forcing band. In Fig. 1(b) we plot the normalized entropy spectra, $E_\theta(k)k^{7/5}$ for the *BO* scaling, and $E_\theta(k)k^{5/3}$ for the *KO* scaling. The figure indicates that the *BO* scaling fits with the numerical data better than *KO* scaling. Along with the spectrum results, we also compute the KE and entropy fluxes, which are plotted in Fig. 1(c). $\Pi_u(k) \sim k^{-4/5}$, and $\Pi_\theta = \text{const}$ (though not in the same wavenumber range), consistent with *BO* scaling. These results show that *BO* scaling is valid for stably stratified flow. We find that the energy supply rate by buoyancy, $F(k) = \Re\langle u_k \theta_k^* \rangle$, is negative, indicating a conversion of kinetic energy to the potential energy[13]. We obtain an approximate dual scaling

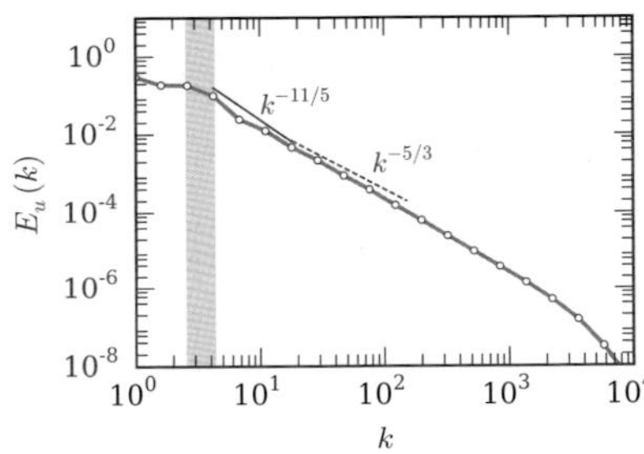

Fig. 2. Stably stratified turbulence with $\text{Pr} = 1$, $\text{Ra} = 10^{10}$ and $\text{Ri} = 0.25$: plot of KE spectrum. The wavenumber range $4 < k < 18$ exhibit $E(k) \sim k^{-11/5}$, and $18 < k < 100$ exhibit $E(k) \sim k^{-5/3}$.

($k^{-11/5}$ and $k^{-5/3}$) for $\text{Ra} = 10^{10}$. For this run we keep $\varepsilon = 10$, and under steady state we achieved $\text{Ri} = 0.25$, $\text{Re} = 2 \times 10^5$, $\epsilon_u = 0.6$, and $\epsilon_\theta = 0.8$. It turns out that the dual scaling occurs for a very narrow set of parameters for which the buoyancy is strong, but not so strong so as to make the flow quasi two-dimensional.

In addition, for small Richardson number or for weak buoyancy, we obtain Kolmogorov's $k^{-5/3}$ energy spectrum. The aforementioned results are consistent with the DNS results of Kumar *et al.*[13].

4.2. *Convective turbulence shell model*

The parameters used for the convective turbulence simulations are listed in Table I. For one of the runs, $\text{Ra} = 10^{16}$ and $\text{Pr} = 1$, we plot the normalized KE spectrum, $E_u(k)k^{5/3}$, and the normalized entropy spectrum, $E_\theta(k)k^{5/3}$ in Fig. 3(a). The figure indicates that Kolmogorov (*KO*) scaling fits quite well with the data for three decades. We cross check our spectrum result with the KE and entropy fluxes, which are plotted in Figs. 3(b) and 3(c) respectively. The KE flux $\Pi_u(k)$ slightly increases with k (see Fig. 3(b)) as shown by dashed red curve and then becomes constant. The entropy flux $\Pi_\theta(k)$ remains constant in the inertial range. These results demonstrate that the Kolmogorov scaling is valid for the convective turbulence.

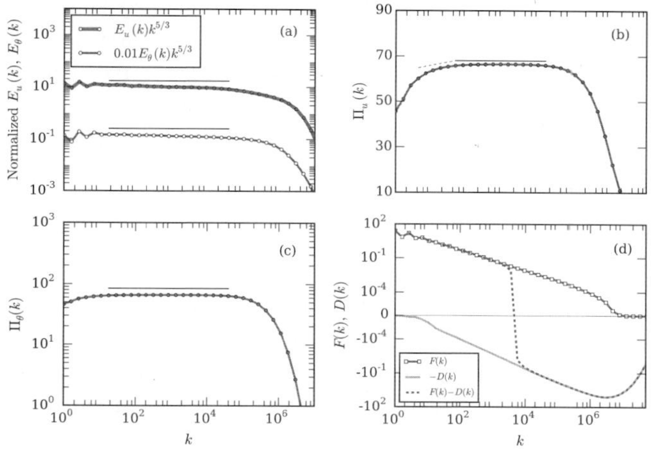

Fig. 3. Convective turbulence with Pr = 1 and Ra = 10^{16}: plots of (a) normalized KE spectra and normalized entropy spectra for Kolmogorov-Obukhov (KO) scaling, (b) KE flux $\Pi_u(k)$ and (c) entropy flux $\Pi_\theta(k)$, and (d) energy supply rate $F(k)$, dissipation rate $D(k)$, and $F(k) - D(k)$.

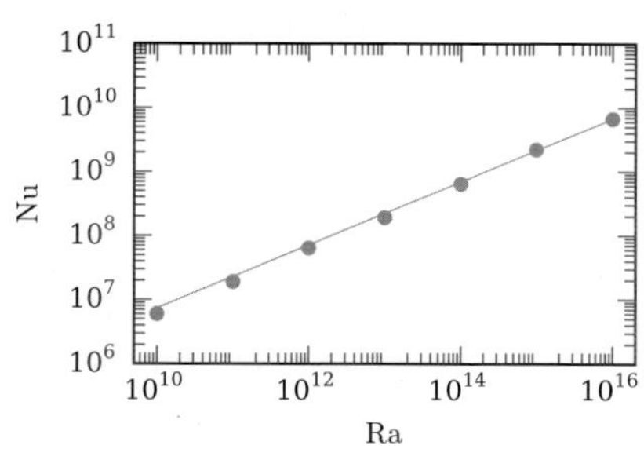

Fig. 4. Plot of the Nusselt number Nu versus Ra. The function Nu $= (86.4 \pm 18.7)\text{Ra}^{(0.49\pm0.01)}$ fits well with the simulation data.

We also compute energy supply rate $F(k) = \Re\langle u_k \theta_k^* \rangle$ and plot it in Fig. 3(d). The plot indicates a positive energy transfer from buoyancy to the kinetic energy[13]. In Fig. 3(d), we also plot the dissipation rate $D(k)$

and $F(k) - D(k)$. $F(k)$ and $D(k)$ cancel each other in the inertial range, and hence $d\Pi_u(k)/dk \approx 0$ or $\Pi(k)$ is approximately constant. This yields a constant KE flux $\Pi_u(k)$ which leads to Kolmogorov's spectrum.

We compute the Richardson number for all convection runs, and observe it to be approximately 0.01. The above number is related to the Péclet number:

$$\begin{aligned}\mathrm{Ri} &= \frac{\alpha g d (d\bar{T}/dz)}{u^2/d} \\ &= \frac{\alpha g d^4 (d\bar{T}/dz)}{\nu\kappa}\frac{\nu}{\kappa}\frac{\kappa^2}{u^2 d^2} \\ &= \frac{\mathrm{RaPr}}{\mathrm{Pe}^2}\end{aligned} \tag{10}$$

where Pe is the Péclet number. Verma *et al.*[20] showed that

$$\mathrm{Pe} = C\sqrt{\mathrm{RaPr}}. \tag{11}$$

In our numerical simulations of shell model for the convective turbulence, we observe that $C \approx 10$, hence $\mathrm{Ri} \approx 0.01$. Since Richardson number is relatively small, the nonlinear term dominates the buoyancy term; the observed Kolmogorov's spectrum is possibly because of this reason.

Since the Prandtl number is unity, the Reynolds number ud/ν is same as the Péclet number, and it is listed in Table I for all the runs. We compute the Kolmogorov constant K_{Ko} using the energy spectrum and the energy flux. We observe that K_{Ko} listed in Table I ranges from 0.9 to 1.2. We compute the Nusselt number using Eq. (9), and plot it in Fig. 4 as a function of Ra. The best fit of our data is $\mathrm{Nu} = (86.4 \pm 18.7)\mathrm{Ra}^{(0.49\pm0.01)}$. This result is consistent with Kraichnan's predication[17], according to which $\mathrm{Nu} \sim \mathrm{Ra}^{1/2}$ for very high Rayleigh numbers (called the ultimate regime).

5. Summary and Conclusions

In summary, we constructed a combined shell models for convective turbulence and stably-stratified turbulence. The difference in these models is in the temperature gradient term of the temperature equation. Using numerical computation of these models we demonstrate the stably-stratified turbulence shows Bolgiano-Obukhov scaling, while the convective turbulence exhibits Kolmogorov spectrum. These results are consistent with the recent numerical results of Kumar *et al.*[13,14] which are based on pseudospectral simulation. We remark that the Bolgiano-Obukhov scaling for stably stratified flow is observed for the isotropic turbulence which is observed when the Froude number is of the order of unity.

6. Acknowledgments

We thank K. R. Sreenivasan for useful comments during the IUTAM meeting, and T. K. Sengupta and the organizers for hosting the wonderful and interactive IUTAM meeting. We also thanks the editors of the proceedings for the useful suggestions on the paper. This work was supported by a research grant SERB/F/3279/2013-14 from Science and Engineering Research Board, India.

References

1. P. Ditlevsen, *Turbulence and Shell Models* Cambridge Univ. Press, Cambridge, UK (2011).
2. L. Biferale, *Ann. Rev. Fluid Mech.* **35**, 441 (2003).
3. A. Brandenburg, *Phys. Rev. Lett.* **69**, 605 (1992).
4. J. Mingshun and L. Shida, *Phys. Rev. E* **56**, 441 (1997).
5. E. S. C. Ching and W. C. Cheng, *Phys. Rev. E.* **77**, 015303 (2008).
6. M. K. Verma, A. Kumar, and A. G. Chatterjee, Under review in IUTAM proceedings (2015).
7. E. Lindborg, *Geo. Res. Lett.* **32**, L01809 (2005).
8. E. Lindborg, *J. Fluid Mech.* **550**, 207 (2006).
9. B. Dutta and M. K. Verma, Unpublished (2015).
10. A. Pandey, M. K. Verma, and P. K. Mishra *Phys. Rev. E* **89**, 023006 (2014).
11. R. Bolgiano, *J. Geophys. Res.* **64**, 2226 (1959).
12. A. N. Obukhov, *Dokl. Akad. Nauk SSSR* **125**, 1246 (1959).
13. A. Kumar, A. G. Chatterjee, and M. K. Verma, *Phys. Rev. E* **90**, 023016 (2014).
14. M. K. Verma, A. Kumar, and A. G. Chatterjee, *Phys Focus* **25**(1), 45 (2015).
15. I. Procaccia and R. Zeitak, *Phys. Rev. Lett.* **62**, 2128 (1989).
16. V. S. L'vov, *Phys. Rev. Lett.* **67**, 687 (1991).
17. R. H. Kraichnan, *Phys. Fluids* **5**, 1374 (1962).
18. V. L'vov, E. Podivilov, A. Pomyalov, I. Procaccia, and D. Vandembroucq, *Phys. Rev. E* **58**, 1811 (1998).
19. A. Kumar and M. K. Verma, *Phys. Rev. E* **91**, 043014 (2015).
20. M. K. Verma, P. K. Mishra, A. Pandey, and S. Paul, *Phys. Rev. E* **85**, 016310 (2012).

Numerical Simulations in Low–Prandtl Number Convection

J. D. Scheel+

Physics Department, Occidental College Los Angeles,
Los Angeles, CA 90041, USA
+*E-mail: jscheel@oxy.edu*
www.oxy.edu/physics

J. Schumacher*

Mechanical Engineering, Technische Universität Ilmenau,
Ilmenau, D-98684, Germany
**E-mail: joerg.schumacher@tu-ilmenau.de*
www.tu-ilmenau.de/tsm

Direct numerical simulations in low–Prandtl number turbulent convection which are based on a spectral element method in a cylindrical cell are discussed. Here, we report results on the structure of the thermal boundary layer and on the statistics of the locally fluctuating Kolmogorov scales.

Keywords: Turbulent convection; Direct numerical simulation; Low Prandtl number.

1. Introduction

Turbulent thermal convection phenomena are found in many systems in nature and technology.[3] Convection depends strongly on the material properties of the working fluid. They are quantified by the Prandtl number which relates the kinematic viscosity of the fluid, ν, to the thermal diffusivity of the temperature field, κ, and is defined as $Pr = \nu/\kappa$. Turbulent convection in the Sun is present at Prandtl numbers $Pr < 10^{-3}$. The Prandtl number in the liquid metal core of the Earth is $Pr \sim 10^{-2}$. Convection in material processing, nuclear engineering, or liquid metal batteries have Prandtl numbers between 3×10^{-2} and 10^{-3}. One reason for significantly fewer low–Pr RBC studies is that laboratory measurements have to be conducted in opaque liquid metals, partly operated at temperatures significantly larger than room temperature. Examples are mercury or gallium at $Pr = 0.021$[4,15] and liquid sodium at $Pr = 0.005$.[5] Thermal convection in liquid sodium provides the lowest Prandtl number which is accessible in

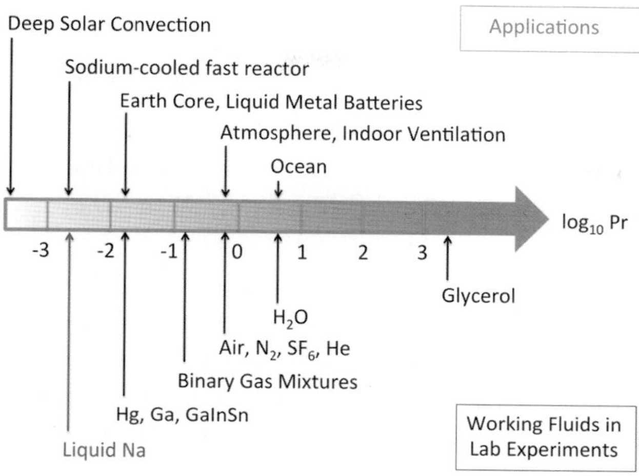

Fig. 1. Typical working fluids in laboratory convection experiments and some applications for different Prandtl numbers are listed. Liquid sodium provides the smallest Prandtl number obtainable in laboratory. It is therefore indicated in grey.

controlled laboratory experiments. In Fig. 1, we summarize Prandtl numbers of working fluids in laboratory experiments and applications. Direct numerical simulations (DNS) have access to the full three-dimensional convective turbulent fields in low-Pr convection. These DNS have been carried out in homogeneous turbulent flows[9], in planar turbulent flows between parallel free-slip or no-slip walls[1,8,17] or in closed cylindrical cells.[2,16] While the turbulent heat transport is reduced in low–Pr convection, vorticity and shear amplitudes are enhanced significantly. In the present work, we want to present some first results of our DNS studies. The outline is as follows. In the next section we describe the equations of motion and the numerical method. We will discuss two aspects of low–Prandtl number convection, the thermal boundary layer structure and the locally fluctuating Kolmogorov scales which are a manifestation of the small-scale intermittency.

2. Boussinesq Equations for Thermal Convection and Numerical Method

We solve the three-dimensional equations of motion numerically in the *Boussinesq approximation*. In this approximation, the density is a linear function of the temperature which results in the last term in Eq. (2). The height of the cell H, the free-fall velocity $U_f = \sqrt{g\alpha\Delta T H}$ and the imposed temperature difference ΔT are used to rescale the equations of motion. The aspect ratio is defined as $\Gamma = D/H$ with the cell diameter D. This results in the following dimensionless form of the equations of motion

$$\nabla \cdot \mathbf{u} = 0\,, \tag{1}$$

$$\frac{\partial \mathbf{u}}{\partial t} + (\mathbf{u} \cdot \nabla)\mathbf{u} = -\nabla p + \sqrt{\frac{Pr}{Ra}}\nabla^2 \mathbf{u} + T\mathbf{e}_z\,, \tag{2}$$

$$\frac{\partial T}{\partial t} + (\mathbf{u} \cdot \nabla)T = \frac{1}{\sqrt{RaPr}}\nabla^2 T\,. \tag{3}$$

where $Ra = g\alpha\Delta T H^3/(\nu\kappa)$ is the Rayleigh number. Throughout the study we set $\Gamma = 1$. At all walls no-slip boundary conditions for the fluid are applied, $\mathbf{u} = 0$. The side walls are adiabatic, i.e., the normal derivative of the temperature field vanishes, $\partial T/\partial \mathbf{n} = 0$. The top and bottom plates are held at fixed temperatures $T = 0$ and 1, respectively. In response to the input parameters Ra, Pr and Γ, a turbulent heat and momentum flux in the cell will be established.

The studies are conducted with the spectral element code (SEM) Nek5000.[10,12] The code employs second order time-stepping, using a backward difference formula. The convective and buoyancy terms are treated explicitly. The remaining linear, symmetric Stokes problem is solved implicitly. This system is split, decoupling the viscous and pressure steps into independent symmetric positive definite subproblems which are solved either by Jacobi (viscous) or multilevel Schwartz (pressure) preconditioned conjugate gradient iteration.[7] Fast parallel solvers based on algebraic multigrid are used for the coarse-grid solve that is part of the pressure preconditioner.

The spectral approximation of a function $u_e \in \mathrm{L}^2_w(\Omega)$ (with $w(x)$ being a positive weight function and $\Omega = [-1, 1]$) can be written as follows

$$u_e(x) = \sum_{k=0}^{N} u(\xi_k)\psi_k(x)\,, \tag{4}$$

where $\psi_k(x)$ is the kth order basis function and the (N+1) points are the nodes of the Gauss-Lobatto quadrature. They are determined by the Gauss-Lobatto integration theorem.[6] For the approximation, one has to take a set of polynomials which form an orthogonal system of the underlying Hilbert space of square-integrable functions $L^2_w(\Omega)$ which will be the Legendre polynomials P_k for the present study. In this case, $w(x) = 1$ and the Gauss-Lobatto-Legendre (GLL) points $\xi_1 \ldots \xi_{N+1}$ in the interval Ω follow from solving the linear system

$$\xi_k P_N(\xi_k) = \frac{N+1}{2N+1} P_{N+1}(\xi_k) + \frac{N}{2N+1} P_{N-1}(\xi_k) \,. \tag{5}$$

Furthermore $\xi_0 = -1$ and $\xi_N = 1$. The basis functions $\psi_k(x)$ in (4) are Lagrangian interpolation polynomials of order N which make now use of the Legendre polynomials and the GLL collocation points. They are given by

$$\psi_k(x) = -\frac{1}{N(N+1)} \frac{(1-x^2)P'_N(x)}{(x-\xi_k)P_N(\xi_k)} \,. \tag{6}$$

In contrast to a classical spectral or pseudospectral method the evaluation of spatial derivatives translates into matrix multiplications. These operations are here highly optimized.

The extension of the framework to the three-dimensional case is straightforward and based on the tensor product formulation of the basis functions (cf. Eq. (4))

$$\mathbf{u}_e(x,y,z) = \sum_{i=0}^{N} \sum_{j=0}^{N} \sum_{k=0}^{N} \mathbf{u}(\xi_i, \xi_j, \xi_k) \; \psi_i(x) \otimes \psi_j(y) \otimes \psi_k(z) \,. \tag{7}$$

The corresponding GLL points have to be mapped on the corresponding deformed elements. We perform a de-aliasing by an over-integration of the convective terms by a factor of $3(N+1)/2$. The code has been used for production runs on up to 131072 MPI tasks. The largest DNS used 2374400 spectral elements, each containing a polynomial expansion at order $N = 13$ in each of the three space dimensions. This sums up to more than 5.2 billion individual mesh cells.

3. Structure of the Boundary Layer

In the range of Rayleigh numbers which we discuss here, the boundary layers of velocity and temperature are not fully turbulent. Recently much effort has been put into a comparison of the experimental and numerical results with those from classical boundary layer theory, such as the laminar boundary layer theory.[11] The theoretical Prandtl-Blasius-Pohlhausen solutions[13] are self-similar solutions for the stream function, Ψ and the temperature T (where ξ is the similarity variable):

$$\frac{d^3\Psi}{d\xi^3} + 0.5\Psi \frac{d^2\Psi}{d\xi^2} = 0, \qquad \frac{d^2T}{d\xi^2} + 0.5Pr\Psi \frac{dT}{d\xi} = 0 \,. \tag{8}$$

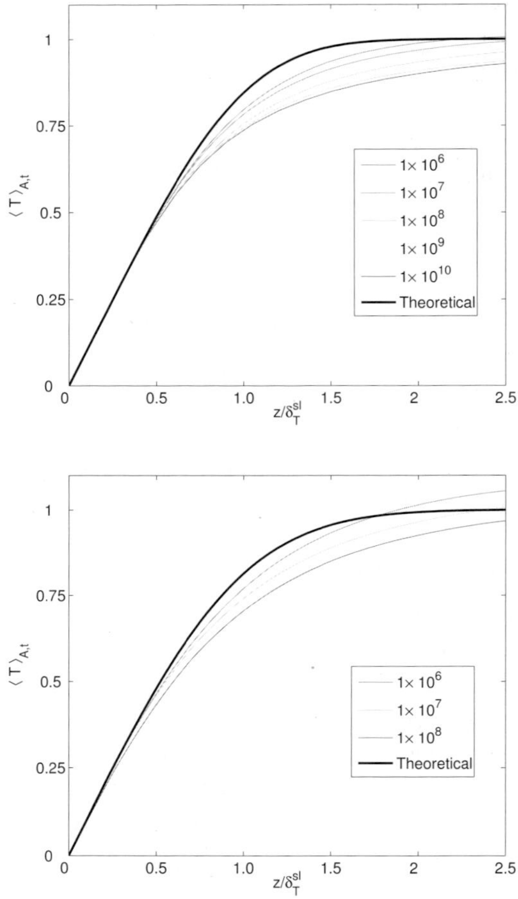

Fig. 2. Boundary layer profiles for (top) Pr=0.7 and (bottom) Pr=0.021. In both cases the thick black line is the theoretical Prandtl-Blasius solution (8) for a laminar boundary layer. Each boundary layer profile is obtained by taking the data for a snapshot, and averaging plane-by-plane, but only for $r < 0.3$ so the effect of the sidewalls (at $r = 0.5$) is eliminated. Then these snapshots, taken at statistically independent times, are averaged as well. The boundary layer thickness δ_T^{sl} is found by finding the slope near the wall and then finding the intersection point of this linear function with the mean temperature in the bulk.

The initial conditions are given by

$$\Psi(0) = 0, \frac{d\Psi}{d\xi}(0) = 0, \frac{d\Psi}{d\xi}(\infty) = 1, \quad (9)$$
$$T(0) = 0, T(\infty) = 1\,.$$

In Fig. 2, we display our numerical findings for the thermal boundary layer. The top figure contains data for a series of DNS at $Pr = 0.7$, the lower panel for $Pr = 0.021$. In all cases we detect deviations from the Blasius profiles. In a next step, we quantify these deviations. For our thermal boundary layer profiles, the shape factor is a ratio of the two quantities: the displacement thickness

$$\delta_d = \int_0^\infty \left(1 - \frac{T(z)}{[T(z)]_{\rm mean}}\right) dz, \tag{10}$$

and the momentum thickness

$$\delta_m = \int_0^\infty \left(1 - \frac{T(z)}{[T(z)]_{\rm mean}}\right) \left(\frac{T(z)}{[T(z)]_{\rm mean}}\right) dz, \tag{11}$$

where $[T(z)]_{\rm mean}$ is the average temperature in the bulk. The shape factor for the thermal boundary layer is then

$$S_T = \frac{\delta_d}{\delta_m}. \tag{12}$$

The value for S_T is found to be 2.58 for Pr=0.7 and 2.47 for Pr=0.021. In Fig. 3, we summarize the shape factors for the whole data record. With increasing Rayleigh number, the value deviates increasingly from the theoretical one. The data for $Ra = 10^7$ and 10^8 in the low–Pr series are found to be close to the corresponding cases for convection in air. The growing deviations are a consequence of the growing fluctuations of the thermal boundary layer structure. Further detailed studies of these intermittent fluctuations, in particular for the velocity boundary layer, are in progress.

4. Local Dissipation Scales

In the second part, we turn from the thermal boundary layer structure to the small-scale turbulence properties in the convection flow. Of central importance is the kinetic energy dissipation rate field. It is known to fluctuate intermittently[14] and is defined as

$$\epsilon(\mathbf{x}, t) = \frac{1}{2}\sqrt{\frac{Pr}{Ra}} \left(\boldsymbol{\nabla u} + \boldsymbol{\nabla u}^T\right)^2 . \tag{13}$$

The mean kinetic energy dissipation rate $\langle\epsilon\rangle$ is related to the mean Kolmogorov scale $\langle\eta_K\rangle$ which is the smallest mean scale when $\nu \leq \kappa$, i.e. $Pr < 1$ as is true in this case. The symbol $\langle\cdot\rangle$ denotes an ensemble average which is calculated in numerical simulations as a volume-time average. We generalize the mean Kolmogorov scale to the local Kolmorgorov scale

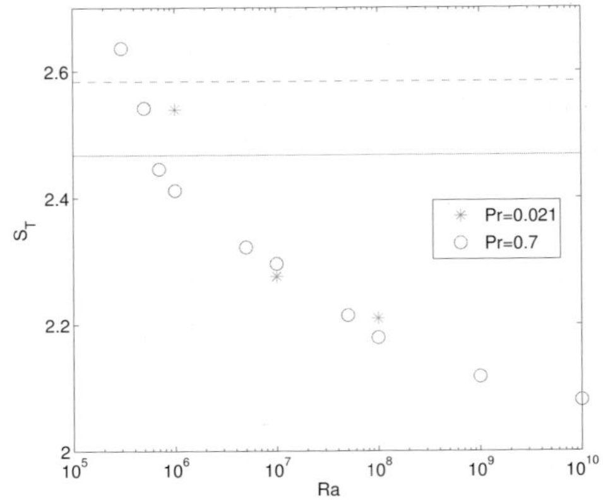

Fig. 3. Shape factors S_T as computed by equation (12) for Pr=0.7 (open circles) and Pr=0.021 (stars) as a function of Rayleigh number. The horizontal lines are the theoretical values for the shape factors as computed from the solution to equation (8).

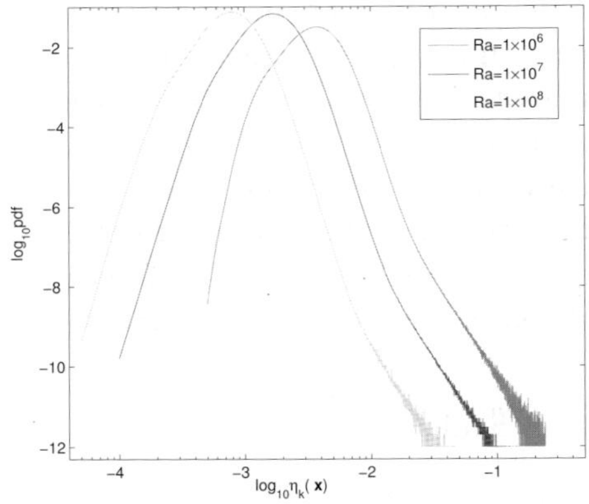

Fig. 4. Pdfs of the local dissipation scale $\eta_k(\mathbf{x})$ as given by (14) for $Pr = 0.021$ and various Rayleigh numbers. Distributions for different Ra are directly comparable in the dimensionless units which we use in the simulations (see section 2).

$\eta_K(\mathbf{x}, t)$ in order to reflect the strong fluctuations of velocity gradients at small scales. In a nutshell, these scales can be thought as measuring the cross section of local shear layers. They are defined as

$$\eta_K(\mathbf{x}, t) = \frac{Pr^{3/8}}{Ra^{3/8}}\, \epsilon(\mathbf{x}, t)^{-1/4} . \tag{14}$$

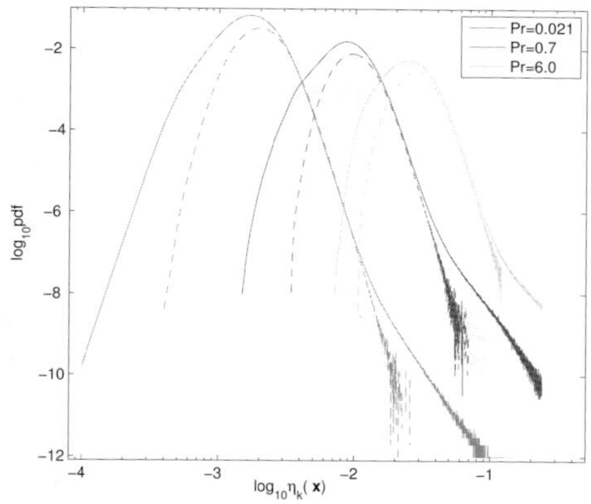

Fig. 5. Pdfs of the local dissipation scale $\eta_k(\mathbf{x})$ as given by (14) for $Ra = 1 \times 10^7$ and different Prandtl number. The solid lines are pdfs taken over the entire volume V and the dashed lines correspond to pdfs taken over the subvolume $V_b = \{\mathbf{x} = (r, \phi, z) \mid 0 \leq r \leq 0.3; 0.2 \leq z \leq 0.8\}$.

In Fig. 4 we display the distributions for the low–Pr series. The distributions are always peaked around the classical Kolmogorov scale. With increasing Rayleigh number the whole distribution is shifted towards smaller scales which indicates that the magnitude of the local shear and thus the magnitude of the kinetic energy dissipation rate field increases. In Fig. 5, we keep the Rayleigh number fixed and vary the Prandtl number. The solid lines display the distribution as obtained for the whole cell, the dashed lines stand for the distribution in a sub volume V_b which is given in the caption of the figure. The finest scales for all three Prandtl numbers are always generated close to the boundary layer which is visible by the comparison of the distributions obtained in V and V_b. Furthermore, we see that the range between maximum and minimum local Kolmogorov scale gets larger as the Prandtl number decreases which is a manifestation of the increasing small-scale intermittency in low–Pr number convection compared to the one at larger Pr. Further detailed studies of the statistics of strain (energy dissipation rate) and vorticity (local enstrophy), such as the differences in the enstrophy production mechanisms in convection at $Pr \ll 1$ compared to $Pr \sim 1$ are currently in progress and will be reported elsewhere.

Acknowledgments

The work is supported by Research Unit 1182 of the Deutsche Forschungsgemeinschaft. We acknowledge supercomputing time at the Jülich Supercomputing Centre provided by grant HIL07 of the John von Neumann Institute for Computing. Furthermore, we acknowledge an award of computer time provided by the INCITE program. This research used resources of the ALCF at ANL, which is supported by the DOE under contract DE-AC02-06CH11357.

References

1. M. Breuer, S. Wessling, J. Schmalzl and U. Hansen, Phys, Rev. E **69**, 026302 (2004).
2. R. Camussi and R. Verzicco, Phys. Fluids **10**, 516 (1998).
3. F. Chillà and J. Schumacher, Eur. J. Phys. E **35**, 58 (2012).
4. S. Cioni, S. Ciliberto and J. Sommeria, J. Fluid Mech. **335**, 111 (1997).
5. S. Cioni, S. Horanyi, L. Krebs and U Müller, Phys. Rev. E **56**, R3753 (1997).
6. M. O. Deville, P. F. Fischer, and E. H. Mund, *High-order methods for incompressible fluid flow*, Cambrigde University Press, Cambridge, 2002.
7. P. F. Fischer, J. Comp. Phys. **133**, 84 (1997).
8. R. M. Kerr and J. R. Herring, J. Fluid Mech. **419**, 325 (2000).
9. P. K. Mishra and M. Verma, Phys. Rev. E **81**, 056316 (2010).
10. P. F. Fischer, J. W. Lottes and S. G. Kerkemeier, http://nek5000.mcs.anl.gov
11. J. D. Scheel, E. Kim and K. R. White, J. Fluid Mech. **711**, 281 (2012).
12. J. D. Scheel, M. S. Emran and J. Schumacher, New J. Phys. **15**, 113063 (2013).
13. H. Schlichting and K. Gersten, *Boundary layer Theory*, 8th ed. Springer, 2000.
14. J. Schumacher, J. D. Scheel, D. Krasnov, D. A. Donzis, V. Yakhot and K. R. Sreenivasan, Proc. Natl. Acad. Sci. USA **111**, 10961 (2014).
15. T. Takeshita, T. Segawa, J. A. Glazier and M. Sano, Phys. Rev. Lett. **76**, 1465 (1996).
16. E. van der Poel, R. A. J. M. Stevens and D. Lohse, J. Fluid Mech. **736**, 177 (2013).
17. M. Verma, P. K. Mishra, A. Pandey and S. Paul, Phys. Rev. E **85**, 016310 (2012).

Effect of Buoyancy on Turbulent Mixed Convection Flow Through Vertical and Horizontal Channels

N. Satish, K. Venkatasubbaiah* and R. Harish

Department of Mechanical and Aerospace Engineering,
Indian Institute of Technology Hyderabad, Hyderabad -502205, India
**E-mail: kvenkat@iith.ac.in*

The effect of buoyancy on turbulent mixed convection flow through vertical and horizontal channels has been numerically studied. Turbulence is modeled by Reynolds averaged Navier -Stokes equations (RANS) with standard $k-\epsilon$ turbulence model. The present analysis is valid when the buoyancy force effects are small compared with the forced convection effects. The mixed convection flow problem is formulated by two dimensional unsteady incompressible flow with the buoyancy term modeled by Boussinesq approximation. The governing equations are solved by high accuracy compact finite difference schemes with four stage Runge-Kutta method for time integration. Results are reported for Reynolds number of 6000 with Richardson number varied from 0 to 0.5. The velocity, temperature profiles and average Nusselt number values are presented. The buoyancy shows significant effect on the flow and heat transfer characteristics in vertical channel in comparison with the horizontal channel. The effects of heated plate velocity relative to fluid velocity in a channel are reported. The present results are matching well with the experimental results available in the literature.

Keywords: Turbulent mixed convection; Assisting flow; Opposing flow; Richardson number.

1. Introduction

The study of turbulent mixed convection flow through a channel is important as it has several engineering applications such as heat exchangers, internal cooling system of turbine blades, cooling of nuclear reactors and electronic devices. Studies on mixed convection flows has received considerable attention in literature. Jackson *et al.*[1] presented a comprehensive review of experimental and theoretical studies on mixed convection flow in vertical tubes. Experiments[2,3] were performed to investigate the effect of buoyancy on turbulent mixed convection flows in a vertical passages. Direct numerical simulations[4] were performed on combined forced and natural turbulent convection in a vertical plane channel. Sekimoto *et al.*[5] numerically investigated the effect of buoyancy on turbulent mixed convection flow in a horizontal square duct and they found that the secondary flow regimes appear with increase in Richardson number. Conjugate turbulent mixed convection heat transfer in a vertical channel were studied using $k-\epsilon$ turbulence model[6]. Turbulent mixed convection of air flow in vertical tubes were studied using a zonal turbulence model[7].

Studies in the past have not emphasized the effect of buoyancy on turbulent forced convection dominated mixed convection flows. The effects of buoyancy on turbulent mixed convection flow through vertical and horizontal channels using high accuracy numerical methods are not available in literature. This has been the motivation for present investigation. Turbulent mixed convection flow through vertical and horizontal channels is numerically investigated. Turbulence is modeled by Reynolds Averaged Navier Stokes (RANS) equations with standard $k-\epsilon$ model. The effects of buoyancy on flow and heat transfer characteristics for assisting and opposing mixed convection flows are reported.

2. Governing Equations

Turbulent mixed convection flow through vertical channel and horizontal channel is described by Reynolds averaged Navier-Stokes equations (RANS). The buoyancy term is modeled by Boussinesq approximation that treats density as a constant value in all equations, except for the buoyancy term in the momentum equation. Turbulence is modeled with standard $k-\epsilon$ model including the contribution of buoyancy force in the turbulent kinetic energy generation and dissipation. Two dimensional unsteady RANS equations are obtained in the stream function ψ and vorticity ω formulation. Advantages of this formulation are added accuracy due to exact satisfaction of mass conservation and reduction in the number of unknowns to two as compared to three unknowns for primitive variable formulation. Pressure is eliminated by taking the curl of the momentum equation to give the vorticity transport equation (VTE).

The dimensionless governing equations are obtained by the following variables.

$$X = \frac{x}{W};\ Y = \frac{y}{W};\ U = \frac{u}{U_i};\ V = \frac{v}{U_i};\ \tau = \frac{tU_i}{W};\ \theta = \frac{T-T_i}{T_s-T_i};\ P = \frac{p-p_i}{\rho U_i^2};$$

$$\Omega = \omega\frac{W}{U_i};\ \Psi = \frac{\psi}{U_i W}\ K = \frac{k}{U_i^2};\ \varepsilon = \frac{\epsilon W}{U_i^3}$$

The non-dimensional governing equations are as follows:

$$\frac{\partial^2 \Psi}{\partial X^2} + \frac{\partial^2 \Psi}{\partial Y^2} = -\Omega \tag{1}$$

$$\begin{aligned}\frac{\partial \Omega}{\partial \tau} + U\frac{\partial \Omega}{\partial X} + V\frac{\partial \Omega}{\partial Y} &= \frac{1}{Re}\frac{\partial^2}{\partial X^2}\left[\left(1+\frac{\nu_t}{\nu}\right)\Omega\right] + \frac{Gr_y}{Re^2}\frac{\partial \theta}{\partial X} \\ &\pm \frac{Gr_x}{Re^2}\frac{\partial \theta}{\partial Y} + \frac{1}{Re}\frac{\partial^2}{\partial Y^2}\left[\left(1+\frac{\nu_t}{\nu}\right)\Omega\right] + \frac{2}{\nu Re}\frac{\partial U}{\partial Y}\frac{\partial^2 \nu_t}{\partial X^2} \\ &- \frac{2}{\nu Re}\frac{\partial V}{\partial X}\frac{\partial^2 \nu_t}{\partial Y^2} + \frac{2}{\nu Re}\left[\frac{\partial V}{\partial Y} - \frac{\partial U}{\partial X}\right]\frac{\partial^2 \nu_t}{\partial X \partial Y}\end{aligned} \tag{2}$$

$$\begin{aligned}\frac{\partial \theta}{\partial \tau} + U\frac{\partial \theta}{\partial X} + V\frac{\partial \theta}{\partial Y} &= \frac{1}{Re}\frac{\partial}{\partial X}\left[\left(\frac{1}{Pr} + \frac{\nu_t}{\nu Pr_t}\right)\frac{\partial \theta}{\partial X}\right] \\ &+ \frac{1}{Re}\frac{\partial}{\partial Y}\left[\left(\frac{1}{Pr} + \frac{\nu_t}{\nu Pr_t}\right)\frac{\partial \theta}{\partial Y}\right]\end{aligned} \tag{3}$$

where $Gr_x = \frac{g_x \beta \Delta T W^3}{\nu^2}$ is the Grashof number in x direction; $Gr_y = \frac{g_y \beta \Delta T W^3}{\nu^2}$ is the Grashof number in y direction; $Pr = \frac{\nu}{\alpha}$ is the Prandtl number; $\nu_t = \frac{C_\mu k^2}{\epsilon}$; Here, length scale (L) is chosen as width of the channel. The quantity $\frac{Gr}{Re^2}$ is known as the Richardson number (Ri). This number quantifies the influence of free convection in comparison with the forced convection. positive and negative signs on Ri refer to assisting and opposing mixed convection flows. In a vertical channel, the upward forced flow is termed as assisted mixed convection flow because the natural convection created by buoyancy is in the same direction as the bulk flow. In contrast, the downward flow is called opposing mixed convection flow based on its direction opposite to the natural convection. Details of standard $k-\epsilon$ equations are given in Wilcox[8].

2.1. *Boundary and Initial conditions*

Uniform velocity and temperature are specified at the inlet of the channel. No-slip velocity boundary conditions are applied on the walls. At the outlet of the channel, fully developed flow condition is applied for both velocity and temperature fields. The left wall of the channel is considered to be insulated and adiabatic boundary condition is applied to it. Constant wall temperature is applied at the right wall.The turbulent kinetic energy and normal gradient of dissipation is set to zero at the walls. The initial conditions for velocity and temperature fields are considered as inviscid solution.

3. Numerical Methods

The governing equations are discretized using finite difference schemes and the solver is developed in Fortran 90. The stream function equation (SFE) is discretized using a second order central difference scheme CD_2. An iteration method of Bi-conjugate gradient is used to solve the SFE. The VTE, energy equation (EE), Kinetic energy and dissipation equations are solved by discretizing the diffusion terms using CD_2 scheme. Time integration is performed explicitly by the four-stage Runge-Kutta (RK4) scheme and non-linear convection terms are discretized by high accuracy compact schemes, whose details are given in[9,10]. A small time step of 10^{-4} is used to avoid numerical instability.

4. Grid Independence and Validation

The grid independence test has been conducted with different grids for Reynolds number Re = 6000. It is noticed that the velocity and temperature profiles are shown grid independent with gird size of 805 × 80. Hence all simulations are reported with this grid size. The accuracy of the present numerical model and the method of solution has been tested with the experimental results available in the literature. The results are validated for pure forced convection flow (Ri = 0) through a channel at the Reynolds number Re = 6000. The average Nusselt number is 14.58 for such flow using the experimental correlation[11]. The average Nusselt number obtained using the present numerical model is 14.39.

5. Results and Discussions

The turbulent mixed convection flow through vertical and horizontal channels are numerically studied and the heat transfer characteristics are reported for different Richardson numbers. The Prandtl number (Pr) for air is taken as 0.72. Here the buoyancy force effects are small compared with the forced convection effects. The vertical channel has height to width aspect ratio of 9.0. The right wall is maintained at constant surface temperature (T_s) and it is greater than the inlet temperature (T_i) of air.

To know the effect of buoyancy, results of steady state vorticity and temperature contours in vertical channel and horizontal channel are shown in Figs. 1 and 2 for Re = 6000 with Ri = 0.3. Figure 1(a) represents the pure forced convection flow (Ri = 0) which depicts the boundary layer formation near the surfaces, the center core vortex and leading edge vortex formations

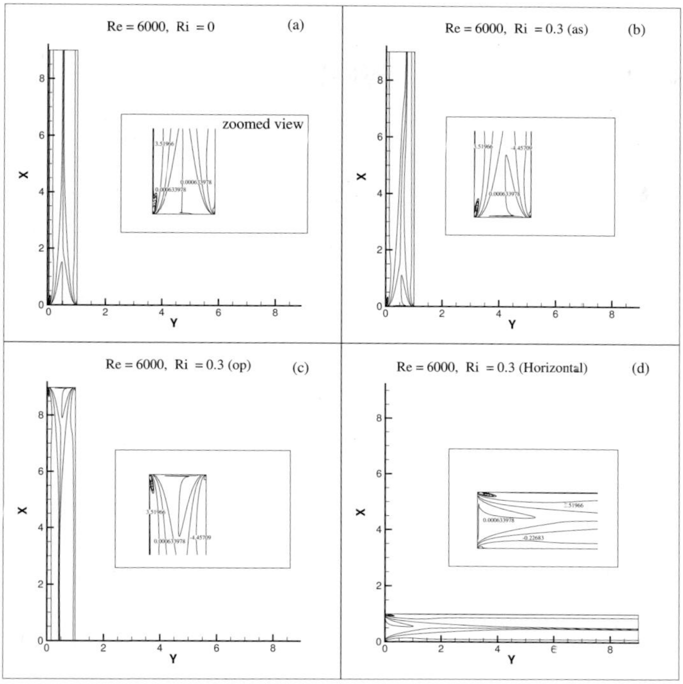

Fig. 1. Vorticity contours in a vertical and horizontal channels.

near the inlet of the channel due to turbulence and viscous effects. Figure 1(b) illustrates the reduction of the vortex formation strength in a vertical channel for assisting mixed convection flow with Ri = 0.3 compared to Ri = 0, due to stabilizing the flow by buoyancy. The strength of the leading edge vortex increases for opposing mixed convection flow in comparison with the assisting mixed convection flow since buoyancy destabilizes the flow, as shown in Fig. 1(c). The leading edge vortex strength has increased in a horizontal channel with Ri = 0.3 compared to Ri = 0 due to buoyancy, as shown in Fig. 1(d). The corresponding temperature contours are shown in Fig. 2 and it is observed that the growth of thermal boundary layer slightly varies with Ri.

To know the effect of buoyancy in a vertical channel, results of the velocity and temperature profiles at the outlet of the channel are shown in Fig. 3 for assisting and opposing mixed convection flows at Re = 6000. The velocity increases and temperature decreases near the heated wall with increase in Ri due to buoyancy aids the flow, as shown in Fig. 3 (a & b). For

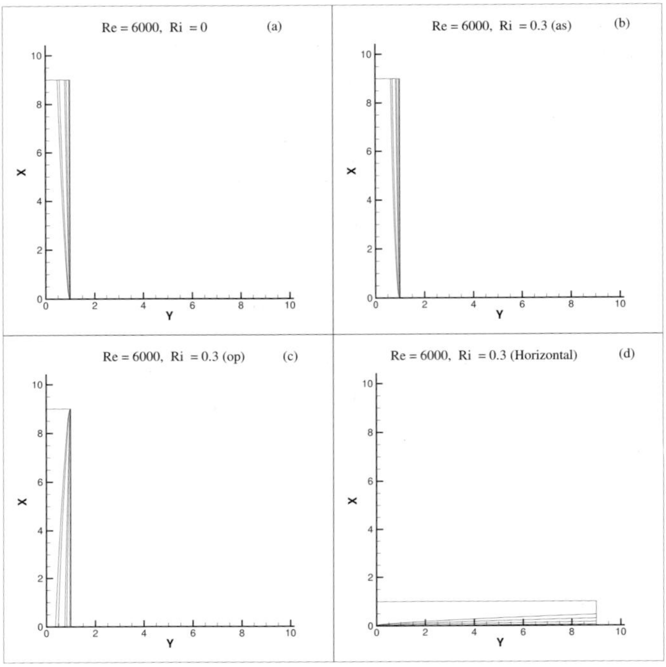

Fig. 2. Temperature contours in a vertical and horizontal channels.

Table 1: Average Nusselt number values for assisting and opposing mixed convection flows in a vertical channel

Type of flow	Re	Ri	Nu_{avg}
Forced convection flow	6000	0	14.39
Assisting flow	6000	0.1	14.63
Assisting flow	6000	0.3	15.05
Assisting flow	6000	0.5	15.45
Opposing flow	6000	0.1	14.22
Opposing flow	6000	0.3	13.84
Opposing flow	6000	0.5	13.51

opposing mixed convection flows, the velocity increases near the left wall and temperature increases with increase in Ri due to buoyancy opposing the bulk flow, as shown in Fig. 3(c & d). The average Nusselt number values with different Richardson numbers are given in Table 1 for assisting and opposing mixed convection flows in a vertical channel at Re = 6000.

For assisting mixed convection flows, the average Nusselt number increases with increase in Ri due to buoyancy aids the flow. The average Nusslet number decreases with increase in Ri for opposing mixed convection flow due to buoyancy opposes the bulk flow. This shows that the buoyancy enhances the heat transfer rate in assisting mixed convection flows and opposite trend for opposing mixed convection flows.

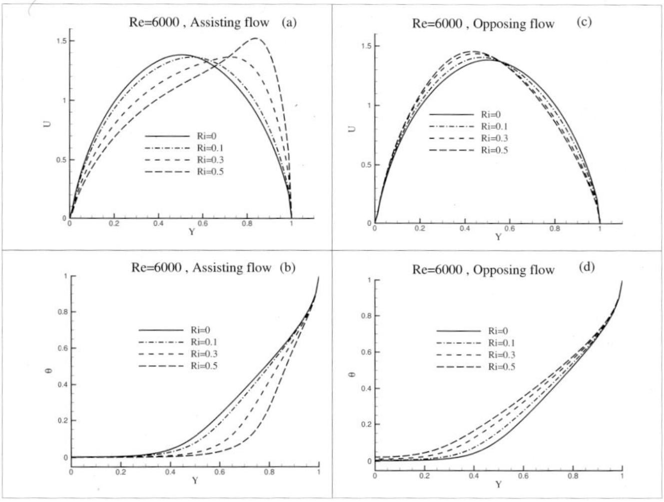

Fig. 3. Effect of buoyancy for assisting and opposing mixed convection flows in a vertical channel (a, c) Velocity and (b, d) Temperature.

To know the effect of buoyancy in a horizontal channel, results of the velocity and temperature profiles at the outlet of the channel are shown in Fig. 4 (a & b) for different Richardson numbers at Re $= 6000$. The velocity slightly increases near the bottom wall and temperature increases with increase in Ri due to buoyancy. The average Nusselt number values with different Richardson numbers are given in Table 2 for mixed convection flow in a horizontal channel at Re $= 6000$. From Table 2, the average Nusselt number decreases with increase in Ri due to buoyancy force acts normal to the flow. This shows that the buoyancy reduces the heat transfer rate in a horizontal channel flows.

To know the effect of heated plate velocity related to fluid velocity in a vertical channel, results of the velocity and temperature profiles at the outlet of the channel are shown in Fig. 4(c & d) for different plate velocities at Re $= 6000$ with Ri $= 0.3$. The velocity increases and temperature de-

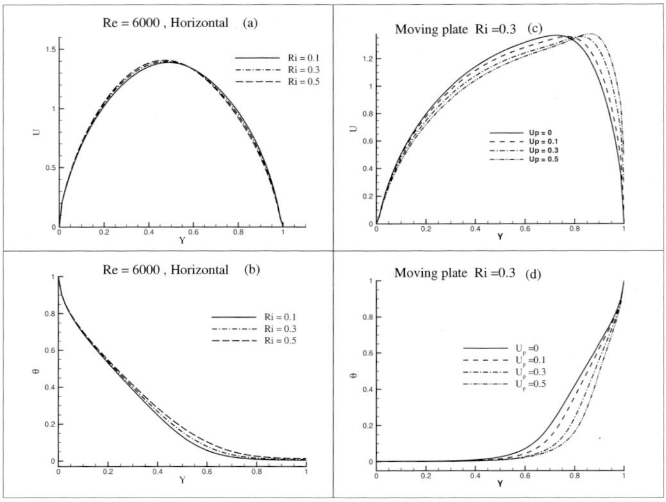

Fig. 4. (a, b)Velocity and Temperature profiles for mixed convection flow in a horizontal channel; (c, d) Velocity and Temperature profiles in a vertical channel for different heated plate velocities.

Table 2: Average Nusselt number values for mixed convection flows in a horizontal channel

S.No	Re	Ri	Nu_{avg}
1	6000	0	14.39
2	6000	0.1	14.35
3	6000	0.3	14.23
4	6000	0.5	14.11

creases near the heated wall with increase in plate velocity. This shows that heat transfer rate increases with increase in plate velocity due to decrease in thermal boundary layer thickness.

6. Conclusions

The effect of buoyancy on flow and heat transfer characteristics of mixed convection flow in a vertical and horizontal channels has been investigated numerically. Results are presented at a Reynolds number of 6000 with different Richardson numbers varied from 0 to 0.5. In a vertical channel, the velocity increases near the heated wall and the convective heat transfer rate increases with increase in Richardson number for assisting mixed convection flow due to buoyancy aiding the flow. For opposing mixed convection flows in vertical channel, the significant decrease in convective heat transfer rate with increase in Richardson number due to buoyancy opposing the bulk flow. The convective heat transfer coefficient slightly decreases with increase in Richardson number for mixed convection flow in a horizontal channel due to buoyancy acts normal to the flow. Buoyancy significantly altered the flow and heat transfer characteristics in a vertical channel compared to horizontal channel. The convective heat transfer rate increases with increase in heated plate velocity due to decrease in thermal boundary layer thickness. Present results would be useful to understand the flow and heat transfer characteristics of mixed convection flows in vertical and horizontal channels.

References

1. J. D. Jackson, M. A. Cotton and B. P. Axcell, Studies of mixed convection in vertical tubes, *Int. J. Heat and Fluid Flow*, 10, 2–15 (1989).
2. Xiaodong Zhang and Sandip Dutta, Heat transfer analysis of buoyancy-assisted mixed convection with asymmetric heating conditions, *Int. J. Heat and Mass Transfer*, 41, 3255–3264 (1998).
3. J. Wang, J. Li and J. D. Jackson, A study of the influence of buoyancy on turbulent flow in a vertical plane passag, *Int. J. Heat and Fluid Flow*, 25, 420–430 (2004).
4. N. Kasagi and M. Nishimura, Direct numerical simulation of combined forced and natural turbulent convection in a vertical plane channel, *Int. J. Heat and Fluid Flow*, 18, 88–89 (1997).
5. A. Sekimoto, G. Kawahara, K. Sekiyama, M. Uhlmann, and A. Pinelli, Turbulence and buoyancy-driven secondary flow in a horizontal square duct heated from below, *Physics of Fluids*, 23, 075103:1–19 (2011).
6. Roy N. Mathews and C. Balaji, Numerical simulation of conjugate, turbulent mixed convection heat transfer in a vertical channel with discrete heat sources, *International Communications in Heat and Mass Transfer*, 33, 908–916 (2006).
7. M. Shahraeeni and M. Raisee, Investigation of turbulent mixed convection of air flow in vertical tubes using a zonal turbulence model, *Int. J. Heat and Fluid Flow*, 31, 179–190 (2010).
8. David C. Wilcox, Turbulence modeling for CFD, *DCW Industries,Inc., USA* (2006).
9. T. K. Sengupta, S. K. Sirkar and A. Dipankar, High accuracy schemes

for DNS and acoustics, J. Sci. Comput. 26, 151–193 (2006).
10. K. Venkatasubbaiah and T. K. Sengupta, Mixed convection flow past a vertical plate: Stability analysis and its direct simulation, Int. J. Therm. Sci. 48, 461–474 (2009).
11. Frank P. Incropera and David P. Dewitt, Fundamentals of Heat and Mass Transfer, *Wiley India (p) Ltd., New Delhi* (2006).

Computation of Boundary Layer Flow over Porous Laminated Flat Plate

Aswathy Nair K.* and A. Sameen

Department of Aerospace Engineering, Indian Institute of Technology Madras, Chennai, Tamil Nadu 600 036, India
**E-mail: ae12s013@smail.iitm.ac.in*

S. Anil Lal

Department of Mechanical Engineering, College of Engineering Trivandrum, Thiruvananthapuram, Kerala 695 016, India

The slip velocity of the flow at the porous-fluid interface is studied using a 2D porous region, modeled as an array of circular cylinders in square and hexagonal arrangements to resemble the structure of real porous media. The Navier-Stokes and continuity equations are solved using a hybrid FEM-FVM solver by applying appropriate macroscopic boundary conditions. The existence of a slip velocity at the porous-fluid interface is demonstrated and it is observed that the slip velocity increases with increase in porosity and flow Reynolds number.

Keywords: porous lamination; boundary layer; slip velocity; similarity profile

1. Introduction

Flow through porous media has been studied extensively due to its relevance in engineering, such as sediment/pollutant transport in hydrology and soil mechanics, nutrient transport/transfusion in bio-medical engineering, catalyst engineering, heat exchangers and coolers, oil permeation in petroleum engineering etc. to mention a few. Coupled flows are fluid flows through and over a porous substrate and have been of interest lately with specific application in flow control. The applications of porous media as a flow control device has not received much acceptance because the flow physics at the porous-fluid interface is inadequately understood. In this computational study, we try to understand the flow characteristics at the porous-fluid interface, most importantly the inter-facial boundary conditions, and then investigate the effectiveness of porous media in passive flow control.

Many theories have been developed for the porous-fluid interface, with each of them valid under certain limitations. The flow in the porous region is governed by the well-known Darcy's law [1], which is a first order equation suitable for low permeability. The flow in the homogeneous fluid region is governed by the second order Navier-Stokes equations. The difference in the order of the two governing equations renders it impossible to deduce boundary conditions from them. Hence, analyses of the flow through and over the porous medium have been made with intuitive assumptions from kinematic and dynamic considerations. Brinkman [2], in his analytical work, modeled the coupled flow as a macroscopic flow with an added diffusion term to the Darcy's equation. Studies on actual porous materials by Beavers and Joseph [3], based on a slip-flow boundary condition at the interface with an experimentally determined parameter, proved inconclusive due to the inability to arrive at a unique empirical relation for the parameter. Later works of Sahraoui and Kaviany [4] and Larson and Higdon [5,6] with a tube bundle/array of cylinders to model 2D porous media, showed that the Brinkman model is valid only for very dilute systems with isotropic structures, which is rarely the practical case. However, Ochoa-Tapia and Whitaker [7,8] developed a boundary condition based on momentum transfer at the porous-fluid interface to match the Brinkman model to the Stokes flow of the homogeneous fluid above the porous region. Nonetheless, the authors suggest further extensive experimental and numerical studies to verify and establish their boundary condition. More recently Breugem *et al.* [9] developed a self-similar solution for the developing laminar flow over a permeable wall, which is limited to low value of permeability κ, defined as $\kappa = v\ \mu\ \Delta x/\Delta P$ where v is the average superficial velocity calculated as if the fluid were the only phase present in the porous medium, μ is the dynamic viscosity of the fluid and $\Delta P/\Delta x$ is the pressure gradient applied in the direction of the flow. Thus, an understanding of slip velocity at moderate values of porosity (ϕ) and lower values of κ remains largely elusive.

The slip velocity and other flow parameters are averaged over a minimal volume of the flow domain to eliminate the local jumps and irregularities in these properties due to the complex and random structure of the porous media. A concept analogous to the control volume approach for homogeneous fluid flows is followed here, namely the Representative Elemental Volume (REV) [10]. Thus, the flow properties are averaged over the REV to eliminate the local singularities. For the present 2D study, a 2D REV or spatial averaging of the required data is performed. The aim of the present study is to understand the flow at the porous-fluid interface for systems with moderate and high porosities (ϕ) over a range of flow Reynolds number, Re_d, which are defined in Sec. 2.

2. Computational Analysis

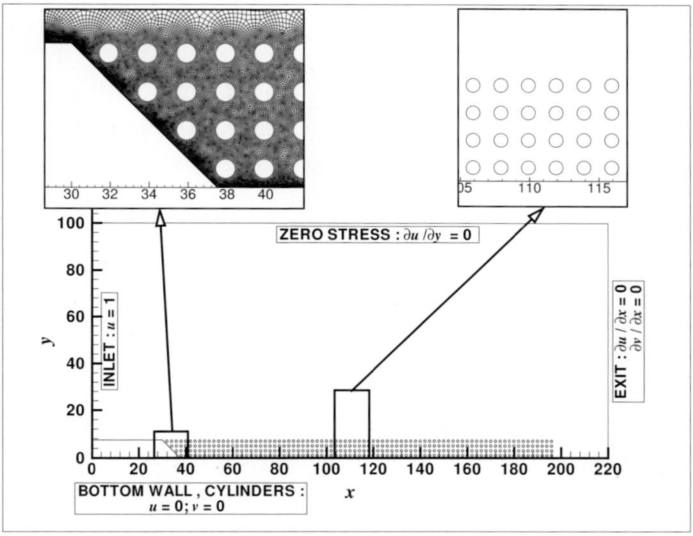

Fig. 1. The domain used for the computational study. Insets:(top left)the zoomed in view of the upstream region with the unstructured mesh; (top right) the zoomed in view of the porous region near the mid-part of the domain.

A direct numerical simulation of the Navier-Stokes equation for the boundary layer flow over the porous laminated flat plate is performed in this study. The flow is assumed incompressible. The porous region is modeled as an array of circular cylinders in square and hexagonal arrangements. The stream-wise and wall-normal distance variables, X and Y respectively, are non-dimensionalised with the cylinder diameter d ($x = X/d$ and $y = Y/d$). The porosity, ϕ, is defined as ratio of void volume to total volume of the REV. The Reynolds number is defined based on d as $Re_d = ud/\nu$, where u is the non-dimensional stream-wise velocity and ν is the kinematic viscosity. The height h of the porous layer is another characteristic dimension used in this analysis. The non-dimensional Navier-Stokes equation along with continuity equation for incompressible flow, Eqns. (1), (2) and (3) where u and v are the non-dimensional stream-wise and wall-normal velocities, are solved using a hybrid FEM-FVM solver[11].

Continuity Equation:

$$\frac{\partial u}{\partial x} + \frac{\partial v}{\partial y} = 0 \tag{1}$$

X-momentum Equation:

$$\frac{\partial u}{\partial t} + u\frac{\partial u}{\partial x} + v\frac{\partial u}{\partial y} = -\frac{\partial p}{\partial x} + \frac{1}{Re_d}\left(\frac{\partial^2 u}{\partial x^2} + \frac{\partial^2 u}{\partial y^2}\right) \tag{2}$$

Y-momentum Equation:

$$\frac{\partial v}{\partial t} + u\frac{\partial v}{\partial x} + v\frac{\partial v}{\partial y} = -\frac{\partial p}{\partial y} + \frac{1}{Re_d}\left(\frac{\partial^2 v}{\partial x^2} + \frac{\partial^2 v}{\partial y^2}\right) \tag{3}$$

Models with square and hexagonal arrays of circular cylinders with $\phi =$ 0.4061, 0.4973, 0.6509, 0.8036 and 0.9127 are used in the computation. A uniform velocity boundary condition of $u = 1.0$ is specified on the inlet boundary. Zero-stress boundary condition, $\partial u/\partial y = 0$, is specified on the top boundary. The exit boundary is specified by a zero normal gradient boundary condition for both velocities i.e. $\partial u/\partial x = 0$ and $\partial v/\partial x = 0$. No slip boundary condition is imposed on the bottom wall and the cylinder surfaces.

The domain for $\phi = 0.8036$ used in the study is shown in Fig. 1. The domain has a total length of $220d$ and height of $100d$, which were decided after domain independence studies. An inlet length of $30d$ is provided upstream of the porous region and that of $20d$ is provided at the end of the porous region. It is discretized using an unstructured mesh into $4,87,226$ quadrilateral elements and $4,92,100$ nodal points, with clustering in the porous region. Other porosities are analyzed using similar domains with dimensions corresponding to the cylinder arrangements within the domain. Domain and grid independence are verified for all the cases.

3. Results and Discussions

The results of the numerical study of the flow over a porous laminated flat plate of porosities $\phi = 0.4061,\ 0.4973,\ 0.6509,\ 0.8036$ and 0.9127 are presented in this section. The u-velocity and z-component vorticity (ω_z) contours for the case with $\phi = 0.4061$ and $\phi = 0.8036$ at $Re_d = 100$ are shown in Fig. 2. A concentration of clockwise vorticity at the fluid-porous interface (just above the top layer of cylinders) can be seen in the ω_z contours, which suggests a wall-like interface. The u contours show a permeation velocity at the porous-fluid interface which indicates the existence of a finite slip velocity U_s. The local, instantaneous velocity profile is plotted in the u contours to qualitatively demonstrate this U_s. The presence of slip further indicates that the porous-fluid interface acts as a wall with a slip velocity.

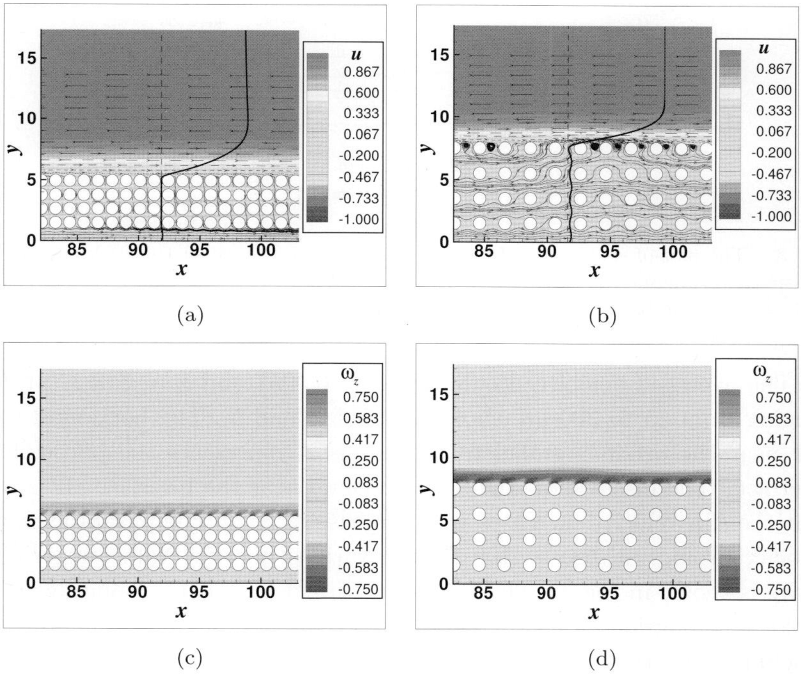

Fig. 2. The local, instantaneous velocity profile (solid line) plotted on the instantaneous u-velocity contour at $Re_d = 100$: (a)$\phi = 0.4061$, (b)$\phi = 0.8036$, and contour of ω_z vorticity : (c)$\phi = 0.4061$, (d)$\phi = 0.8036$.

The local variations in the structure of the porous media and the porous-fluid interface gives rise to local jumps and disparities in the flow parameters, as described earlier in section 1. Figure 3 shows the plot of spatially and temporally averaged U_s ($\overline{< U_s >}$, where $< >$ denotes spatial averaging and overbar denotes temporal averaging) against the REVs chosen for the spatial averaging, for various Re_d. It was found that the variation in

$\overline{< U_s >}$ was less than 5% when the averaging length or the dimension of the REV was greater than or equal to $4d$. Thus an REV of dimension greater than $4d$ was chosen to smooth out the local disparities.

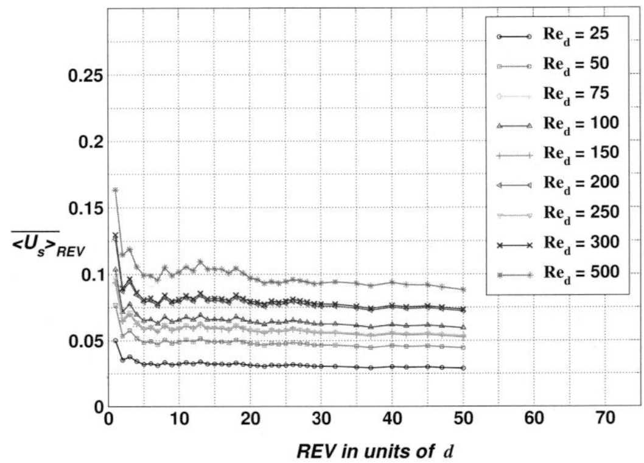

Fig. 3. The plot of spatially and temporally averaged U_s against the REVs chosen for the spatial averaging for various Re_d. It is seen that there is less than 5% variation in $\overline{< U_s >}$ when the REV dimension is greater than or equal to $4d$.

The averaged velocity profiles for $\phi = 0.4061$ and $\phi = 0.9127$ at Re_d ranging from 25 to 300 are shown in Figs. 4 and 5. The horizontal solid line near the origin, parallel to the abscissa, represents the porous-fluid interface. The insets show the zoomed views near the interface. It is seen that the slip velocity increases with increase in Re_d. The plot of averaged velocity profiles at $Re_d = 100$ for the various porosity values mentioned above is shown in Fig. 6. The Blasius profile for flow over a solid flat plate is plotted alongside for comparison. From the figure, it can be qualitatively concluded that the value of slip velocity increases with the porosity as well.

4. Conclusion

The boundary layer flow over a porous laminated flat plate was investigated numerically for a range of Reynolds numbers, Re_d. The existence of a finite slip velocity at the porous-fluid interface was demonstrated for the range of porosities from 0.4 to 0.9 and flow Reynolds numbers, $Re_d : 25 - 300$. The flow parameters were spatially averaged over a nominally sufficient REV, chosen after a careful analysis. It was found that the slip velocity, U_s at the porous-fluid interface increases with increase in flow Reynolds number, Re_d, and the porosity.

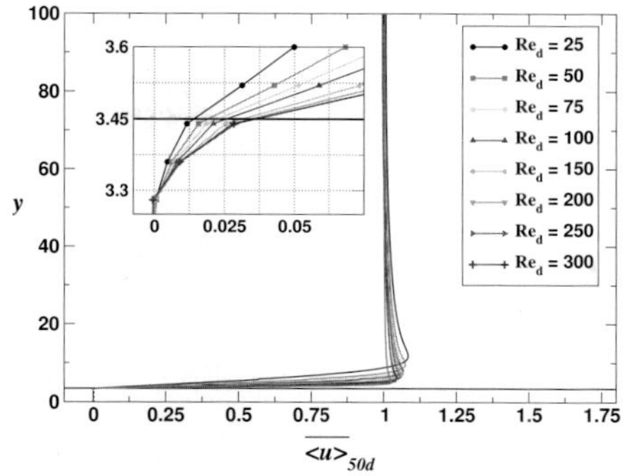

Fig. 4. The averaged velocity profiles with REV= $50d$ for $\phi = 0.4061$ at various Re_d. The horizontal line represents the porous-fluid interface. The insets show the zoomed in views near the interface. It is seen that the slip velocity increases with increase in Re_d.

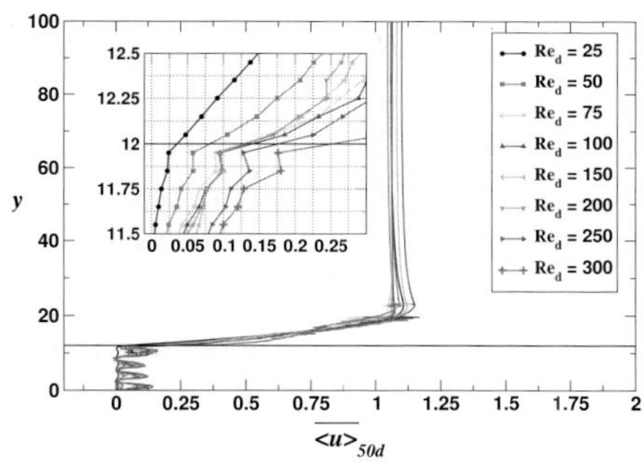

Fig. 5. Same as Fig. 4 for $\phi = 0.9127$ at various Re_d.

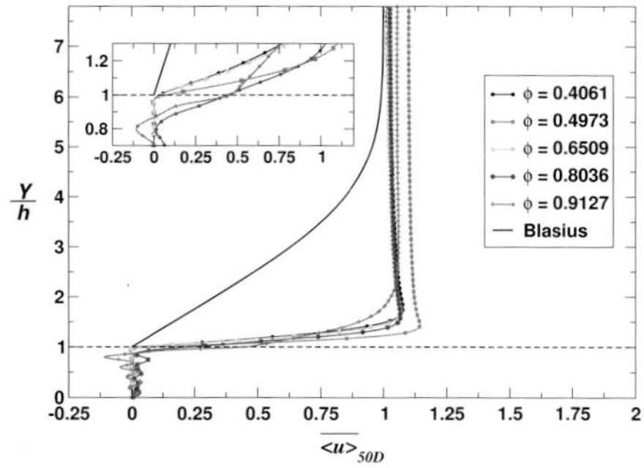

Fig. 6. The averaged velocity profiles for various porosities at $Re_d = 100$. It can be noted that the slip velocity increases with increase in porosity.

References

1. H. Darcy, *Les Fontaines Publiques de la Ville de Dijon.* (Dalmont, Paris, 1856).
2. H. C. Brinkman, A calculation of the viscous force exerted by a flowing fluid on a dense swarm of particles, *App. Sci. Res.* **A1**, 27 (1947).
3. G. S. Beavers, D. D. Joseph, Boundary condition at a naturally permeable wall. *J. Fluid Mech.* **30**, 197 (1967).
4. M. Sahraoui, M. Kaviany, Slip and no-slip velocity boundary conditions at interface of porous, plain media. *Int. J. Heat Mass Transfer* **35**(4), 927 (1992).
5. R. E. Larson, J. J. L. Higdon, Microscopic flow near the surface of two-dimensional porous media, Part 1: axial flow. *J. Fluid Mech.* **166**, 449 (1986).
6. R. E. Larson, J. J. L. Higdon, Microscopic flow near the surface of two-dimensional porous media, Part 2: transverse flow. *J. Fluid Mech.* **178**, 119 (1987).
7. J. A. Ochoa-Tapia, S. Whitaker, Momentum transfer at the boundary between a porous medium and a homogeneous fluid - I: theoretical development. *Int. J. Heat Mass Transfer* **38**(4), 2635 (1995).
8. J. A. Ochoa-Tapia, S. Whitaker, Momentum transfer at the boundary between a porous medium and a homogeneous fluid - II: comparison with experiment. *Int. J. Heat Mass Transfer* **38**(4), 2647 (1995).
9. W. P. Breugem, B. J. Boersma, R. E. Uittenbogaard. The laminar boundary layer over a permeable wall. *Trans. Porous Med.* **59**, 267 (2005).
10. S. Whitaker, *The method of volume-averaging*, **13**, Springer (1999).
11. S. Anil Lal, E. Jabir. A hybrid finite element finite volume method for the analysis of incompressible flow through complex geometries using mixed grids. *Proc. I. Mech E, Part G: J. Aero. Engg.* **224**(1), 23 (2010).

Boundary Condition Development for an Adverse Pressure Gradient Turbulent Boundary Layer at the Verge of Separation

V. Kitsios*,[a], C. Atkinson[a], J. A. Sillero[b], G. Borrell[b], A. G. Gungor[c], J. Jiménez[b] and J. Soria[a,d]

[a] *Laboratory For Turbulence Research in Aerospace and Combustion, Department of Mechanical and Aerospace Engineering, Monash University, Clayton 3800, AUSTRALIA*

[b] *School of Aeronautics, Universidad Politécnica de Madrid, E-28040 Madrid, SPAIN*

[c] *ITU Faculty of Aeronautics and Astronautics, Department of Astronautical Engineering, Maslak 34469 Istanbul, TURKEY*

[d] *Department of Aeronautical Engineering, King Abdulaziz University, Jeddah 21589, KINGDOM OF SAUDI ARABIA*

* *E-mail: vassili.kitsios@monash.edu*

Details are provided on the development of a farfield boundary condition for the direct numerical simulation of an adverse pressure gradient (APG) turbulent boundary layer (TBL) at the verge of separation. The APG TBL achieves a region constant pressure velocity to freestream velocity ratio over a momentum thickness based Reynolds number range of $Re_\theta \approx 2000$ to 6000. Mean velocity deficit and Reynolds stress profiles are presented under both friction velocity and pressure velocity based scaling within this range. The pressure velocity scaling demonstrates a collapse of the statistical profiles.

Keywords: Adverse Pressure Gradient; Turbulent Boundary Layer; Direct Numerical Simulation.

1. Introduction

The performance of many engineering systems relies on fluid flows remaining attached to aerodynamic surfaces, and the occurrence of flow separation can potentially result in catastrophic consequences or at best energy efficiency degradation. The study of fluid flow separation is, therefore, of utmost importance. Real world examples include the flow over aerofoil geometries including aircraft wings, wind turbine blades, and turbo-machinery. The accurate prediction of TBL separation remains a significant challenge for engineering design. One of the complexities of these aerofoil geometries, however, is that the pressure gradient is constantly changing in the streamwise direction, as in the large eddy simulation of Kitsios *et al.*[1].

In the present study we focus our efforts on the canonical flow configuration of a self-similar TBL subjected to an APG such that the TBL is at the verge of separation, akin to the configuration in the experimental study of Skate & Krogstad[2]. A self-similar APG TBL is defined as having a constant ratio of friction velocity (u_τ) to freestream velocity (u_∞), and also a constant ratio of pressure velocity (u_P) to u_∞[3]. Previous DNS of TBL APG include the non-self-similar separated flow of Gungor *et al.*[4], and self-similar APG cases at relatively low Reynolds numbers in Lee & Sung[5]. In the present study we undertake simulations of the incipient separation case on a flat surface using the TBL DNS code of Simens *et al.*[6] and Borrell *et al.*[7], with a modified farfield APG boundary condition (BC). This effectively decouples the effect of upstream flow history and surface curvature from the influence of the local pressure gradient. The maximum momentum number based Reynolds number in present simulations is $Re_\theta = 6000$.

An overview of the original ZPG TBL DNS code is presented in section 2, with the modifications to the BC required to generated the self-similar APG TBL presented in section 3. In section 4, profiles of first and second order statistical moments from the DNS of the APG and ZPG TBLs are presented scaled on the basis of the friction and pressure velocities. Finally concluding remarks are made in section 5.

2. Direct Numerical Simulation Details

The TBL DNS code adopted within solves the Navier-Stokes equations in a three-dimensional rectangular volume, with constant density (ρ) and kinematic viscosity (ν). The three flow directions are the streamwise (x), wall-normal (y) and spanwise (z), with respective instantaneous velocity components in these directions of u, v and w. Note throughout this paper the time averaged velocity components are denoted by $(\overline{u}, \overline{v}, \overline{w})$, where the overline is the time averaging operation, with associated fluctuating components of (u', v', w'). A fractional-step method[8,9] is used to solve the governing equations for the velocity and pressure (p) fields. Fourier decomposition is used in the periodic spanwise direction, with compact finite difference[10] in the aperiodic wall-normal and streamwise directions. The equations are stepped forward in time using a modified three sub-step Runge-Kutta scheme[6].

The code utilizes MPI and openMP parallelisation. For each MPI process the physical domain is decomposed into streamwise regions containing all spanwise and wall-normal points. The physical sub-domain is further decomposed into wall normal planes for each openMP thread[7]. All spatial derivatives in the spanwise and wall-normal direction can then be calculated with no MPI message passing. To calculate the streamwise derivatives the data is rearranged into streamwise oriented lines[7].

The boundary conditions of the original ZPG version of the TBL DNS code are as follows. The bottom surface is a flat plate with a no-slip (zero velocity) BC. The spanwise boundaries are periodic. Due to the TBL growing in height as it develops in the streamwise direction, a downstream streamwise normal plane is copied, and mapped to the inlet BC[11]. At the farfield boundary the spanwise vorticity is zero, and the wall normal suction velocity is given by

$$v_{ZPG}(x) = \frac{d\delta^*(x)}{dx} u_{ZPG} \,, \tag{1}$$

where u_{ZPG} is the constant freestream streamwise velocity, and δ^* is the displacement thickness[12].

3. Adverse Pressure Gradient Boundary Condition Development

In order to generate the desired self-similar APG TBL flow, the farfield wall-normal velocity BC must be modified. From Mellor & Gibson[3] the freestream streamwise velocity $u_{APG}(x) \propto x^m$ where $m = -0.23$ for the incipient separation APG TBL ($u_\tau \to 0$). The wall normal suction velocity $v_{APG}(x)$ is deduced from $u_{APG}(x)$ via the boundary layer streamfunction solution in the farfield region to be

$$v_{APG}(x) = -\frac{\partial u_{APG}(x)}{\partial x}\left[y_{BC} - \delta^*(x)\right] + u_{APG}(x)\frac{\partial \delta^*(x)}{\partial x} \,, \tag{2}$$

where y_{BC} is the wall normal position of the farfield boundary[3]. Note for the case of a constant streamwise velocity (zero streamwise derivative), as in the ZPG TBL, (2) becomes equivalent to (1).

The structure of the complete farfield wall normal BC, $v_\infty(x)$, is as follows. In the APG TBL DNS, to allow the rescaling necessary for the inlet BC an initial ZPG TBL is simulated up until the streamwise position $x_s = 100\delta_{99,I}$ (located after the recycling plane) by applying $v_{ZPG}(x)$ as defined in (1). Note $\delta_{99,I}$ is the boundary layer thickness at the inlet. Downstream of the position $x_f = 140\delta_{99,I}$ the wall normal velocity $v_{APG}(x)$ is applied at the farfield boundary as given by (2), which imparts the desired deceleration and hence expansion of the boundary layer. From x_s to x_f

the velocity $v_{APG}(x)$ is gradually introduced using a smoothing function. Finally the farfield velocity is transitioned from suction ($v_\infty(x) > 0$) at $x_o = 760\delta_{99,I}$ to blowing ($v_\infty(x) < 0$) at the outlet to reduce the number of instantaneous reversed flow events, such that numerical stability of the outflow boundary condition is maintained. The structure of the ZPG and APG farfield boundary conditions, $v(x)/u_{\infty,I}$, is illustrated in Fig. 1, where $u_{\infty,I}$ is the freestream streamwise velocity at the inlet. This farfield BC was first implemented and tested in two-dimensional Reynolds Averaged Navier Stokes (RANS) simulations, yielding the appropriate self-similar velocity profiles, and subsequently implemented into the DNS solver.

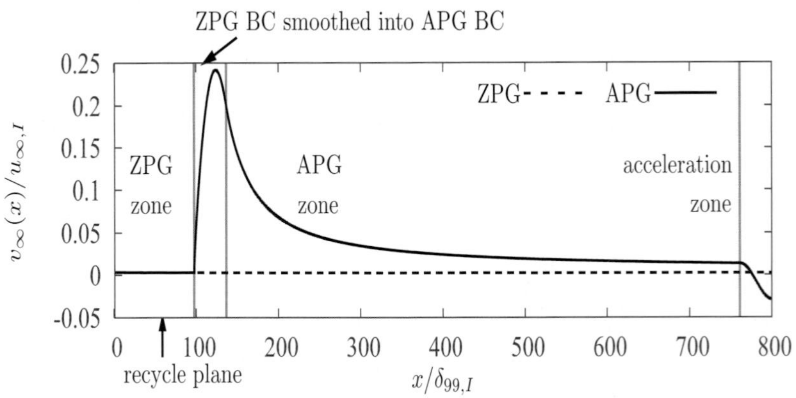

Fig. 1. Farfield wall normal velocity boundary condition in the adverse pressure gradient (solid line) and zero pressure gradient (dashed line) direct numerical simulation.

Table 1. Numerical details of the APG and ZPG TBL DNS.

	$(L_x, L_y, L_z)/\delta_{99,I}$	$N_x \times N_y \times N_z$	$(\Delta x, \Delta y_{wall}, \Delta y_{\ infty}, \Delta z)/\delta_{99,I}$
ZPG	$(801, 38, 134)$	$8193 \times 315 \times 1362$	$(0.098, 0.0026, 0.17, 0.098)$
APG	$(801, 70, 134)$	$8193 \times 500 \times 1362$	$(0.098, 0.0026, 0.17, 0.098)$

The numerical details of the ZPG and APG simulations are summarized in Table 1. The number of grid points in each of the three flow directions is given by (N_x, N_y, N_z). The extents of the computational domain in the respective directions (L_x, L_y, L_z) are also presented, along with the constant grid spacings in streamwise (Δx) and spanwise directions (Δz)

and the wall normal grid spacing at the farfield boundary (Δy_∞) and at the wall (Δy_{wall}). Note the APG simulation has a larger wall normal domain (L_y) and more points in this direction (N_y). This is due to the fact that the APG TBL expands more quickly in the streamwise direction than the ZPG TBL, as discussed in the following section. In both simulations the Courant number was set to unity. The length scales in Table 1 are non-dimensionalised by $\delta_{99,I}$ as viscous scaling is inappropriate for the APG flow.

4. Scaling of the First and Second Order Statistical Moments

The ZPG and APG boundary layers are first compared on the basis of Re_Θ, and the friction and pressure velocity scales. The mean velocity deficit and Reynolds stress profiles from various streamwise locations are then presented in both friction and pressure velocity scaling. In all of the following figures the dashed and solid lines represent the ZPG and APG cases respectively.

The momentum thickness based Reynolds number illustrated in Fig. 2(a), increases in the APG TBL more rapidly than the ZPG TBL, as the former expands more rapidly (hence larger Θ) in the streamwise direction as it decelerates. This deceleration of the flow also reduces the friction velocity $u_\tau = \sqrt{\tau_w/\rho}$, where τ_w is the mean shear stress at the wall. In Fig. 2(b), u_τ of the APG case is less than that of ZPG TBL as the former is decelerated more than the latter. However, the APG TBL DNS has not yet attained the desired $u_\tau \to 0$ condition, representative of the incipient separation case. Further fine tuning of the BC is required. To achieve a self-similar boundary layer the ratio of the pressure velocity (u_P) to u_∞ must be constant. The pressure velocity, defined by $u_P = \sqrt{(\partial p/\partial x)\delta^*/\rho}$, is a velocity scale based on the streamwise pressure gradient $\partial p/\partial x$. As illustrated in Fig. 2(c) a near constant ratio of u_P/u_∞ is achieved over a streamwise domain from $180\delta_{99,I}$ to $650\delta_{99,I}$.

Mean streamwise velocity deficit profiles ($u_\infty - \overline{u}$) are now presented at the streamwise locations indicated by the arrows in Fig. 2(c). In Fig. 3(a) the deficit profiles are non-dimensionalised by u_τ and plotted against y/δ_{99}, where δ_{99} is the boundary layer thickness. The dots in this figure represent results from the previous ZPG DNS of Jiménez *et al.*[13], which agree with the present ZPG simulation. When scaled by u_τ, the non-dimensional velocity profiles do not collapse, but in fact increase in the downstream

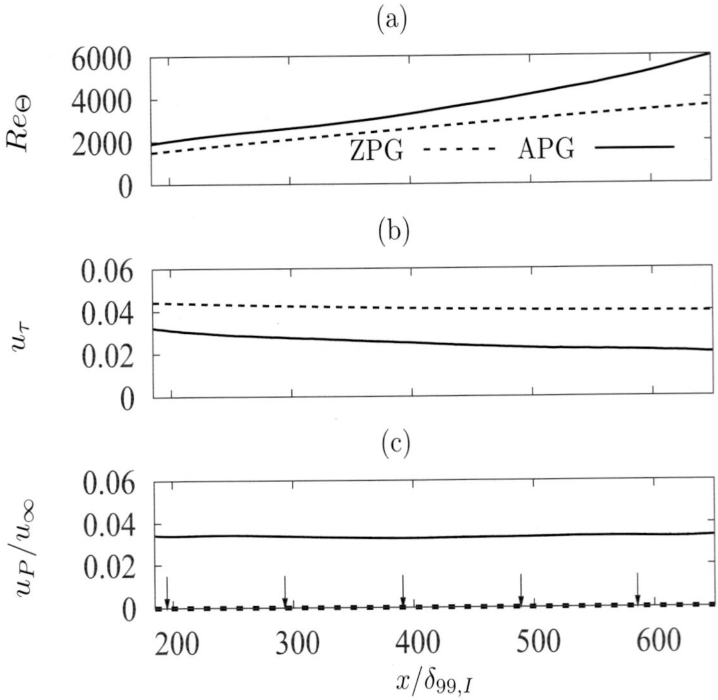

Fig. 2. Boundary layer properties of the adverse pressure gradient DNS (solid line) and the zero pressure gradient DNS (dashed line): (a) momentum thickness Reynolds number $Re_\Theta = u_\infty \Theta/\nu$; (b) friction velocity $u_\tau = \sqrt{\tau_w/\rho}$; (c) pressure velocity $u_P = \sqrt{(\partial p/\partial x)\delta^*/\rho}$ divided by the freestream velocity u_∞, with arrows indicating the positions of the APG TBL velocity profiles illustrated in Fig. 3 and Fig. 4.

direction – indicated by the arrow – as u_τ decreases. However, the profiles do collapse when scaled by u_P as illustrated Fig. 3(b). Note the ZPG profiles are not included in this figure as $u_P = 0$ in this case.

As undertaken for the velocity deficit profiles, the second order statistical moments are now presented scaled on the basis of both the wall shear stress and pressure gradient. Under the former scaling the Reynolds stress profiles are nondimensionalized using a velocity scale of u_τ and length scale of ν/u_τ. Under the latter scaling the pertinent velocity and length scales are u_P and δ^*. Profiles of the variance of the fluctuating component of the streamwise velocity $(\overline{u'u'})$ are plotted in friction velocity scaling in Fig. 3(c), which also increase as u_τ decreases in the downstream direction. Additionally a second outer peak is evident, which becomes more evident in the

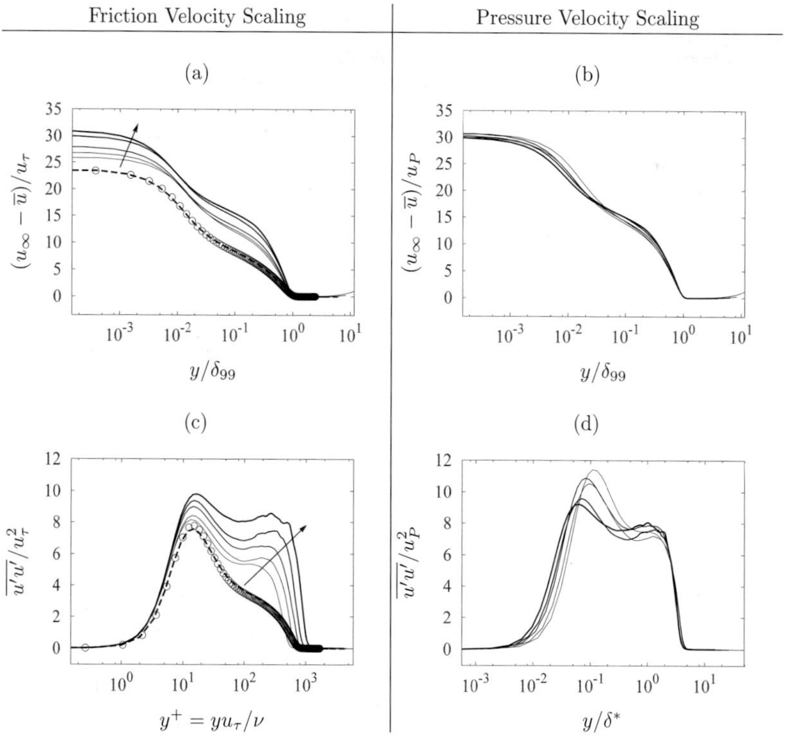

Fig. 3. Mean velocity deficit profiles nondimensionalized by: (a) friction velocity (u_τ) and boundary layer thickness (δ_{99}); and (b) pressure velocity (u_P) and displacement thickness (δ^*). Reynolds stress profile $\overline{u'u'}$ nondimensionalized by: (c) u_τ and viscous length scale ν/u_τ; and (d) u_P and δ^*. ZPG TBL DNS of Jiménez *et al.*[13] - dots; ZPG TBL DNS current simulation - dashed line; APG TBL DNS from current simulation at different streamwise locations - solid lines, of thickness increasing with downstream position. Arrows in the left column indicate the direction of increasing x position. Positions of the APG TBL profiles are illustrated in Fig. 2(c).

downstream direction. Similar observations are also made concerning the wall normal velocity variance, $\overline{v'v'}$, spanwise velocity variance, $\overline{w'w'}$, and Reynolds stress, $\overline{u'v'}$, plotted under friction velocity scaling in Fig. 4(a), Fig. 4(c), and Fig. 4(e) respectively. All of these Reynolds stresses increase in magnitude as u_τ decreases in the downstream direction, with a prominent second outer peak becoming stronger downstream. This outer peak collapses for all of the streamwise stations when plotted in pressure velocity scaling as illustrated for $\overline{u'u'}$, $\overline{v'v'}$, $\overline{w'w'}$ and $\overline{u'v'}$ in Fig. 3(d), Fig. 4(b), Fig. 4(d), and Fig. 4(f) respectively.

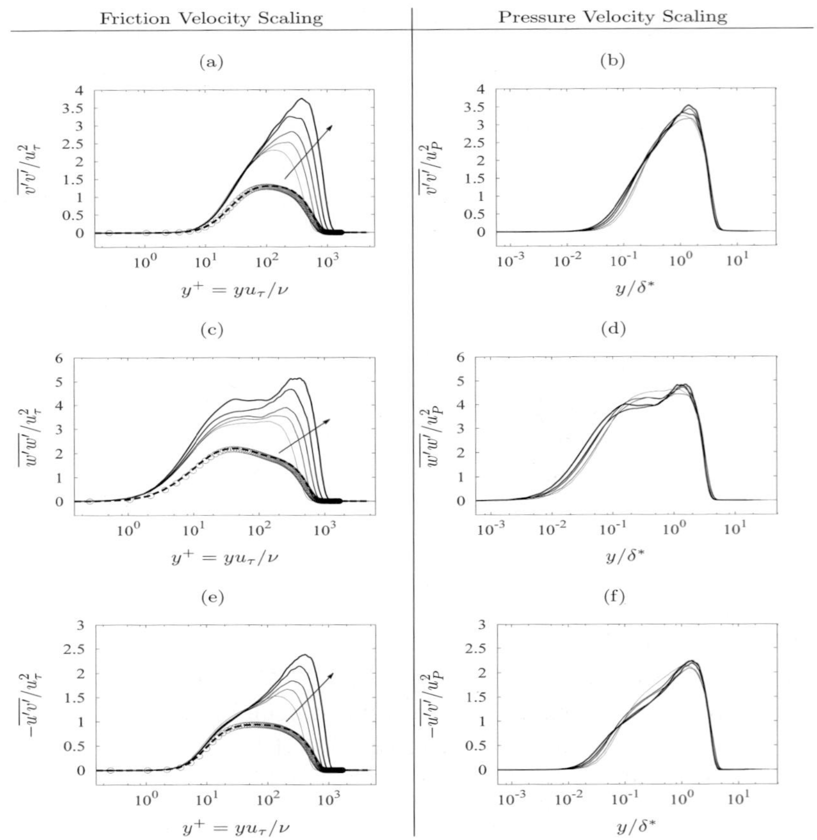

Fig. 4. Reynolds stress profiles: (a) $\overline{v'v'}$ nondimensionalized by friction velocity (u_τ) and viscous length scale ν/u_τ; (b) $\overline{v'v'}$ multidimensionality by pressure velocity (u_P) and displacement thickness (δ^*); (c) $\overline{w'w'}$ multidimensionality by u_τ and ν/u_τ; (d) $\overline{w'w'}$ multidimensionality by u_P and δ^*; (e) $\overline{u'v'}$ multidimensionality by u_τ and ν/u_τ; and (f) $\overline{u'v'}$ multidimensionality by u_P and δ^*. ZPG TBL DNS of Jiménez *et al.*[13] - dots; ZPG TBL DNS current simulation - dashed line; APG TBL DNS from current simulation at different stream wise locations - solid lines, of thickness increasing with downstream position. Arrows in left column indicate the direction of increasing x position. Positions of the APG TBL profiles are illustrated in Fig. 2(c).

5. Concluding Remarks

Development of a farfield boundary condition for the direct numerical simulation of an adverse pressure gradient turbulent boundary layer at the verge of separation has been presented. A region of constant ratio of pressure velocity to freestream velocity is attained over a momentum thickness based Reynolds number range of $Re_\theta \approx 2000$ to 6000. Mean velocity deficit profiles were shown to collapse under pressure velocity scaling, but not in friction velocity scaling. The Reynolds stresses also exhibit a second outer peak, which collapses under pressure velocity scaling.

Acknowledgements

The authors would like to acknowledge the research funding from the Australian Research Council and European Research Council, and the computational resources provided by the Australian National Computational Infrastructure, iVEC and PRACE. Julio Soria gratefully acknowledges the support of an Australian Research Council Discovery Outstanding Researcher Award fellowship.

References

1. V. Kitsios, L. Cordier, J.-P. Bonnet, A. Ooi and J. Soria, On the coherent structures and stability properties of a leading edge separated airfoil with turbulent recirculation, *J. Fluid Mech.* **683**, 395 (2011).
2. P. Skåre and P.-A. Krogstad, A turbulent equilibrium boundary layer near separation, *J. Fluid. Mech.* **272**, 319 (1994).
3. G. Mellor and D. Gibson, Equilibrium turbulent boundary layers, *J. Fluid. Mech.* **24**, 225 (1966).
4. A. Gungor, M. Simens and J. Jiménez, Direct numerical simulation of wake-perturbed separated boundary layers, *J. Turb* **134**, 061024 (2012).
5. J.-H. Lee and J. Sung, Effects of an adverse pressure gradient on a turbulent boundary layer, *Int. J. Heat Fluid Fl.* **29**, 568 (2008).
6. M. P. Simens, J. Jiménez, S. Hoyas and Y. Mizuno, A high-resolution code for tubulent boundary layers, *J. Comp. Phys.* **228**, 4128 (2009).
7. G. Borrell, J. Sillero and J. Jiménez, A code for direct numerical simulation of turbulent boundary layers at high Reynolds numbers in BG/P supercomputers, *Comp. Fluids* **80**, 37 (2013).
8. F. Harrow and J. Welch, Numerical calculation of time-dependent

viscous incompressible flow of fluid with free surface, *Phys. Fluids* **8** (1965).
9. J. Perot, An analysis of the fractional step method, *J. Comput. Phys.* **108**, 51 (1993).
10. S. Lele, Compact finite difference schemes with spectral-like resolution, *J. Comput. Phys.* **103**, 16 (1992).
11. J. Sillero, J. Jiménez and R. Moser, One-point statistics for turbulent wall-bounded at Reynolds numbers up to $\delta^+ \approx 2000$, *Phys. Fluids* **25**, 105102 (2013).
12. J. Sillero, High Reynolds number turbulent boundary layers, PhD thesis, Universidad Politénica de Madrid (2014).
13. J. Jiménez, S. Hoyas, M. Simens and Y. Mizuno, Turbulent boundary layers and channels at moderate Reynolds numbers, *J. Fluid. Mech.* **657**, 335 (2010).

Some Interesting Features of Flow Past Slotted Circular Cylinder at Re = 3500

Suryanarayana G K[†], Vidyadhar Y Mudkavi, Rajan Kurade and Naveen K M

CSIR National Aerospace Laboratories
P B 1779, Kodihalli, Bangalore 560017, India
[†]*E-mail: surya@nal.res.in*

The features of flow past a cylinder with a vertical slot at Reynolds number of 3500 based on high-speed visualizations of smoke-flow and computations are presented. The pressure differential across the slot causes natural periodic suction and blowing across the cylinder. The 'dead-water' region that forms behind the basic circular cylinder is replaced by a highly dynamic and complex vortex dominated periodic base flow. The separated shear layer on one side rolls up into a large vortex and reattaches at the base. The reattaching vortex entrains the opposite shear layer and promotes extended regions of attached flow featuring multiple separations and reattachments. Computations show qualitative agreement with experiments and clearly bring out the occurrence of periodic wall jets transverse to the cylinder, caused by oscillatory flow across the slot excited by the external shedding. These wall jets lead to the formation of multi-pole vortices in the shear layers and periodic ejection of vorticity leading to 'aerodynamic tripping' of the shear layers. It seems likely that the separated shear layers undergo bypass transition to turbulence and reattach onto the base, resulting in formation of periodic separation bubbles around the base of the slotted cylinder. Variation of instantaneous static pressures around the slotted cylinder shows oscillations with increasing amplitude from the stagnation point, which tend to coalesce towards the beginning of the slot. Soon after the slot, large amplification of pressure occurs due to the wall-jet. Time-averaged pressure distributions indicate increased pressures in the entire separated flow region of the slotted cylinder as compared to that of the basic cylinder. The slotted cylinder shows dramatic reductions of more than 80% in the mean and fluctuating drag and lift forces. The present studies show that the vortex-induced vibrations of a circular cylinder could be passively controlled using natural ventilation through a slot across the cylinder.

Keywords: Slotted cylinder; Drag reduction; Natural ventilation, Passive flow control

1. Introduction

The first study on flow past a slotted cylinder[1] reported three flow-regimes based on orientation of the slot (β) with respect to the mean flow. In regime I ($0^o < \beta < 40^o$), the slot resulted in injection of flow into the wake, base pressure rise and drag reduction. In regime II, ($40^o < \beta < 60^o$), there was practically no effect of the slot. In a finer resolution of flow features in regime III ($60^o < \beta < 90^o$), intermittent boundary layer suction with asymmetric flow ($62^o < \beta < 75^o$) and periodic blowing and suction ($78^o < \beta < 90^o$) at the shedding frequency was reported[2]. In this regime, generation of strong and alternate vortices is observed. Under Regime I, the use of a slotted cylinder as a bluff-body flame holder was reported[3]. A detailed review of the effectiveness of various passive devices in suppression of vortex shedding from cylinders has been presented[4]. Considering the entry at the front of the cylinder as 'Windward Suction' and exit at the base as 'Leeward Blowing' (energetic technique), the streamwise-aligned slot can suppress Vortex Induced Vibrations (VIV) at Re = 500 and 1000 is demonstrated computationally[5]. Considering the natural ventilation at $\beta = 0^o$, suppression of VIV and optimal slot width at Re = 500 through numerical simulations of incompressible 3-D Navier Stokes equations is reported[6]. They also show that the optimal slot width for suppression of VIV reduces at Re = 1000. Following Roshko's[7] concept of a flow-meter based on assumption of constant Strouhal number S_t (= fd/U where f is the vortex shedding frequency, d is the diameter and U the freestream velocity) in the range $40 < Re < 10^4$, Karman vortex flow meters for measurement of mass flow rate in pipe flows have been developed[7,8,9], which rely on vortices shed by a 2-D circular cylinder. In such flow-meters, the generation of strong and periodic vortices significantly improves the signal to noise ratio essential for reliable measurements. From this perspective, under Regime III[1], increased amplitude and periodicity of velocity fluctuations measured using a hot wire are reported[11]. They have concluded that vertical slot increases the strength of shed vortices and that a concave notch at the base stabilizes the vortex shedding. Computational results[6] shows the effect of bleeding the flow from stagnation region at various circumferential locations using a thin outer shell surrounding the cylinder, and concludes that $\beta = 90^o$ offered the strongest vortices and amplified VIV. Various innovative combinations of bluff bodies separated by a short distance, including the vertically slotted cylinder for developing low-loss flow-meters and optimized the combination of bluff bodies are reported[12] which shows high sensitivity and linearity of shedding frequency with the freestream velocity in pipe flows. Flow visualizations[13,14] show that a divergent slot brings about a dramatic reduction in wake width in addition to suppression of vortex shedding, as compared to a parallel vent. They have also shown the possible occurrence of sound production in addition to generation of large vortices at high Re by a vertically slotted cylinder. Nevertheless, the flow past a vertically slotted circular cylinder has not been explored in great detail. Limited studies have shown very interesting flow features at high Re[13,14]. The present work is mainly to explore the flow features of a vertically slotted cylinder at Re = 3500 through experiments and computations and in addition, to compute the flow features at Re = 150, 1000 and 15000.

2. Experimental Set Up and Details of the Model

Experiments were carried out on a 30 mm diameter (d) cylinder with a slot width (s) of 3 mm (s/d = 0.1) in the 0.55 m low speed wind tunnel at the Experimental Aerodynamics Division of the National Aerospace Laboratories at Bangalore. Experiments for slot angles β = 0°, 15°, 30°, 45°, 60°, 75° and 90°. Studies related to β = 90° alone are reported in this paper. Details of the model are shown in Fig. 1. A Photron high-speed camera was used to record the images at 250 fps with resolution of 1 MP. Phantom software was used to extract instantaneous images.

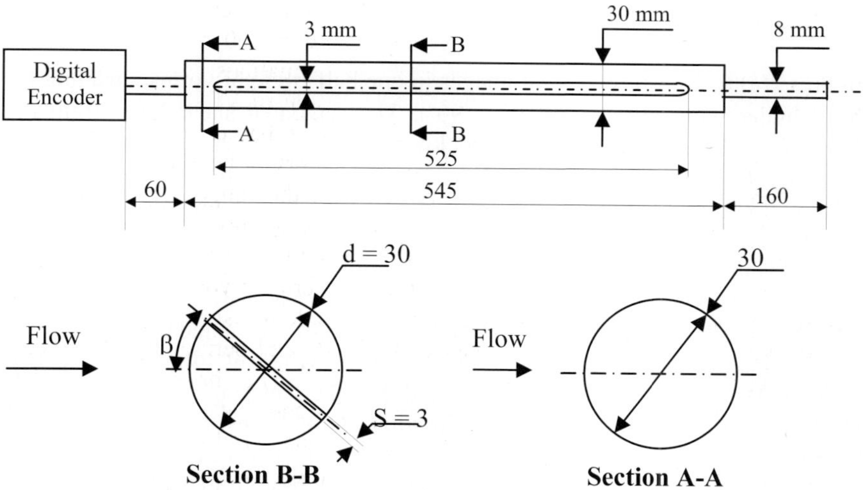

Fig. 1. Details of the slotted cylinder and the test set-up.

3. Computations

Computations were carried out to solve RANS equations using turbFOAM solver of OpenFOAM (http://www.openfoam.com) with Spalart-Allmaras (SA) turbulence model. Velocity was specified at the inlet of the domain while pressure was specified at the outlet. Far-field boundary condition was imposed on the top and bottom boundaries. On the cylinder surface and inside the vent, wall condition was applied. More than 75 thousand cells were used for vented cylinder. For all the simulations, y+ value for first point was about 30. The Pressure-Implicit with Splitting of Operators algorithm was used to solve the pressure-velocity coupling of the governing equations. For time discretization, second order Euler scheme was employed and time step was set at 0.0001s for Re of 150 to 15000. Simulations were carried out for Re = 150 (without any turbulence model), 1000, 3500 and 15000. The upstream domain extends to 15 R while the downstream domain extends to 40 R.

4. Results and Discussions

The visualizations reported here correspond to a time resolution (Δt) of 4 ms or 250 frames per second, corresponding to a non-dimensional time step ($\tau = U\Delta t/D$) of 0.2747. Hence, while the phenomenon of global vortex shedding (low frequency) is well captured at this frame rate, shear layer instabilities, which require higher frame rate for proper imaging, may not be properly observed.

4.1. Experimental results (β = 90o)

Figures 2(a) to 2(f) show typical images in one half-cycle of vortex shedding for the cylinder with vertical slot ($\beta = 90^o$). The separated shear layer from the upper side rolls up into a large anticlockwise vortex (A^+) very close to the base, resulting in reattachment at the base ($\theta \sim 180^o$) and further roll up. The combination of A^+ and the clockwise vortex (C^-) shed in the previous cycle leads to a strong entrainment of the external flow from the upper side (E^U) due to the induced velocity of the vortex pair (C^-:A^+). Formation of clockwise vortex (C^+) of the current cycle on the lower side (Fig. 2(b)) and its pairing with A^+ leads to entrainment of shear layer (E^L) from the lower part of the vented cylinder.

E^U gets sucked towards the wake centerline where it bifurcates into two portions (B_1), with one part entering C^- and the other entering A^+. As A^+ rolls up and traverses downstream, C^+ starts growing and entrains the shear layer and external flow from the lower side (E^L) due to the induced velocity of the vortex pair (C^+:A^+), with streamlines traversing around the base from the lower side. E^L travels upwards in the wake from the lower side and bifurcates (B_2) with a part merging into A^+ and the rest rolling into C^+. A^+ grows bigger and traverses downstream, with simultaneous growth of C^+. In time, C^+ causes reattachment of the flow at the base ($\theta \sim 180^\circ$) and starts entraining the shear layer and the external flow from the upper side. The average frequency of shedding, based on image analyses of 135 cycles is 10.103 Hz, corresponding to St of 0.147. In contrast, similar analyses for the basic cylinder indicated St of 0.2747, suggesting significant reduction in the shedding frequency with the slot.

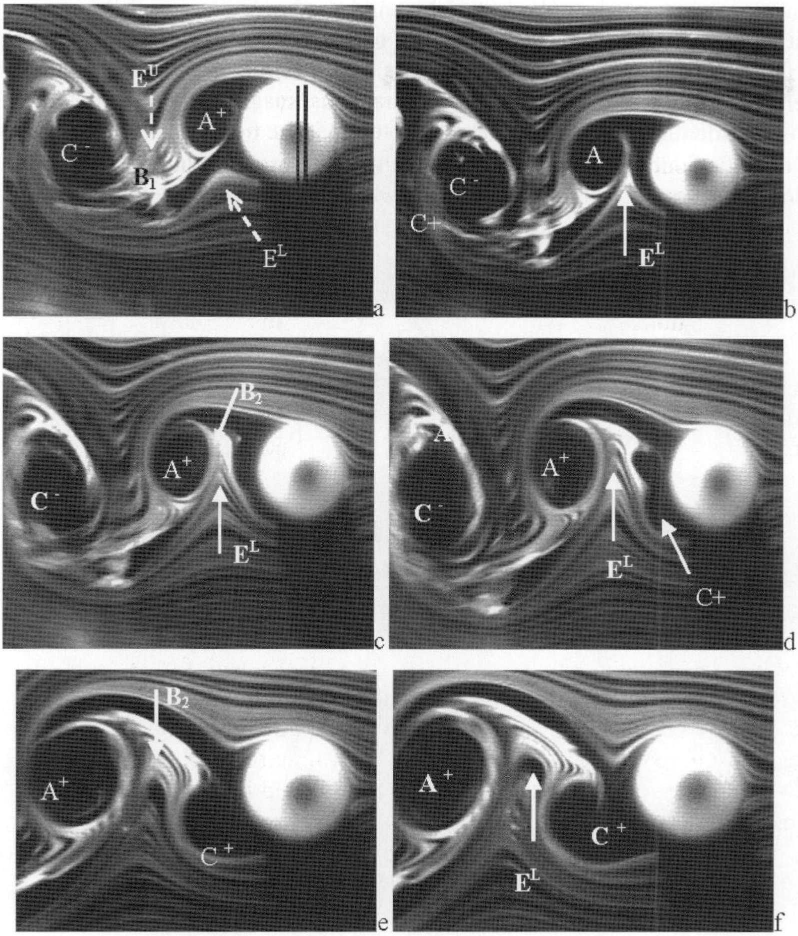

Fig. 2. Smoke flow visualization of streamlines around slotted circular cylinder: $\beta = 90^{\circ}$.

4.2. *Computed results*

4.2.1. Streamlines

Figures 3(a) to 3(i) show the computed streamlines in one period of vortex shedding. Extensive regions of reattached flow may be seen, with the reattached streamlines separating at nearly 250° from the stagnation point at some instants. Figures 4 show two instantaneous streamline patterns in one half-cycle of shedding. It may be noted that the external periodic flow excites an oscillatory vertical flow across the slot, leading to occurrence of alternate suction and blowing across the slot. The larger dynamic head of the external flow bends the exiting slot flow as a wall jet, which traverses nearly tangential to the rear surface of the slotted cylinder. Instances of extensively reattached flow can be seen. Figure 5 compares a typical instantaneous image with the computed result showing identical features. It may be noted that the occurrence of periodic wall jet at ±90° from the stagnation point is equivalent to introducing periodic slip velocity across a circular cylinder, which is similar to forced rotational oscillation of the cylinder at the shedding frequency.

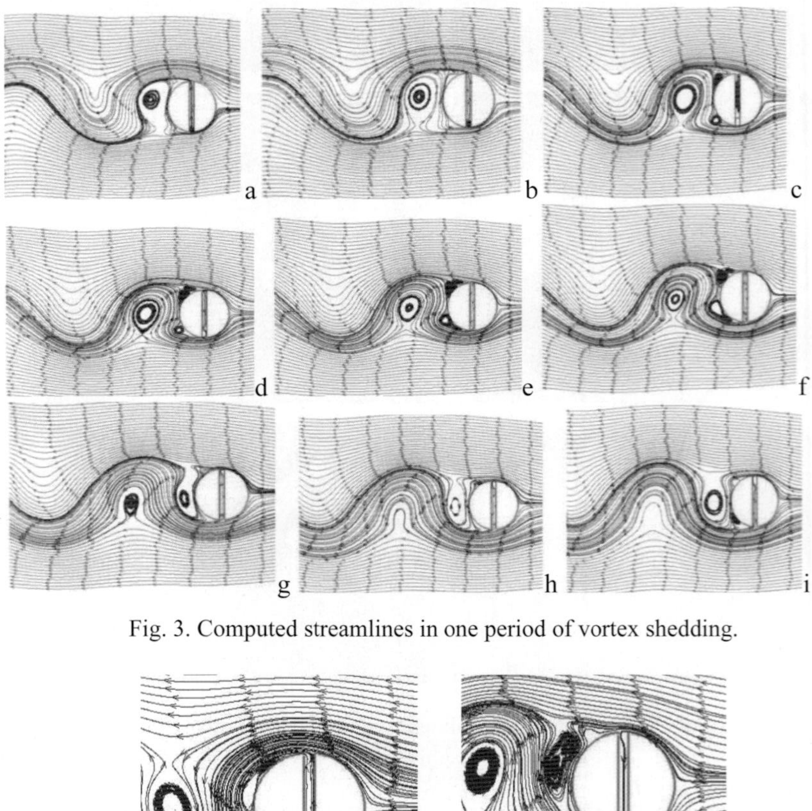

Fig. 3. Computed streamlines in one period of vortex shedding.

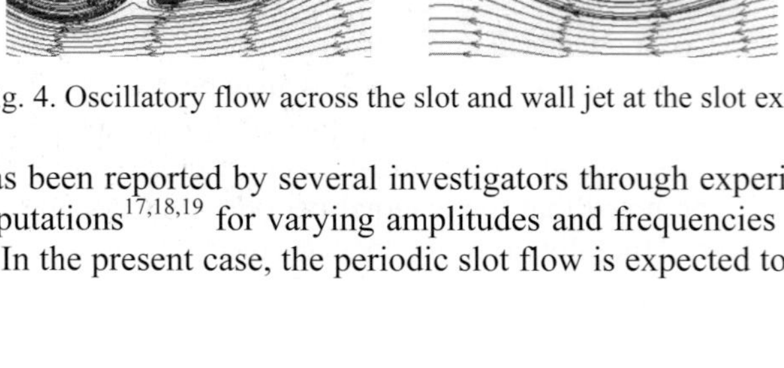

Fig. 4. Oscillatory flow across the slot and wall jet at the slot exit.

The latter has been reported by several investigators through experiments[15,16] as well as computations[17,18,19] for varying amplitudes and frequencies of rotational oscillations. In the present case, the periodic slot flow is expected to occur at the vortex shedding frequency of the cylinder. Nevertheless, by analogy, some of the features of the flow reported in the literature may be expected in the present case also.

4.2.2. *Variation of instantaneous Cd and Cl*

Time-variations of C_d and C_l at Re = 3500 and the corresponding spectra are compared in Figs. 6 and 7. The slotted cylinder causes significant reductions in fluctuations of both the quantities. Large reduction in the magnitude at the dominant frequency is indicated, but no change in frequency of fluctuations is indicated: however, experiments clearly indicate reduction in the shedding frequency (S_t from 0.275 to 0.174 from visual observations). It is interesting to note that some of the earlier studies have suggested enhanced Vortex Induced Vibration (VIV)[6,11] due to increased lift fluctuations, while the current study show significant decrease. Thus, experimental validations of the present computed results are essential.

4.2.3. *Variation of mean drag coefficient (Cd)*

Figure 8 compares the time-averaged C_d from present computations with experimental data of Roshko[7], experiments of forced rotational oscillations of Tokumaru & Dimotakis[15] and computed results of Shiels & Leonard[18]. The present computations of slotted cylinder indicate large drag reduction as well as invariance with Re (Table 1), suggesting very large change in the overall flow features as compared to the basic cylinder.

Table 1. Comparison of mean C_d of basic and slotted cylinders.

Re	*Cd Basic*	*Cd Slotted*
150	1.619	0.141
1000	1.603	0.169
3500	0.935	0.161
15000	0.523	0.142

Very significant reductions in C_d, of as much as 80% (Re =150) to 75% (Re =15000) are indicated, with C_d nearly independent of Re for the slotted cylinder.

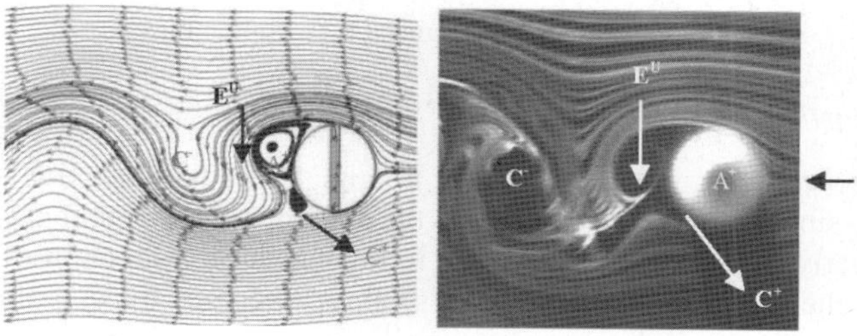

Fig. 5. Comparison between computed and experimental streamlines.

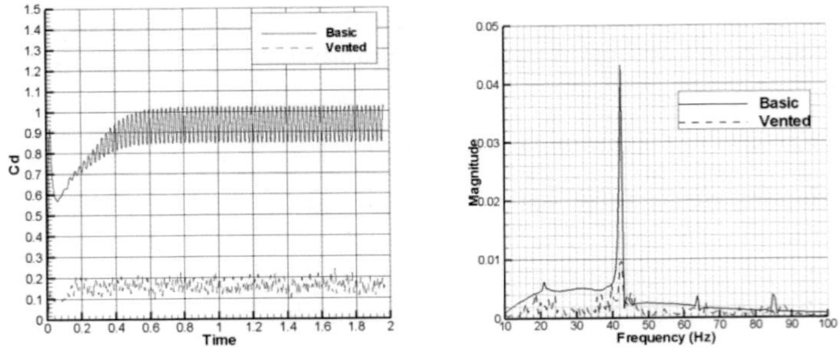

Fig. 6. Comparisons of time-dependent C_d and spectra.

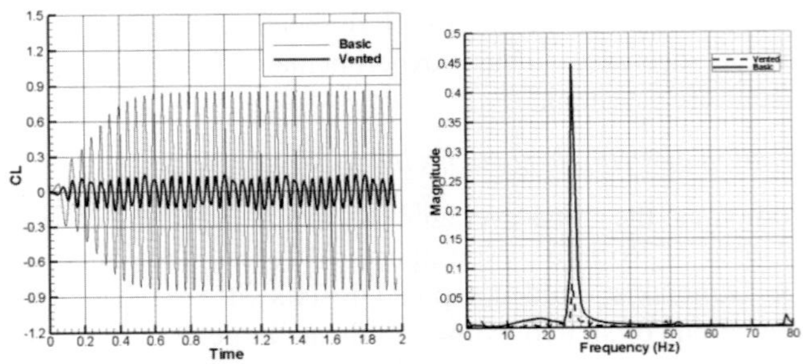

Fig. 7. Comparisons of time-dependent C_l and spectra.

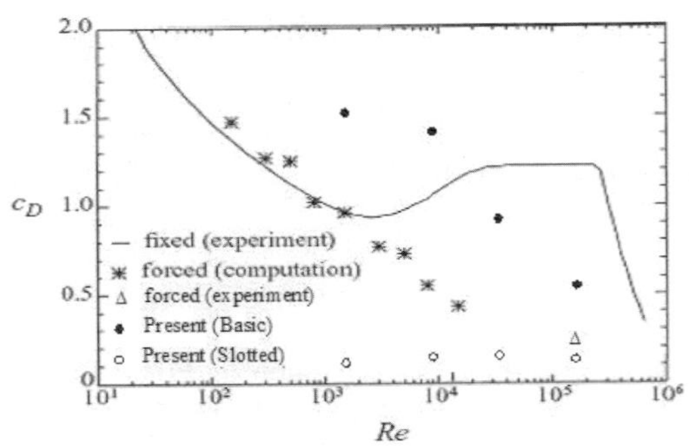

Fig. 8. Comparison of mean C_d with literature.

4.2.4. Vorticity

Typical iso-vorticity lines are shown in Fig. 9. Radial injection of vortical structures, similar to those reported in the case of forced rotational oscillation can be clearly observed. It seems that following the introduction of additional vorticity, the separating shear layers may undergo bypass transition to turbulence due to 'aerodynamic tripping'[19]. Pressure gradients induced by the passing vortices and proximity to the base may lead to multiple reattachments and separations of the streamlines, as shown in Fig. 3.

Fig. 9. Typical iso-vorticity lines.

4.2.5. Pressure distributions

Figures 10 show instantaneous and time-averaged pressure distributions around the slotted cylinder within one cycle of shedding. The first quarter of the cylinder is not greatly influenced by ventilation. However, as the slot is approached, effect of the slot begins to dominate. At some instants, laminar separation (S_1) around $\theta = 96^o$, constant pressure plateau indicating separated flow, turbulent reattachment of the shear layer (R_1) around $\theta = 125^o$, turbulent separation (S_2) around $\theta = 135^o$, turbulent reattachment (R_2) around $\theta = 155^o$ and ultimate turbulent separation (S_3) around $\theta = 165^o$ may be observed. At all locations downstream of the slot, the time-averaged pressures are higher, suggesting that base pressure increase is the cause for drag reduction.

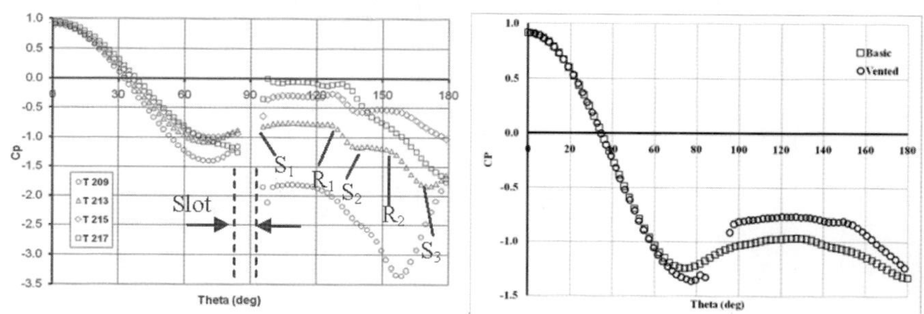

Fig. 10. Instantaneous and time-averaged pressure distributions, Re = 3500.

Figures 11(a) to 11(g) show the time-variations of pressure at various locations around the slotted cylinder. Based on similarity of patterns, the distributions may be grouped into specific regions upstream and downstream of the slot. The region upstream of the slot may be subdivided into regions (a) $0 < \theta < 72$ and (b) $72 < \theta < 85$. Likewise, the region downstream may be divided into five regions (c) $95 < \theta < 120$ (d) $120 < \theta < 132$ (e) $132 < \theta < 152$ (f) $152 < \theta < 168$ and (g) $168 < \theta < 180$. In region (a) upstream of the slot, small amplitude oscillations of a single frequency are initially seen (Fig. 11(a)). As the slot is approached (region b, Fig. 11(b)), the frequency of oscillations gets modulated. The amplitudes seem to coalesce towards the values corresponding to the last location upstream of the slot. As soon as the slot is crossed, large amplification occurs (region c, Fig. 11(c)) as a result of the inflow / outflow through the slot, depending on the phase of the flow. In region d (Fig. 11(d)) steady state oscillations are seen. In region e (Fig. 11(e)), variations in the peak pressures are observed. In the region f (Fig. 11(f)), the pressure peaks appear sharper and variations in the location of the peak pressure occur. The region g (Fig. 11(g)) shows similar pressure variations with reduction in the peak values towards the base. All the variations indicate the existence of a dominant frequency and systematic modulations originating from specific locations.

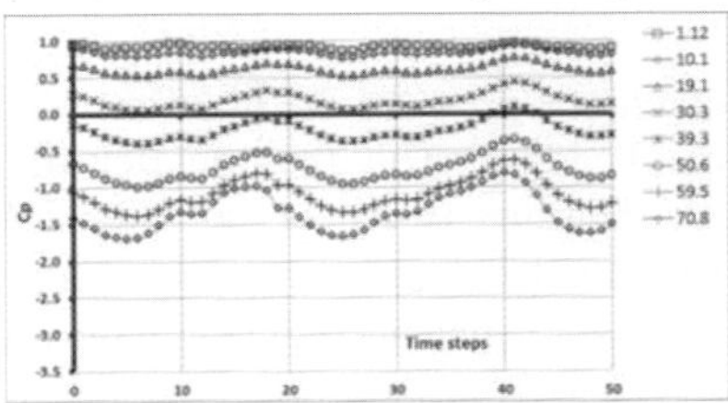

(a) 0 < θ < 72

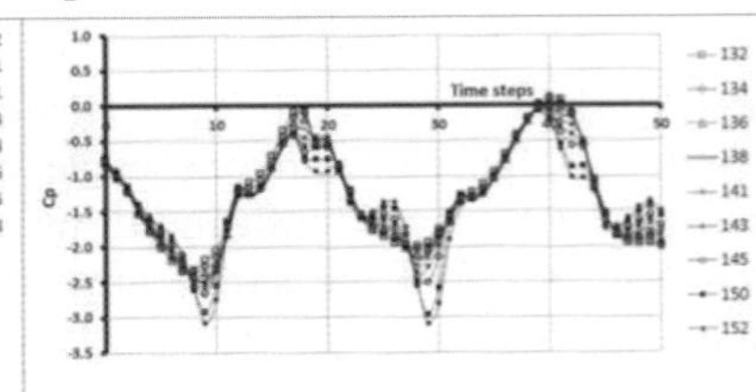

(b) 72 < θ < 85

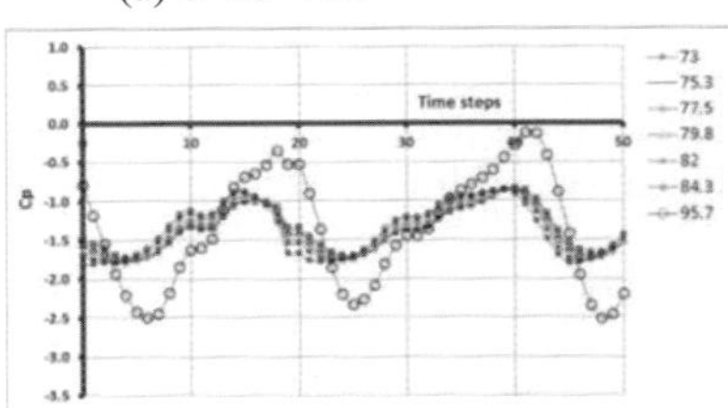

(c) 95 < θ < 120

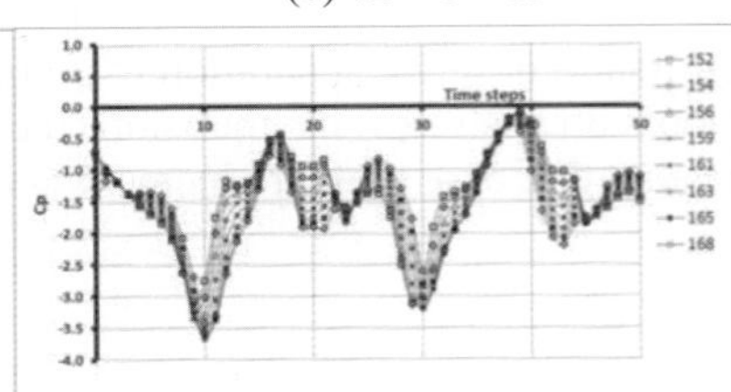

(d) 120 < θ < 132

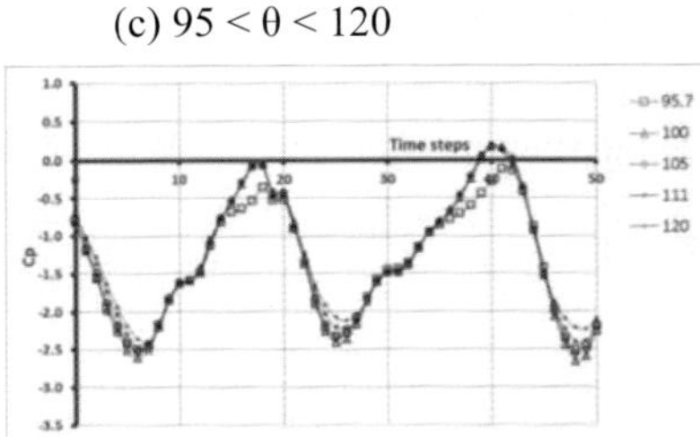

(e) 132 < θ < 152

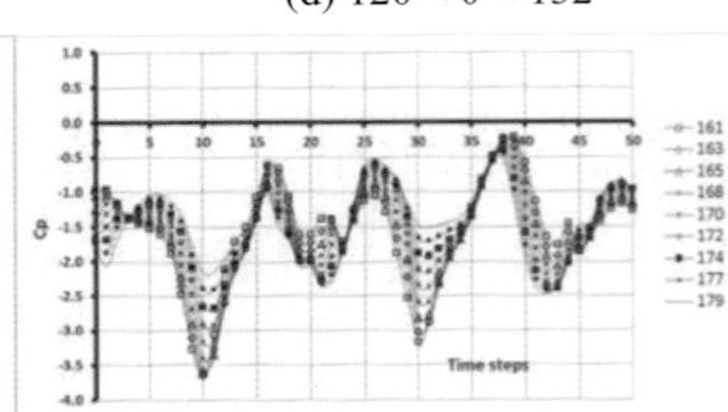

(f) 152 < θ < 168

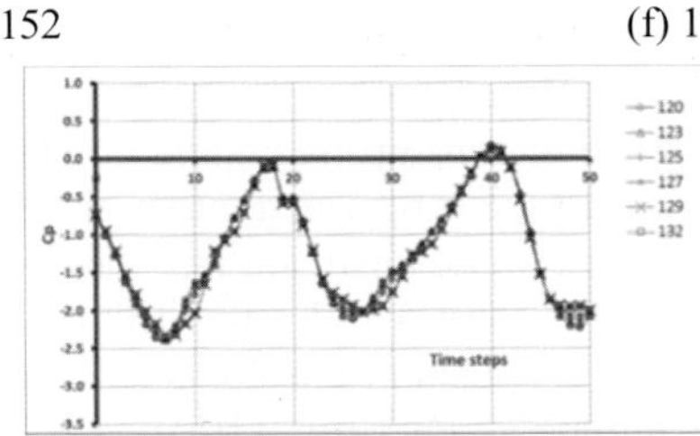

(g) 168 < θ < 180

Fig. 11. Pressure fluctuations at various locations around the slotted cylinder.

It seems likely that while the dominant frequency is due to the global shedding, the modulations are due to the multi-pole vortices, which have different time scales and specific locations on the rear of the vented cylinder.

Concluding Remarks

Visualizations of flow past a vertically slotted cylinder show the formation of large vortices very close to the base, replacing the dead-water region of the basic cylinder by a dynamic and vortex dominated flow at Re = 3500. Even though flow-visualizations suggest a large reduction in the vortex shedding frequency compared to the basic cylinder, RANS computations based on OpenFOAM do not indicate any change; However, the computed streamlines match well with the experiments. The flow features seems to be analogous with the flow past a rotationally oscillating cylinder wherein significant drag reduction has been earlier reported. The computations predict very significant reductions in mean and fluctuating drag as well substantial reductions in fluctuating lift, in contrast to some previous computations wherein enhanced VIV has been reported. The present study shows low values of drag nearly invariant with Re in the range 150-15000, underlining the need for experimental verification of the predictions made using OpenFOAM.

References

1. T. Igarashi, Flow characteristics around a circular cylinder with a slit, Bulletin of the JSME, **21(**154), 6641 (1978).
2. T. Igarashi, Flow characteristics around a circular cylinder with a slit, Bulletin of the JSME, **25**(207), 1389–1937 (1982).
3. Filippi, F. and Fabrovich-Mazza, L, 8th Symposium on Combustion, Williams and Wilkins, **9**, 956 (1961)
4. Zdravkovich M. M., Review and classification of various aerodynamic means for vortex suppression, *J. Wind Engineering & Industrial Aerodynamics*, 7, 145–189 (1981).
5. S. Dong, G. S., Triantafyllou and G. E. Karniadakis, Elimination of vortex streets in bluff body flows, *Phys. Rev. Lett.,* PRL 100, 204501 (2008).
6. H. Baek and G. E. Karniadakis, Suppressing vortex-induced vibrations via passive means, *J. Fluids and Structures* **25**, 848–866 (2009).
7. A. Roshko, On the wake and drag of bluff bodies, *J. Aeronautical Sciences*, **22**, 124–132 (1955).
8. S. J. Bailey, Two new flowmeters have no moving parts, *Control Eng*. **16**, 73–77 (1969).
9. T. H. Burgess, Flow measurement using vortex principles, in: Proceedings of the Flow Conference, Brighton, 321–339 (1977).
10. R. Blickey and J. George, Vortex flowmeters provide higher accuracy lower pressure drops, *Control Engg*. **42**, 745–749 (1995).

11. Olsen, J. F. and Rajagopalan S., Vortex shedding behind modified circular cylinders, *J. Wind Engg and Industrial Aerodynamics*, **86**, 55–63 (2000).
12. J. Peng, X. Fu, and Y. Chen, Flow measurement by a new type of vortex flowmeter of dual triangulate bluff body, *J. Sensors and Actuators A* **115**, 53–59 (2004).
13. G. K. Suryanarayana, *Aerodynamic drag reduction of bluff bodies – Application of natural ventilation,* ISBN 978-3-8383-1103-6, Lambert Academic Publishers, Germany, Germany (2009).
14. G. K. Suryanarayana, G. E. A. Meier and H. Pauer, Vented circular cylinder as a vortex generator, J. Visualization, **9**, 197 (2005).
15. Tokumaru, P. T. and Dimotakis, P. E.: Rotary oscillation control of a cylinder wake, *J. Fluid Mech.* **224**, 77–90 (1991).
16. Thiria, B. Goujan–Durand S. and Wesfreid, J. E., The wake of a circular cylinder performing rotary oscillations, *J. Fluid Mech.* **560**, 123–147, (2006).
17. Chou, Mo Hong: Synchronization of vortex shedding from a cylinder under rotary oscillation, *Comput. and Fluids* **26**(8), 755–774, (1997).
18. Shiels, D. & Leonard, A.: Investigation of drag reduction on a circular cylinder in rotary oscillation, *J. Fluid Mech.* **431**, 297–322, (2001).
19. T. K. Sengupta, K. Deb, S. B. Talla: Control of flow using generic algorithm for a circular cylinder executing rotary oscillation, *Comput. & Fluids* **36**, 578–600, (2007).

A High-Resolution Compressible DNS Study of Flow Past a Low-Pressure Gas Turbine Blade

Rajesh Ranjan, S. M. Deshpande and Roddam Narasimha*

Engineering Mechanics Unit
Jawaharlal Nehru Centre for Advanced Scientific Research
Bangalore, Karnataka 560064, India
**E-mail: roddam@jncasr.ac.in*
www.jncasr.ac.in

Direct numerical simulation (DNS) studies of compressible flow past low pressure (LP) turbine blades have been carried out using a new compressible code ANUROOP. The code has been developed at JNCASR and validated for both free-shear and wall-bounded flows. The mid-section of the Pratt and Whitney LP turbine blade T106A has been chosen for detailed analysis. The set-up enables direct comparison with both experimental and computational data available on the blade. The Reynolds number(Re) of the flow is 51,831 based on inlet velocity and axial chord of the blade.

The grid size has been varied from 25 to 160 million, and shows that the flow near transitional Re depends crucially on resolution. The separation bubble at the leading edge of the suction side present at low resolutions disappears at the highest resolution, with consequent changes in blade fluid dynamics that are striking. Around the leading edge, the high curvature demands a correspondingly near wall resolution, which plays a major role in determining the pressure distribution.

1. Introduction

Flow past turbine blades remains a challenging problem for the fluid dynamicist as well as the design engineer because of its complex nature. Operating in a harsh environment, these blades may exhibit a wide range of flow phenomena such as wake-boundary layer interaction, boundary layer separation, transition, relaminarization, retransition etc.

Recently Direct Numerical Simulation (DNS) has emerged as a promising tool to investigate complex fluid-dynamical flows. Because of its ability to solve the problem without any additional physical assumptions, DNS provides a powerful tool complementary to experimental data, especially as the numerical experiments generally offer complete three-dimensional datasets, whereas datasets obtained in the laboratory are still limited by

Table 1. DNS on LP turbine blades.

Authors	Blade	Re	$N_x \times N_y \times N_z$	Grid Size	Δx^+	Δy^+	Δz^+
Wu & Durbin[1] (2001)	T106A	1.48×10^5	$1152 \times 384 \times 128$	56×10^6			
Michelassi *et al.*[2] (2002)	T106A	51,800	$1014 \times 266 \times 64$	17×10^6	10	0.8	3
Kalitzin *et al.*[3] (2002)	T106A	1.48×10^5	$1152 \times 576 \times 128$	85×10^6	28	2.3	19
Wissink *et al.*[4] (2003)	T106A	51,831	$1,014 \times 260 \times 64$	17×10^6			
Wissink & Rodi[5] (2006)	T106A	51,800	$1014 \times 260 \times 64$	17×10^6			
Wissink & Rodi[6] (2006)	MTU	72,000	$1254 \times 582 \times 128$	93×10^6	25	1.5	15
Rajesh *et al.* (2014)	T106A	51,831	$12,57,162 \times 128$	160×10^6	2.2	0.1	2.4

the available instrumentation.

The present work describes flow past a low pressure (LP) turbine blade using a compressible DNS code with a very fine hybrid grid. This new code has been developed for simulating compressible flow past complicated geometrical shapes. Table 1 lists the DNS studies on LP turbine blades available in the published literature and also the present study with the finest grid to-date, along with Reynolds number (Re) and grid size used; Δx^+, Δy^+ and Δz^+ represent the distance of nearest grid point from the boundary in wall units in streamwise, wall-normal and spanwise directions at the mid-blade section on the suction side, respectively[a].

All these simulations listed in Table 1 (except the present study) have been performed assuming flow to be incompressible, and hence compressibility effects are ignored. Also in many works, the inadequate resolution of the grid near the wall, as a consequence of using relatively inflexible elliptic schemes, has remained a concern for detailed boundary layer investigations[3]. The present work is an attempt to fill these gaps and to throw light on fluid-dynamical issues hitherto not considered in the turbine blade literature.

2. The Code ANUROOP

In the present work, 3D compressible Navier-Stokes equations are solved in 3D space and time using a code called ANUROOP (meaning resembling or

[a]Empty columns in Table 1 indicate that corresponding values are not reported in the literature. The last entry in the table does not provide N_x, N_y separately but gives the total no. of elements on the face, because the grid topology in our work is different.

concordant in Sanskrit) developed at JNCASR.

ANUROOP is developed in the finite-volume framework with a 2nd order central differencing scheme for spatial discretization of convective and diffusive fluxes.

The finite-volume approach is a natural choice for compressible flow simulations as it solves Navier-Stokes equations in conservative form. The flux-discretization scheme used here follows Jameson[7], and guarantees conservation of kinetic energy in a discrete space. It has been found to be more robust and stable than a regular central differencing scheme.

A total-variation-diminishing (TVD) scheme is preferred for time discretization as it does not allow kinetic energy to grow rapidly. The Shu-Osher based Runge-Kutta 3 scheme has therefore been used here for time-marching. The Green-Gauss method has been adopted for calculating the requisite gradients at cell faces as required for handling the viscous terms.

Inlet, outlet, no-slip wall and periodic boundary conditions between the blades are implemented in the code. All the boundary conditions are imposed by adding ghost cells.

For inlet and outlet, a Riemann invariant boundary condition has been used. Incoming and outgoing characteristics are calculated using values at infinity and in the interior. Primitive flow variables at the boundaries are then constructed using this information. For the wall, a mirror boundary condition has been used, where the variables in the interior cells next to the boundary are mirrored appropriately to corresponding ghost cells and an average is taken to get the value at the wall face. For periodic boundary conditions, the values at the first interior cells of the first boundary are copied to corresponding ghost cells in the other periodic boundary and vice versa.

The code has been parallelized using the standard Message Passing Interface (MPI). The open-source package METIS from Karypis Lab has been used for breaking the computational domain into subdomains.

3. Validation Studies

ANUROOP has been verified and validated against a number of standard test cases, both free-shear as well as wall-bounded. Given below are comparisons of results for Taylor-Green flow and supersonic turbulent channel flows, against the results available in literature.

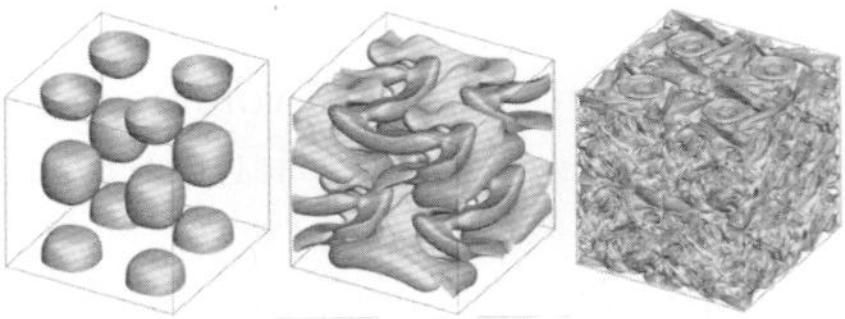

Fig. 1. Iso-surfaces of vorticity magnitude at non-dimensional time $t = 0.5, 2.0$ and 13.0 (from left to right).

3.1. *Taylor-Green flow*

The Taylor-Green case has been selected as one of the challenge problems for DNS schemes in the High-Order Workshop, 2013 (Ref. 8). This case is used to verify the accuracy of the numerical scheme employed, and evolution of the flow with time. A 2-D incompressible flow problem proposed by G. I. Taylor and George Green has been extended to 3-D in a weakly compressible limit. The flow inside a periodic box is initialized by simple perturbations of sines and cosines with zero-mean representing counter-rotating vortices, which decay with time at a rate governed by viscosity.

The initial flow field of a Taylor-Green vortex is a 2D, single-mode velocity field,

$$u = U_{\text{ref}} \sin x \cos y \cos z$$
$$v = -U_{\text{ref}} \cos x \sin y \cos z$$
$$w = 0$$

The flow is computed within a periodic cube given by $0 \leq x, y, z \leq 2\pi$. The initial density and pressure are

$$\rho = \rho_{\text{ref}}$$
$$p = p_0 + \frac{\rho_{\text{ref}} U_{\text{ref}}^2}{16} (\cos 2x + \cos 2y)(\cos 2z + 2)$$

Flow parameters set for the simulation are $Re = 1600$ and $M = 0.1$, with both U_{ref} and ρ_{ref} are taken as 1.0. The initial Mach number M gives the value of p_0. The grid chosen for the current simulation is 256^3 and flow has been simulated till $t = 13.0$. Several test simulations have been made over a range of Re to check the stability of the present scheme. The simulations progressed without any difficulty for all the cases, thus confirming the robustness of ANUROOP.

Figure 1 shows iso-surfaces of vorticity magnitude as it evolves in time. It illustrates the cascade process, in which multiple scales are generated

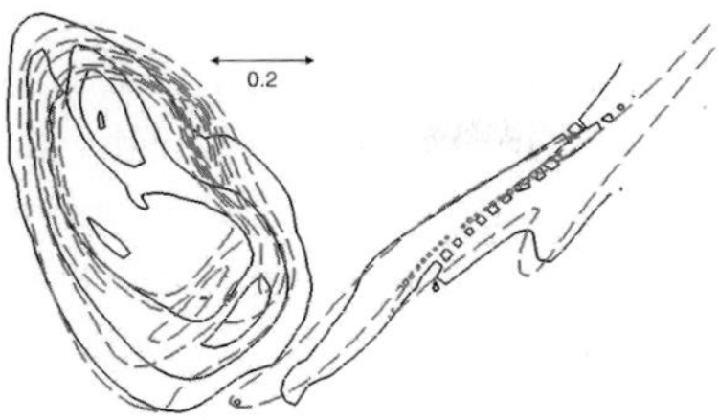

(a)

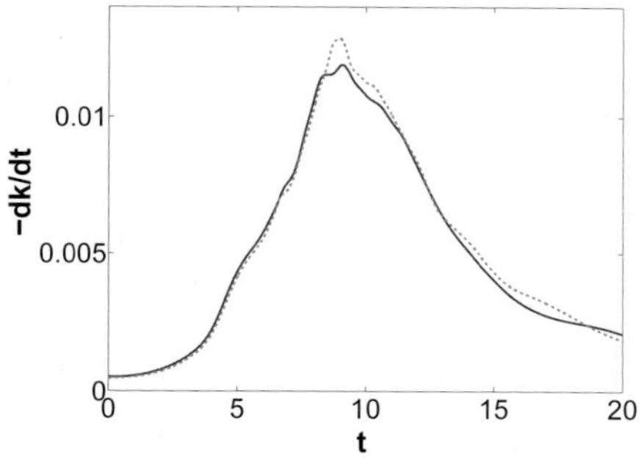

(b)

Fig. 2. (a) Contours of non-dimensional vorticity magnitude of 5, 10 and 15 at $t = 8.0$ on one of the periodic faces ($x = 0$) (b) Kinetic energy dissipation rate, compared with a 512^3 pseudo-spectral simulation[8] (dashed lines).

from a single initial large scale into very small scales due to non-linear interaction of eddies. Figure 2(a) shows contours of non-dimensional vorticity magnitude norm on one of the periodic faces at $t \approx 9.0$ when dissipation is near maximum. The structures are strikingly similar to those obtained by the reference pseudo-spectral code[8].

The kinetic energy of the system decays due to viscous dissipation. Figure 2(b) shows rate of dissipation as compared with the results of a 512^3 psuedo-spectral simulation[8]. Despite the smaller resolution used in our simulation, compared to the reference pseudo-spectral simulation, the comparison is good. This confirms that ANUROOP correctly simulates the evolution of kinetic energy in time. This is an important requirement for the codes used for DNS studies.

3.2. *Compressible channel flow*

DNS of compressible channnel flow has been performed to test the code against wall-bounded flows at finite Mach number. Past studies[9] of an isothermal supersonic channel flow offer a good case for validation of the present code. Two plates separated by a width of $2H$ constitute the channel in which the fluid flows. The flow is statistically homogeneous in the streamwise as well as spanwise directions, and is driven by a body-force that keeps the mass-flux constant. The parameters chosen here for the validation are $M = 1.5$ and $Re = 3000$ (based on bulk velocity, viscosity at wall temperature, and H).

The sides of the computational domain are $L_x = 2\pi, L_y = 2, L_z = 4\pi/3$. There are 100 grid points chosen in each direction to make the grid. The grid is uniform in streamwise (x) and spanwise (z) directions but stretching is used along the wall-normal direction y to obtain a sufficiently fine grid to capture the boundary layer.

Favre-averaged streamwise velocity is compared with DNS of Coleman[9] as shown in Fig. 3(a). It shows a good agreement between the two simulations. Table 2 shows quantitative comparisons of various parameters obtained in the current simulation against Coleman[9]. The mean centreline velocity is 1.178, which is very close to 1.175 as reported in the earlier DNS.

Table 2. Comparison with earlier DNS.

	M_c	M_τ	Re_c	Re_τ	$\langle\rho_c\rangle$	$\langle\rho_w\rangle$	$\langle T_c\rangle$
Coleman[9]	1.502	0.082	2760	222	0.98	1.355	1.378
Present	1.498	0.081	2746	216	0.979	1.343	1.391

Figure 3(b) shows the turbulent normal stresses plotted against those obtained in earlier DNS. Even with moderate grid size, these second order quantities compare reasonably well.

4. Flow Past LP Turbine Blade: Numerical Parameters

DNS has been performed on flow past a T106A blade, which is representative of blades used in a low-pressure turbine. The blade profile is shown in Fig. 4(a) along with the computational domain adopted for simulations. The Reynolds number, based on axial chord length L and inlet velocity U, is $Re = 51,831$ for this particular case. The flow enters the domain at an angle $\alpha = 45.5^0$ with the axial chord (x-axis, consistent with the value recommended by Stadtmuller[10]). For the present compressible simulation, the exit Mach number is around 0.4.

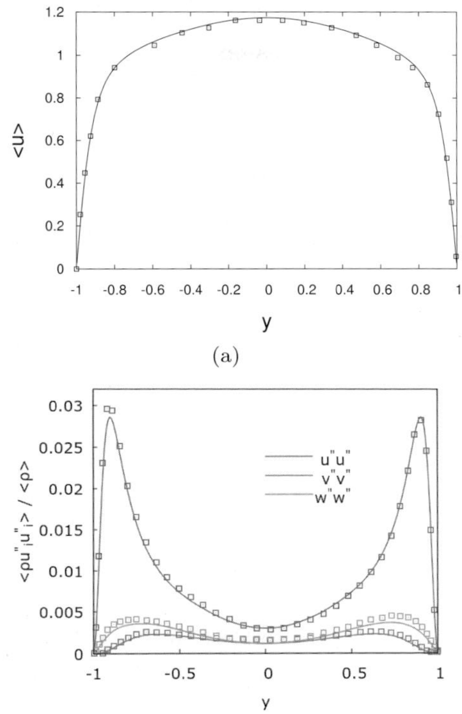

(a)

(b)

Fig. 3. (a) Mean velocity profile (b) Turbulent normal stresses ($\langle \rho u'u' \rangle$). Squares are values from earlier DNS[9].

Computational domain and grid

Figure 4(a) shows the computational domain for the flow around a turbine blade in a cascade. Inlet is $0.5L$ ahead of the leading edge and outlet is $1L$ behind the trailing edge. Pitch between the blades is $P = 0.9306L$, and the spanwise dimension is $S = 0.2L$. No-slip boundary condition has been applied at the blade wall, and characteristics-based boundary conditions have been used at inlet and outlet. The flow is periodic on the rest of the boundaries.

The grid used here is hybrid in nature, with very fine boundary layer grid near the body and an unstructured grid in the outer region. Using hybrid grid gives the flexibility to use a proper orthogonal grid in the boundary layer. Figure 4(a) indicates the inner and outer regions used for different grid topologies. Figure 4(b) shows the boundary layer grid near the leading

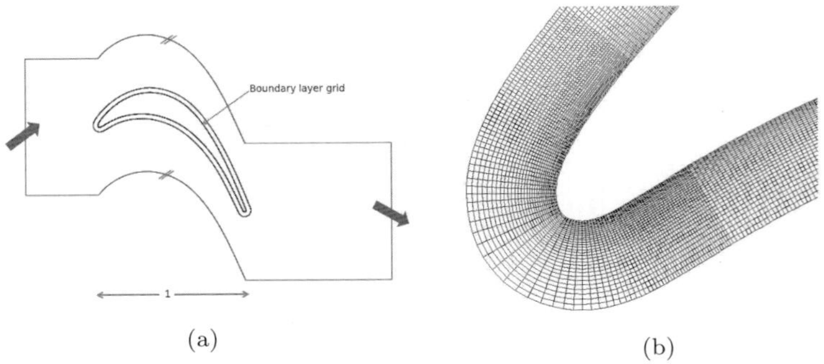

Fig. 4. (a) Computational domain (b) Boundary layer grid near leading edge.

Table 3. Overview of the simulations.

Simulation	N_f	N_z	Grid Size
A	384,164	64	25×10^6
B	740,088	64	47×10^6
C	740,088	128	95×10^6
D	1,257,162	128	161×10^6

edge of the suction side. A grid is first created for 2D xy domain (Fig. 4(a)) and then extruded in the spanwise(z) direction. Simulations have been performed at different grid sizes to study grid independence (see Table 3 for details). The number of grid elements used for the face mesh (N_f), and the intervals used in the spanwise direction (N_z), are also given. The streamwise resolution along the blade has been changed significantly among A, B and D as shown in Fig. 5(b). In simulation C, the face mesh has been kept same as B but the spanwise resolution (N_z) has been doubled. This did not seem to have any effect on the development of the flow.

It is to be noted that the highest resolution used earlier for simulation, performed with a 2nd-order accurate numerical method at the present Re, was 17 million (see Table 1). Also for the finest mesh (SIM D), Δy^+ in wall units was below 0.1 throughout the blade.

5. Results and Discussion

The present simulations have been carried out on the PARAM Yuva-II supercomputer of C-DAC using Intel Xeon E5-2670 8C processors. The non-dimensional time step employed corresponds to $CFL \approx 1.0$. The simulations have been performed without any free-stream turbulence or inlet wake. However for the finest grid (D), simulation has also been performed with 10% free-stream turbulence, but it didn't have any appreciable difference in the overall pressure distribution.

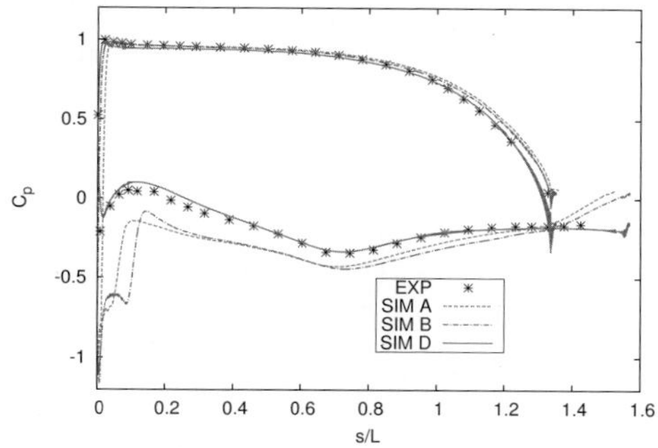

(a)

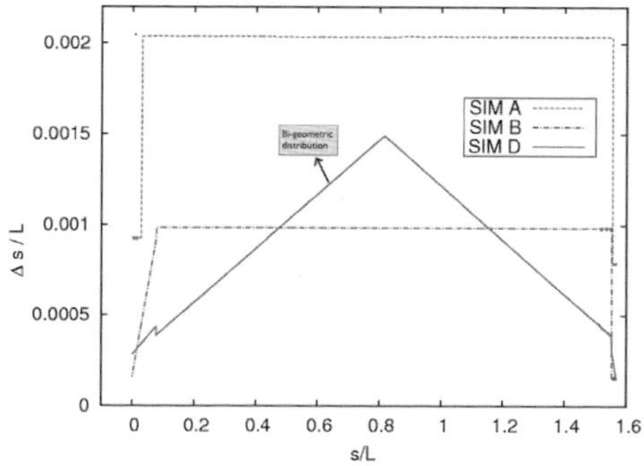

(b)

Fig. 5. (a) Co-efficient of pressure. (b) Mesh spacing in suction side boundary layer (streamwise).

The plot in Fig. 5(a) shows a comparison of mean static pressure coefficients (C_p) as obtained by the present simulation against experimental results (Stadtmuller[10]). On the pressure side, the results match very well for all the grids; however on the suction side, where the flow is rather complex, the results are not in such good agreement for grids other than the finest (D). It is to be remembered that the flow conditions used for all the cases are the same.

The results on the pressure distribution are of great interest for two reasons. First, previous DNS/LES work has been unable to match experimental data near the leading edge of the suction side. Second (as preliminary results show), downstream flow on the suction side is extremely sensitive to the pressure distribution at the leading edge, and so is also strongly affected by resolution. It is interesting that such a fundamental bulk parameter as pressure distribution can be so sensitive to resolution, highlighting the need to ensure grid-independence in simulations of flow past highly curved bodies such as turbine blades.

6. Acknowledgments

We would like to thank the Gas Turbine Enabling Technology (GATET) Program for funding this project, and in particular Dr. S. Kishore Kumar for his enthusiasm. We also thank C-DAC, Pune for providing us the necessary computational resources to perform simulations, and Prof. Wolfgang Rodi, Prof. Jan Wissink and Mr. Stephan Stotz for their help with respect to the data.

References

1. X. Wu and P. A. Durbin, Evidence of longitudinal vortices evolved from distorted wakes in a turbine passage, *J. Fluid Mech.* **446**, 199 (2001).
2. V. Michelassi, J. Wissink and W. Rodi, Analysis of DNS and LES of flow in a low pressure turbine cascade with incoming wakes and comparison with experiments, *Flow, Turbulence and Combustion* **69**, 295 (2002).
3. G. Kalitzin, X. Wu and P. A. Durbin, DNS of fully turbulent flow in a LPT passage, *Int. J. Heat and Fluid Flow* **24**, 636 (2003).
4. J. Wissink, DNS of separating, low Reynolds number flow in a turbine cascade with incoming wakes, *Int. J. Heat and Fluid Flow* **24**, 626 (2003).
5. J. Wissink and W. Rodi, Direct numerical simulations of transitional flow in turbomachinery, *J. Turbomachinery* **128**, 668 (2006).
6. J. G. Wissink and W. Rodi, Direct numerical simulation of flow and heat transfer in a turbine cascade with incoming wakes, *J. Fluid Mech.* **569**, 209 (2006).

7. A. Jameson, Formulation of kinetic energy preserving conservative schemes for gas dynamics and direct numerical simulation of one-dimensional viscous compressible flow in a shock tube using entropy and kinetic energy preserving schemes, *J. Sci. Comput.* **34**, 188 (2008).
8. K. Hillewaert, Direct numerical simulation of the Taylor-Green vortex at Re = 1600, 2nd international high order cfd workshop On the WWW, (2012).
9. G. N. Coleman, J. Kim and R. D. Moser, A numerical study of turbulent supersonic isothermal-wall channel flow, *J. Fluid Mech.* **305**, 159 (1995).
10. P. Stadtmuller, *Investigation of wake-induced transition on the LP turbine cascade T106A-EIZ*, Tech. Report (2002).

Numerical Simulation of Impulsive Supersonic Flow from an Open End of a Shock Tube: A Comparative Study

T. Murugan* and S. De

CSIR-Central Mechanical Engineering Research Institute, Durgapur, West Bengal 713209, India

**E-mail: murugan.thangadurai@gmail.com*

V. Thiagarajan

Scientist, Defence Research & Development Laboratory, Hyderabad, 500 058, India

E-mail: trajan306@yahoo.co.in

Impulsive transient jet originating from the open end of a short driver section shock tube has been simulated numerically by solving the unsteady Navier-Stokes equation in axisymmetric form using AUSM+ scheme for a driver section pressure ratio of 10. The simulations are performed using second order central and fifth order upwind schemes with different spatial resolutions. The effect of higher order schemes and spacial resolution on the accuracy of the solution is studied in detail. It has been observed that the results obtained from the fifth order scheme with lower spacial resolution and denser grids along the free shear layer matched close with experimental smoke flow visualization.

Keywords: Shock tube, AUSM+, Higher order schemes, Impulsive flow.

1. Introduction

Impulsive flow emanating from the open end of a shock tube after the diffraction of incident shock for higher Mach numbers (M>1.6) is complex and unsteady due to propagation of expansion waves inside the shock tube, supersonic expansion of under-expanded jet which consists of Mach disc, reflected shock and the unstable shear layer besides the rolling up and formation of the compressible vortex ring. The presence of diffracted incident shock ahead of the primary compressible vortex ring make this flow different from the conventional supersonic jet. The increase in pressure behind the incident shock (ahead of the vortex ring) plays a major role in the jet flow development and the characteristics of the vortex ring.

Knowledge in the detailed flow field of impulsive jet is helpful in understanding the noise associated with the muzzle blast, shock-vortices interaction, the large and small scales turbulence interaction[1]. Though, the qualitative results of shock cell structure of the impulsive jet obtained from optical techniques (shadowgraph, schlieren and interferometry) existing in the literature[2–5] could be used for calculating the pressure up to a certain extent, the velocity, temperature and density data are essential for understanding the noise associated with the flow. The accurate velocity and vorticity data of impulsive jet are difficult to obtain for higher M even from particle image velocimetry (PIV) due to the presence of shock cell structures and embedded shock within the core of vortex ring. The deceleration of flow behind the strong shock causes accumulation of a large number of particles behind the shock which causes dispersion in PIV results[6,7].

The complete flow field of impulsive supersonic jet which includes rapid spatial and temporal variations of turbulent eddies, free shear layer and unsteady shock cell structures can be resolved accurately using direct numerical simulation (DNS). The DNS of shock tube flow using the time dependent Navier-Stokes (N-S) equations is beyond the capabilities of our computing power. In large eddy simulation (LES), the bigger size eddies are explicitly resolved and small scale eddies are modeled using the filtered N-S equations. However, it also needs substantial computational power.

The Euler simulation is used for predicting the near-field supersonic jet noise, as the physical viscosity is negligible compared to the numerical dissipation[8]. The shock cell structures of the supersonic jet can be predicted well with Euler solver. Ishii *et al.*[9] predicted the formation of starting vortex and shock cell structure for a wide range of jet strength and it matched well with experimental results. However the evolution of starting vortex was not simulated well at a later stage as the vortex motion is affected by gas viscosity. Hence Murugan *et al.*[7] had simulated the flow inside the shock tube and at the muzzle by N-S solver where the axisymmetric unsteady N-S equations are solved using AUSM+ scheme.

The solver was explicit and time stepping was performed with a four-stage Runge-Kutta time integration scheme developed by Arnone *et al.*[10]. Viscous fluxes were calculated by standard second-order accurate central differencing formulas. The centre line velocity of the jet and the velocity across the primary vortex were matched with the PIV results except behind the embedded shock and Mach disc. However, the shear layer vortices originating at the trailing jet and their entrainment into the primary vortex core observed in smoke flow visualization[11] during the initial stage were not predicted well in the simulation.

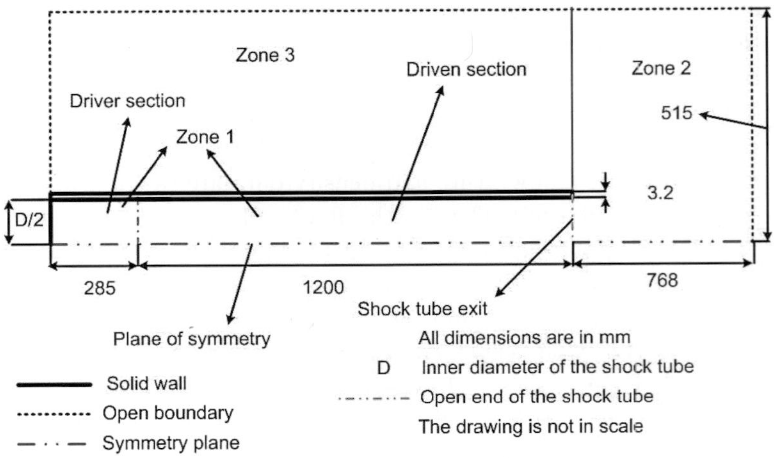

Fig. 1. The computational domain.

Table 1. Grid details of the numerical simulation.

Case no	Scheme	Zone 1 grid points	Zone 2 grid points	Thickness region grid points	Total number of grids (in millions)
1	2nd order	201 X 49	600 X 400	4	0.32
2	2nd order	301 X 99	1200 X 800	7	1.2
3	5th order upwind	201 X 49	600 X 400	4	0.32
4	5th order upwind	301 X 99	1200 X 800	7	1.2
5	5th order upwind	221 X 100	900 X 520	20	0.58

Murugan *et al.*[12] showed the formation mechanism of counter rotating vortex ring and its evolution for various impulse amplitude and duration using a 2nd order scheme. The action of a limiter makes it a first order upwind scheme near shocks. They showed that NS simulation differs with Euler simulation at latter stages as the vortices generated by Kelvin-Helmholtz (K-H) instability grow. It is shown that a certain amount of impulse amplitude is necessary for the creation of counter rotating vortex ring (CRVR) along with the pulse duration. Dora *et al.*[13] found the critical value of non-dimensional impulse for the formation of single and multiple CRVR as 1.2 and 2, respectively, for the impulse amplitude of $P/P_{atm} >$ 1.8 using AUSM+ scheme and fifth-order upwind differencing formula.

The growth rate of the slipstream originated from the Mach disc has followed the model given by Rikanati *et al.*[14]. The formation of shear layer and CRVRs obtained from simulation were matched well with the flow visualization and PIV results. However, the formation of shear layer vortices originating at the trailing jet during the initial phase of vortex ring evolution was not seen. In the present study, the effect of higher order schemes and spatial resolution on the numerical results of impulsive flow is studied and compared with experimental flow visualization.

2. Computational Procedure

Figure 1 shows the computational domain of the flow originating from the shock tube. Numerical simulation is performed for one half of the shock tube with the flow assumed to be axisymmetric. Driver and driven sections length of the shock tube are 285 mm and 1200 mm and are represented by Zone 1. The inner diameter (D) of the shock tube is 64 mm. The pressure in the driver section is 10 bar and everywhere else ambient condition (1 bar) is assumed. The temperature is 300 K everywhere. No-slip condition is applied on shock tube walls. Open boundaries of the domain are treated by the Navier-Stokes characteristic boundary condition. The NS equations in the axisymmetric form in finite volume formulation have been solved along with continuity, energy and state equations using the AUSM+ scheme.

Second and fifth order *Monotonic Upstream Centered Scheme for Conservation Laws* (*MUSCL*) approach has been employed to interpolate variables to the cell faces for use in convective terms. The minmod limiter has been applied to limit the interpolated values as sharp gradients are present in the form of shocks. The solver is explicit and time stepping is performed with a six-stage Runge-Kutta time integration scheme. Viscous fluxes are calculated by standard second-order-accurate central differencing formulae. Further details about the governing equations, and numerical scheme can be found elsewhere[7,12,13]. Numerical simulation of five cases shown in Table are analysed and presented in the subsequent section.

3. Results and Discussion

Figure 2 shows the vorticity field of starting jet obtained using the second order scheme with 0.32 and 1.2 million grids at t=480 μs and 880 μs for M=1.62. The precursor shock wave formed inside the shock tube moves towards the exit of the shock tube. Here, at t=0 denotes the time when the precursor shock is at the exit. Though the Mach disc, embedded shock and shear layer originating from the triple point look similar at t=480 μs, the shear layer broke down to form different vortical structures at t=880 μs. The $K-H$ vortices originating at the trailing jet are seen at the later stage with higher grid resolution.

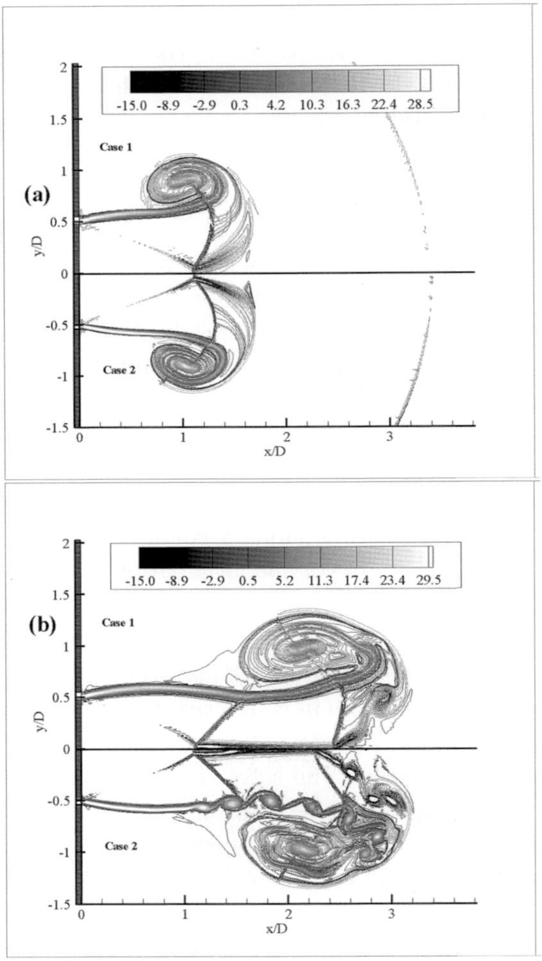

Fig. 2. Comparison of second order scheme results at (a) t= 480 μs and (b) 880 μs.

Figure 3 shows the vorticity field obtained from the fifth order scheme with 0.32 and 1.2 million grid points at t=480 μs and 880 μs for the same M. The fifth order scheme has resolved the vortex induced shock with low resolution at t=480 μs as compared to the second order scheme. It has resolved the K-H vortices to a certain extent and shown the K-H vortices entering into the primary vortex core which was seen from experimental flow visualizations[11] compared to the second order scheme at low grid resolution.

Further, it has also resolved more K-H vortices at later stage with higher spatial resolution. However, the K-H vortices originating at the trailing jet near the exit of the shock tube are not resolved. In order to capture these vortices near the shock tube exit, the grid points are increased along the shear layer region.

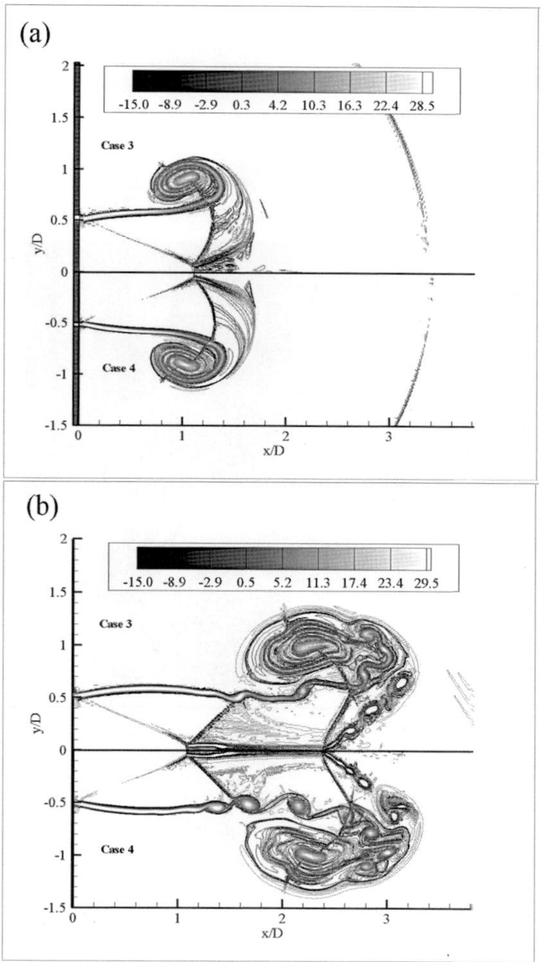

Fig. 3. Results of 5th order upwind scheme at (a) t= 480 μs and (b) 880 μs.

In case 5, the number of grid points has increased from 7 to 20 along the axial direction in the thickness region of the shock tube. In Fig. 4, the high speed smoke flow visualization of the impulsive jet obtained at t=880 μs is shown in one half of the shock tube exit. The vorticity contour obtained from grid 5 is shown in the other half. The K-H vortices originating at the trailing jet near the shock tube exit are resolved much better compared to the other cases, though the total grid points are reduced to 0.58 million. However, the K-H vortices seem to be slightly larger in size compared to those in the visualization image and the K-H vortices entered into the vortex core are clearly seen in the simulation which are not seen in the flow visualization due to lack of smoke particles.

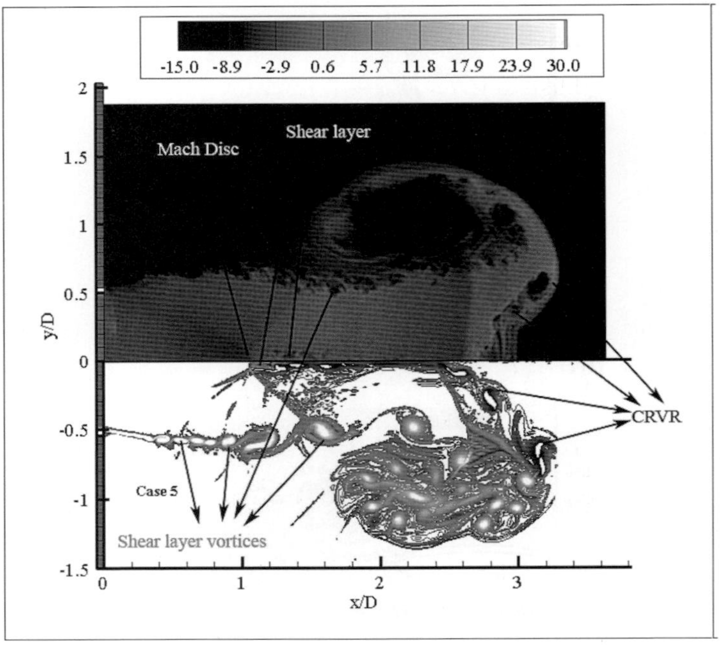

Fig. 4. Comparison of case 5 results at t=880 μs with experimental flow visualization[8].

4. Concluding Remarks

Impulsive jet originating from the shock tube for shock Mach number 1.62 has been studied numerically by solving the axisymmetric Navier-Stokes equation using the AUSM+ scheme with different spatial higher order upwind schemes. The fifth order scheme resolves well the Kelvin-Helmholtz (K-H) vortices originating at the free shear layer at lower spatial resolution as compared to the second order scheme. The embedded shock, vortex induced shock and shear layer vortices originating at the Mach disc are also resolved quite well at high spatial resolution. Use of denser grid along the free shear layer resolves the K-H vortices formation near the shock tube exit. However, the K-H vortices are larger in size compared to the size obtained from the experimental flow visualization. Implementation of a turbulence model in $3D$ simulation with higher order scheme may bring the gap closer to the experimental result as the flow is turbulent at M>1.6 and it will be done in future.

5. Acknowledgments

The authors would like to thank Dr. Debopam Das, associate professor and Mr. C. L. Dora, doctoral student, Department of Aerospace Engineering, IIT Kanpur for providing the experimental flow visualization images for comparing the numerical results.

References

1. M. J. Lighthill, On sound generated aerodynamically. ii. turbulence as a source of sound, *Proc. R. Soc. London, ser. A* **222**, 1 (1954).
2. K. C. Phan, J. L. Stollery, The effect of suppressors and muzzle brakes on shock wave strength, *Proc. 14th Int. Symp. on Shock Tubes and Waves*, August 1922, Sydney (1984).
3. J. P. Baird, Supersonic vortex rings. *Proc. R. Soc. Lond. Ser. A* **409**, 59 (1987).
4. M. Brouillette, C. Hebert, Propagation and interaction of shock generated vortices. *Fluid Dyn. Res.* **21**, 159 (1997).
5. T. Murugan and D. Das, Characteristics of counter-rotating vortex rings formed ahead of a compressible vortex ring, *Exp. Fluids* **49**, 1247 (2010).
6. H. Zare-Behtash, K. Kontis, N. Gongora-Orozco, Experimental investigation of compressible vortex loops. *Phys. Fluids* **20**, 126105 (2008).
7. T. Murugan, S. De, C. L. Dora and D. Das, Numerical simulation and PIV study of formation and evolution of compressible vortex ring, *Shock Waves* **22**, 69 (2012).
8. R. Mankbadi, R. Hixon, S. Shih and L. Povinelli, Use of linearized Euler equations for supersonic jet noise prediction, *AIAA J.* **36(2)**, 140 (1998).

9. R. Ishii, H. Fujimoto, N. Hatta and Y. Umeda, Experimental and numerical analysis of circular pulse jets, *J. Fluid Mech.* **392**, 129 (1999).
10. A. Arnone, M. S. Liou and L. A. Povinelli, Multigrid time-accurate integration of Navier–Stokes equations, *NASA TM* **106373** (1999).
11. T. Murugan, Flow and acoustic characteristics of high Mach number vortex rings during evolution and wall-interaction: an experimental investigation, *Doctoral thesis*, I. I. T. Kanpur (2008).
12. T. Murugan, S. De, C. L. Dora, D. Das and P. Prem Kumar, A study of the counter rotating vortex rings interacting with the primary vortex ring in shock tube generated flows, *Fluid Dyn. Research* **45**, 025506 (2013).
13. C. L. Dora, T. Murugan, S. De and D. Das, Role of slipstream instability in formation of counter rotating vortex rings ahead of a compressible vortex ring, *J. Fluid Mech.* **753**, 29 (2014).
14. A. Rikanati, O. Sadot, G. Ben-Dor, D. Shvarts, T. Kuribayashi and K. Takayama, Shock-wave Mach-reflection slip-stream instability: a secondary small-scale turbulent mixing phenomenon, *Phys. Rev. Lett.* **96**, 174503 (2006).

Green's Function Analysis of Pressure-Strain Correlations in a Supersonic Pipe, Nozzle and Diffuser

Somnath Ghosh

Aerospace Engineering,
Indian Institute of Technology, Kharagpur, India
E-mail: sghosh@aero.iitkgp.ernet.in

Rainer Friedrich

Lehrstuhl fuer Aerodynamik und Stroemungsmechanik,
Technische Universitaet Munich, Germany
E-mail: r.friedrich@lrz.tum.de

1. Introduction

Compressible turbulence occurs in nature and in engineering applications. Kovasznay[9] examined the linearized equations for compressible turbulence and showed the existence of three basic modes: acoustic, vorticity and entropy. In the absence of viscous effects and mean shear these modes are independent of each other. If viscous effects are taken into account, then coupling exists between the acoustic and entropy modes. If only mean shear is considered, then the vorticity and acoustic modes are coupled. In the presence of both mean shear and viscous effects, the three modes are coupled to each other[1]. Sarkar *et al.*[13] studied isotropic turbulence at low turbulent Mach numbers (M_t), and showed that the acoustic mode can be effectively isolated. In such flows, the effects of compressibility on vorticity is much smaller than that on acoustics. The analysis given in Sarkar *et al.*[13] yields a parameter which represents the partition of internal and kinetic energy of the acoustic mode. It was shown that this parameter evolves to a constant value of unity which denotes a state of acoustic equilibrium i.e. the equipartition of kinetic and potential energies of the acoustic mode. Lee *et al.*[10] studied decaying isotropic compressible turbulence over a range of turbulent Mach numbers M_t and found eddy shocklets in the flow at large M_t. They found that the probability of occurrence of shocklets increases at larger M_t and higher Reynolds numbers. Near the shocklets, the compressible dissipation (which depends on dilatation fluctuations) was found to be large. Shocklets, and hence explicit compressibility effects related to dilatation fluctuations, were also observed at moderate M_t in the DNS of isotropic compressible turbulence by Pirozzoli and Grasso[11] and Samtaney *et al.*[12].

DNS studies of supersonic channel flows with isothermal walls[3,4] have revealed that explicit compressibility effects are negligible in these flows for the range of supersonic Mach numbers studied. 'Compressibility' effects in these flows manifest themselves as mean density and temperature variations in the near-wall region. The mean density variation leads to a reduction of pressure-strain correlations at supersonic Mach numbers and in turn to an increase in Reynolds stress anisotropy[4]. These observations were also made in DNS of supersonic pipe flow with isothermal wall[5]. Effects of mean dilatation and extra rates of strain add further complications to supersonic wall-bounded flows and lead to changes in the turbulence structure which cannot be explained only by mean property variations. Such effects were described by Bradshaw[2] and numerically predicted in LES and DNS of canonical supersonic nozzle and diffuser flows where fully developed supersonic pipe flow serves as inflow[6,7]. It was observed that the Reynolds stresses decrease dramatically in the nozzle and increase in the diffuser. The pressure-strain correlations were found to play a pivotal role in changing the Reynolds stresses in these flows. Hence, it appeared important to gain insights into the behaviour of pressure-strain correlations in these flows and an elegant way of doing this is a Green's function analysis based on DNS data. Kim[8] used Green's function to solve the Poisson equation for pressure fluctuations based on DNS data of an incompressible channel flow. He found that the contribution of the slow part (nonlinear source terms of the Poisson equation) to the pressure fluctuations was greater than that of the rapid part (linear source terms) in the channel, except very close to the walls. Foysi *et al.*[4] applied Green's function to analyse pressure-strain correlations using supersonic channel flow DNS data and also found the contribution of the slow terms to be greater than that of the rapid terms. The wave equation that governs the pressure fluctuations in a compressible flow was converted into a Poisson equation by neglecting the unimportant terms containing density fluctuations. Ghosh *et al.*[5] carried out a similar study in cylindrical coordinates with DNS data of a supersonic pipe flow. In the present paper, we extend the Green's function analysis to DNS data of a supersonic nozzle and a diffuser with circular cross-sections where the cross-sectional area varies slowly in the axial direction. We also demonstrate the usefulness of well-resolved LES in modelling pressure-strain correlations by performing a Green's function analysis of LES data of supersonic pipe flow.

Mathematical formulation and numerical technique

The equation for pressure fluctuations in a compressible flow is :

$$\nabla^2 p' = -\nabla\cdot\nabla\cdot(\rho\vec{u}.\vec{u} - \overline{\rho\vec{u}.\vec{u}}) + \frac{\partial^2\rho'}{\partial t^2} + \nabla\cdot\nabla\cdot\underline{\underline{\tau}}' \tag{1}$$

where the overbar and dash have their usual meanings. The first term on the RHS can be expanded into terms containing the mean density and density fluctuations. The second term can be expressed as a substantial derivative in order to preserve Galilean invariance of the above equation. Foysi[4] has shown that the terms containing density fluctuations can be neglected at supersonic speeds and therefore equation (1) can be reduced to a Poisson equation. The exact form of Poisson's equation in cylindrical coordinates applicable to pipe, nozzle and diffuser flows is given in Ghosh *et al.*[5,6]. In the pipe flow, the periodicity in axial and azimuthal directions allows us to do Fourier transforms and the Poisson equation reduces to the following ordinary differential equation

$$\frac{d^2\hat{p}}{dr^2} + \frac{1}{r}\frac{d\hat{p}}{dr} - (k_x^2 + \frac{1}{r^2}k_\theta^2)\hat{p} = \bar{\rho}\hat{f}, \tag{2}$$

where f denotes the source terms. This equation is solved subject to the boundary condition $\frac{\partial p'}{\partial r} = 4/3(\bar{\mu}\frac{\partial^2 u'}{\partial r^2} + \frac{\partial\bar{\mu}}{\partial r}\frac{\partial u'}{\partial r})$ at the pipe wall ($r = 1$). Equation (2) can be solved using a Green's function having the following form[5]:

$$G_<(k_x, k_\theta, r, r_o) = -\frac{I_{k_\theta}(k_x r_o) I_{k_\theta}(k_x r) K'_{k_\theta}(k_x)}{I'_{k_\theta}(k_x)} + K_{k_\theta}(k_x r_o) I_{k_\theta}(k_x r)$$

$$G_>(k_x, k_\theta, r, r_o) = -\frac{I_{k_\theta}(k_x r_o) I_{k_\theta}(k_x r) K'_{k_\theta}(k_x)}{I'_{k_\theta}(k_x)} + I_{k_\theta}(k_x r_o) K_{k_\theta}(k_x r) \tag{3}$$

where I, K are modified Bessel functions of the first and second kind, respectively. The solution to Eq. (2) is then given as:

$$\hat{p}(k_x, k_\theta, r) = \int_0^1 G(k_x, k_\theta, r, r_o)\bar{\rho}(r_o)\hat{f}(k_x, k_\theta, r_o) r_o dr_o \tag{4}$$

In the nozzle/diffuser flow, periodicity (and hence Fourier transform) can only be applied in the azimuthal direction and the Poisson equation can be written as:

$$\frac{\partial^2 \hat{p}}{\partial r^2} + \frac{1}{r}\frac{\partial \hat{p}}{\partial r} + \frac{\partial^2 \hat{p}}{\partial x^2} - \frac{1}{r^2}k_\theta^2 \hat{p} = \bar{\rho}\hat{f} \tag{5}$$

This equation is solved along with homogeneous Neumann boundary conditions at the wall and at $x = 0, L$. Using a series expansion form of the delta function, we can express the Green's function for Eq. (5) as

$$G(x, x_o, k_\theta, r, r_o) = G_o(k_\theta, r, r_o) + \sum_{n=1}^{\infty} G_n(k_\theta, r, r_o) cos(n\pi x_o/L) cos(n\pi x/L) \tag{6}$$

where

$$G_{n,<}(k_\theta, r, r_o) = \left(-\frac{I_{k_\theta}(k_x r_o) I_{k_\theta}(k_x r) K'_{k_\theta}(k_x)}{I'_{k_\theta}(k_x)} + K_{k_\theta}(k_x r_o) I_{k_\theta}(k_x r)\right)/\pi L$$

$$G_{n,>}(k_x, k_\theta, r, r_o) = \left(-\frac{I_{k_\theta}(k_x r_o) I_{k_\theta}(k_x r) K'_{k_\theta}(k_x)}{I'_{k_\theta}(k_x)} + I_{k_\theta}(k_x r_o) K_{k_\theta}(k_x r)\right)/\pi L$$

and

$$G_{o,<}(k_\theta, r, r_o) = \Big(- \frac{r_o^{k_\theta} . r^{k_\theta}}{2k_\theta} - \frac{r_o^{-k_\theta} . r^{k_\theta}}{2k_\theta} \Big)/2\pi L$$

$$G_{o,>}(k_\theta, r, r_o) = \Big(- \frac{r_o^{k_\theta} . r^{k_\theta}}{2k_\theta} - \frac{r_o^{k_\theta} . r^{-k_\theta}}{2k_\theta} \Big)/2\pi L$$

Here, $k_x = n\pi R/L$, R being the radius. The solution of Eq. (5) is then given by

$$\hat{p}(x, k_\theta, r) = \int_0^L \int_0^1 G(x, x_o, k_\theta, r, r_o) \bar{\rho}(x_o, r_o) \hat{f}(x_o, k_\theta, r_o) r_o dr_o dx_o \tag{7}$$

The pressure fluctuations in physical space are obtained by inverse Fourier transform of the above equation,

$$p'(x, \theta, r) = \int_0^L \int_0^1 \bar{\rho}(x_o, r_o) G * f'(x_o, \theta, r_o) r_o dr_o dx_o \tag{8}$$

and the pressure-strain correlations (here the axial component is shown) for the nozzle/diffuser flow are obtained by

$$PS_{xx} = \overline{p' \frac{\partial u''}{\partial x}} = \int_0^L \int_0^1 \bar{\rho}(x_o, r_o) \overline{G * f'(x_o, \theta, r_o) \frac{\partial u''}{\partial x}} r_o dr_o dx_o \tag{9}$$

The corresponding equation for the pipe flow is

$$PS_{xx} = \overline{p'\frac{\partial u''}{\partial x}} = \int_0^1 \bar{\rho}(r_o)\overline{G * f'(x,r,\theta,r_o)\frac{\partial u''}{\partial x}}r_o dr_o \tag{10}$$

The abovementioned Green's function method is used to analyse DNS database of supersonic pipe, nozzle and diffuser and also LES data of supersonic pipe. 5th order low-dissipation compact upwind schemes have been used in the DNS for the convective terms and 6th order central schemes for the molecular transport terms[5]. 6th order compact central schemes are used for LES along with an explicit filtering variant of the approximate deconvolution method as subgrid scale model[7]. Time integration is done using a 3rd-order Runge-Kutta method. The pipe flow has a centerline Mach number of 1.5 and a friction Reynolds number of 245. $256 \times 128 \times 91$ points have been used in the axial, spanwise and radial directions respectively in the DNS. $64 \times 64 \times 50$ points were used in the LES. This pipe flow acts as inflow to the nozzle simulation which is similarly discretized. The incoming pipe flow for the diffuser simulation has a friction Reynolds number of 300 and a centerline Mach number of 1.8. The diffuser domain is discretized with $384 \times 256 \times 140$ points in the DNS and with $64 \times 64 \times 50$ points in the LES. The walls are isothermal in all the simulations. The modified Bessel functions are evaluated using GNU Scientific Library (GSL) software. In the nozzle/diffuser, 100 terms of the series in Eq. (6) were included in the analysis to obtain accurate results.

2. Results

In supersonic pipe flow with isothermal wall, there is a rise in mean temperature from the wall to the core region. This results in a drop in mean density from its wall value to that in the core region, since the mean pressure does not vary much in the radial direction[5]. Acceleration in the supersonic nozzle flow with isothermal wall results in the axial decrease of mean temperature in the core region from its inlet value. This leads to an axial increase in the ratio of core to wall density. Opposite effects are observed in the decelerating flow in the diffuser where the temperature in the core region increases from its inlet value (see Fig. 1). The acceleration of the flow in the nozzle results in attenuation of turbulence intensities (see Ghosh *et al.*[6]). Opposite effects are observed in the decelerating flow in the diffuser. The pressure-strain correlations decay in the nozzle and are magnified in the diffuser and these terms are the ones which eventually control the levels of turbulence intensities in the nozzle and diffuser. The Green's function method provides an accurate prediction of pressure-strain correlations as evidenced by the following comparison with DNS data. Figure 2 (left) shows the comparison for the axial pressure-strain correlation in the pipe flow and we find that the Green's function solution follows the DNS profile very closely. Also, we see that the slow source terms in the Poisson equation are the major contributors in this case. The contribution of the rapid terms is lower. In Fig. 2 (right), we observe an accurate estimate of the azimuthal pressure-strain component as well. Interestingly, for this component, the rapid terms are the major contributors as opposed to the slow terms for the axial component. For the radial component, the slow terms are the major contributors as seen in Fig. 4 in the pipe and nozzle flows. Figure 3 shows the Green's function solution at a particular axial location in the supersonic nozzle. We note that the Green's function provides very good predictions for the axial and azimuthal pressure-strain correlations. As in the pipe flow, the slow terms are the major contributors in the axial component and the rapid terms are dominant in the azimuthal component. Similar observations are made in the diffuser flow as shown in Fig. 5 where the axial and azimuthal pressure-strain correlations are plotted. It is interesting to compare the axial pressure-strain correlations obtained from LES of supersonic pipe flow with DNS data. As shown in Fig. 6 (left), the LES result lies a little above the one from the DNS because of absence of the effects of subgrid scales in the LES. The Green's function predictions however prove to be accurate in the LES as in the DNS which provides confidence to future applications of such analysis using LES data. As demonstrated in Fig. 6 (left), the Green function solution in the LES taking into account only the slow terms differs from the corresponding solution in the DNS while the solution taking into account only the rapid terms follows the corresponding DNS solution closely. Thus the difference between pressure-strain correlation obtained in the LES and the DNS can be attributed to an underprediction of the effects of the slow terms in the LES.

3. Conclusions

We have performed Green's function analyses of pressure-strain correlations in supersonic pipe, nozzle and diffuser flows based on DNS data. In all cases, the Green's function solutions follow the DNS results very closely. The contributions of slow terms in the Poisson equation are found to be

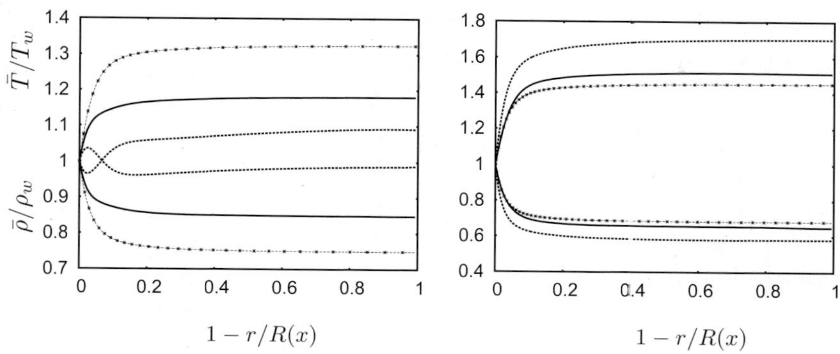

Fig. 1. Radial profiles of mean temperature and density, normalized with their wall values. Stations at $x/L = 0(- * - * -)$,0.2(—),0.85(.....) in the supersonic diffuser (left) and at $x/L = 0.0(- * - * -)$, 0.45(—),0.8(.....) in the supersonic nozzle (right). (courtesy Ghosh *et al.*[6])

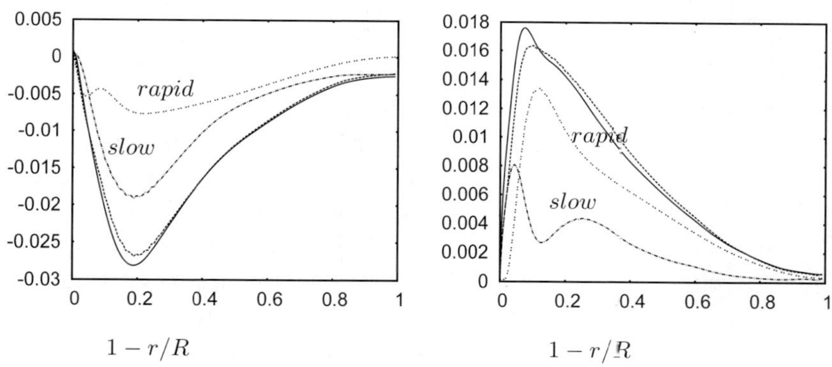

Fig. 2. Comparison of Green's function solution (...) and DNS (—) data for axial (left) and azimuthal (right) pressure-strain correlations in supersonic nozzle at $x/L = 0.25$. Contributions of slow (-.-.-), rapid (.. .. ..) source terms are shown. All terms are normalized with the local $\tau_w^2/\bar{\mu}$.

dominant in the axial and radial pressure-strain correlations while those of the rapid terms are dominant in the radial component. We have also performed similar analysis using LES data of supersonic pipe flow and found the Green's function approach to be successful there as well. Similar analysis is underway for LES of nozzle and diffuser flows.

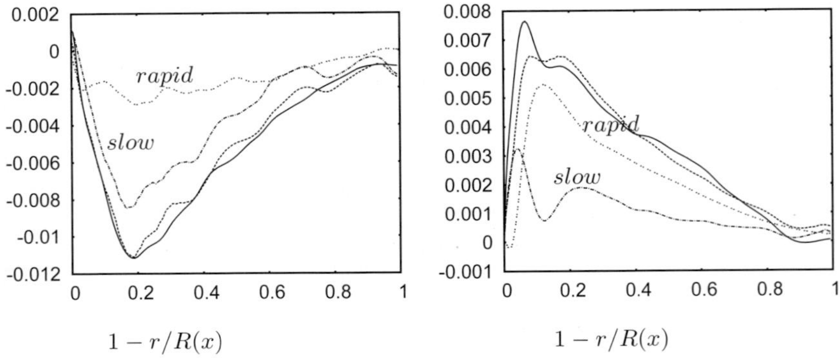

Fig. 3. Comparison of Green's function solution (...) and DNS (—) data for axial (left) and azimuthal (right) pressure-strain correlations in supersonic nozzle at $x/L = 0.25$. Contributions of slow (-.-.-), rapid (.. .. ..) source terms are shown. All terms are normalized with the local $\tau_w^2/\bar{\mu}$.

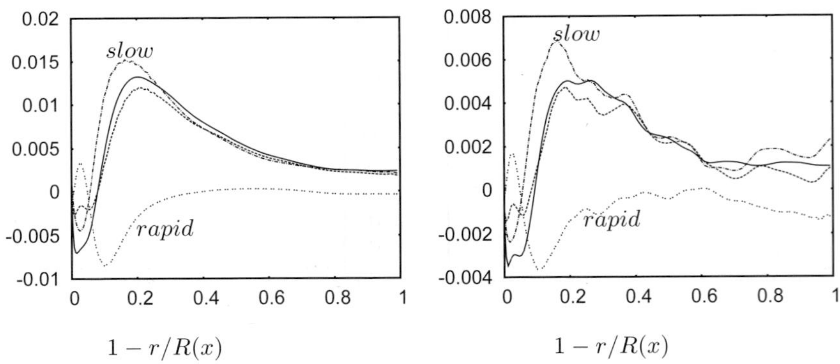

Fig. 4. Comparison of Green's function solution (...) and DNS (—) data for radial pressure-strain correlations in supersonic pipe (left) and in supersonic nozzle at $x/L = 0.25$ (right). Contributions of slow (-.-.-), rapid (.. .. ..) source terms are shown. All terms are normalized with the local $\tau_w^2/\bar{\mu}$.

References

1. Blaisdell, G. A., Mansour N. N. and Reynolds, W. C., Compressibility effects on the growth and structure of homogeneous turbulent shear flow, *J. Fluid Mechanics* **256**, 443–485 (1993).
2. Bradshaw, P., The effect of mean compression or dilatation on the

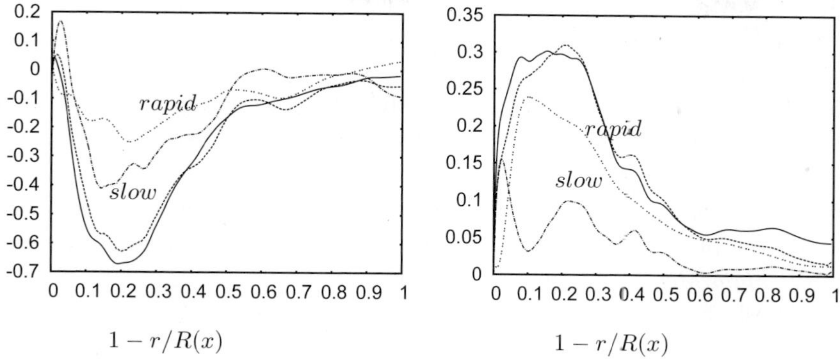

Fig. 5. Comparison of Green's function solution (...) and DNS (—) data for axial (left) and azimuthal (right) pressure-strain correlations in supersonic diffuser at $x/L = 0.3$. Contributions of slow (-.-.-), rapid (.. .. ..) source terms are shown. All terms are normalized with the local $\tau_w^2/\bar{\mu}$.

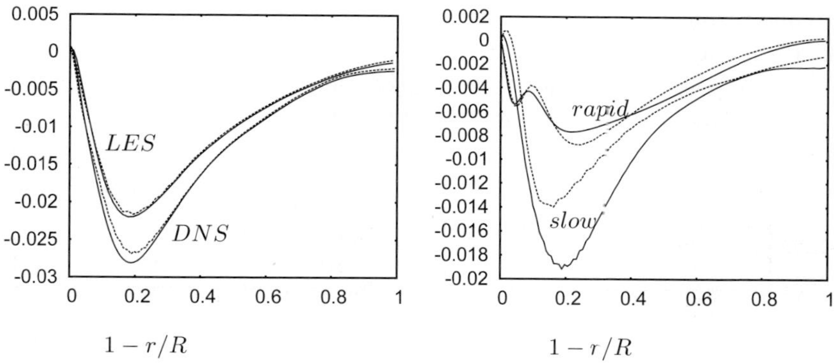

Fig. 6. (left) Comparison of axial pressure-strain correlation in supersonic pipe flow from DNS and LES data. The corresponding Green's function solutions are shown by dashed lines. (right) Slow and rapid parts of axial pressure-strain correlations obtained using Green's function based on DNS (solid line) and LES (dashed line). All terms are normalized with $\tau_w^2/\bar{\mu}$.

turbulence structure of supersonic boundary layers. *J. Fluid Mech.* **63**, 449–464 (1974).

3. Coleman, G. N., Kim, J. and Moser, R. D, A numerical study of turbulent supersonic isothermal-wall channel flow, *J. Fluid Mech.* **305**, 159–183 (1995).

4. Foysi, H., Sarkar, S. and Friedrich, R.. Compressibility effects and turbulence scalings in supersonic channel flow, *J. Fluid Mech.* **509**, 207–216 (2004).
5. Ghosh, S., Foysi, H. and Friedrich, R., Compressible turbulent channel and pipe flow: similarities and differences, *J. Fluid Mech.* **648**, 155–181 (2010).
6. Ghosh, S. and Friedrich, R., Effects of distributed pressure gradients on the pressure-strain correlations in a supersonic nozzle and diffuser. *J. Fluid Mech.* **742**, 466–494 (2014).
7. Ghosh, S., Sesterhenn, J. and Friedrich, R., Large-eddy simulation of supersonic turbulent flow in axisymmetric nozzles and diffusers. *Int. J. Heat and Fluid Flow* **29**, 579–590 (2008).
8. Kim, J., On the structure of pressure fluctuations in simulated turbulent channel flow. *J. Fluid Mech.* **205**, 421–451 (1989).
9. Kovasznay, L. S. G., Turbulence in supersonic flow. *J. Aeronaut. Sci.* **20**, 657–682 (1953).
10. Lee, S., Lele, S.K. and Moin, P., Eddy shocklets in decaying compressible turbulence, *Phys. Fluids A* **3**, 657 (1991).
11. Pirozzoli, S. and Grasso, F., Direct numerical simulations of isotropic compressible turbulence: Influence of compressibility on dynamics and structures, *Phys.Fluids* **16**, 4386 (2004).
12. Samtaney, R., Pullin, D. I. and Kosovic, B., Direct numerical simulation of decaying compressible isotropic turbulence and shocklet statistics, *Phys. Fluids* **13**, 1415 (2001).
13. Sarkar, S., Erlebacher, G., Hussaini, M. Y. and Kreiss, H. O., The analysis and modelling of dilatational terms in compressible turbulence, *J. Fluid Mech.* **227**, 473–493 (1991).

The Structure of Turbulence in Poiseuille and Couette Flow at Computationally High Reynolds Number

S. Pirozzoli*, M. Bernardini and P. Orlandi
Dipartimento di Ingegneria Meccanica e Aerospaziale,
Università di Roma La Sapienza,
via Eudossiana 18, 00184, Roma, Italia
**E-mail: sergio.pirozzoli@uniroma1.it*

We study internal wall-bounded flows by means of direct numerical simulations of canonical Poiseuille and Couette flow at computationally high Reynolds number. The two flows exhibit a number of similarities, in that they both feature a near log layer in the mean velocity profile and in the wall-parallel velocity variances, and exhibit excess energy production over dissipation in the outer layer. When closely inspected, the mean velocity profiles exhibit systematic deviations from the alleged log law, which are more severe in Couette flow. The latter also exhibits clearer outer-layer activity associated with the formation of quasi-streamwise rollers, and which implies the emergence of a secondary outer-layer peak of the streamwise turbulence intensity.

Keywords: Wall turbulence, Direct Numerical Simulation, high-Reynolds-number effects

1. Introduction

The study of turbulent wall-bounded flows has great importance in engineering applications, but it is also a subject of substantial intrinsic academic interest. In the commonly accepted scenario, turbulence in the near-wall layer is approximately universal when quantities are scaled in wall units (the friction velocity $u_\tau = \sqrt{\tau_w/\rho}$, and the viscous length-scale $\delta_v = \nu/u_\tau$). On the other hand, far from the wall, the significant length scale is the thickness of the wall layer (in channels, the half-width h), whereas the relevant velocity scale is still u_τ. The universal nature of wall turbulence has been put into question by a series of studies which have highlighted violations of the canonical structure, the most evident being perhaps the variation of the near-wall peak of the streamwise turbulence intensity, whose amplitude is found to grow logarithmically with the friction Reynolds number ($Re_\tau = h/\delta_v$)[5]. This behavior has been traced to the influence of outer-scaled eddies, which mainly contain wall-parallel, inactive motions[8]. The

Table 1. List of parameters for turbulent Poiseuille and Couette flows, including friction Reynolds number, collocation points and mesh spacings. The time averaging window T is given in terms of eddy turnover times h/u_τ.

Flow case	Re_τ	N_x	N_y	N_z	Δx^+	Δz^+	Tu_τ/h	Line style
P1	546	1024	256	512	10.0	6.7	36.3	Dashed
P2	999	2048	384	1024	9.2	6.1	26.9	Dash-dot
P3	2012	4096	768	2048	9.3	6.2	14.9	Dash-dot-dot
P4	4079	8192	1024	4096	9.4	6.2	8.54	Solid
C1	171	1280	192	896	7.55	4.80	113.9	Dashed
C2	260	2048	256	1280	7.18	5.10	72.2	Dash-dot
C3	507	4096	384	2560	7.00	4.99	74.9	Dash-dot-dot
C4	986	8192	512	5120	6.80	4.84	54.1	Solid

importance of such large-scale structures on the dynamics of wall-bounded turbulent flows is now well recognized thanks to extensive research conducted in the last years, which has contributed to characterizing the role of large- and very large-scale motions (LSMs,VLSMs) observed in the core of channel and pipe flows and of the superstructures found in the log region of turbulent boundary layers. Large-scale motions are known to exert an influence on the self-sustained near-wall cycle through an imprinting mechanism, consisting of the superposition of low-wavenumber energy, coming from a secondary energy site located in the outer layer, and through non-linear mechanisms of amplitude modulation[9]. Both imprinting and modulation effects increase with Reynolds number, becoming visible/appreciable only for $Re_\tau \geq 1000$, and it is generally believed that, to be able to probe high-Reynolds-number effects, friction Reynolds numbers of the order of a few thousands should be explored, the consensus threshold to observe an order-of-magnitude logarithmic variation of the mean velocity being $Re_\tau \approx 4000$[7].

Studying wall-bounded turbulent flows (boundary layers, channels, pipes) in the moderate-to-high Reynolds-number regime is then a challenging task both for experiments, because of the stringent requirements on the velocity probes size, and even more for DNS. The latter offers the advantage of easier access to the full three-dimensional flow fields, but it is seriously hampered by the need of huge computational resources. This restriction makes the numerical simulation of wall-bounded flows at high Reynolds number (in the operational sense given above) quite challenging. It is the purpose of this contribution to investigate the structure of wall turbulence at moderately high Reynolds number, with specific reference to the structure of the outer layer, whose features are relatively less known than those of the inner layer. For this purpose novel DNS databases[4,14] of incompressible Poiseuille flow and Couette flow are exploited.

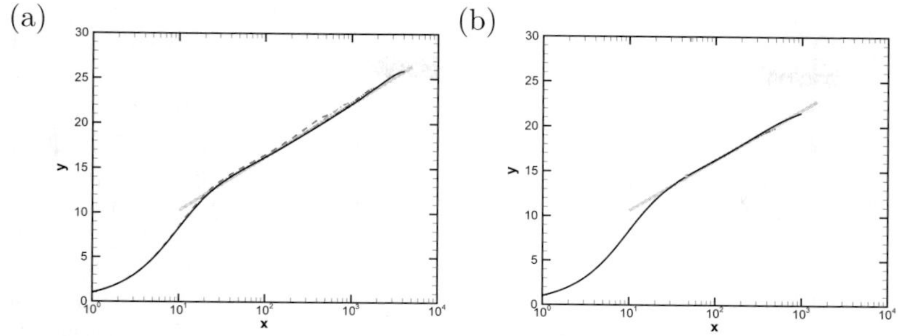

Fig. 1. Mean velocity profiles for Poiseuille flow (a) and Couette flow (b). The thick grey lines indicate logarithmic fits $u^+ = 1/k \log y^+ + C$, with the sets of constants given in the text.

2. Numerical method

The numerical method is based on the use of second-order finite-difference approximations for the solution of the Navier-Stokes equations for a divergence-free velocity field[12]. The equations are discretized in an orthogonal coordinate system ($x, y, z/u, v, w$ denote the streamwise, wall-normal and spanwise directions/velocities) using staggered central finite-difference approximations, which guarantee that kinetic energy is globally conserved in the limit of inviscid flow. Time advancement is achieved by a hybrid third-order low-storage Runge-Kutta algorithm coupled with the second-order Crank-Nicolson scheme, combined in the fractional-step procedure, whereby the convective and diffusive terms are treated explicitly and implicitly, respectively. The Poisson equation for the pressure field stemming from the incompressibility condition is efficiently solved using a direct solver based on Fourier transform methods. Details on the computational mesh and on the flow parameters for the DNS are provided in Table 1. The simulations have been carried out in a moving reference frame in which the bulk velocity is zero[3], in large computational domains to obtain correct representation of the largest outer-layer flow structures. For that purpose, the box size is $6\pi h \times 2\pi h$ for Poiseuille flow, and $18\pi h \times 6\pi h$ for Couette flow, the difference being dictated by the greater importance of the core eddies in the latter case.

3. Results

The results of Poiseuille and Couette flow are here compared in terms of the mean velocity profiles and of the velocity fluctuations statistics. When plotted in a conventional semi-log representation (see Fig. 1), the velocity

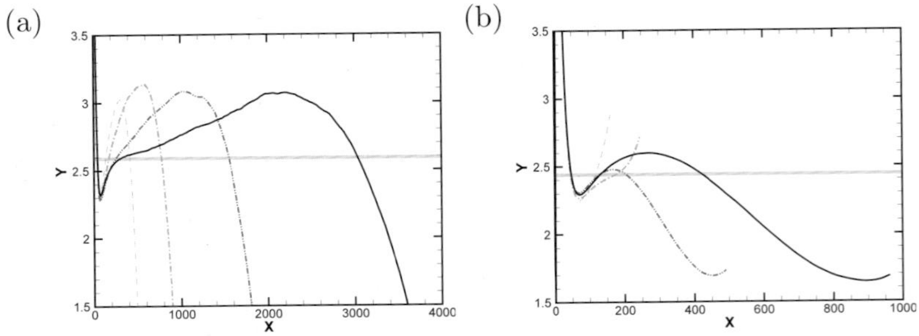

Fig. 2. Diagnostic function for Poiseuille flow (a) and Couette flow (b). The thick grey lines indicate the alleged von Karman constants for the two flows.

profiles are shaped consistently with the common wisdom, in that extended logarithmic layers appear which become wider as Re_τ is increased. The same statement applies to both flows, although the set of log-law constants which can be extrapolated based on fitting the velocity curves is slightly different. Specifically, in Poiseuille flow we find $k = 0.386$, $C = 4.3$, not too far from recent experimental estimates[11], and in Couette flow we find $k = 0.41$, $C = 5.0$, which is closer to the canonical values quoted in text-books[15]. However, the scenario is much less reassuring if the log-law diagnostic function, defined as $\Xi = y^+ \mathrm{d}u^+/\mathrm{d}y^+$, is monitored, as in Fig. 2. In Poiseuille flow the diagnostic function does not exhibit any significant plateau, which would correspond to a genuine log layer. On the other hand, the distributions of Ξ in about half of the channel seems to exhibit a linear variation, with slope decreasing in inner units. This behavior is consistent with refined overlap arguments[1,4], which imply that a genuine log layer only appears in the infinite Reynolds number limit. The situation is even more intricate in the case of Couette flow, in which no genuine log layer is observed, nor any systematic tendency to it is found. This is particularly disappointing in that Couette flow is the only flow in which the total stress is exactly constant across the wall layer, hence the best candidate for log layers to emerge. This odd behavior is then likely dictated by the impact of intense core eddies on the mean flow behavior, and which are much less energetic in Poiseuille flow.

Velocity fluctuations statistics are shown in inner coordinates in Fig. 3 for Poiseuille and Couette flows, and compared with available experimental data. The longitudinal stresses clearly lack universality when reported in inner scaling even in the near-wall region, with fluctuation amplitudes

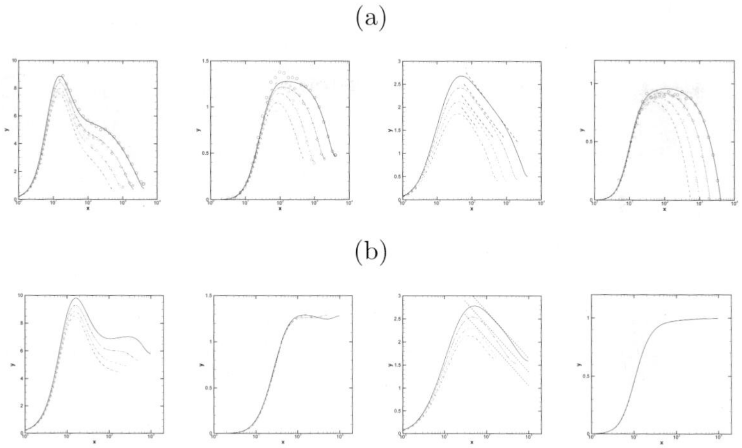

Fig. 3. Reynolds stress components for Poiseuille (a) and Couette (b) flow. Open symbols indicate channel experimental data from Schultz and Flack [16], at $Re_\tau = 1010$ (squares), $Re_\tau = 1956$ (diamonds), $Re_\tau = 4048$ (circles). The dashed lines indicate data fit obtained according to (1).

increasing nearly logarithmically with Re_τ. This behavior is due to the increasing influence of inactive outer-layer modes on the near-wall dynamics [10]. No clear range with logarithmic decrease of the streamwise stress is observed, as predicted by the attached eddy hypothesis [17]. Interestingly, a secondary outer peak of $\overline{u'^2}$ is found in the Couette flow at $y^+ \approx 300$ ($y/h \approx 0.33$). The existence of a secondary peak of u' has been the subject of strong debate in recent years [2,6], and it has been speculated that not outer peak should arise in Poiseuille flow up to $Re_\tau \approx 15000$. Apparently, the nature of the Couette flow is different, and high-Reynolds number effects occur earlier than in other wall-bounded flows. The formation of a logarithmic layer is rather evident in the spanwise velocity variances, and we find evidence for a universal scaling

$$\overline{w'^2}/u_\tau^2 \approx B_3 - A_3 \log(y/h), \tag{1}$$

with slope $B_3 \approx 0.95$, and $A_3 \approx 0.44$ for Poiseuille flow, $A_3 \approx 1.65$ for Couette flow (the fits are shown with dashed lines in Fig. 3). The wall-normal velocity fluctuations and the turbulent shear stress exhibit universality in inner scaling in a wider part of the near-wall region. especially for the Couette flow, with the formation of an outer plateau consistent with Townsend's attached-eddy hypothesis. However a continuing increase of the peak of $\overline{v'^2}$ is observed in the Poiseuille flow, which is probably connected with the slow

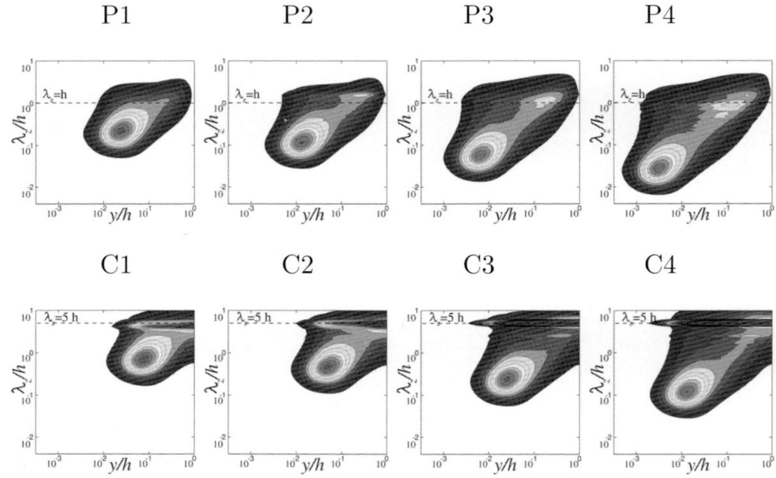

Fig. 4. Pre-multiplied spanwise spectral densities (in wall units) of streamwise velocity with wall distance for Poiseuille (top) and Couette (bottom) flow cases.

tendency to saturation of the shear stress.

The behavior of the streamwise Reynolds stress can be analyzed in greater detail through the pre-multiplied power spectral densities of u', shown in Fig. 4. The figure highlights the presence of a main energetic inner-layer site at $y^+ \approx 15$, corresponding to inner-layer streaks with a typical spanwise wavelength of about 100 wall units, which is roughly universal across the Reynolds number range, for both Poiseuille and Couette flows. The spectra also exhibit a secondary energetic site in the outer layer, whose intensity increases with Re_τ when expressed in wall units, and which has been traditionally associated with boundary layer superstructures[7], and with large-scale and very-large-scale motions in internal flows[8]. In the case of Poiseuille flow the energetic outer motions have a typical spanwise size $\lambda_z \approx h$, and their effects is felt, albeit weakly, all the way down to the wall. In the case of Couette flow the spectral footprint of the outer-layer organized structures consists of a nearly pure tone at the wavelength $\lambda_z \approx 5h$, with a weaker secondary harmonic spectral line visible in the C3 and C4 flow cases. The intensity of the outer peak in Couette flow appears to be much higher than in Poiseuille flow at the same Reynolds number, with stronger imprint on the near-wall layer. It is noteworthy that the spectral signature in the core part of the flow is consistent with the presence of rollers which occupy the entire channel. In fact, inviscid models of rollers[18] predict $u(y, z) \sim \cos(\pi/2\eta)\cos(\beta z)$, where $\beta = 2\pi/\lambda_z$ is the roll wavenumber

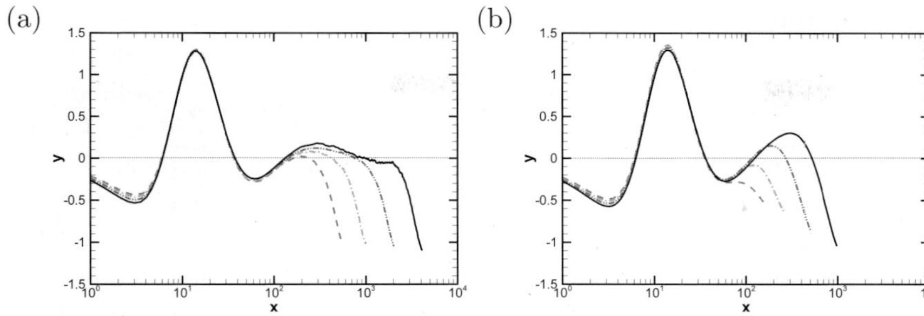

Fig. 5. Pre-multiplied turbulence kinetic energy production excess for Poiseuille flow (a) and Couette flow (b).

in the spanwise direction.

The observed structural flow changes with Re_τ can be interpreted by monitoring the local excess of kinetic energy production ($P = -(\mathrm{d}\overline{u}/\mathrm{d}y)\overline{u'v'}$) over its dissipation rate ($\varepsilon = 2\nu\overline{s'_{ij}s'_{ij}}$). This is shown in Fig. 5, where we report the distributions of $P - \varepsilon$ in pre-multiplied form, so that equal areas underneath the curves correspond to equal integral contributions. Figure 5 confirms the presence of a strong excess of production centered at $y^+ \approx 15$ for both flows, which corresponds to the near-wall turbulence self-sustainment cycle, and which is universal in wall units. The figure also shows the onset of a secondary spot with positive excess production, whose intensity increases with the Reynolds number, and whose position moves outwards in wall units. While this effect is observed in both flows, it is certainly more apparent in Couette flow. In that case, at the highest Reynolds number achieved in the present study, the outer peak accounts for an integrated production excess which is about 20% as the inner production site. Hence, its effect on the overall wall dynamics is certainly not negligible, and it is observed to increase with Re_τ. Notably, in the C4 flow case, the peak location well corresponds with the position of the secondary peak of the streamwise velocity variance (see Fig. 3(a)).

4. Conclusions

Canonical Poiseuille and Couette flow in the moderate-to-high Reynolds number regime highlight a number of (more or less) standard common features, as well as a number of differences, whose full explanation will certainly require further efforts. From a practical standpoint, the standard log law (with slightly different sets of coefficients) provides a good approximation of the mean velocity profiles of both flows, as well as good prediction of the wall friction coefficient (the data are not shown here). Deviations from the log law are found to be systematic in Poiseuille flow, where data suggest that it will eventually only show up in the infinite Reynolds number limit. In Couette flow, deviations from the log law do not appear to have a definite trend, suggesting that strong core eddies alter the mean flow to an extent not recoverable at any finite Re. The Reynolds stress components have a similar behavior in the two flows, however an additional outer-layer peak for the streamwise component emerges in Couette flow, which is found to be associated with a clear tone in the spectral maps. The typical outer-layer eddies are found to be wide about one channel half-height in Poiseuille, and about five channel half-heights in Couette, in which case they can be identified with rollers spanning the whole channel. Further considerations about similarities and differences between Poiseuille and Couette flow will be provided in a forthcoming publication[13].

References

1. N. Afzal and K. Yajnik, Analysis of turbulent pipe and channel flows at moderately large Reynolds number, *J. Fluid Mech.*, **61**, 23–31 (1973).
2. P. H. Alfredsson, A. Segalini and R. Örlü, A new scaling for the streamwise turbulence intensity in wall-bounded turbulent flows and what it tells us about the 'outer' peak. *Phys. Fluids* **23** 041702 (2011).
3. M. Bernardini, S. Pirozzoli, M. Quadrio and P. Orlandi, Turbulent channel flow simulations in convecting reference frames. *J. Comput. Phys.* **232** 1–6 (2013).
4. M. Bernardini, S. Pirozzoli and P. Orlandi, Velocity statistics in turbulent channel flow up to $\mathrm{Re}_\tau = 4000$. *J. Fluid Mech.* **742** 171–191 (2014).
5. D. B. DeGraaff and J. K. Eaton, Reynolds-number scaling of the flat-plate turbulent boundary layer. *J. Fluid Mech.* **422** 319–346 (2000).
6. M. Hultmark, M. Vallikivi, S. C. C. Bailey and A. J. Smits, Turbulent pipe flow at extreme Reynolds numbers. *Phys. Rev. Lett.* **108** 094501 (2012).
7. N. Hutchins and I. Marusic, Evidence of very long meandering features in the logarithmic region of turbulent boundary layers. *J. Fluid Mech.* **579** 1–28 (2007).
8. K. C. Kim and R. J. Adrian, Very large-scale motion in the outer layer. *Phys. Fluids* **11** 417–422 (1999).

9. R. Mathis, N. Hutchins and I. Marusic, Large-scale amplitude modulation of the small-scale structures in turbulent boundary layers. *J. Fluid Mech.* **628** 311–337 (2009).
10. M. M. Metzger and J. C. Klewicki, A comparative study of near-wall turbulence in high and low Reynolds number boundary layers. *Phys. Fluids* **13** 692–701 (2001).
11. H. M. Nagib and K. A. Chauhan, Variations of von Kármán coefficient in canonical flows. *Phys. Fluids* **20** 101518 (2008).
12. P. Orlandi, *Fluid Flow Phenomena: A Numerical Toolkit.* Kluwer, Netherlands (2000).
13. P. Orlandi, M. Bernardini and S. Pirozzoli, Poiseuille and Couette flows in the transitional and fully turbulent regime. *J. Fluid Mech.* (2015). accepted for publication.
14. S. Pirozzoli, M. Bernardini and P. Orlandi, Turbulence statistics in Couette flow at high Reynolds number. *J. Fluid Mech.* **758** 327–343 (2014).
15. H. Schlichting and K. Gersten, *Boundary layer theory.* 8th edn. Springer-Verlag, Germany (2000).
16. M. P. Schultz and K. A. Flack, Reynolds-number scaling of turbulent channel flow. *Phys. Fluids* **25** 025104 (2013).
17. A. A. Townsend, Equilibrium layers and wall turbulence. *J. Fluid Mech.* **11** 97–120 (1961).
18. F. Waleffe, On a self-sustaining process in shear flows. *Phys. Fluids* **9** 883–900 (1997).

A New Reynolds Stress Damping Function for Hybrid RANS/LES with an Evolved Functional Form

J. Weatheritt* and R. D. Sandberg**

*Institute for Complex Systems Simulation, **Aerodynamics and Flight Mechanics,
University of Southampton, United Kingdom
*E-mail: jw22g11@soton.ac.uk
www.southampton.ac.uk

A new methodology for finding hybrid RANS/LES damping functions is proposed. The Flow Simulation Methodology framework is modified from Speziale's original formulation [AIAA 36,173 1998] to allow for fewer grid points away from solid boundaries. The proposed functional form is regressed using an evolutionary algorithm. The solution is trained by considering DNS data of a separated flow. This sophisticated approach to building the damping function allows for the development of a hybrid methodology adaptable to any level of required modelling. The regressed model is compared to highly resolved reference LES on the classical two dimensional periodic hills case, for which the agreement is very good.

Keywords: Hybrid RANS/LES; FSM; Evolutionary Optimisation; Symbolic Regression.

1. Introduction

The cost of Large Eddy Simulation (LES) is too great for industrially relevant flows, whereas Reynolds-Averaging the Navier-Stokes (RANS) equations is too inaccurate in many cases. Instead Hybrid RANS/LES methods have appeared that aim to be cheaper turbulence resolving approaches by leaning on a RANS model in some way.

The Reynolds-averaging and low-pass spatial filtering of the (incompressible) Navier-Stokes equations form structurally similar transport equations,

$$\partial_t u_i + u_j \partial_{x_j} u_i = -\partial_{x_i} p + \nu \partial^2_{x_j x_j} u_i + \partial_{x_j} \tau_{ij}^{MOD} \tag{1}$$

where u_i is the resolved velocity. This similarity makes it possible to construct closures τ_{ij}^{MOD} that are a blend of both τ_{ij}^{RANS} and τ_{ij}^{LES}.

Flow Simulation Methodology (FSM) is an adaptable hybrid approach for which

$$\tau_{ij}^{MOD} = F(\Delta, ...) \cdot \tau_{ij}^{RANS}. \tag{2}$$

τ_{ij}^{RANS} models the turbulence of the flow field and the function F automatically damps this to ensure that the overall τ_{ij}^{MODEL} is representative of the local sub-grid scales (SGS). F is defined between 0 and 1, the DNS and RANS limits for total resolution or complete turbulence modelling. A non-linear stress-strain relation is often used for SGS anisotropy on coarse meshes. The damping function depends locally on the grid spacing Δ and also on some turbulent length scales in order to estimate local grid resolution. The success of a given hybrid closure crucially depends on the formulation of F, however it is not trivial to derive a functional form. Instead, authors have generally defined F on an *ad-hoc* basis, by either making assumptions or considering a certain philosophy. In the latter, we mean a hybrid approach has been formulated by *implicitly* defining F. In this case, no explicit thought is necessarily given to the RANS and DNS limiting regimes.

For example, Speziale[1] defined F as a function of the Kolmogorov microscale ℓ_k,

$$F = \left(1 - e^{-\beta \frac{\Delta}{\ell_k}}\right)^n, \tag{3}$$

where $\beta = 0.001$ and $n = 1$. These coefficients are set so F behaves well in the DNS limit. However as $\Delta/\ell_k >> 1$, it is unclear whether this ratio is a suitable parameter for the required level of modelling and worse the coefficients dictate such a slow response to lower mesh resolutions that the RANS limit, or even coarse LES regimes are unrecoverable. Rather FSM acts as a poorly calibrated implicit LES which lacks the required model dissipation to be numerically stable.

Alternatively, Spalart *et al.*[2] formulated a hybrid approach known as Detached-Eddy Simulation (DES) — which has become extremely popular and has met great success at predicting flows where the geometry determines the unsteadiness. The concept, which has recently been adopted by the developers of FSM,[3] is to damp the length scale ℓ_t in the RANS dissipation term,

$$D_{hyb} = \frac{k^{3/2}}{\ell_{hyb}}, \quad \ell_{hyb} = F(\Delta, ...) \cdot \ell_t. \tag{4}$$

By considering the Boussinesq approximation, Eq. 2 and Eq. 4 are almost identical in method and so damping functions are interchangeable between the two approaches. For DES, F is defined as,

$$F = \min(\frac{d}{\ell_t}, C_{DES}\frac{\Delta}{\ell_t}), \tag{5}$$

where the d is the wall distance and $C_{DES} = 0.65$. Away from solid boundaries, the value of C_{DES} makes the recovery of the unsteady RANS regime impossible. Consider that for this limit, $F \approx 1$ which implies $\Delta > \ell_t$. Therefore the model will provide too little dissipation. This is again another issue with constructing the damping function on an *ad-hoc* basis.

This paper outlines a more rigorous construction of a new damping function F. A data driven approach is taken to formulate F applicable to a wider range of flows. Whilst the data set used in this paper is not yet big enough, the methodology is introduced and a new damping function F is proposed, more as a proof of concept. Subsequent studies will expand on the data employed here, in order to produce a truly adaptable damping function.

2. Construction of the Damping Function

2.1. *Process*

The essential concept behind the new formulation methodology is to use a regression analysis, with large datasets representing idealised forms of F. To use a regression analysis, training data is required to fit potential solutions. Here DNS data is used to calculate the idealised damping function, then a functional form can be found via symbolic regression. To evaluate the idealised form, the DNS flow field must be made to represent a hybrid RANS/LES flow field. Hybrid RANS/LES flow fields are under-resolved due to the effective filter created by large grid scales Δ and time steps Δ_t. To mimic this, filters are applied to DNS snapshots removing the small scales. The large scales are those that would still be resolved by a hybrid model and the small scales removed by the filter are those that must be picked up by τ_{ij}^{MOD}.

This decomposition creates a hybrid flow field without having to perform an under resolved DNS. This is undesirable not just for time constraints but also for non-physical dissipation levels which would impact on the quality of the data. Filtered DNS (FDNS) also allows for the creation of a pseudo hybrid flow field without having to run any hybrid model. This is beneficial for several reasons. Firstly, FDNS is free from any model assumptions, reducing bias. Secondly, FDNS allows testing without having to repeatedly run hybrid simulations to assess potential candidates.

The filtering for the FDNS is performed locally, such that the strength of the filter depends on the grid size of the DNS. The number of points used in the filter is kept constant, so that the filter width varies. In other words the pseudo hybrid grid is still fine near solid boundaries and in coarser regions the filter is stronger. This is desired because it means that the size of the filter mimics a grid used in a hybrid simulation. The transfer function T is inspected for a given filter and the hybrid grid width is defined by the point where $T = 0.5$.

The ideal form of the damping function describes the amount of resolved over total stress, derived[4] from Eq. 2,

$$F = 1 - \frac{\tau_{ij}^{RANS} T_{ij}}{\tau_{mn}^{RANS} \tau_{mn}^{RANS}}, \tag{6}$$

where $T_{ij} = \langle u_i u_j \rangle - \langle u_i \rangle \langle u_j \rangle$ is the time averaged resolved stress. Due to this time average Eq. 6 is not suitable for actual CFD because of the increase in computational time. However, Eq. 6 can be evaluated for a range of FDNS datasets, using a series of snapshots to build a picture of the damping function for different flow fields.

Equation 6 is free from any explicit length scale dependence, allowing freedom in the choice of independent variables. These are calculated from the FDNS flow field. Only variables that are available to a RANS model at runtime are chosen to increase the portability of the resulting function F. The independent variables reflect those presently used in damping functions in the literature.[5] Also, the Taylor micro-scale is added, as this is the length scale of the smallest resolved eddies. It is easily calculated and has not been previously considered for hybrid closures of the type defined by Eq. 4. The length scales used in the regression are,

$$\begin{aligned}
L_k &= \frac{\Delta}{\ell_k}, & \ell_k &= \left(\frac{\nu^3}{\varepsilon}\right)^{\frac{1}{4}} & &\text{Kolmogorov length scale}\\
L_t &= \frac{\Delta}{\ell_t}, & \ell_t &= \frac{k^{\frac{3}{2}}}{\varepsilon} & &\text{Integral length scale}\\
L_\lambda &= \frac{\Delta}{\ell_\lambda}, & \ell_\lambda &= \sqrt{\frac{10k\nu}{\varepsilon}} & &\text{Taylor micro-scale}
\end{aligned}$$

$$L_{TRRANS} = \max\left(\left(\frac{S}{1.25\Omega}\right), 1\right)^{-2}, \quad L_{OES_S} = \frac{kS}{\varepsilon}, \qquad L_{OES_\Omega} = \frac{k\Omega}{\varepsilon}. \tag{7}$$

Note, $L_{TRRANS} = L_T$, $L_{OES_S} = L_S$ and $L_{OES_\Omega} = L_\Omega$ is used interchangeably. See, for example, the book[5] by Sagaut *et al.* for their original uses in TRRANS, OES, FSM, PANS and DES.

All six scales are non-dimensional, which means that the non-dimensionality of F is trivially preserved. As a result the regression process can be handled by a state-of-the-art tool, without the need to enforce extra constraints. Gene Expression Programming[6] (GEP) is chosen for this study. GEP is an optimisation method which creates and evolves a population of functional forms in a mimicry of biological survival of the fittest. Using such a technique comes with the benefit of being able to treat the evolutionary process in an ensemble sense and thus ask deeper questions about the effectiveness of each of the variables listed in Eq. 7. For this study a look at the ensemble frequency, across the evolution of many disjoint populations, of each variable L_i is considered when deciding the optimal shape of F.

2.2. *Formulation using backward facing step flow data*

The flow behind a backward facing step is considered to formulate the shape of the function F. This flow presents a number of challenges for a locally formulated turbulence model, as the step induces non-local effects. That is for models with limited history effects, the propogation of the flow state from the step is challenging to capture correctly. The Reynolds number is $Re = 3,000$ based on step height. A previous study[7] looked at a simple attached flow. This study is aiming to look at the performance of the methodology using more complex training data. The reattachment position of the DNS is $x/h = 6.7$, where the step is located at $x/h = 0$. The plane $x/h = 6.5$, just before reattachment, is a region where the normal stress components are significant and $x/h = 10$ is a slice of the recovering boundary layer. These two locations present a reduction in the number of data-points to consider, yet still capture in essence the features of the flow field and therefore make good training data for the regression.

Figure 1 shows the plots of the length scales L_i for the FDNS at the locations $x/h = 6.5$ and $x/h = 10$ using fifty snapshots and spanwise averaging. Note L_k has been factored by 0.04 in order to fit conveniently into the plots. $y = 0$ corresponds to the lower surface and $y = 1$ is the height of the step.

Figure 1 also contains a plot of the idealised damping function at $x/h = 6.5$ and $x/h = 10$. This represents the training data for the regression. The echo of the shear layer can be seen — the location where F reaches its peak for the $x/h = 6.5$ profile. This peak has gone from the profile at $x/h = 10$ as the boundary layer has undergone some recovery.

Thirty independent runs of the evolutionary process, using both loca-

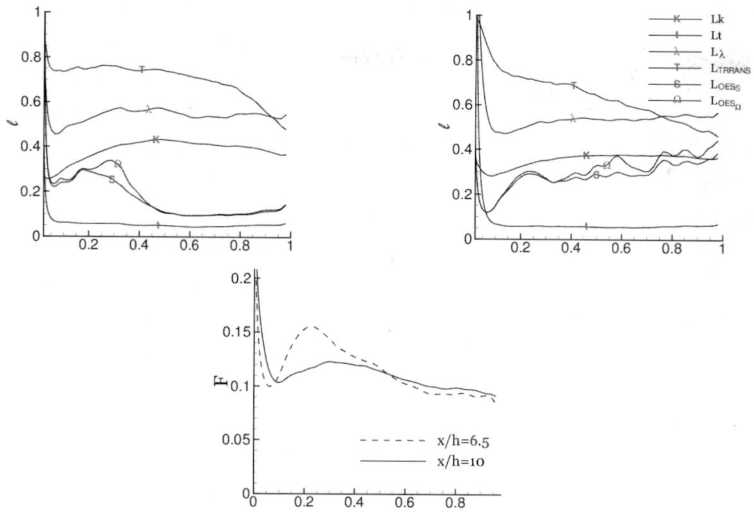

Fig. 1. Length scales for the backward facing step flow. Left: $x/h = 6.5$, Right: $x/h = 10$.

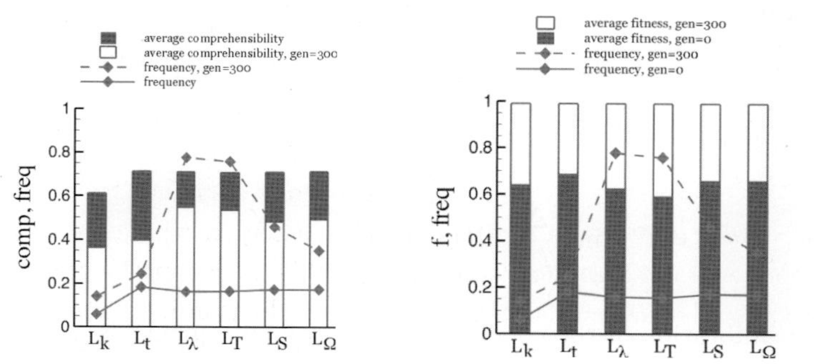

Fig. 2. Ensemble statistics for evolutionary system. Left: comprehensibility, Right: fitness.

tions as training data, are performed to gather ensemble statistics for the final state of the evolutionary system. During the evolutionary process, each candidate solution in the population is given a fitness and comprehensibility score bound between 0 and 1. The fitness is a measure of how closely an individual function matches the training data (1 is an exact correspondence), whilst the comprehensibility is a measure of how long its mathematical expression is (1 is a function consisting of just one mathematical symbol). The top 2% of the population based on fitness, is plotted in Fig. 2. Bars correspond to either the fitness or comprehensibility, given that a function contains L_i. L_λ and L_T are the most frequent and produce the simplest functions (not mutually exclusive phenomenon). Evident from the average fitness after 300 generations, the algorithm is capable of finding good functional forms for F from any of the variables L_i. However, good solutions containing L_k, L_t, L_S and L_Ω are much rarer, hinting at an underlying complexity required in the functional form.

The simplicity and major preference of the variables L_λ and L_T are the driving forces behind an optimisation including only these variables. An optimal solution from evolutionary runs including only L_T and L_λ is,

$$F = CL_\lambda L_T = C\frac{\Delta}{\ell_\lambda}\max\left[\left(\frac{S}{1.25\Omega}\right), 1\right]^{-2}. \tag{8}$$

This is not the final form every time, rather a function is produced that easily simplifies to this form by cancelling negligible terms.

2.3. *Implementation to form a hybrid closure*

In order to turn Eq. 8 into a fully working hybrid model, it must be modified to ensure proper boundedness between 0 and 1. The lower bound is trivially maintained, but to ensure $F \leq 1$, Eq. 8 is modified,

$$F = C\min\left(\frac{\Delta}{\ell_\lambda}, 1.0\right)\max\left[\left(\frac{S}{1.25\Omega}\right), 1\right]^{-2}. \tag{9}$$

Before the calibration of the coefficient, near wall treatment[3] is added,

$$F = \min\left(\frac{\tilde{F}}{1 - f_b}, 1\right), \tag{10}$$

where $\tilde{F}$ is the damping function Eq. 9. f_b is the damping function from the SST turbulence model[8] designed to be 0 at the edge of the boundary layer and 1 at the wall. This achieves RANS at the wall and LES outside of the boundary layer. This is vital to the success of the hybrid simulation, not only for near wall cost requirements, but also for predictive performance.[5] The underlying RANS model chosen is the explicit algebraic stress model (EASM) of Speziale, Sarkar and Gatski[9] (SSG). The coefficient was calibrated using a channel flow of friction Reynolds number 550. By matching the shear stress component the coefficient is found to be $C = 0.8$.

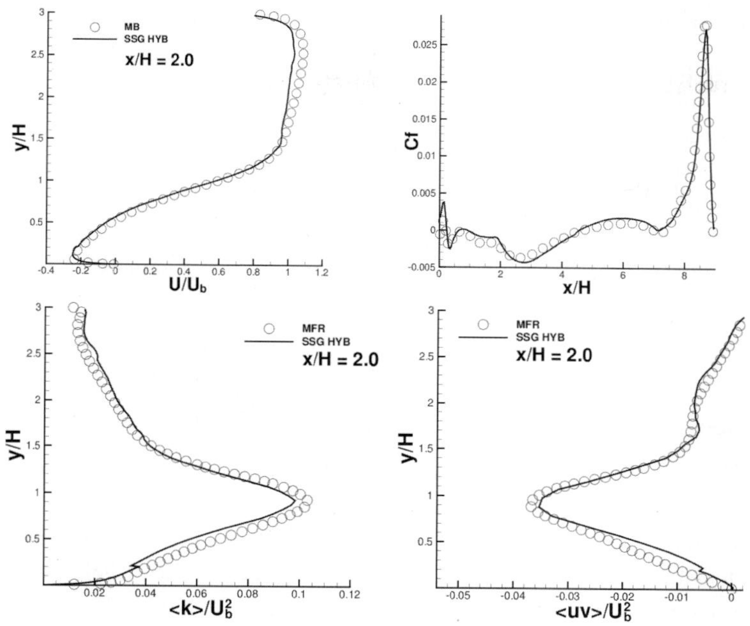

Fig. 3. 2D Periodic Hills. Top left: Mean streamwise velocity. Top right: Friction Coefficient. Bottom left: Total turbulent kinetic energy. Bottom right: Total shear stress. Symbols denote reference LES data.[10,11]

3. Application to Periodic Hills

The new closure has been tested on the classical periodic hills geometry. This is a well studied case and is a good validation study for new turbulence closures. In Fig. 3, results on a relatively coarse grid of 800, 000 cells, are shown against two highly resolved LES datasets (MB)[10] and (MFR)[11] which agree well with experimental values.[10] The location $x/h = 2.0$ is particularly challenging as this profile crosses the reverse flow region, free shear layer and the bulk flow. In general the hybrid methodology with the new damping function is in excellent agreement with the reference data for every variable.

One cause for concern is the kink present in the second order statistics in the lower surface boundary layer. This could be caused by an unforeseen interaction with the shielding function which effectively forces a boundary condition on F. This should be investigated and incorporated into the regression process.

4. Conclusion and Outlook

A new damping function for hybrid RANS/LES built using a novel approach for turbulence modelling has been presented and applied to the well known periodic hills test case, with good agreement to the reference data.

This method can be improved by simultaneously considering more datasets that include complex flow structures with differing order filters so that a damping function suitable for a wide range of flow problems can be found. Further, specialised damping functions can be generated for flow problems that contain a dominant feature such as transitional, massively separated or turbo-machinery flows.

References

1. C. G. Speziale, Turbulence modeling for time-dependent RANS and VLES: A review, *AIAA J.* **36**, 173 (1998).
2. P. R. Spalart, W.-H. Jou, M. Strelets and S. R. Allmaras, Comments on the feasibility of LES for wings, and on a Hybrid RANS/LES approach, in *Advances in DNS/LES*, eds. C. Liu and Z. Liu, Gayden Press (1997).
3. M. Weinmann, R. D. Sandberg and C. Doolan, Tandem cylinder flow and noise predictions using a hybrid RANS/LES approach, *International J. Heat and Fluid Flow* **50**, 263 (2014).
4. M. Germano, Comment on "Turbulence Modeling for Time-dependent RANS and VLES: A Review", *AIAA J.* **36** (1998).
5. P. Sagaut, S. Deck and M. Terracol, *Multiscale and multiresolution approaches in turbulence* World Scientific (2013).
6. C. Ferreira, Gene Expression Programming: A New Adaptive Algorithm for Solving Problems, *Complex Systems* **13**, 87 (2001).
7. J. Weatheritt and R. D. Sandberg, Use of Symbolic Regression for construction of Reynolds-stress damping functions for Hybrid RANS/LES, in *53rd AIAA Aerospace Sciences Meeting*, AIAA SciTech, American Institute of Aeronautics and Astronautics, January (2015).
8. F. R. Menter, Zonal Two Equation k-ω Turbulence Models for Aerodynamic Flows, *AIAA*-93-2906 (1993).
9. C. G. Speziale, S. Sarkar and T. B. Gatski, Modelling the pressure strain correlation of turbulence: an invariant dynamical systems approach, *J. Fluid Mech.* **227**, 245 (1991).
10. M. Breuer, N. Peller, C. Rapp and M. Manhart, Flow over periodic hills - Numerical and experimental study in a wide range of Reynolds numbers, *Comput. and Fluids* **38**, 433 (2009).

11. J. Fröhlich, C. Mellen, W. Rodi, L. Temmerman and M. Leschziner, Highly resolved large-eddy simulation of separated flow in a channel with streamwise periodic constrictions, *J. Fluid Mech.* **526**, 16 (2005).

Direct Numerical and Large Eddy Simulations of Helicity-Induced Stably Stratified Turbulent Flows

A. Rahimi and A. J. Chandy*

Department of Mechanical Engineering, University of Akron, Akron, OH 44325-3903, USA
**E-mail: achandy@uakron.edu*

Subgrid-scale (SGS) turbulence models for large eddy simulations (LES) are quantitatively assessed for the strongly-stratified helical transition-to-turbulence, Arnold-Beltrami-Childress (ABC) problem through comparisons to in-house direct numerical simulations (DNS). LES predictions using the classical non-dynamic and dynamic Smagorinsky models (SSM and DSM, respectively) are compared to a modified Smagorinsky model (MSM) designed specifically for natural convection. DSM predicts the best results when compared to DNS, although none of the SGS models are able to predict the peak of nonlinearities.

Keywords: Turbulence; Stratification; Helicity; large eddy simulation

1. Introduction

The study of decay of stratified turbulent flows is relevant to many oceanic and atmospheric flows. Stratification is due to he spatial variations in density. Stable density stratification can be found in the intermediate scales of atmosphere and ocean. In the presence of stable density stratification, fluids undergo slow quasi-horizontal vortical motion and fast internal gravity waves[1]. Buoyancy forces that arise due to spatial variations in density break the symmetry of the flow by favoring the direction in which the gravitational force acts. They tend to damp turbulence in the case of stable stratification, i.e. when lighter fluid overlays heavy counterparts. In stable stratification, a fluid parcel displaced from equilibrium oscillates vertically with frequency $\mathcal{N} = \sqrt{-\frac{g}{\rho_0}\frac{\partial \rho}{\partial z}}$, where $\mathcal{N}$ is called Brunt-Vaisala frequency and ρ_0 is the ambient reference density. Furthermore, a mean non-zero helicity is also present in any geophysical environment[2]. Previous geophysical simulations[3,4] have demonstrated the important role of helicity in storm systems for instance. The volume integrated helicity introduced by

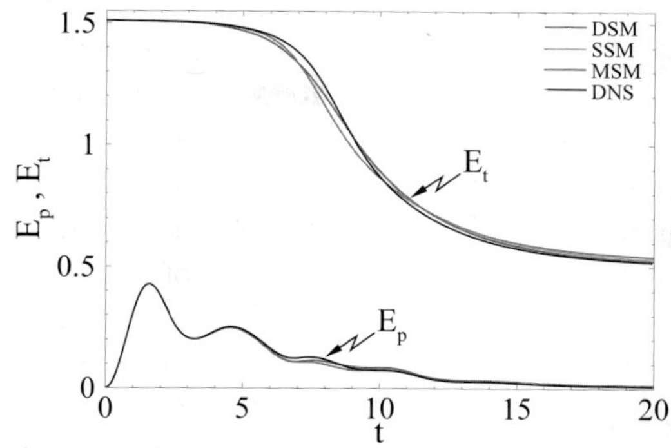

Fig. 1. Temporal evolution of energy budget for DNS and LES using three SGS models.

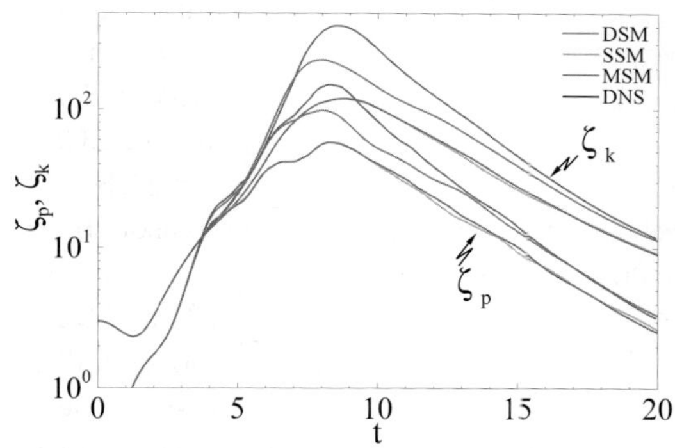

Fig. 2. Temporal evolution of enstrophy for DNS and LES using three SGS models.

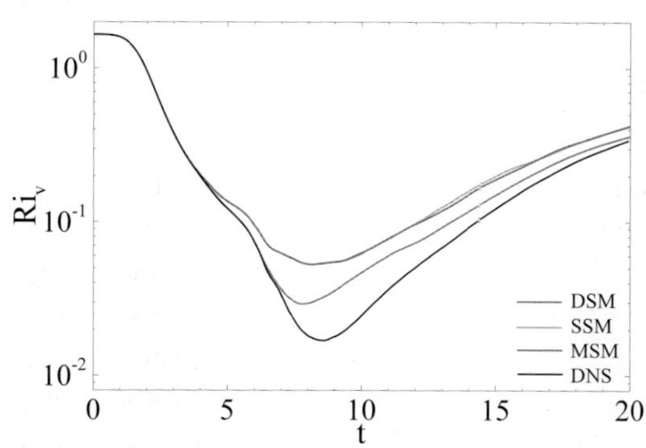

Fig. 3. Temporal evolution of volume-averaged Richardson number, Ri_v, for DNS and LES using three SGS models.

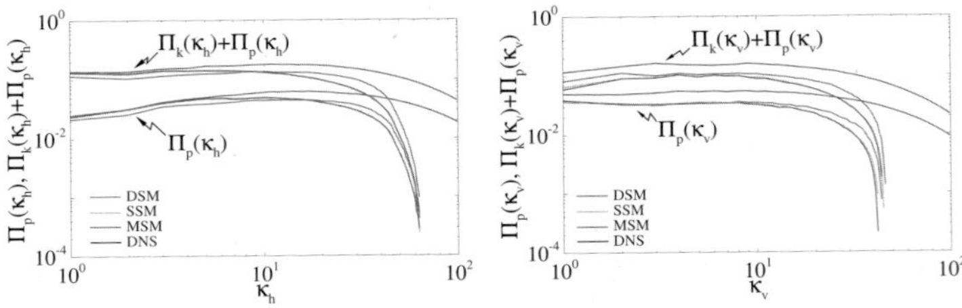

(a) (b)

Fig. 4. Spectrum of (a) potential and total energy, and (b) helicity at the peak of DNS enstrophy for DNS and LES using three SGS models.

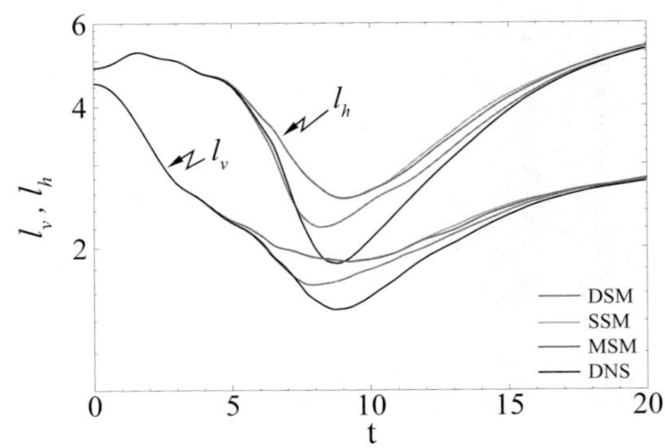

Fig. 5. Temporal evolution of horizontal and vertical length scales for DNS and LES using three SGS models.

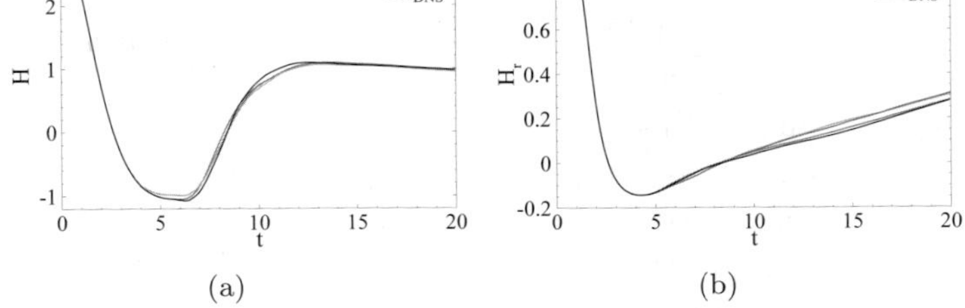

(a) (b)

Fig. 6. Temporal Evolution of (a) helicity and (b) relative helicity for DNS and LES using three SGS models.

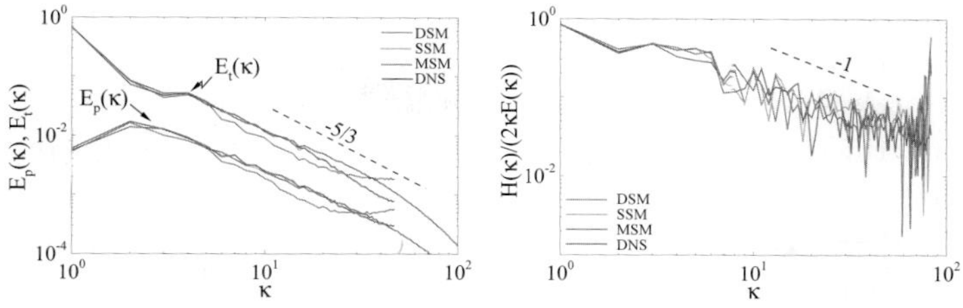

(a) (b)

Fig. 7. Spectrum of (a) potential and total energy, and (b) helicity at the peak of DNS enstrophy for DNS and LES using three SGS models.

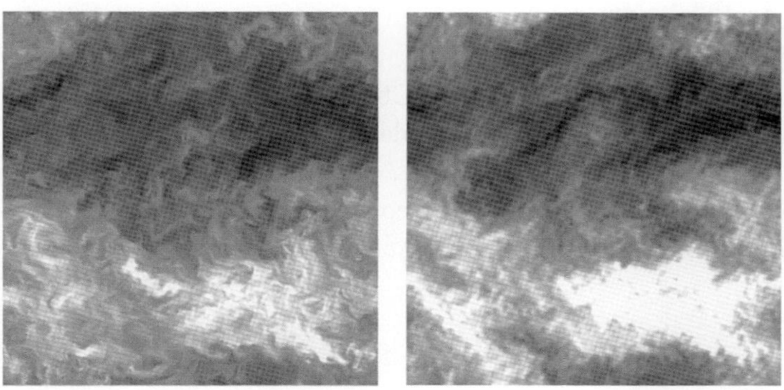

(a) (b)

Fig. 8. Horizontal velocity field for (a) DNS and (b) DSM.

References

1. B. Cushman-Roisin and J.-M. Beckers, *Introduction to geophysical fluid dynamics: Physical and numerical aspects* Academic Press (2011).
2. S. Moiseev and O. Chkhetiani, Helical scaling in turbulence, *J. Experimental and Theoretical Phys.* **83**, 192 (1996).
3. D. K. Lilly, Stratified turbulence and the mesoscale variability of the atmosphere, *J. the Atmospheric Sciences* **40**, 749 (1983).
4. D. K. Lilly, The structure, energetics and propagation of rotating convective storms. part ii: Helicity and storm stabilization, *J. atmospheric sciences* **43**, 126 (1986).

5. H. K. Moffatt, The degree of knottedness of tangled vortex lines, *J. Fluid Mech.* **35**, 117 (1969).
6. C. Rorai, D. Rosenberg, A. Pouquet and P. Mininni, Helicity dynamics in stratified turbulence in the absence of forcing, *Phys. Rev. E* **87**, 063007 (2013).
7. S. B. Pope, *Turbulent flows* Cambridge univ. press (2000).
8. S.-H. Peng and L. Davidson, Comparison of subgrid-scale models in les for turbulent convection flow with heat transfer, *Turbulent Heat Transfer* **2**, 5 (1998).
9. V. I. ARNOLD, Sur la topologie des bcoulements stationnaires des fluides parfaits, *C. R. Academy of Science* **261**, 17 (1965).
10. S. Remmler and S. Hickel, Direct and large eddy simulation of stratified turbulence, *International J. Heat and Fluid Flow* **35**, 13 (2012).
11. E. Lindborg, The energy cascade in a strongly stratified fluid, *J. Fluid Mech.* **550**, 207 (2006).
12. C. Yu, Z. Xiao, Y. Shi and S. Chen, Joint-constraint model for large-eddy simulation of helical turbulence, *Phys. Rev. E* **89**, 043021 (2014).
13. S. Kurien, M. A. Taylor and T. Matsumoto, Cascade time scales for energy and helicity in homogeneous isotropic turbulence, *Phys. Rev. E* **69**, 066313 (2004).

Comparison of RANS and DNS for Transitional Flow Over WTEA-TE1 Airfoil

*Pramod Bagade[a], Éric Laurendeau[b], Ashish Bhole[a], Nidhi Sharma[a], T. K. Sengupta[a]

[a] *Department of Aerospace Engineering, IIT Kanpur, Kanpur, U.P., 208 016, India*
**E-mail: pmbagade@iitk.ac.in*

[b] *Department of Mechanical Engineering*
École Polytechnique de Montréal
E-mail:eric.laurendeau@polymtl.ca

Here an attempt has been made to demonstrate relative strengths and weaknesses of various methods of simulating transitional flows. Flow simulations have been done using compressible RANS and DNS for compressible and incompressible formulations. Flow over the WTEA-TE1 airfoil is simulated for free-stream Mach no. $M_\infty = 0.1815$ and angle of attack $\alpha = -0.0328°$. In particular, the pressure and skin friction distributions are compared. In additions RMS signals for the DNS calculations are made, including Fourier Transforms of pressure signals over specific experimental measurement stations.

1. Introduction

Computing transitional flows is one of the tougher challenges of direct numerical simulation (DNS). Here, DNS of a low Mach number flow is computed by incompressible and compressible formulations and are compared with the results obtained by compressible RANS formulation with a well calibrated empirical model for transition. Here, we have computed flow past WTEA-TE1 airfoil for a Mach number of 0.1815 to characterize transition occurring over the airfoil, which is tested for morphing wing applications[1]. However, in the present case we only focus upon fixed geometry configuration, while the aerodynamic performance of this morphing model was optimized in Mamou *et al.*[1] for transitional delay. General review of morphing aircraft wing, from the point of view of transitional delay by shape deformation, has been described by Sofla *et al.*[2]. Other notable references in the area of adaptive wing and flow control are due to Stanewsky, Popov and others[3–6]. The works in Popov and Mamou[1,4] are of greater relevance considering the fact that the present study focuses on transition prediction by various computing formulations and methodologies.

Results using RANS methodology with $\gamma - Re_\theta$ transition model due to Langtry and Menter[7] are reported in Robitaille *et al.*[8]. We have compared these results with DNS results obtained by compressible flow solver on structured orthogonal grid using high accuracy compact scheme[9,10]. For this incompressible flow, we have also solved the incompressible Navier-Stokes equations in orthogonal grid formulation[11]. The geometry of the airfoil is as shown in Fig. 1, which also indicates the locations of Kulite transducers for unsteady pressure used in the experiment[1]. Airfoil geometry has a surface discontinuity at the junction of flexible and rigid part as shown in the inset of Fig. 1.

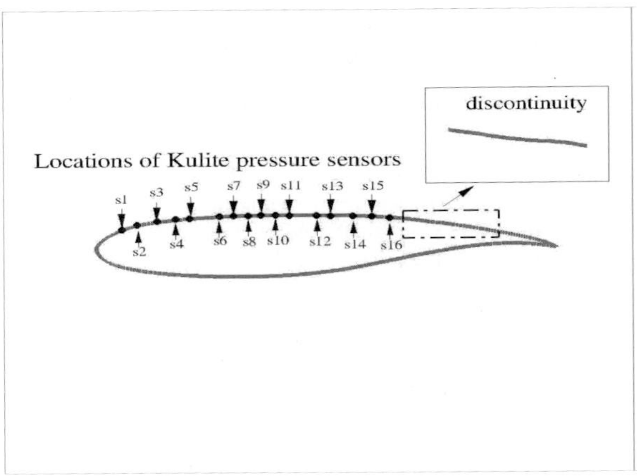

Fig. 1. WTEA-TE1 airfoil geometry and location of Kulite pressure sensors.

2. Experimental Data

The experimental data are collected by the research team of the CRIAQ 7.1 project[1], whose objective is to study the possibility of using morphing wing airfoil surface to delay the onset of laminar-turbulent transition to reduce drag. In the experiments, the transition point has been measured by the use of two techniques: First, Kulite pressure sensors were positioned at several locations along the upper surface of the airfoil, to obtain unsteady pressure measurements. The root mean square (RMS) value of the signals indicate whether the flow is turbulent or not. Secondly, infrared imagery was used to visualize the flow over the airfoil. Infrared photograph in Fig. 2

shows the abrupt change in heat transfer rate at the juncture of laminar and turbulent flow, which gives rise to variation in wall temperature.

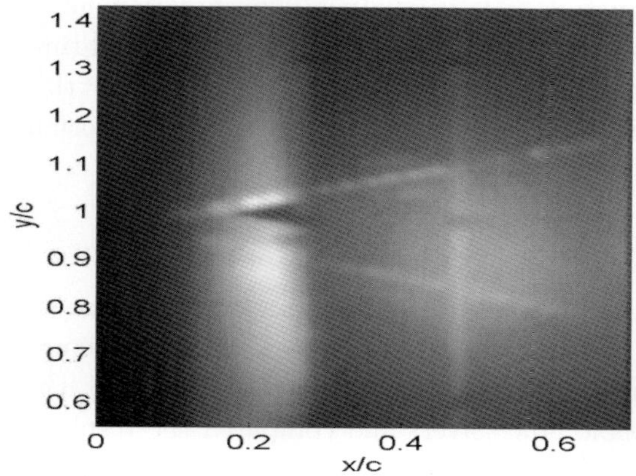

Fig. 2. Infrared photograph of flow over the WTEA-TE1 airfoil showing transition location.

3. Governing Equations and Numerical Methods

Details of governing equations, boundary conditions for compressible DNS solver and numerical methods used are described in Sengupta *et al.*[9] High resolution, dispersion relation preserving (DRP) numerical method has been used in the DNS solver, where convective derivatives are evaluated using optimized upwind compact scheme (OUCS3). Diffusive terms are evaluated using second order central difference (CD2) method. While artificial diffusion due to Jameson *et al.*[12] by the JST model has been used earlier[9], here in evaluating convective derivatives 2D filter has been used in an adaptive manner[14]. For time integration, optimized three-stage Runge-Kutta (ORK3) method[10] has been used. An orthogonal grid is generated using hyperbolic grid generation technique[10] on the WTEA-TE1 airfoil section with near wall grid resolution (ΔS_η) equals to 3.072×10^{-5} and outer boundary located at a distance of 10 chords (c).

A bi-dimensional multi-block structured RANS solver (NSCODE) of École Polytechnique de Montréal uses an explicit five-stage Runge-Kutta time integration routine along with JST model[8]. Apart from using local time stepping, multi-grid and residual smoothing, the transitional turbulence model[7] was used on C-grid with 624×128 points and the wall spacing set to 1×10^{-6} with the outer boundary located at $50c$.

For the DNS solver of incompressible Navier-Stokes equation formulation, stream function (ψ)- vorticity (ω) formulation has been used with spatial discretization performed by OUCS3 scheme and time integration by ORK3 methods[10]. We have also used 2D filters[14] for the incompressible solver, using the same grid used for the compressible formulation.

4. Results

Flow over the WTEA-TE1 airfoil is simulated for free-stream Mach no. $M_\infty = 0.1815$ and angle of attack $\alpha = -0.0328°$. Figure 3 shows mean C_p-distribution plotted against x/c, obtained by DNS using compressible and incompressible flow solvers and compared the results with the RANS solver[8]. In this figure, all C_p-distributions match well with each other, showing a kink in the mean C_p-distribution corresponding to the slope discontinuity location shown in Fig. 1. This is due to airfoil surface discontinuity to allow wing morphing. The kink is prominent in all the cases of RANS and DNS solvers. However, if one uses JST model in compressible DNS solver, the kink is diffused. Also, shown in the figure is the experimental data measured only on the top surface, up to the point of surface discontinuity. One notices that the incompressible DNS overpredicts the lift significantly as compared to the RANS and compressible DNS results.

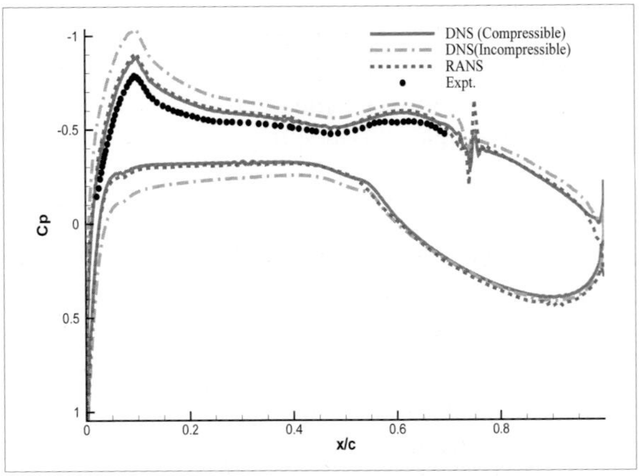

Fig. 3. Mean Cp-distribution plotted *vs* x/c, obtained using DNS compressible and incompressible solver as well as RANS solver.

In Fig. 4, mean coefficient of skin friction (C_f) is shown against x/c for all the three solvers, which helps one to identify the transition location. In Fig. 4(a), mean C_f-distribution is plotted for the upper surface and it can be seen that for RANS, incompressible DNS and compressible DNS, transition is occurring at $0.45c$, $0.48c$ and $0.46c$, respectively.

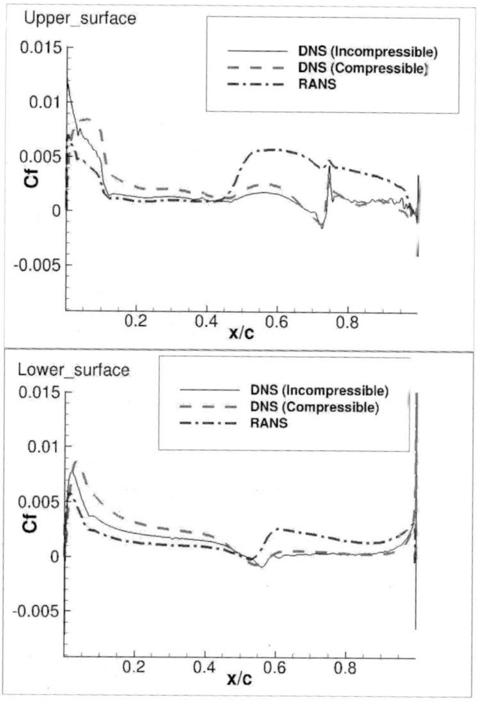

Fig. 4. Mean Cf-distribution plotted *vs* x/c, obtained using DNS compressible, DNS incompressible and RANS solver for (a) upper surface and (b) lower surface.

As is usual, the incompressible DNS displays a suction peak at the leading edge. While the compressible DNS solver shows progressive acceleration of the flow near the leading edge. For the RANS solver, once the transition is predicted by the empirical model, the flow is switched to fully developed turbulent flow with higher skin friction coefficient and such a flow is also insensitive to the surface discontinuity. However, the DNS solvers are more capable in capturing the actual skin friction variation, along with the rapid fluctuation near the kink.

In Fig. 4(b), mean C_f-distribution is shown for lower surface and it is seen that for RANS shows earlier and stronger transition, compressible and incompressible DNS shows milder transition. Like the top surface, once the flow transition occurs for RANS solver, subsequently the skin friction values remain higher.

One of the major purposes of the present work is to evaluate the transition location and the property of the transitional flow. From Fig. 4 we have noted that RANS is quite capable of predicting the transitional location due to the $\gamma - Re_\theta$ model. However, subsequent higher drag due to the turbulence model makes the estimate more conservative. The property of the transitional flows evinced by c_f distribution shows that DNS can be a very valuable tool.

In Fig. 5, root mean square (RMS) value of stream-wise component of velocity obtained using compressible and incompressible DNS solvers are plotted as function of streamwise distance, for more accurate transition location. Figure 5(a) shows RMS of stream-wise component of velocity for the upper surface and it is observed that the transition location indicated by the localized peak of the value occurs at identical location. Pre-transition dip also is noted for both the methods, also at same location. The transitional flow following the peak show that the incompressible formulation has consistently higher RMS value, implying higher turbulence level. RMS of stream-wise component of velocity for the lower surface of the airfoil is shown in Fig. 5(b). While the pretransition dip occurs at almost identical position for both the DNS formulation, the incompressible DNS is characterized by higher value of turbulence in the aft portion of the airfoil, as compared to compressible formulation.

In Fig. 6, Fourier transform of pressure time series obtained by compressible DNS at the even numbered locations of Kulite pressure sensors are shown for $t = 22.60$ to 38.93 with a sampling rate of 0.001, for the non-dimensional computational time. It is seen that the fundamental frequency for the spectra is at 8.4627 kHz and instead of discrete higher harmonics one observes wide-band spectra. It is noted that with JST model (results not shown) there are only distinct higher harmonics (five in total), while the use of high fidelity filter the spectrum is more realistic and is therefore recommended.

On the right column corresponding experimental pressure spectrum is shown where the sampling rate of 15kHz was used. Thus, one can resolve up to only 7.5kHz as shown in these frames. A direct comparison with DNS would reveal that this resolution is not adequate which results in significant aliasing in the experimental data at the higher ranges of frequency.

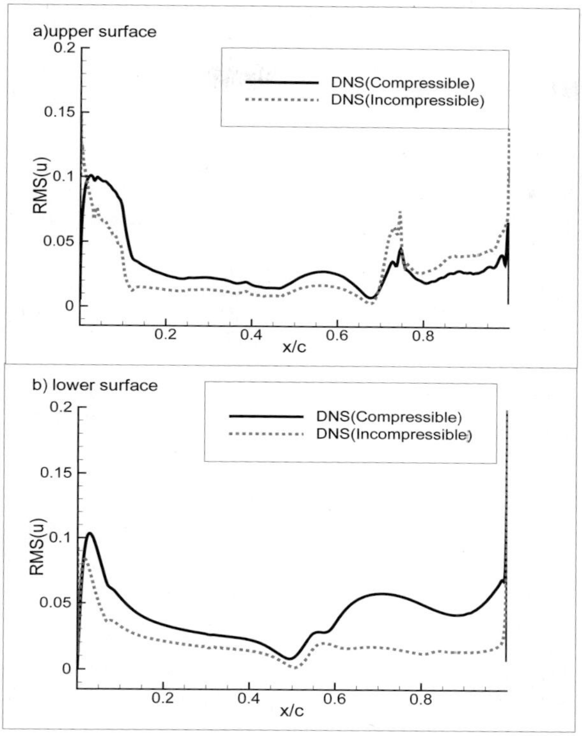

Fig. 5. Distribution of root mean square of stream-wise component of velocity obtained by DNS solvers for (a) upper surface and (b) lower surface.

5. Summary and Conclusions

Present exercise demonstrate relative strengths and weaknesses of various methods of simulating transitional flows. Without a doubt, the RANS formulation is a fast method and provides lift value as effectively as the DNS methods using compressible and incompressible formulation for a low Mach number flow ($M_\infty = 0.1815$). When it comes to drag prediction, the efficiency of the RANS method is good in predicting the transition location. However, post-transition the results obtained by DNS are able to capture more physics.

6. Acknowledgement

The authors thank all partners of the Consortium for Research and Innovation in Aerospace in Qubec (CRIAQ) 7.1 project for allowing to use some

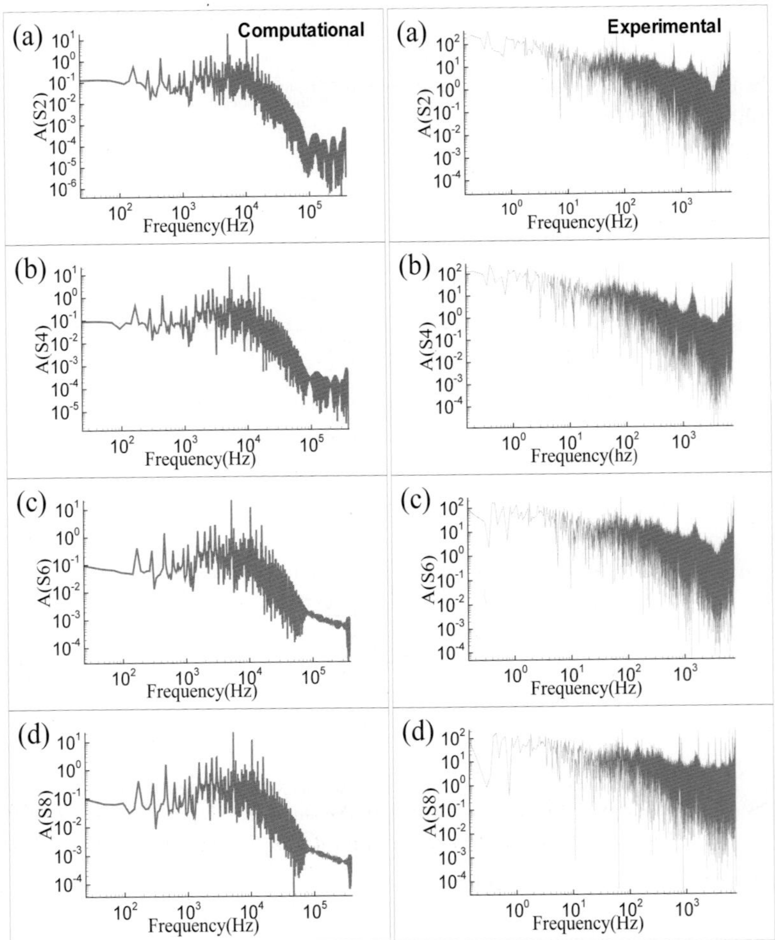

Fig. 6. Fourier transform of pressure signals obtained by compressible DNS solver (taken over $t = 22.60to38.93$), at the even numbered locations of Kulite pressure sensors: (a)-(h) corresponds to 2nd-16th Kulite pressure sensors, respectively.

of the project results.

References

1. Mamou M., Mébarki, Y., Khalid M., Genest M., Coutu D., Popov A. V., Sainmont C., Georges T., Grigorie L., Botez R. M., Brailovski V., Terriault P., Paraschivoiu I., Laurendeau E., Aerodynamic performance optimization of a wind tunnel morphing wing model subject

to various cruise flow conditions, 27th *International Congress Of The Aeronautical Sciences, ICAS*, 2010.

2. Sofla A. Y. N., Meguid S. A., Tan K. T., and Yea W. K., Shape morphing of aircraft wing: status and challenges, *J. Materials and Design*, Vol.-31, pp 1284–1292, 2010.
3. Stanewsky E., Adaptive wing and flow control technology, *Progress in Aerospace Sciences*, Vol. 37, pp 583–667, 2001.
4. Popov A. V., Botez R. M., Michel Labib, Transition point detection from the surface pressure distribution for controller design, *J. Aircraft*, Vol. 45, No. 1, 2008.
5. Popov A. V., Grigorie L. T., Botez R. M., Mébarki, Y., Mamou, M., Modeling and testing of a morphing wing in open-loop architecture, *J. Aircraft*, Vol. 47, No. 3, 2010.
6. Popov A. V., Grigorie L. T., Botez R. M., Mébarki, Y., Mamou, M., Closed-loop control validation of a morphing wing using wind tunnel tests, *J. Aircraft*, Vol. 47, No. 4, 2010.
7. Langtry R., Menter F., Transition modeling for general CFD applications in aeronautics *AIAA*, 43rd *AIAA Aerospace Sciences Metting and Exhibit, Reno, Nevada*, Vol. 522, 2005.
8. Robitaille M., Mosahebi A., Laurendeau E., Verification and validation of the NSCODE implementation of the $\gamma - Re_\theta$ transition model. In *Advanced Aero Concepts, Design and Operations* conf. of Royal Aero. Soc. Bristol, UK (July, 2014).
9. Sengupta T. K., Bhole A., Sreejith N. A., Direct numerical simulation of 2D transonic flows around airfoils, *Computers and Fluids*, Vol. 88, pp 19–37, 2013.
10. Sengupta T. K., *High Accuracy Computing Methods*, Cambridge University Press, New York, USA, 2013.
11. Sengupta T. K., Bhumkar Y. G., Direct numerical simulation of transition over a natural laminar flow airfoil, *Frontiers in Aerospace Engineering*, Vol. 2(1), pp 39–52, 2013.
12. Jameson A., Schmidt W., Turkel E., Numerical solution of the Euler equations by finite volume methods using Runge-Kutta time-stepping schemes, *AIAA 19811259*, AIAA 14th Fluid and Plasma Dynamic Conference, June 2325, 1981, Palo Alto, California.
13. Menter F. R., Two-equation eddy-viscosity turbulence models for engineering applications, *AIAA J.*, Vol. 32, No. 8, pp 1598–1605, 1994.
14. Bhumkar Y. G., Sengupta T. K., Adaptive multi-dimensional filters, *Computers and Fluids* Vol. 49, pp 128–140, 2011.

Extracting Coherent Structures to Explore the Minimum Jet Noise

Z. Fu and A. Agarwal*

Department of Engineering, University of Cambridge,
*Cambridge, CB2 1PZ, UK, *E-mail: aa406@cam.ac.uk*

A. V. G. Cavalieri

Divisão de Engenharia Aeronáutica, Instituto Tecnológico de Aeronáutica,
São José dos Campos, 12228-900, Brazil

P. Jordan

Département Fluides, Thermique, Combustion, Institut Pprime,
CNRS-Université de Poitiers-ENSMA, Poitiers CEDEX, 86036, France

This paper examines the lower bound of the sound that a circular turbulent cold jet (Mach number 0.88) radiates to the far field, for the given jet exit velocity and nozzle size. By applying discrete wavelet transform (DWT) and Fourier transform in azimuth, the jet flow is decomposed into wave packets (WP), coherent eddies (CE), and incoherent structures (IS). With the time-averaged component of the hydrodynamic near field unchanged, WP are filtered out using wavelets. It is shown that the remaining IS, which has a Gaussian PDF by definition, is de-correlated. Using the same approach, the acoustic far field is filtered in the low azimuthal modes, leading to 9-15dB lower overall far-field sound compared with the original jet at different axial angles. Sound spectra show that the high-frequency component dominates and remains unaffected by the filtering process. After IS are removed, the values of the correlations between the axisymmetric modes of the near and far field are significantly higher than those reported for flow-acoustic correlations of the axisymmetric mode. It is suggested that the sound reduction after all modes are filtered is the best that can be achieved in practice, given the proposed constraints.

Keywords: Jet turbulence; Coherent structures; Jet noise.

1. Introduction

Coherent structures play a significant role in the dynamics of jet turbulence and jet noise radiation[1,2]. For jet noise control, it is important to suppress these acoustically efficient structures. In this paper we perform numerical experiments to artificially remove the coherent structures and explore the lower limit of jet noise. The lower limit suggests the best-case scenario for jet noise suppression, and the approach to this limit gives researchers hints as to what could be done to mitigate jet noise in practice.

In order to remove the coherent structures (CS), we first need to define them. We do this indirectly by defining incoherent structures (IS). Following the previous studies[3,4], we define IS as random fluctuations. This paper aims to examine jet noise reduction after the acoustically efficient CS are removed. The study is motivated by the empirical jet noise control related to these structures. After CS are removed, the sound radiation from the remaining structures would suggest the capability of empirical jet noise control. We hope that the current study will provide guidance for future empirical investigations in jet noise reduction and enhance the understanding of physical mechanisms of jet noise generation.

2. LES Database

In this study, we use both numerical and experimental databases. The numerical database is from a large-eddy simulation (LES) of a circular turbulent cold jet recently performed by the research group in Whittle Laboratory, University of Cambridge[5]. The three-dimensional unsteady compressible Navier-Stokes (NS) equations are solved for the jet flow. The Mach number of the jet is 0.88, based on the jet exit velocity at the centreline (U) and ambient speed of sound (c); the Reynolds number is 2×10^5, based on U, jet exit diameter (D) and ambient kinematic viscosity of air (ν). The details of the scheme can be found in the previous work[5]. The acoustic field is calculated based on Ffowcs Williams-Hawkings formulation. The near field and the far field data has been validated[5].

An experimental database is also used, and the details can be found in the original literature[1].

3. Flow Decomposition and Filtering Process

3.1. *Introduction*

It is tentatively proposed that jet turbulence can be decomposed into the following three parts: (i) wave packets (WP), characterised by low azimuthal wavenumbers m and low Strouhal numbers ($m = 0$, 1, 2, and $St \leq 0.9$); (ii) coherent eddies (CE), having higher azimuthal wavenumbers; (iii) incoherent structures (IS) going from CE to the smallest structures. In this paper by CS we mean CE + WP.

Figure 1 illustrates the decomposition of the hydrodynamic near field and acoustic far field. As shown in Fig. 1(a), using Fourier transform in azimuth, a hydrodynamic variable $q(x, r, \theta, t)$ is first decomposed into azimuthal modes, where x, r, θ, and t are axial and radial coordinates, azimuthal angle, and time, respectively. Then the flow is filtered through discrete wavelet transform (DWT) to educe CS, $q_C(x, r, m \leq 2, t)$ containing mostly WP or $q_C(x, r, m > 2, t)$ containing CE. A similar decomposition is also conducted in the acoustic far field, $p'(R, \varphi, \theta, t)$, where p' is acoustic pressure. We note that $p'(R, \varphi, m \leq 2, t)$ marked by block B in Fig. 1(b) constitutes most of the far-field acoustic energy and is mostly radiated by WP, which are marked by block A in Fig. 1(a).

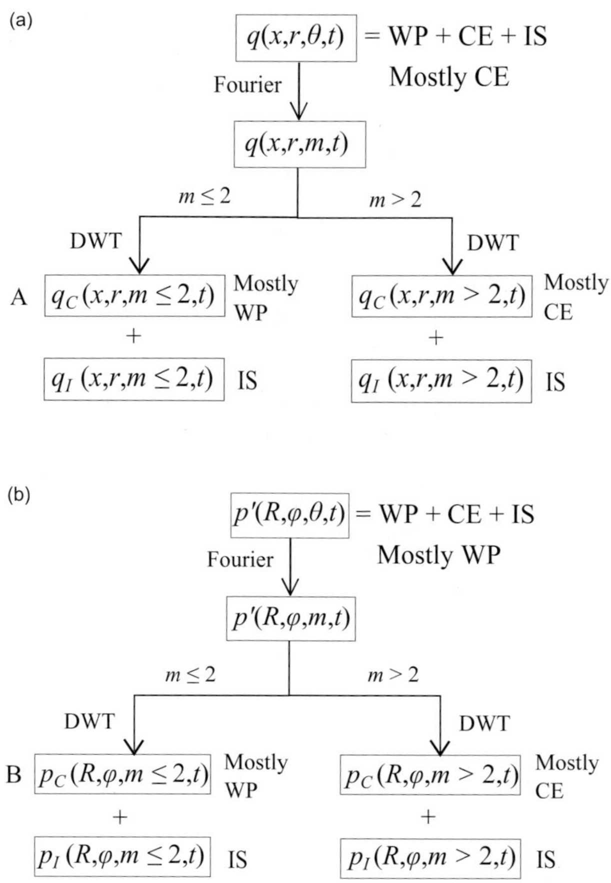

Fig. 1. Block diagrams showing the decomposition of the (a) near field and (b) far field.

3.2. *Wavelet filtering*

The discrete wavelet transform (DWT)[6] is employed to analyse the temporal structures of the flow field in the current study. Following the previous studies[3,4], the current study defines the incoherent component of the flow as random fluctuations with Gaussian probability density functions (PDF). An arbitrary variable $q(\mathbf{x}, t)$ in the time-dependent flow field can be decomposed as

$$q(\mathbf{x}, t) = \bar{q}(\mathbf{x}) + q_C(\mathbf{x}, t) + q_I(\mathbf{x}, t), \tag{1}$$

where $\bar{q}$ is the time-averaged mean value, q_C is the coherent component of the fluctuations, and q_I is the incoherent component. The time-dependent fluctuations (i.e. $q_C + q_I$) are considered for the filtering process.

CS have been proposed to be extracted by filtering the wavelet coefficients by setting a threshold[4]. The threshold (ϵ) for identifying a Gaussian white noise is given by Ruppert-Felsot *et al.*[4]

$$\epsilon = \sqrt{2\sigma_I^2 \ln N}, \tag{2}$$

where σ_I^2 is the variance of the incoherent part (with Gaussian PDF) and N the resolution at which the time series is sampled.

4. Results and Discussions

4.1. *Hydrodynamic near field*

The mean centreline velocity of the LES agrees well with the experiment[7], and the maximum error is around 5%. The root-mean-square value of the streamwise velocity perturbations on the centreline also agrees well with the experiment[7], and for $2 < x/D < 10$ the maximum error is within 5%. We do not show these results here in the interest of brevity.

4.1.1. *Velocity spectra*

The power spectral density (PSD) of the velocity fluctuations in the axial direction is shown in Fig. 2, where PSD (S_{xx}) is obtained by,

$$S_{xx}(x, \omega) = \langle \hat{u}_x(x, \omega)\hat{u}_x^*(x, \omega)\rangle / U^2, \tag{3}$$

and $\omega = 2\pi St$, where u_x is streamwise velocity fluctuation, and St is based on jet diameter and jet exit velocity. The averaging between Fourier transforms of segments of the time series is implicit, and the cylindrical coordinates r and θ have been dropped for conciseness.

The data in Fig. 2 is extracted at $x/D = 4$, 5, and 6, respectively, from left to right. The PSD has been extracted on the jet lipline, where the flow contains high fluctuation energy. It can be seen that, at low Strouhal numbers, the coherent component (marked by the dashed lines), i.e. WP + CE, shares the same PSD with the original flow (marked by the solid lines), but the incoherent component, i.e. IS (marked by the dotted lines), has negligible energy. However, IS account for almost all the PSD of the flow for St higher than 1.5 ~1.7.

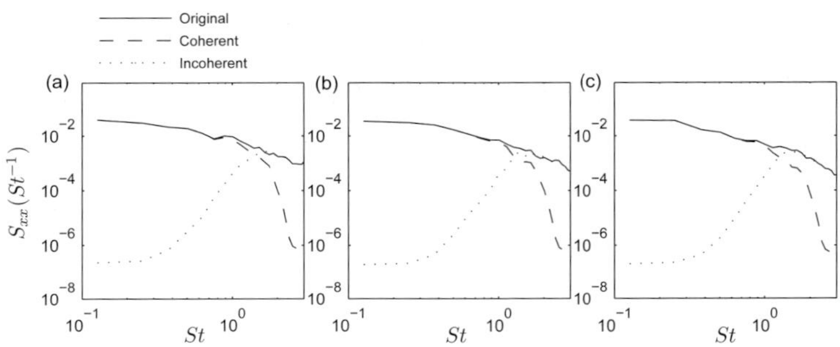

Fig. 2. PSD of the original u'_x (solid lines), the coherent (dashed lines) and incoherent components (dotted lines) of u'_x: (a) - (c) $x/D = 4$, 5, 6, respectively, on the jet lipline.

4.1.2. *Spatial coherence*

The spatial coherence function of velocity fluctuations is defined as

$$C(x, \Delta x, \omega) = \frac{|W(x, \Delta x, \omega)|^2}{S_{xx}(x, \omega) S_{xx}(x + \Delta x, \omega)}, \tag{4}$$

with W being the cross spectral density, given by

$$W(x, \Delta x, \omega) = \langle \hat{u}_x(x, \omega) \hat{u}_x^*(x + \Delta x, \omega) \rangle / U^2. \tag{5}$$

The coherence of the velocity perturbations on the lipline is shown in Fig. 3(a), where $\Delta x/D = 0.1$, and x/D ranges from 1 to 8. It can be seen in Fig. 3(a) that in general the coherence of IS declines as x/D increases, which suggests the coherence length scale of IS could also drop as x/D increases. The decay with x/D is expected due to the increasing turbulence scales with x/D.

The coherence length scales (L_C/D) are determined by fitting the values of the coherence function given by Eq. 4. For a given St at a certain location

(x/D), L_C/D is fitted as

$$C(x, \Delta x, St) = \frac{A}{w} e^{-\frac{(\Delta x/D)^2}{w^2}} + C_0, \tag{6}$$

where A, w and C_0 are constants. The presence of C_0 in the equation above results from the numerical oscillations for low coherence values. The coherence length scale (L_C) is then given by

$$L_C/D = w\sqrt{\ln \frac{2A}{w(C(\Delta x = 0) - C_0)}}, \tag{7}$$

where $C(\Delta x = 0) = 1$. The results are shown in Fig. 3(b). It indicates that the coherence length scale drops in the original unfiltered flow as x/D increases. In the incoherent flow, the length scales are below 0.05 for $x/D > 3$. This confirms that, after both WP and CE are eliminated, IS are decorrelated and spatially compact (for instance, for $St = 1.5$ the acoustic wavelength is $0.76D$). The low L_C for $x/D > 3$ suggests the peak for the coherence of IS for $x/D > 6$ seems minor and may result from numerical error.

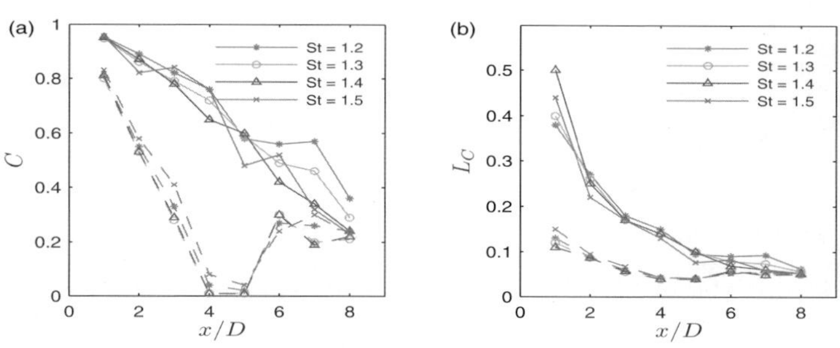

Fig. 3. (a) Coherence function C of the original and the remaining incoherent flow for x_0/D ranging from 1 to 8 and $x/D = 0.1$; (b) the coherence length scale as a function of x_0/D; The solid lines indicate the results of the original flow, and the dashed lines represent the data of the incoherent flow.

4.2. *Acoustic far field*

Figure 4(a) shows the overall sound pressure level (OASPL) of the filtered far field. As can be seen, the sound directivity of LES agrees well with the experiment measurement[8]. After the far field is filtered in the low azimuthal modes (i.e. $m = 0$, 1 and 2) through azimuthal decomposition and reconstruction, the OASPL is significantly reduced for a wide range of polar angles from the jet axis. With the CS of the acoustic field removed in the azimuthal modes up to 2, the OASPL is decreased by 9 to 15dB.

The sound spectra are shown in Fig. 4(b). It shows that the LES captures the trend of sound spectrum compared with the experiment[8]. After the CS have been removed, the sound pressure level (SPL) of the low-frequency component decreases significantly. When the filter is applied to the azimuthal modes up to order 2, the highest SPL occurs at $St \approx 1.6$. This Strouhal number is consistent with the frequency at which the IS capture the PSD of the original flow in the hydrodynamic near-field (see Fig. 2).

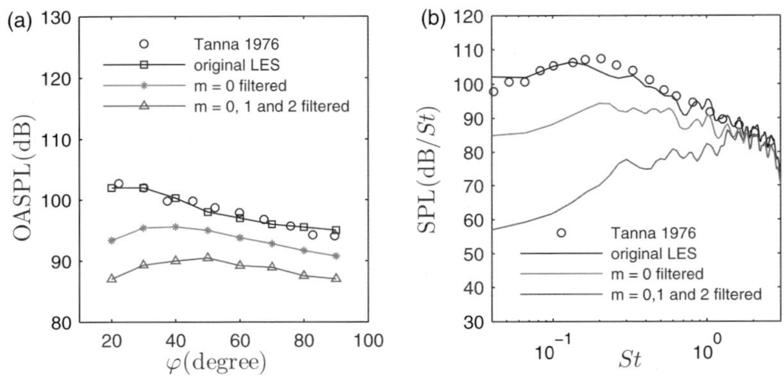

Fig. 4. Far-field sound spectra at $R = 120D$. (a) OASPL of the experiment measurement[8], the LES result with $m = 0$ filtered, $m = 0$, 1 and 2 filtered, and all m filtered;(b) SPL at $\varphi = 30^0$: experiment[8], the original flow, $m = 0$ filtered, $m = 0$, 1 and 2 filtered.

4.3. *Correlation between the near-field velocity and far-field pressure*

We calculate cross correlations between near-field axial velocity perturbations and far-field pressure before and after the filtering process. Specifically, we are going to examine (i) the correlation between $u_x(x, r, m = 0, t)$ and $p(R, \varphi, m = 0, t)$, and (ii) the correlation between $u_{xC}(x, r, m = 0, t)$ and $p_C(R, \varphi, m = 0, t)$. We use near-field PIV and far-field microphone data[1]. The cross correlations between near-field axisymmetric mode axial velocity and far-field axisymmetric mode pressure ($C^0_{u_x,p}$) is defined as

$$C^0_{u_x,p}(\tau) = \frac{1}{T}\int_0^T u_x(m = 0, t)p(m = 0, t + \tau)dt. \quad (8)$$

The normalised correlations between the original signals are presented as solid lines in Fig. 5 reproducing the results in Fig. 13(b) and (c) of the original work[1]. We see that the correlations reach the order of 10% at the average propagation time. Using the filtering technique described in Section 3, we isolate the CS and IS of the near-field and far-field data. This enables us to calculate the correlations between the CS in the flow (which in this case are supposed to be mostly WP) with the associated far-field sound, as shown by the dashed lines in Fig. 5. It can be seen that the normalised correlation increases to 15%, which suggests strong correlation between the CS extracted by the filter. The correlation between the IS in the near and far field is also calculated, as presented by the dash-dotted lines in Fig. 5. It is clear that these correlations are low without any obvious peak over the noise level.

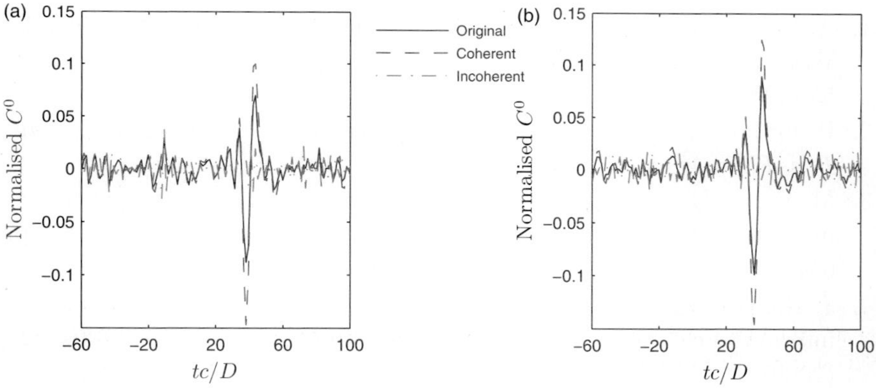

Fig. 5. Cross correlations between axisymmetric mode axial velocity at $r = D/4$ and axisymmetric mode far-field pressure at $\varphi = 20^0$ for the $M = 0.6$ jet[1]: (a) $x/D = 5$; (b) $x/D = 6$. The solid, dashed and dash-dotted lines indicate cross correlations between the original signals, CS and IS. The average propagation time without flow-acoustic interaction (the sound wave arrives at the far field) is approximately $tc/D = 40$.

5. Conclusions

In the current study we decompose the near field of a turbulent jet and the acoustic far field into wave packets (WP), coherent eddies (CE) and incoherent structures (IS). The decomposition is achieved by a combination of a Fourier transform in azimuth and discrete wavelet transform (DWT) after assuming that IS comprise of random fluctuations. The WP are identified as the most acoustically-efficient coherent structures (CS) in the jet and dominate the fluctuation energy at low Strouhal numbers (≤ 1.5). The IS account for almost all the fluctuation energy in the high Strouhal number range. The correlations between the CS in the axisymmetric modes of the near and far field have significantly higher values than those previously reported for flow-acoustic correlations of the axisymmetric mode. This suggests a link connecting the near-field and far-field CS.

The current study suggests that flow control measures to minimise acoustically-efficient CS in the near field should target low azimuthal modes (≤ 2) and Strouhal numbers (≤ 1.5). As a result, a significant reduction of overall sound pressure level could potentially be achieved.

References

1. A. V. G. Cavalieri, D. Rodríguez, P. Jordan, T. Colonius and Y. Gervais, Wavepackets in the velocity field of turbulent jets, *J. Fluid Mech.* **730**, 559 (2013).
2. P. Jordan and T. Colonius, Wave packets and turbulent jet noise, *Annu. Rev. Fluid Mech.* **45**, 173 (2013).
3. M. Farge, G. Pellegrino and K. Schneider, Coherent vortex extraction in 3D turbulent flows using orthogonal wavelets, *Phys. Rev. Lett.* **87**, p. 054501 (2001).
4. J. Ruppert-Felsot, M. Farge and P. Petitjeans, Wavelet tools to study intermittency: application to vortex bursting, *J. Fluid Mech.* **636**, 427 (2009).
5. Z.-N. Wang, I. Naqavi, M. Mahak, P. Tucker, X. Yuan and P. Strange, Far field noise prediction of subsonic hot and cold jets using large-eddy simulation, in *Proceedings of ASME Turbo Expo 2014: Turbine Technical Conference and Exposition*, (Düsseldorf, Germany, 2014).
6. M. Farge, Wavelet transforms and their applications to turbulence, *Annu. Rev. Fluid Mech.* **24**, 395 (1992).
7. J. Bridges and M. P. Wernet, Establishing consensus turbulence statistics for hot subsonic jets, in *Proceedings of 16th AIAA/CEAS Aeroacoustics Conference*, (Stockholm, Sweden, 2010).
8. H. K. Tanna, *The generation and radiation of supersonic jet noise*, tech. rep., Air Force Aero Propulsion Laboratory (1976).

Synchronized Large-Eddy Simulations for Sound Generation Analysis

S. Unnikrishnan and Datta V. Gaitonde

Mech. and Aerosp. Engineering, The Ohio State University, Columbus, OH 43210

We employ a new technique comprised of multiple simultaneous Large-Eddy Simulations (LES) to analyze the dynamics of turbulence, with emphasis on jet noise. Unlike methods which seek to use correlation of signals at different locations, here we isolate and tag perturbations in different regions of interest and examine the manner in which they are modulated and filtered by the turbulence. The technique essentially uses two (or more) simultaneous, synchronized Large Eddy Simulations (LES). One, denoted baseline, provides the natural perturbations existing in the jet. These are processed and injected into a twin to provide a perturbation boost, which can then be tracked in the twin. Several simulations on a Mach 1.3 cold jet demonstrate the effectiveness of the approach in determining the self and mutual interactions in core turbulence that ultimately yield the near acoustic field. Detailed 1-D and 2-D statistical analyses yield unique insight into the nature of intermittency, spatio-temporal correlations, wave-packet generation and directivity.

Keywords: Jet noise; Synchronized LES; Wave-packets; Correlation analysis.

1. Introduction

The manner in which turbulence evolves is fundamental to almost all aerodynamic applications. A particularly difficult problem is that of determining how jet turbulence is processed into noise. Although it is known that the turbulent core of a jet plays an important role in creating its near- and far-field noise signatures, the underlying physics of this process linking the non-linear core to the linear acoustic far-field is not well understood. Correlations have been employed to gain some insight, (*e.g.*, Ref. 1), but the details of how high-energy broad-spectrum spatio-temporal turbulence is modulated and filtered through highly non-linear processes to create the small but highly directive far-field signal remain unknown[2].

Recent evidence suggests that the dominant component of the far field signal originates in intermittent events[3]. Several efforts[4,5] analyze the noise generation process using wave-packet models. Although these models explain the importance of intermittent events, the manner in which they are generated remains unexplained. In this work, we develop a method using LES to explore the role of specific regions of the core in the far-field noise signature. The approach essentially uses a baseline LES to generate native (also denoted natural or organic) perturbations to boost the signal in a parallel LES by a very small amount. The time accurate difference between the two solutions can then be employed to analyze the turbulent dynamics.

2. Synchronized LES Strategy

The basic LES approach has been proven for jets in multiple prior publications[6,7]. Briefly, the inviscid fluxes are simulated with a third-order upwind biased method combined with the compressive harmonic limiter, which enforces monotonicity and provides sub-grid closure. Viscous effects are treated with centered differencing. Time integration is based on an implicit diagonalized Beam-Warming approach, which facilitates larger time traces for statistical analysis. The synchronized LES approach may be summarized as follows. First, the baseline LES of the case under examination is marched to statistically stationary state and the mean flow is obtained. A second, twin LES is started from this initial condition. At each step, perturbations in the baseline (derived from a difference between instantaneous and mean) from a chosen area of interest are extracted, windowed, scaled and injected into the twin. The two solutions are marched in this fashion in a lock-step manner. At subsequent time instants, differences between the baseline and the twin yield the effect of the perturbation boost, which can be analyzed to understand signal propagation and the synthesis of the near field. In this manner, the process by which specific regions of the jet modulate the natural perturbation to yield the acoustic near-field can be determined. The method ensures that the injected perturbations have the same acoustic, vortical and entropic components as the original signals already present in the jet and hence do not induce any new instabilities in the twin. The scaling of the perturbation boost by a small value ensures that the evolving perturbation boost field remains linear as shown in Ref. 8. Once the near-field signals are obtained, the effect of the region under study on the far-field acoustic signature of the jet can be determined through usual linear propagation analyses.

The case considered to explore this approach, a cold Mach 1.3 jet issuing from a circular nozzle, has been discussed extensively in the literature[6,7,9]. A structured cylindrical grid is employed with $422 \times 242 \times 105$ points in the axial, radial and azimuthal directions respectively (Fig. 1(a)). Verification and validation studies (not shown for brevity) in the above references demonstrate that the mean and fluctuating quantities are reproduced quite accurately, and that the flowfield is comprised of coherent structures with broad spatio-temporal spectra.

3. Results

3.1. *Cases investigated*

For discussion purposes, Fig. 1(b) shows the mean flow field obtained from time-averaging the baseline flow. Disturbances are separately extracted from three point locations, designated A and B on the centerline and C on the lip-line, and injected into the twin to yield Cases A, B and C. Case AB introduces perturbations simultaneously at A and B while Case AC does the same at A and C. Details of these cases including corresponding source locations and scaling parameter (ϵ) employed for scaling the perturbation boost amplitudes are presented in Table 1.

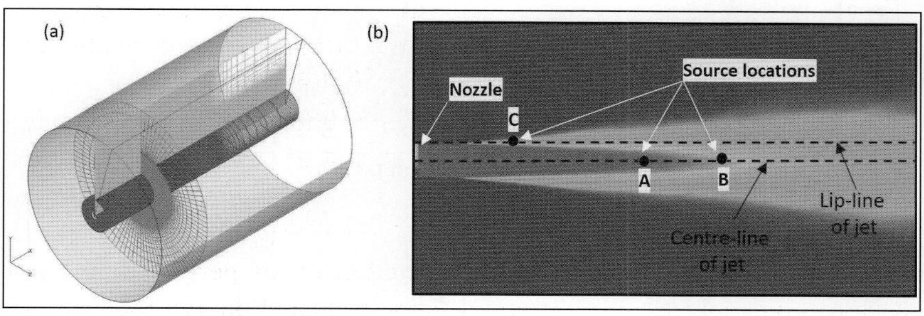

Fig. 1. Computational grid and source locations used for synchronized LES.

Table 1. Source configuration and scaling for various cases.

Case	Point source location	Amplitude scaling
Case A	(6.897, 0, 0)	$1e^{-6}$
Case B	(9.242, 0, 0)	$1e^{-6}$
Case C	(2.988,0.5, 0)	$1e^{-6}$
Case AB	A+B	$1e^{-6}$
Case AC	A+C	$1e^{-6}$

Basic tests confirm that the mean flow field is unaltered by the perturbation boost, and remains axisymmetric. The potential core of the jet extends to approximately 7 jet diameters in the down-stream direction.

3.2. *Unsteady perturbation field*

Sub-figures (a) to (d) in Fig. 2 show snapshots of perturbation boost pressure field in Case A at different time instants marked. The initial symmetric

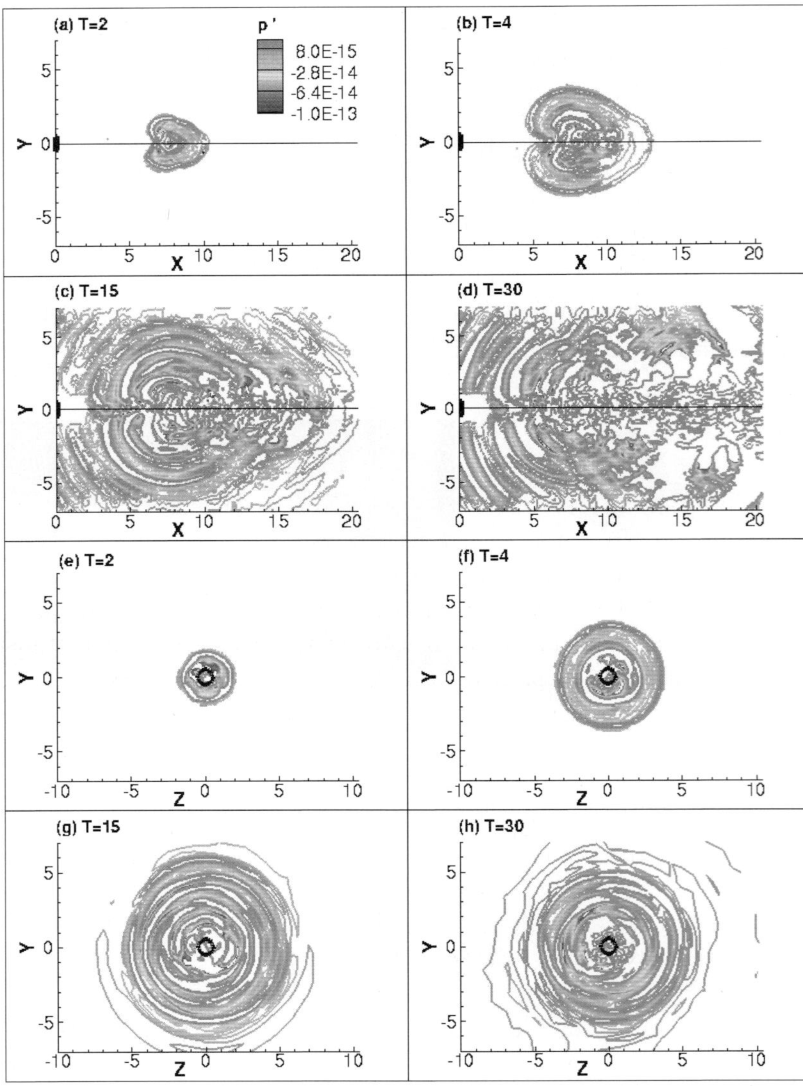

Fig. 2. Instantaneous pressure perturbation field for Case A at indicated nondimensional time instances. The axial planes are taken at the source location in Case A (Table 1). The contour levels for all the subplots are indicated in (a).

wave fronts grow outward, and form intermittent wave-packets that propagate into the near acoustic field. At later time instances, $t = 15$ and 30 in Fig. 2, the perturbations are amplified to higher values towards the center of the jet at further downstream locations. These also then start emitting outward propagating wave-packets at locations which are downstream from the previous set. This trend is consistent in both the centerline source cases A and B. The upstream propagating wave fronts are also observed to be more intact and well defined than downstream propagating waves, which are subjected to the more turbulent flow.

Sub-figures (e) to (h) in Fig. 2 visualize the corresponding snapshots from the axial direction. The axial plane is located at the source location of Case A. The manner in which the nature of perturbations vary as they propagate from the core to the near field is clear in these images. The perturbation field appears to be disorderly and turbulent within the core of the jet, but it assumes a recurrent wave-packet nature towards the near field region. It also becomes increasingly axi-symmetric as the core turbulence decays. The amplitudes of the propagated field at various axial locations at a radial distance of $7D$ from the jet centerline are found to be one to two orders of magnitude lesser than the scaled perturbations at the source location. This indicates the effect induced due to the filtering and modulation of natural signals within the core of the jet by the turbulence. The perturbation field is axi-symmetric in a statistical sense due the symmetry of the source. The presence of the recurrent wave-packets in the near field is an important feature of jet noise. The characteristic feature of such localized packets is that, although they contain relatively lesser amount of energy compared to the turbulent hydrodynamic fluctuations within the core, they are persistent over longer distances which results in the long range influence of acoustic signature of the jet. This analysis indicates that, although the natural perturbations are boosted only at a point source, they are still capable of producing localized wave-packets. They collectively determine the intermittency and acoustic characteristics of the total signal in the near field, which is a combined result of natural perturbations emanating form the entire core.

Figure 3 shows perturbation boost pressure contours for Case C at various nondimensional time instances. The approximate end of the potential core is indicated by a solid circle in each snapshot. The snapshot at $t = 5$ indicates the roll-up of perturbation boost structures in the shear layer. Once they are past the end of the potential core, these structures are channeled to the centerline, their amplitudes rise and they emit stronger

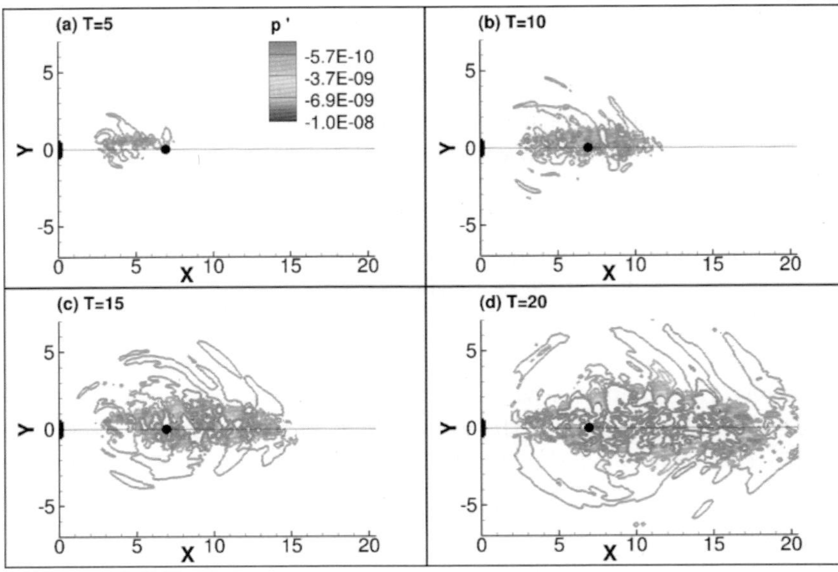

Fig. 3. Shear layer instabilities in Case C. The contour values for all the subplots are indicated in (a).

wave-packets predominantly along downstream angles as seen in Fig. 3(d). Centerline cases A and B result in relatively lesser amplification of the perturbations relative to Case C. In a previous effort[8], we have also shown that upstream centerline sources (*e.g.*, Case A) have relatively lesser upstream influence compared to downstream sources (*e.g.*, Case B) on RMS values of the perturbation boost.

In Cases C and AC, perturbations are amplified in the twin simulation to a much higher level compared to Cases A, B and AB because of the presence of the source at point C in the twin (which lies on the nozzle lip-line) where shear layer instabilities grow and roll up into structures which convect downstream. Figure 4 shows the instantaneous pressure perturbation boost field for Case AC where the centerline and lip-line sources are activated simultaneously. Development of the symmetric centerline and the asymmetric lip-line perturbation fields are evident in this figure. It is also evident that the perturbation field for the lip-line case has higher contour values compared to the centerline case. Once the leading edge of the lip-line perturbation field crosses the location of the downstream centerline source, the influence of the centerline source on the resultant perturbation field is masked as the lip-line perturbation field amplifies to values four to five

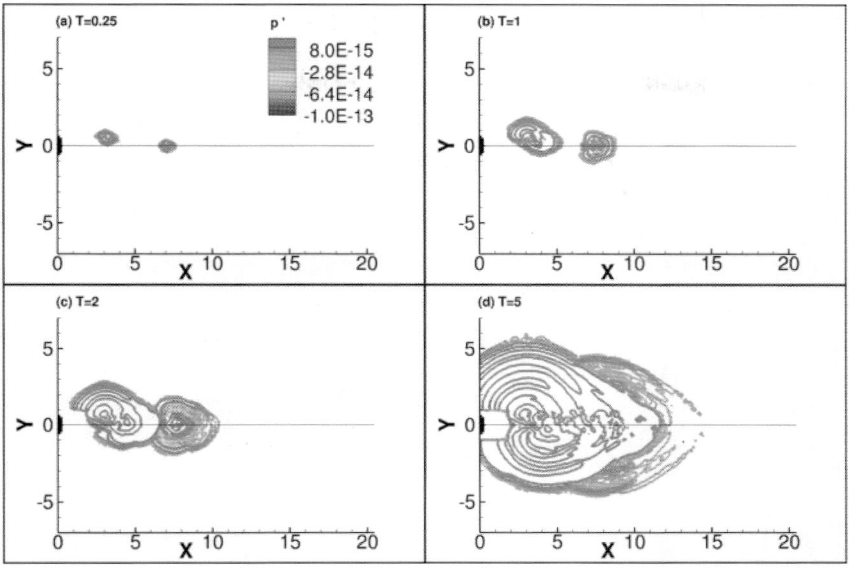

Fig. 4. Instantaneous pressure perturbation field for Case AC at indicated nondimensional time instances. The contour values for all the subplots are indicated in (a).

orders of magnitude grater than the centerline perturbation field.

3.3. *Correlation analysis*

Spatio-temporal correlations provide insight into quantitative aspects of the intermittent wave-packets observed in the perturbation boost field. The process is significantly different from previous studies in that it facilitates investigation of how the region of interest correlates with the actual source in establishing near-acoustic field intermittency. Figure 5 shows two point correlations with respect to source point A. Frame (a) represents the time signal at the source while Frame (b) shows the symmetric auto-correlation, *cc*, of this signal plotted at various lag values. The vertical dotted line at a time lag of about $3.5e^{-5}s$ indicates the local time scale of the flow. Frame (c) shows the time signal at a point vertically above the source point, at $y = 4.424$ and $z = 0$ while frame (d) shows the correlation between this signal and the source signal. It can be seen that *cc* exhibits an oscillating profile, indicating the presence of recurring similarity between the source and the destination point. This is the key signature of wave-packets created by the core through modulating the time varying source signal, similar to those observed experimentally.

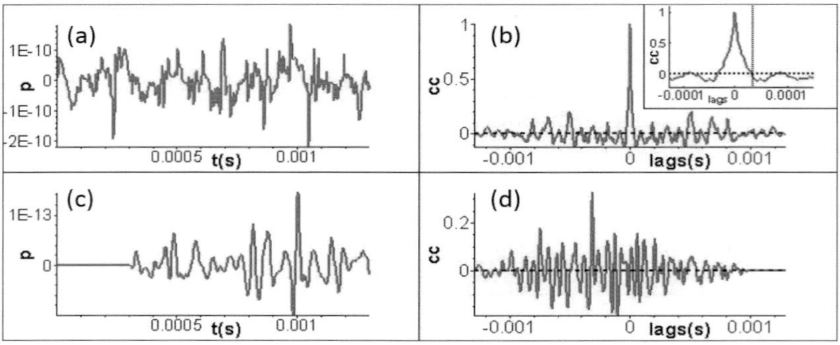

Fig. 5. Auto-correlation and cross-correlation at source of Case A.

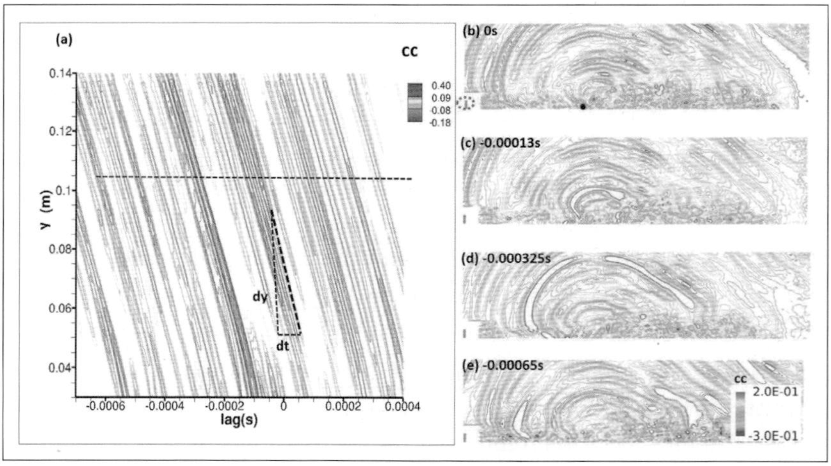

Fig. 6. 1-D and 2-D spatio-temporal correlation plots for Case A.

Figure 6 presents 1-D and 2-D spatio-temporal correlations for Case A. Crests and troughs of the recurring wave-fronts appear in the 1-D analysis, Fig. 6(a) as adjoining streaks of opposite signs. Blank spaces indicate the silent periods between the fronts. The slopes of the streaks represent the convection velocity of the perturbation boost (370 m/s). Similar analysis conducted for Cases B and C yields values of 360m/s and 335m/s respectively. Figures 6(b)-(e) show the 2-D spatio-temporal correlation for Case A at the indicated lags. The signature of well-defined recurring wave fronts propagating along upstream angles, lesser coherent structures emitted at low aft-angles, and finer structures produced along the downstream axial direction are clearly visible. Progressing from zero to negative values of lag in the contour plots of 2D spatio-temporal correlation is equivalent to moving from right to left along the horizontal axis in Fig. 6(a). Similar observations are made in the analysis of Case B and its spatio-temporal correlation contours at zero lag is shown in Fig. 7(a). The corresponding

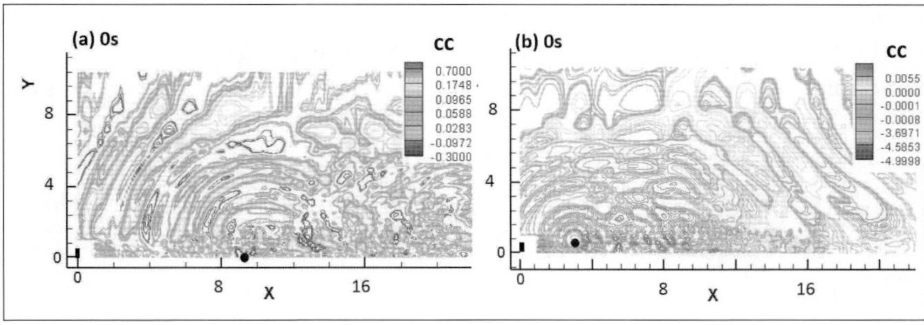

Fig. 7. 2-D spatio-temporal correlation at zero lag for Case B and C. The dots mark the corresponding source location in each case.

analysis carried out for Case C with the lip-line source indicates the signature of rolled up shear layer instabilities leading to strong radiation along the downstream angles as shown in Fig. 7(b).

Figures 8(a)-(d) show the 2-D spatio-temporal correlations of Case A on an axial plane at the source location. These figures correspond to the

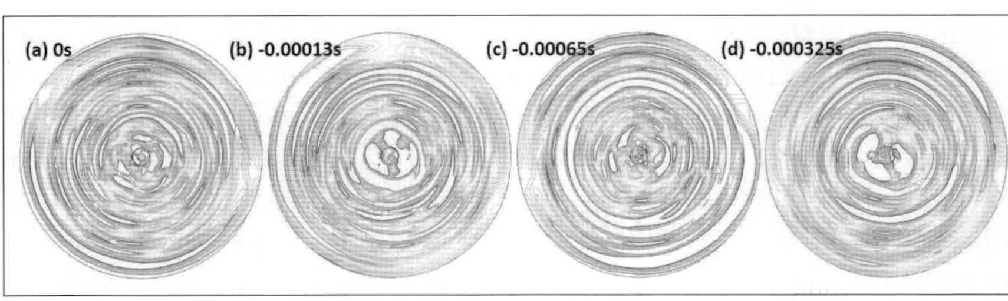

Fig. 8. 2-D spatio-temporal correlation plots on axial plane at source location for Case A.

lag values indicated in Figs. 6(b)-(e). The outward propagation of near concentric recurring perturbation field is evident in these figures. These are the foot prints of the outward propagating wave-packets discussed in Fig. 2. Alternating adjacent sign values are visible in these figures also indicating the axysymmetric nature of the crests and troughs of wave-packets. The

2-D spatio-temporal analysis thus highlights the intermittency induced by the signal, as well as the coherence of associated perturbations.

Correlation analysis also provides estimates of local timescales present in the flow. This information is useful in determining the directivity induced by the core of the jet. The local time scale of the flow obtained from the autocorrelation of pressure perturbation boost signals are shown in Fig. 5(b). This is the lag value at which the autocorrelation changes sign for the first time. This value is obtained at all the points on the $Z = 0$ half plane for Case A and its $Log10$ value is presented in Fig. 9. It can be seen

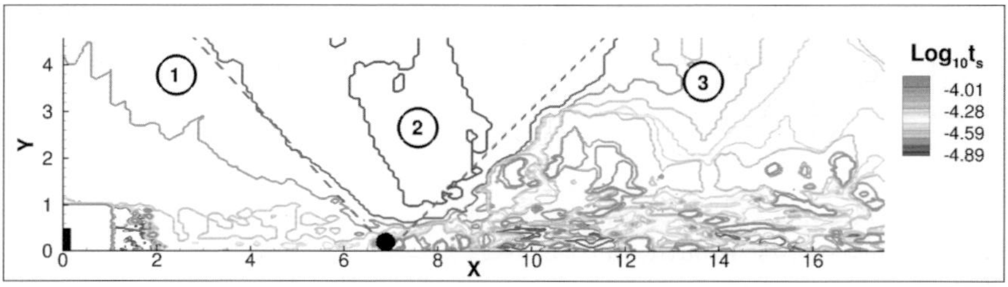

Fig. 9. Integral timescale of the perturbation boost field for Case A obtained from the autocorrelation of pressure perturbation boost. The dot indicates the source location. The dashed lines demarcate three major timescale regions in the nearfield as identified by the circled numbers.

that this contour can be broadly demarcated into three zones, 1, 2 and 3. The dotted inclined lines in Fig. 9 align with the approximate boundaries which separate these three zones. Zone 3 represents the highest values of integral time scales present in the perturbation boost field. This corresponds to the larger flow features that convect downstream and propagate at low aft angles from the jet axis. Zone 1 represents an intermediate range of time scale in the flow and these are observed as well preserved wavefronts propagating in the upstream direction at intermediate frequencies. Zone 2 represents the lowest time scales in the perturbation boost field and hence the components with the highest frequencies. This indicates that the high frequency components of perturbations propagate in a direction perpendicular to the jet axis from the source. This trend was found to be consistent with other source configurations also.

4. Conclusion and Future Work

Simultaneous LES have been used to analyze disturbances in turbulent flows, and to understand their dynamics under the influence of viscous and inviscid non-linear mechanisms. Application to jet flows shows that disturbances originating in the initial jet shear layer show relatively higher degree of disturbance amplification. However, the perturbations are first channeled to the centerline, in the vicinity of the collapse of the potential core. The intermittency generated by the turbulent core is analyzed by forward correlation analysis. Spatio-temporal correlations identify and quantify intermittent wave-packets and the degree of coherence of the emitted wave fronts at different locations. Integral timescales present in the perturbation boost indicate that the high frequency components of the field propagate in a direction perpendicular to the jet axis from the source, whereas the lower frequencies are emitted along low aft angles in the downstream direction. These results hence provide insight into the directivity of the signal spectrum.

References

1. C. Tam, K. Viswanathan, K. Ahuja and J. Panda, The sources of jet noise: experimental evidence, *J. Fluid Mech.* **615**, 253 (2008).
2. J. Bridges and M. Wernet, Validating large eddy simulations for jet aeroacoustics, *J. Prop. Power* **28**, 226 (2012).
3. M. Kearney-Fischer, A. Sinha and M. Samimy, Intermittent nature of subsonic jet noise, *AIAA J.* **51**, 1142 (2013).
4. P. Jordan and T. Colonius, Wave packets and turbulent jet noise, *Ann. Rev. Fluid Mech.* **45**, 173 (2013).
5. R. Reba, S. Narayanan and T. Colonius, Wave-packet models for large-scale mixing noise, *Int. J. Aeroacoustics* **9**, 533 (2010).
6. D. Gaitonde and M. Samimy, Coherent structures in plasma-actuator controlled supersonic jets: Axisymmetric and mixed azimuthal modes, *Phys. Fluids* **23** (2011).
7. R. Speth and D. Gaitonde, Parametric study of a Mach 1.3 cold jet excited by the flapping mode using plasma actuators, *Comput. Fluids* **84**, 16 (2013).
8. S. Unnikrishnan and D. Gaitonde, Synchronized large-eddy simulations to track native perturbations in a turbulent jet, in *AIAA Sci. Tech. 53rd Aerospace Sciences Meeting*, AIAA-2015-0505 (2015).
9. D. Gaitonde, Analysis of the near field in a plasma-actuator-controlled supersonic jet, *J. Prop. Power* **28**, 281 (2012).

DNS of a Turbulent Jet Issuing from an Acoustically Lined Pipe

R. D. Sandberg*,[1] and B. J. Tester[2]

[1] *Aerodynamics and Flight Mechanics Group* / [2] *Institute of Sound and Vibration Research, University of Southampton, Southampton, SO17 1BJ, U.K.*

* *E-mail: sandberg@soton.ac.uk*

www.southampton.ac.uk

A series of direct numerical simulations (DNS) were conducted of turbulent jets issuing from acoustically lined pipe. The inclusion of the pipe in the simulations with a fully turbulent flow inside ensures that all possible noise generation mechanisms are represented. Earlier results from a similar pipe/jet configuration, albeit without acoustic liner treatment, showed contamination of the far field noise field by interior noise from the pipe. Therefore, here two key modifications were made. Firstly, the interior pipe walls were acoustically lined using an impedance condition. Secondly, the turbulent pipe inflow boundary condition was modified to reduce spurious noise introduced into the axisymmetric mode. The sound radiation from the pipe/jet configuration was analysed using a phased array source breakdown technique. It is demonstrated that the modification of the inflow boundary results in a strongly reduced contribution of the interior noise component to the farfield noise in the axisymmetric mode, while the acoustic liner is effective in reducing the interior noise contribution to farfield noise in the higher azimuthal modes. This enables the source breakdown analysis to extract the jet mixing noise contributions to the farfield much more clearly.

Keywords: DNS, Jet Noise, Aeroacoustics, Acoustic Liner, Phased-Array

1. Introduction

Jet engines have over the last decades become dramatically quieter, mainly through a continuous increase in bypass ratio. However, further progress on noise reduction now depends on a detailed understanding of the exact noise generation mechanisms. To get further physical insight into jet noise generation mechanisms high-fidelity data of the hydrodynamic and acoustic fields of jets are desirable. Several different sources contribute to the overall sound radiation from subsonic jets: (i) large scale structures mainly occurring close to the potential core region, (ii) breakdown of large scale structures into fine-scale turbulence near the end of the potential core,

(iii) fine-scale turbulence within the initial shear layers of fully turbulent jets, and (iv) trailing-edge noise resulting from the interaction between flow and the solid wall at the nozzle exit. The contribution of each individual source is known to vary with the direction from the jet downstream axis and frequency, e.g. large scale structures predominantly radiate low frequency sound in the downstream direction, whereas fine-scale turbulence (shear layers, breakdown of the potential core) is mainly responsible for the high-frequency sound that is considered to be less directional[1].

In order to include all possible noise sources, it is essential to include the nozzle in the simulation. This was first done using direct numerical simulation (DNS) by Sandberg *et al.*[2], who simulated turbulent flow exiting a pipe into a laminar co-flow, serving as a well defined turbulent inlet condition for the study of sound radiation from fully turbulent jets. The sound radiation from this configuration was studied for three subsonic jet Mach numbers and varying co-flows. However, this study unexpectedly found that the velocity scaling of mean square far-field pressure at 90 degrees was similar for all cases over the range of Mach numbers considered and not following the scaling law expected for an isothermal jet. In an attempt to explain this behaviour, Sandberg and Tester[3,4] investigated the source characteristics of the pressure field decomposed into its azimuthal modes with the aid of a phased array technique. This technique separated out the noise sources associated with jet mixing from additional noise source mechanisms that could be attributed to noise generated in the nozzle or by turbulent flow interacting with the nozzle lip, depending on frequency and azimuthal mode number. The strongest radiation levels were due to azimuthal mode $m = 0$ and it was shown that this was dominated by internal noise. Olivetti *et al.*[5] conducted DNS of the application of an acoustic liner model in a pipe terminated with a non-reflecting boundary condition, aimed at suppressing the internally generated noise. The time domain model developed by Tam and Auriault[6] was implemented in the DNS solver and showed that the liner was effective a removing acoustic fluctuations within the pipe without affecting the turbulent flow field. In the current paper, a new set of DNS of a fully turbulent pipe/jet configuration using an acoustic liner boundary condition for suppression of internally generated noise are presented. The first analysis of this data has shown that the change in inflow boundary condition is largely responsible for the reduction of internal noise of the axisymmetric mode while the main role of the acoustic liner is the absorption of internal noise for higher azimuthal modes[7]. The main focus of the current contribution is on using a source breakdown technique, based on a phased-array approach, to evaluate the effect of the internal noise suppression on the different noise sources.

2. Numerical Setup

2.1. *Direct numerical simulations*

The compressible Navier–Stokes equations are solved in cylindrical coordinates using a 4th-order central difference scheme with Carpenter boundary stencils for the spatial discretization of the radial and streamwise directions. A Fourier method is used for discretization in the azimuthal direction, enabling an axis treatment that exploits parity conditions of individual Fourier modes[8]. The explicit time marching is performed with an ultra low-storage 4th-order Runge–Kutta scheme and stability of the code is enhanced by a skew-symmetric splitting of the nonlinear terms. In addition, an 11 point wave-number optimized filter is used after each full Runge–Kutta cycle with a weighting of 0.2 to remove possible grid-to-grid-point oscillations. At the pipe inlet, turbulent fluctuations obtained by using a digital filter technique[9] with parameters specified from precursor periodic pipe simulations, were superposed onto mean flow values in order to generate a spatially developing turbulent pipe flow. However, in contrast to some of our earlier studies, the inflow perturbation velocities for the axisymmetric mode $m = 0$ are explicitly set to zero. This avoids mass flow fluctuations in the pipe that result in strong acoustic radiation of the axisymmetric mode[3] and was found difficult to suppress with a liner at low Strouhal numbers. Removing velocity fluctuations of $m = 0$ has been found to have no effect on the developed pipe flow downstream. More details about the code and validation can be found in Sandberg *et al.*[10].

A time-domain impedance boundary condition based on a mass-spring-damper analogy has been used, as defined by Tam and Auriault[6]. It can be written in the form where the generic resistance parameter $R0$ is identified as the dissipative term of the mass-spring-damper model and the two reactance parameters are identified as mass-reactance $X1$ and stiffness $X2$, which are chosen to produce resonance at the required frequency or Strouhal number. In the current study, $R0 = 4$, $X1 = 1.177$ and $X2 = -33.49$.

2.2. *Simulation geometry and grid*

The computational domain is composed of seven blocks, as shown in Fig. 1: flow inside the pipe (block 1), jet development downstream of the pipe exit (blocks 2, 3, 4 and 6), and coflow and acoustic field upstream of the pipe exit (blocks 5 and 7).

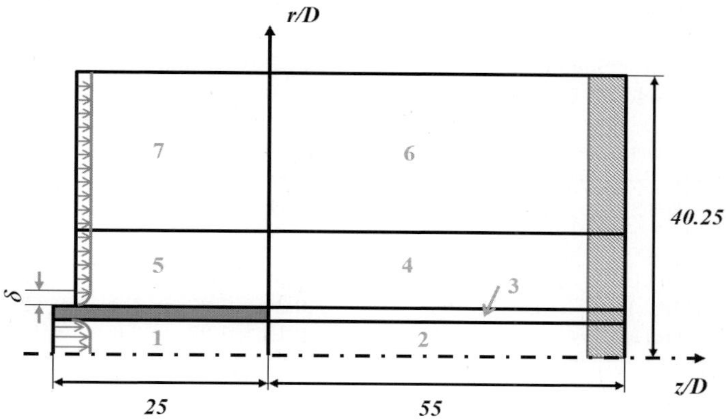

Fig. 1. Sketch of the computational domain; the shaded region at the outflow denotes the region in which the zonal characteristic boundary condition[11] was applied.

In the azimuthal direction 64 or 8 Fourier modes corresponding to 130 or 18 collocation points were used in the turbulent flow (blocks 1 to 5) and acoustic regions (blocks 6 and 7), respectively, resulting in a total of 225×10^6 grid points. Figure 1 also shows the dimensions of the computational domain, determined from preliminary turbulent pipe and jet simulations[10]. The finest grid spacing was $\Delta z/D = 0.0045$ and $\Delta r/D = 0.00129$ at the nozzle exit and the upper bounds of $\Delta z/D = 0.0515$ and $\Delta r/D = 0.0711$ were chosen to resolve acoustic waves up to Strouhal number $St_D \approx 2$ (based on the jet velocity and diameter) with at least 10 grid points.

A laminar boundary layer (Blasius solution) was prescribed at the inflow boundary on the pipe outside, reaching $\delta/D = 0.025$ at the jet exit. At the outflow boundary a zonal characteristic boundary condition was applied[11], while characteristic boundary conditions were used at the upper freestream boundary.

Three different cases were conducted (Tab. 1) to study the effect of the liner and the effect of co-flow Mach number for a given jet exit Mach number. The changes in density and temperature were less than 15% of the wall values, with the wall set to be isothermal at the adiabatic temperature of 1.141. All DNS were run for 150 nondimensional time units (based on diameter and bulk velocity inside the pipe) to allow the initial transients to leave the domain. Each case was then continued for at least 350 time units to achieve statistical convergence. The spatial and temporal resolution of the simulations were rigorously assessed and were found to be adequate[10].

Table 1. Simulation parameters; M_{jet}, M_{co}, u_{co} and u_{jet} based on bulk velocity in pipe, local velocity excess λ at pipe exit, Re_{jet} based on u_{jet}, pipe diameter and ν at pipe wall. Mach numbers based on reference speed of sound using wall temperature.

Case	Liner	M_{jet}	M_{co}	u_{co}	u_{jet}	λ_{exit}	Re_{jet}
M84c2L	yes	0.84	0.2	0.24	1.27	4.3	8,522
M84c2nL	no	0.84	0.2	0.24	1.27	4.3	8,529
M84c1L	yes	0.84	0.1	0.12	1.27	9.6	8,492

3. Results

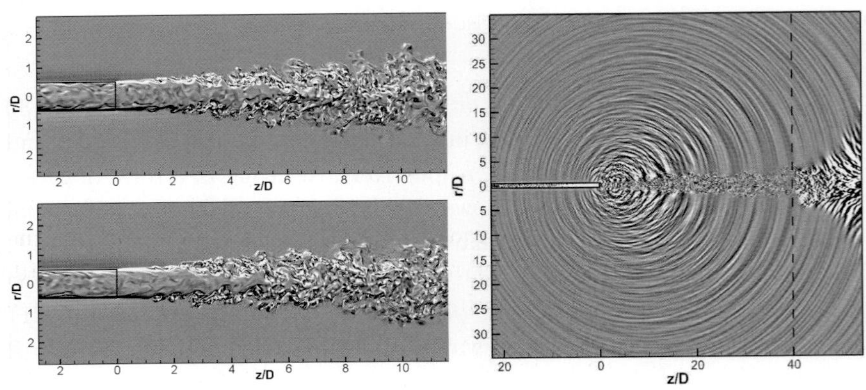

Fig. 2. Instantaneous contours of azimuthal vorticity component for lined (top) and un-lined (bottom) case near the nozzle (left) and dilatation field for azimuthal plane $\Theta = 0°, 180°$, with levels $[-5 \times 10^{-4}; 5 \times 10^{-4}]$ (right); the dashed line at $z/D = 39.5$ denotes the onset of the zonal characteristic boundary condition[11]; Cases M84c2L and M84c2nL.

In Fig. 2 (left) instantaneous contours of the azimuthal vorticity component are shown for the near-nozzle region of cases M84c2L and M84c2nL, illustrating qualitatively the fully developed turbulent pipe flow exiting the nozzle and rapidly developing into a jet. It can be observed that the initial shear layers of the jet are already turbulent and do not first have to undergo a laminar-turbulent transition. Comparing the lined and unlined cases does not reveal any qualitative differences, indicating that the inclusion of the liner does not appear to alter the flow field exiting the nozzle. The similarity of the two cases was shown more rigorously by comparing turbulence spectra obtained within the hydrodynamic jet field. The spectra, both in amplitude and shape, were found to be very similar for the lined and unlined cases, providing evidence that the turbulent flow exiting the pipe is unaffected by the presence of a liner[7]. It was also previously shown that the turbulence statistics at the nozzle exit could be collapsed with profiles of fully developed turbulent pipe flow[10], thus the flow exiting constitutes a well defined turbulent upstream condition suitable for direct noise computations.

For a qualitative impression of the resulting acoustic field, instantaneous contours of the dilatation field $\partial u_i/\partial x_i$ in the entire domain are also shown in Fig. 2. Sound waves can be observed emanating from the nozzle corner and from the jet core. Importantly, no interference from boundary reflections can be detected, which is particularly encouraging given the very small contour levels chosen ($\pm 5 \times 10^{-4}$). The noise radiation shows the highest intensity at roughly $\theta = 40°$, where θ is defined with respect to the streamwise direction. In addition, upstream radiating noise emanating from the nozzle lip, although more weakly, can be detected. When compared to the earlier DNS data without liner and unmodified inflow boundary condition, not shown here for conciseness, two observations can be made that indicate a reduction in nozzle-based noise radiation: i) the dilatation field inside the pipe is significantly weaker for the new case, in particular for $z/D \geq -10$, and ii) the acoustic waves emanating from the nozzle and near-jet regions are considerably more directional.

Sandberg[7] compared the PSDs of individual azimuthal modes of acoustic pressure in the farfield of the lined and unlined simulations at different angles. At the main radiation angle of $\theta = 40°$ it was observed that the liner has a negligible effect on the jet mixing noise component. However, at $\theta = 90°$ the unlined case exhibits cut-on of higher modes at higher frequencies while the lined case does not display this behavior, implying that the internal noise has successfully been removed by the acoustic liner. In this study, the far field noise generated by the turbulent jet was investigated with a phased array technique in order to quantitatively separate the jet mixing noise from any other sources of noise that might be present, e.g. nozzle lip noise. Although such techniques have been available for many years, successful application to LES and DNS data has often been prevented by a combination of two factors: the lack of far-field data and the lack of sufficient time history (record length) to obtain adequate accuracy in the required statistical quantities, such as the spectral density and cross-spectral density of the radiated unsteady pressure. In the current study, a polar array of virtual microphones has been located at 20 jet nozzle diameters from the nozzle, having ascertained that the field is obeying the inverse square law at this radius, without any significant reflections from the DNS domain boundaries. The DNS record length, in excess of 350 convective time units, permits the number of averages for the cross-spectral density to be of order 100 with a filter separation, expressed as a Strouhal number based on the nozzle diameter, of 0.1. A microphone spacing of 0.5 degrees is used over an aperture of 140 degrees to 10 degrees to the jet axis (i.e. 40 to 170 degrees to the 'intake' axis), although even finer spacing is available if required.

A recently developed phased array code, named AFINDS and described in more detail in Tester *et al.*[12], was applied to the DNS data, decomposed into its constituent azimuthal modes. This phased array technique, in addition to the usual imaging capability, is able to extract a 'noise source breakdown' using a model in which the jet mixing noise is represented by one or two distributed axial sources. In the current work only one jet-noise source is used plus two nozzle-based sources, each represented by a simple point-source.

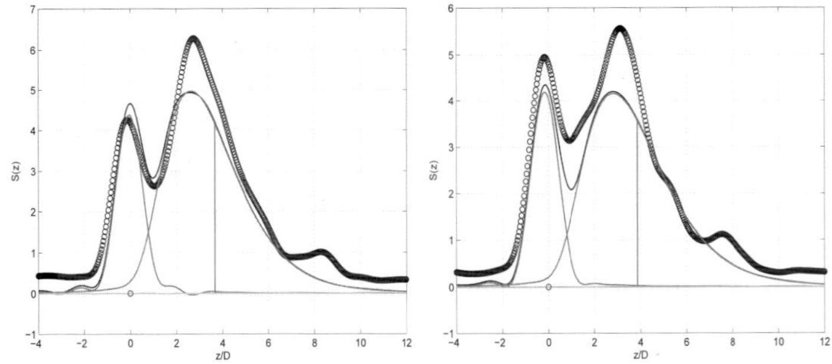

Fig. 3. Axial source image at 90° of DNS data and AFINDS source breakdown; cases M84c2nL (left) and M84c2L (right); $St_D = 1.0$, $m = 0$; open circles represent image derived from DNS data, solid black line is the AFINDS fit to data, the line with solitary peak at origin is one nozzle-based source, the line with two maxima the main jet mixing noise source and the line exhibiting the smallest amplitudes the second nozzle-based source.

In Fig. 3 the noise source breakdowns of the axisymmetric mode $m = 0$

are shown for cases M84c2L and M84c2nL at a Strouhal number $St_D = 1$. The images are in the form of source intensity per unit axial length, $S(z)$, along the jet centreline, the area under each curve being proportional to the radiated sound intensity of that source at 90°. Only this angle is considered here as the main contribution of the nozzle based sources were found here, whereas the smaller angles were dominated by jet mixing noise and not as strongly affected by internal noise sources.

The AFINDS source breakdown is able to fit the general shape of the source intensity, although it does not succeed in capturing the amplitudes perfectly. The overall source image is very similar for both lined and unlined cases, both in terms of magnitude and shape, confirming our earlier findings that the liner boundary condition does not significantly affect the axisymmetric component ($m = 0$) of the far field noise. For both cases, the source breakdown indicates that the overall far field noise measured at 90° is composed of a nozzle based source, located at the nozzle exit ($z/D = 0$), and the main jet noise source centroid at $z/D \approx 4$. Both these sources exhibit very similar intensity. The second nozzle-based source is located just downstream of the nozzle, but with a negligible amplitude. Importantly in these new simulations, the jet noise source can be clearly identified in the noise source breakdown, in contrast to the earlier DNS data, for which only the nozzle based sources, centred at the nozzle exit could be found at high Mach number[3].

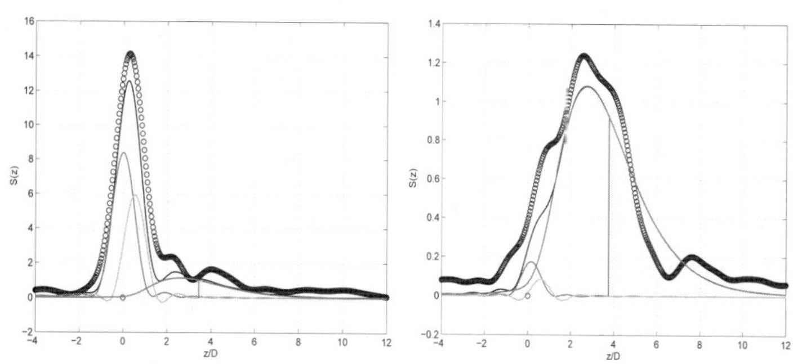

Fig. 4. Axial source image at 90° of DNS data and AFINDS source breakdown; cases M84c2nL (left) and M84c2L (right); $St_D = 1.0$, $m = 1$; lines and symbols as described in Fig. 3.

Figure 3 shows the noise source breakdown for the first azimuthal Fourier mode $m = 1$. In contrast to the axisymmetric mode, the images differ considerably between the lined and unlined cases. Without liner boundary condition, the two nozzle based sources dominate the image, with the first source centred at the nozzle exit and the additional source located downstream of the nozzle. The jet source can be found in the image at the expected position, although at much smaller amplitude. When using the liner, the nozzle based sources become quite small and the source image is dominated by the jet mixing noise source, exhibiting its intensity centroid at $z/D \approx 4$, in good agreement with experimental data at $St_D = 1$. Hence, this is clear evidence that the liner boundary condition is able to significantly reduce nozzle based sources for the higher azimuthal modes and, thus, jet mixing noise can be studied in detail using this high Mach number DNS data set.

4. Conclusions

Direct numerical simulations of fully turbulent flow exiting a long pipe were conducted at target Mach and Reynolds numbers, based on pipe-bulk velocity, of $M = 0.84$ and $Re_j = 8,500$, respectively, for varying co-flow Mach number. To suppress possible internal noise sources emanating from the pipe exit and contaminating the acoustic far field, an acoustic liner model was applied to the interior nozzle walls. In addition, the turbulent inflow generation technique was modified to be free of velocity fluctuations of the axisymmetric mode.

Using a source breakdown code it was possible to elucidate the effect of the acoustic liner model on the contributions from the nozzle based and the jet mixing noise sources. The results obtained from the noise source breakdown suggest, consistent with spectra of individual azimuthal Fourier modes of far field pressure, that the liner only marginally affects the axisymmetric mode while being very effective at reducing nozzle-based noise contributions in the higher azimuthal modes.

References

1. C. Tam, K. Viswanathan, K. Ahuja and J. Panda, The sources of jet noise: experimental evidence, *J. Fluid Mech.* **615**, 253 (2008).
2. R. D. Sandberg, V. Suponitsky and N. Sandham, DNS of fully turbulent jet flows in flight conditions including a canonical nozzle, *AIAA*-2011–2918 (2011).
3. R. D. Sandberg and B. J. Tester, Application of a Phased Array

Technique to Fully Turbulent Subsonic Jet Data Obtained with DNS, *AIAA*-2012–2613 (2012).

4. B. J. Tester and R. D. Sandberg, Application of a phased array technique to DNS-generated turbulent subsonic jet data: source identification and comparison with an analytic model, *AIAA*-2013–2235 (2013).
5. S. Olivetti, R. D. Sandberg and B. J. Tester, Direct Numerical Simulation of Turbulent Flow with an Impedance Condition, *J. Sound and Vibration*, **344**, 28 (2015).
6. C. K. Tam and L. Auriault, Time-domain impedance boundary conditions for computational aeroacoustics, *AIAA J.* **34**, 917 (1996).
7. R. D. Sandberg, DNS of Turbulent Round Jets Using Acoustically Lined Canonical Nozzles, *Proceedings of the AFMC, Paper 143*, 19th Australasian Fluid Mechanics Conference (2014).
8. R. D. Sandberg, An axis treatment for flow equations in cylindrical coordinates based on parity conditions, *Comput. and Fluids* **49**, 166 (2011).
9. E. Touber and N. Sandham, Large-eddy simulation of low-frequency unsteadiness in a turbulent shock-induced separation bubble, *Theor. Compt. Fluid Dyn.* **23**, 79 (2009).
10. R. Sandberg, V. Suponitsky and N. Sandham, DNS of compressible pipe flow exiting into a coflow, *Int. J. Heat Fluid Fl.* **35**, 33 (2012).
11. R. D. Sandberg and N. D. Sandham, Nonreflecting zonal characteristic boundary condition for direct numerical simulation of aerodynamic sound, *AIAA J.* **44**, 402.
12. B. J. Tester and T. Gabard, G. ad Ricoup, Extracting engine noise source levels from phased array measurements with improved internal source models and evaluation against DNS-generated jet noise data, *AIAA*-2012–2272 (2012).

Decomposition of Radiating and Non-Radiating Linear Fluctuating Components in Compressible Flows

Paul Stegeman*

LTRAC, Dept. of Mech and Aero Eng., Monash University, Clayton, Victoria 3800, Australia
E-mail: paul.stegeman@monash.edu

Andrew Ooi

Department of Engineering, The University of Melbourne, Parkville, Victoria, Australia

Julio Soria

LTRAC, Deptartment of Mechanical and Aerospace Engineering, Monash University, Clayton, Victoria 3800, Australia
&
Department of Aeronautical Engineering King Abdulaziz University Jeddah, Kingdom of Saudi Arabia

Decomposition of radiating and non-radiating fluctuating components are important to the study of sound generation and acoustic feedback mechanisms present in a variety of free shear flows. Prior work primarily focused on the development of a decomposition operator for subsonic flows based on the acoustic dispersion relationship which is invariant to the position in time and space, such as found in Goldstein and Sinayoko *et al.* This paper presents the development of an operator that approximately decomposes linear fluctuating flow variables into their radiating and non-radiating components locally in time and space using the wavelet and curvelet transforms as a basis. The operator has been applied to a large-eddy simulation of a highly compressible under-expanded impinging jet to study its radiating and non-radiating fields.

Keywords: Compressible jet flows; Acoustic decomposition; Large-Eddy simulation; Curvelets

1. Flow Decomposition

Decomposition of radiating and non-radiating fluctuating components are important to the study of sound generation and acoustic feedback mechanisms present in a variety of free shear flows. Prior work primarily focused on the development of a decomposition operator for subsonic flows based on the acoustic dispersion relationship which is invariant to its position in time and space[1–3].

The compressible Navier-Stokes equations for a perfect gas flow are given as

$$\frac{\partial \rho}{\partial t} + \nabla \cdot (\rho \mathbf{u}) = 0, \tag{1}$$

$$\frac{\partial \rho \mathbf{u}}{\partial t} + \nabla \cdot (\rho \mathbf{u} \otimes \mathbf{u} + p\mathbb{I} - \mu \mathbb{T}) = 0, \tag{2}$$

and

$$\frac{\partial \rho e}{\partial t} + \nabla \cdot (\mathbf{u}(\rho e + p) - k\nabla T - \mu \mathbb{T} \cdot \mathbf{u}) = 0. \tag{3}$$

where the deviatoric stress tensor ($\mathbb{T}$) is defined as

$$\mathbb{T} = \left[\nabla \mathbf{u} + (\nabla \mathbf{u})^T + \left(\frac{\mu_b}{\mu} - \frac{2}{3}\right)(\nabla \cdot \mathbf{u})\,\mathbb{I}\right]. \tag{4}$$

The equation of state is

$$p = \rho R T,. \tag{5}$$

ρ is the density, $\mathbf{u}$ is the velocity vector, $\rho e = 1/2\,\|\mathbf{u}\|^2 + p/(\gamma - 1)$ is the total energy per unit volume, p is the pressure, T is the temperature, μ is the viscosity, μ_b is the bulk viscosity, k is the thermal conductivity, R is the gas constant and $\mathbb{I}$ is the identity tensor. Performing a plane-wave analysis on the linearised form of equations (1) to (3) yields the dispersion relations for hydrodynamic (non-radiating) and acoustic (radiating) waves in uniform flow are defined respectively as

$$\omega = \bar{\mathbf{u}} \cdot \mathbf{k} \tag{6}$$

and

$$\omega = \bar{\mathbf{u}}x \cdot \mathbf{k} \pm \bar{a}\sqrt{\mathbf{k} \cdot \mathbf{k}} \tag{7}$$

where a is the speed of sound and ω and $\mathbf{k}$ are the temporal frequency and wavenumber vector of the fluctuations, respectively[4]. For spatially non-uniform flow, equations 6 and 7 approximate the dispersion relations for waves with wavelengths smaller than that of the variation in base flow speed of sound and velocity. In this paper, we do not undertake a analysis of the error in decomposition due to this approximation.

Given equations 6 and 7 the separation between acoustic and hydrodynamic waves in the wavenumber-frequency domain for a given local region of space is $\pm\bar{a}\sqrt{\mathbf{k}\cdot\mathbf{k}}$. This separation allows an approximate decomposition of a flow variable into its mean and fluctuating components

$$q\left(\mathbf{x},\mathbf{t}\right)=\bar{q}\left(\mathbf{x},\mathbf{t}\right)+q'\left(\mathbf{x},\mathbf{t}\right) \tag{8}$$

where q is the flow variable to be decomposed, $\bar{\cdot}$ denotes the temporal mean and $\cdot'$ denotes the fluctuating component. Following Sinayoko *et al.*[3] an operator $\mathscr{L}$ is defined that decomposes $\mathbf{q}'$ into its radiating components

$$q'_{\text{ac}}=\mathscr{L}q' \tag{9}$$

and non-radiating components

$$q'_{\text{hd}}=\left(\mathscr{I}-\mathscr{L}\right)q' \tag{10}$$

using the dispersion relations given by equations (6) and (7). For incompressible flows $\mathscr{L}$ has no dependency on $\mathbf{x}$ due to the assumptions that $\mathbf{u}\ll a$ and a is constant, Therefore, it is equivalent to the operator defined by Sinayoko *et al.*[3]. However in the compressible flow case $\mathbf{u}$ may be comparable to a such that the local mean flow effect on the dispersion of hydrodynamic and acoustic waves must now be taken into account. Therefore the decomposition operator must be as compact as possible, limited by the uncertainty principle, in both the space-time and wavenumber-frequency domain.

Temporal decomposition is achieved through the dyadic wavelet transformation

$$c^{\text{temporal}}\left(\mathbf{x},t,i\right)=\int_{\mathbb{R}}q\left(\mathbf{x},\tau\right)\overline{\varphi_{i,\tau}^{\text{wavelet}}\left(\mathbf{x},t\right)}d\tau \tag{11}$$

for a given temporal scale 2^i where the wavelet basis function is defined as

$$\varphi_{i,\tau}^{\text{wavelet}}\left(\mathbf{x},t\right)=\frac{1}{\sqrt{2^i}}\varphi^{\text{motherwavelet}}\left(\frac{t-\tau}{2^i}\right) \tag{12}$$

in which the Haar mother wavelet is defined as

$$\varphi^{\text{motherwavelet}}\left(t\right)=\begin{cases}1 & 0\leq t<1/2\\-1 & 1/2\leq t<1\\0 & \text{otherwise}\end{cases} \tag{13}$$

The original variable may then be reconstructed as

$$q\left(\mathbf{x},t\right)=\sum_i\int_{\mathbb{R}}c^{\text{temporal}}\left(\mathbf{x},\tau,i\right)\varphi_{i,\tau}^{\text{wavelet}}\left(\mathbf{x},t\right)d\tau \tag{14}$$

A spatial decomposition of the temporal wavelet coefficients (c^{temporal}) is then performed using curvelets

$$c(\mathbf{x}, t, i, j, l) = \int_{\mathbb{R}^N} c(\mathbf{y}, t, i)\, \overline{\varphi_{j,l,\mathbf{y}}^{\text{curvelet}}(\mathbf{x})} d\mathbf{y} \tag{15}$$

for a given scale of 2^{-j} and orientation θ_l in wavenumber space where the basis function is defined as[5]

$$\varphi_{j,l,\nu}(\mathbf{x}) = \varphi_j\left(R_{\theta_l}\left(\mathbf{x} - \mathbf{y}^{(j,i)}\right)\right). \tag{16}$$

The original temporal decomposition can be reconstructed independently as

$$c^{\text{temporal}}(\mathbf{x}, \tau, i) = \sum_{j,l,\mathbf{y}} c(\mathbf{x}, t, i, j, l)\, \varphi_{j,l,\mathbf{y}}^{\text{curvelet}}. \tag{17}$$

Where $c(\mathbf{x}, t, i, j, l)$ represent the projection of the flow field variable onto a set of wavelet and curvelet basis functions with position $\mathbf{x}$, time t, temporal scale 2^i, spatial scale 2^j and orientation l. The combination of the wavelet and curvelet basis functions provide a compact window in the wavenumber-frequency domain with a frequency width of 2^{-i}, wavenumber width of 2^{-j} by $2^{-j/2}$ and an orientation l.] Therefore a decomposition operator $\mathscr{L}$ can be constructed by setting all values of $c(\mathbf{x}, t, i, j, l)$ to zero where the acoustic dispersion relation does not intersect the basis function windows in the wavenumber-frequency domain

$$\mathscr{L}q := \sum_i \int_{\mathbb{R}} \left[\sum_{j,l,\mathbf{y}} g(\mathbf{x}, t, i, j, l)\, \varphi_{j,l,\mathbf{y}}^{\text{curvelet}}\right] \varphi_{i,\tau}^{\text{wavelet}}(\mathbf{x}, t)\, d\tau \tag{18}$$

where

$$g(\mathbf{x}, t, i, j, l) = \begin{cases} c(\mathbf{x}, t, i, j, l) & \text{intersects} \\ 0 & \text{otherwise.} \end{cases} \tag{19}$$

That is, the operator $\mathscr{L}$ preserves only the wavenumber-frequency content near the surface defined approximately by the uniform acoustic dispersion relation at the given position in the flow. An overview of the decomposition process is shown in Fig. 1.

This technique not only allows for a decomposition of compressible flows but it allows for an efficient decomposition over a range of frequencies rather than a single frequency as in Sinayoko *et al.*[3].

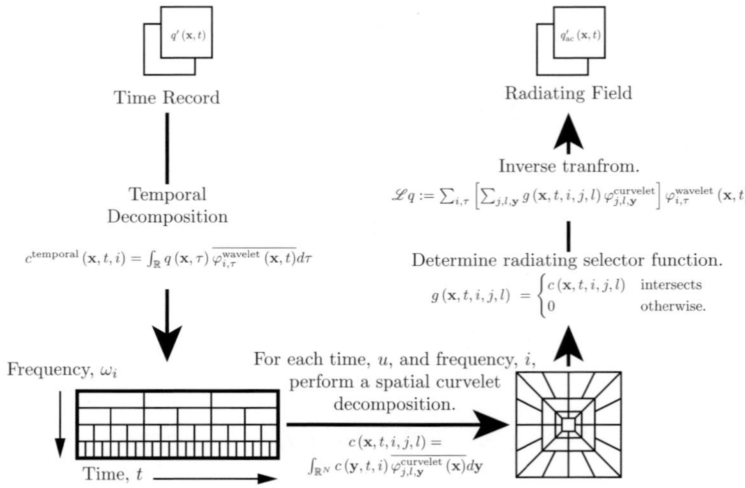

Fig. 1. An overview of the decomposition process. Where $\mathbf{x}$ is the position, t the time 2^i the temporal scale, 2^j the spatial scale and l the orientation.

2. Effects of the Local Mean Flow

The operator's ability to distinguish between the hydrodynamic (Eq. 6) and the acoustic (Eq. 7) dispersion relations is also limited at lower wavenumbers and in regions of the wavenumber-frequency space when the local flow speed is supersonic. Figure 2 illustrates the uniform dispersion relations as a function of the local Mach vector on the separation between the acoustic and hydrodynamic waves in wavenumber-frequency space for a 2D compressible flow. Figures 2(a), (b), (c) and (d) are representative of no local flow, subsonic local flow, nearly supersonic flow and supersonic flow, respectively. This shows that an increase in the local flow speed decreases the separation of the acoustic and hydrodynamic waves at larger spatial and time scales. Thus the decomposition operator is unable to distinguish between the acoustic and hydrodynamic waves at these larger scales in high speed regions of the flow. However, this does not however, detract from its ability to separate the smaller spatial and temporal scales.

3. Application to an Underexpanded Impinging Jet

An in-house developed three dimensional high fidelity compressible LES solver for non-uniform structured meshes was used to generate the underexpanded impinging jet database used to test the decomposition operator.

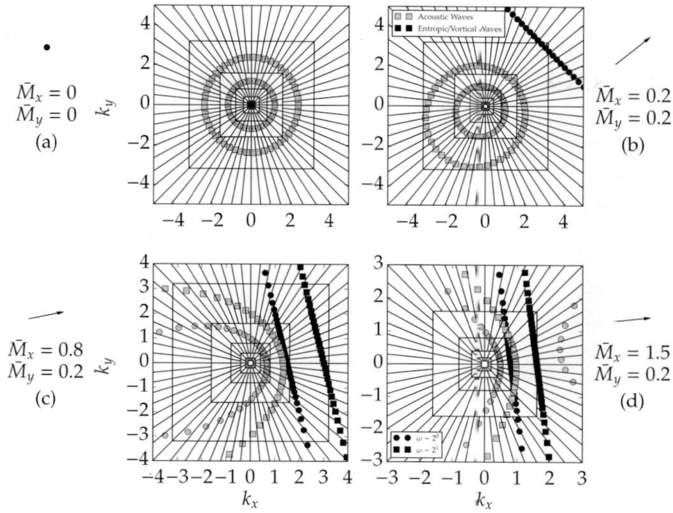

Fig. 2. Effects of the local velocity on the separation between the acoustic and hydrodynamic waves in wavenumber-frequency space for a 2D compressible flow. Grey and black symbols denote the acoustic (Eq. 7) and hydrodynamic (Eq. 6) dispersion relations, respectively. Circle and cross symbols denote two different temporal frequencies with the crosses denoting a frequency that is twice that of the circles.

Important parameters pertaining to this type of flow is the stand-off distance (z/d) between the nozzle outlet and the impingement surface, nozzle pressure ratio (NPR), exit Mach number (M_e), and jet Reynolds number (Re_D) which for this simulation is $z/d = 5$, $NPR = 3.4$, $M_e = 1$, and $\mathrm{Re}_D = 50,000$. For spatial differentiation the hybrid solver employs a sixth order central finite difference scheme for the smooth regions and a fifth order weighted essentially non-oscillatory scheme with local Lax-Friedrichs flux splitting in the discontinuous regions. Temporal integration is performed using a fourth order five step Runge-Kutta scheme. The sub-grid scale (SGS) terms were computed using Germano's dynamic model with the adjustments made by Lilly[6]. The domain consists of approximately 30 million nodes with the spatial extent of $12d$ in the radial direction. Locally one-dimensional inviscid compressible boundary conditions are utilised for the outflow regions[7].

Axisymmetric temporal mean fields of the velocity magnitude and velocity fluctuation magnitude are shown in Fig. 3. Due to length constraints the acoustic spectrum is not shown. However the dominant instability of the jet has a frequency of $St = 0.5$ with the first harmonic at $St = 1.0$ which is in line with prior experimental results at the given stand-off distance and a similar nozzle pressure ratio[8].

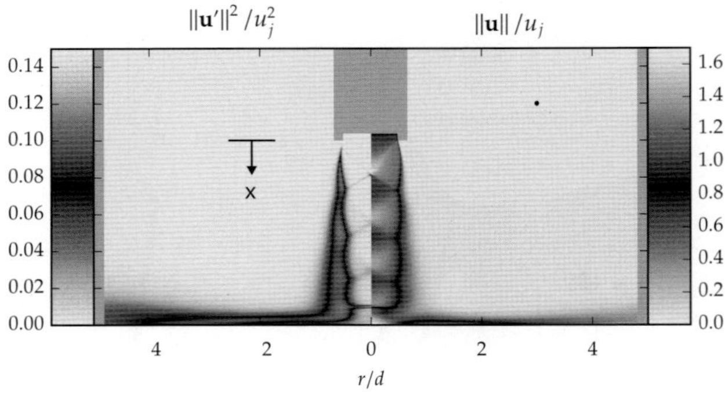

Fig. 3. Mean of the velocity magnitude (right half) and mean square of the velocity fluctuations (left half).

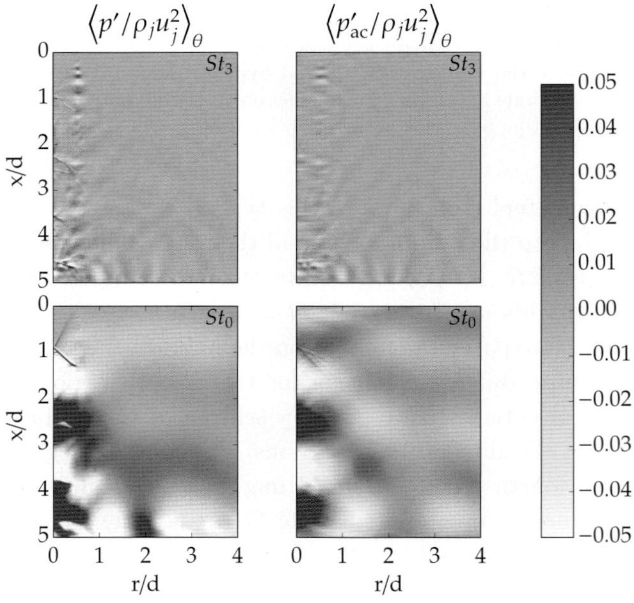

Fig. 4. Effects of the spatial decomposition portion of the operator for the St_0 and St_3 windows. Left hand sub-plots are of the total fluctuating pressure and right hand sub-plots are of the acoustic fluctuating pressure.

Due to the aforementioned the decomposition of the jet time series is only performed for frequencies above $St = 0.24$. Wavelet coefficients corre-

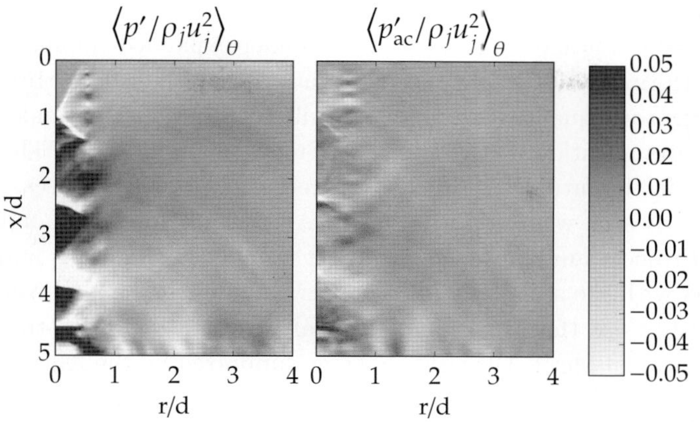

Fig. 5. Instantaneous plot of the total fluctuating pressure (left) and acoustic fluctuating pressure (right).

sponding to frequencies below $St = 0.24$ are thus set to zero. This provides four temporal dyadic windows of $0.24 \leq St < 0.48$, $0.48 \leq St < 0.96$, $0.96 \leq St < 1.92$, and $1.92 \leq St < 3.84$ which are denoted by St_0, St_1, St_2, and St_3, respectively. The effects of the spatial decomposition portion of the operator for the St_0 and St_3 windows are shown in Fig. 4. In the higher frequency window, St_3 the linear region of the shear layer near the nozzle lip and the wall jet region the results are qualitatively similar to those of the subsonic jet studied by Sinayoko *et al.*[3]. As expected near the nozzle lip the fluctuating acoustic wavelength is roughly twice that of the total fluctuating wavelength. Acoustic features such as the waves generated at the first shock reflection point are also more pronounced in the acoustic fluctuation field. In the lowest frequency window the decomposition does not produce accurate results due to the non-linear components of the acoustic dispersion relation for large scale acoustic waves. Further analysis needs to be undertaken to determine the magnitude of the error at these larger scales. Finally, an instantaneous fluctuating pressure field with the corresponding acoustic fluctuating pressure field generated from the full decomposition is presented in Fig. 5.

4. Conclusion

An efficient methodology for decomposing acoustic (radiating) and non-acoustic (non-radiating) flow fields in numerical or experimental data has been outlined. A qualitative comparison of the total fluctuating pressure and acoustic fluctuating pressure for a under-expanded impinging jet has been performed. For brevity a quantitative analysis has not been shown, and a rigorous quantitative analysis will require a better understanding of the decomposition methodology and relevant parameters. This may be achieved in the future via the development of a priori test cases.

This research was supported under Australian Research Council's Discovery Projects funding scheme (project number DP1096474) and was undertaken with the assistance of resources provided at the NCI National Facility systems at the Australian National University through the National Computational Merit Allocation Scheme supported by the Australian Government.

References

1. M. E. Goldstein, A generalized acoustic analogy., *J. Fluid Mech.* **488**, 315 (2003).
2. M. E. Goldstein, On identifying the true sources of aerodynamic sound, *J. Fluid Mech.* **526**, 337 (2005).
3. S. Sinayoko, A. Agarwal and Z. Hu, Flow decomposition and aerodynamic sound generation, *J.f Fluid Mech.* **668**, 335 (2011).
4. P. C. Stegeman, M. E. Young, J. Soria and A. Ooi, Analysis of the anisotropy of group velocity error due to spatial finite difference schemes from the solution of the 2d linear Euler equations., *International J. Numerical Methods in Fluids* **71** (2013).
5. L. Demanet, Curvelets, wave atoms, and wave equations, PhD thesis, California Institute of Technology (2006).
6. D. Lilly, A proposed modification of the germano subgrid-scale closure method *Phys. Fluids* **4**, 633 (1992).
7. T. Poinsot and S. K. Lele, Boundary conditions for direct simulations of compressible viscous flows *J. Comput. Phys.* **101**, 104 (1992).
8. B. Henderson, The connection between sound production and jet structure of the supersonic impinging jet, *The J. the Acoustical Society of America* **111**, 735 (2002).

Toward Control of Compressible Shear Flows: Investigation of Possible Flow Mechanisms

Gaurav Kumar, Rebecca Bertsch, Vishnu Venugopal and Sharath S. Girimaji*

Department of Aerospace Engineering, Texas A&M University,
College Station, TX 77843, USA
**E-mail: girimaji@tamu.edu*
http://engineering.tamu.edu/aerospace

We aim to demonstrate a proof-of-concept of thermodynamics-acoustics-based strategies for controlling the growth of instability in high-speed compressible shear flows. While compressible flows pose many challenges, they also offer novel opportunities for flow control. We show that the equipartition of energy between the potential energy of pressure field and dilatational kinetic energy can be exploited for design of novel thermodynamics-based flow control strategies.

Keywords: Decaying turbulent flows, homogeneous shear flows

1. Introduction

Design of control strategies for high speed and highly compressible flows, especially hypersonic flows, poses novel challenges. Compressibility introduces new flow-physics rendering it more complex than incompressible flows: (i) relaxation of the incompressibility constraint, $\nabla \cdot \mathbf{u} = 0$, (ii) introduction of additional degrees of freedom due to dilatational fluctuations, especially shocks, (iii) change in the behavior/action of pressure, and (iv) appearance of additional interaction mechanisms such as pressure-dilatation interactions especially in regions of shocks and shocklets. On the other hand, compressibility also offers new avenues for novel control strategies. In compressible flows, the acoustic timescale becomes comparable to the flow timescale, enabling control strategies based on pressure/acoustics. Also, a strong coupling between energy and momentum equation permits thermodynamics-based control strategies. Important physical tendencies such as the equipartition of energy between the dilatational kinetic energy and potential energy resident in pressure fluctuations can be used in design of control mechanisms. The aim of the current work is to demonstrate a proof-of-concept for thermodynamics/acoustics based strategies for flow control in hypersonic applications.

One of the prominent features of compressibility is its stabilizing effect on the growth of perturbations. In experiments of high speed mixing layers, the spreading rate (normalized by incompressible spreading rate) decreases with increasing relative Mach number. Stabilizing effect of compressibility has also been observed in experiments of high speed (hypersonic) boundary layer transition wherein the transition zone is elongated in comparison to incompressible flows[1–4]. There is clear evidence that velocity-pressure interactions play a prominent role in this region. In simple canonical flows, like homogeneous shear flows, wherein the underlying physics is very similar to that in mixing and boundary layers, a similar stabilizing effect is observed. The three distinct stages of perturbation growth shown in Fig. 1(a) are: **Stage-1**: pressure-released phase where pressure effects are negligible compared to inertia and growth rate (of kinetic energy) is faster than in incompressible case in shear time units; **Stage-2**: acoustic/wave character stage where acoustic/wave nature of pressure is evident and growth rate slows down and eventually becomes lower than incompressible case; **Stage-3**: final stage where the growth rate of kinetic energy is similar to the incompressible counterpart. In Stage-2, where acoustic timescale becomes comparable to flow timescale, the dynamics is dominated by acoustic waves. Therefore pressure plays a critical role in the evolution of kinetic energy in Stage-2 manifesting via the pressure-strain correlation and pressure dilatation terms.

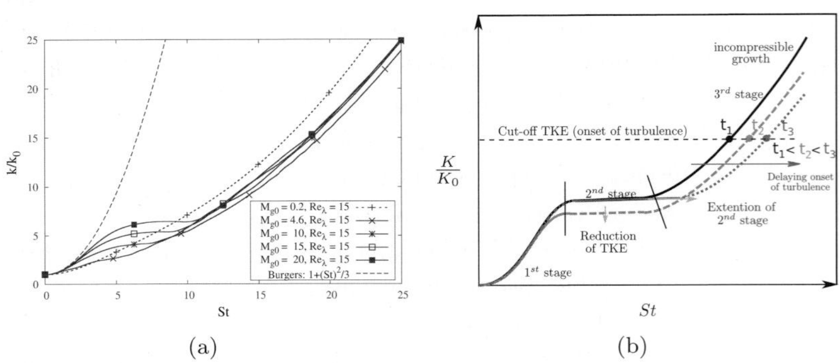

Fig. 1. Evolution of normalized kinetic energy (k/k_0) for collective behavior in shear flows in: a)shear time, b) schematic diagram showing different ways of delaying the onset of turbulence.

The key role played by pressure in Stage-2 provides new avenues for the formulation of novel flow control strategies. Such control strategies can have foundations in the energy redistribution dynamics and exploit the strong coupling between momentum and energy balance equations. In Fig. 1(b), we explore different ways of controlling Stage-2 for delaying the onset of turbulence. In compressible homogeneous shear flows, an equipartition of energy is observed between the dilatational kinetic energy and the potential energy of fluctuations. The so-called potential energy of turbulence that resides in the fluctuating dilatational pressure ($\overline{p'p'}$) and interacts with dilatational kinetic energy ($\overline{u'^d u'^d}$) through the pressure dilatation term $\overline{p'd'}$. In Fig. 2 we describe the equipartition and ways in which it can be exploited for controlling the onset of turbulence through a *two-tank* analogy. Compressibility manifesting through pressure-dilatation term is like a valve controlling the flow between two separate tanks containing potential energy of pressure (fluctuating internal energy) and dilatational kinetic energy. In incompressible flows, the valve is fully closed and there is no exchange of energy. In highly compressible regimes, the valve is fully open, thus allowing a free exchange of energy between the two tanks. The pressure-dilatation works to maintain equal level of energy in both tanks. Now, if we devise a control mechanism to drain energy from either one of the tanks, the level in other tank would also come down. The internal energy and therefore potential energy of turbulence can be removed through endothermic reactions while kinetic energy can be reduced through the body forces. We can mimic the effect of chemical reactions draining the internal energy through a change in transport coefficient namely, the Prandtl number. When internal energy is dissipated, $\overline{p'p'}$ decreases. The pressure dilatation transfers energy from kinetic mode ($\overline{u'^d u'^d}$) to compensate the reduction in $\overline{p'p'}$. This leads to a reduction in dilatation kinetic energy. Since, the production of turbulence in homogeneous shear is highly dependent on the u'_2 component of velocity field, a reduction in dilatational kinetic energy reduces the overall kinetic energy and its subsequent growth.

The objective of this study is to demonstrate thermodynamics-based novel flow control strategies in highly compressible shear flows and provide a modal-analysis-based insight in the flow physics. Towards this end, we perform direct numerical simulations to study the effect of change in Prandtl number on the growth of kinetic energy especially in acoustic dominated second stage.

2. Governing Equations and Numerical Setup

We perform DNS for studying the evolution of turbulent fluctuations in homogeneous shear at different Prandtl numbers. The parameters character-

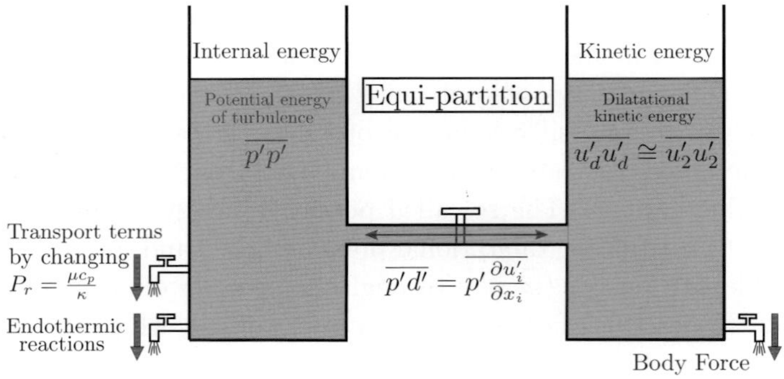

Fig. 2. Schematic for possible control mechanisms which are based on the 'equipartition of energy'.

izing a homogeneous shear flow setup are the initial gradient Mach number (M_{g0}), initial turbulent Mach number (M_{t0}) and initial Taylor micro-scale Reynolds number ($Re_{\lambda 0}$) defined as:

$$M_{g0} = \frac{Sl_0}{\sqrt{\gamma R T_0}}, \quad M_{t0} = \frac{\overline{u'_i u'_i}}{\sqrt{\gamma R T_0}}, \quad Re_{\lambda 0} = \frac{\overline{u'_i u'_i}}{\mu \sqrt{\omega' \omega'}} \tag{1}$$

where S is the shear-rate of the mean flow field, l_0 is the relevant lengthscale of the problem. The gradient of the mean velocity field is given by $\bar{U}_{i,j} = S\delta_{i1}\delta_{2j}$. The lengthscale l_0 is related to initial integral lengthscale of the fluctuating field u'.

The Navier-Stokes equations governing the flow are

$$\frac{\partial \rho}{\partial t} + \frac{\partial(\rho u_i)}{\partial x_i} = 0, \tag{2}$$

$$\frac{\partial(\rho u_i)}{\partial t} + \frac{\partial(\rho u_i u_j + p\delta_{ij})}{\partial x_j} = \frac{\partial \sigma_{ij}}{\partial x_j}, \tag{3}$$

$$\frac{\partial E}{\partial t} + \frac{\partial[(E+p)u_i]}{\partial x_i} = \frac{\partial(\sigma_{ij} u_j)}{\partial x_i} - \frac{\partial}{\partial x_i}\left(\kappa_t \frac{\partial T}{\partial x_i}\right). \tag{4}$$

The viscous stress tensor σ_{ij} is given by a constitutive relation

$$\sigma_{ij} = \mu \left[\frac{\partial u_i}{\partial x_j} + \frac{\partial u_j}{\partial x_i} - \frac{2}{3}\delta_{ij}\frac{\partial u_k}{\partial x_k} + \eta \delta_{ij}\frac{\partial u_k}{\partial x_k}\right], \tag{5}$$

where μ is the dynamic viscosity coefficient, κ_t is the thermal conductivity, η is the bulk viscosity and γ is the ratio of specific heats.

The evolution equation for turbulent kinetic energy in homogeneous shear flows is derived by Favre averaging[5,6] the instantaneous momentum equation (3):

$$\frac{dK}{dt} = P - \epsilon_s - \epsilon_c + \frac{\overline{p'd'}}{\bar{\rho}}, \tag{6}$$

where K is the Favre averaged turbulent kinetic energy (TKE), P is the production of TKE, ϵ_s is the solenoidal dissipation rate, ϵ_c is the compressible/dilatational dissipation rate and $\overline{p'd'}$ is the pressure dilatation. Here $d' = u'_{i,i}$ is the fluctuating velocity dilatation, $p' = \left(\rho'\overline{T} + \rho'T' + \overline{\rho}T' - \overline{\rho'T'}\right) R$ is the fluctuation in pressure and R is the gas constant.

The computational domain is a cubic box of dimension $l = 2\pi$. The box is discretized into $N_x \times N_y \times N_z$ cells with $N_x = 256, N_y = 512$ and $N_z = 256$ along the x, y and, z directions respectively. Periodic boundary conditions are applied for boundaries normal to the stream-wise (x_1) and span-wise/stream-normal (x_3) directions. For boundaries normal to the direction of shear (x_2), we apply shear-periodic boundary condition[7–10]. An isotropic velocity field is generated according to the procedure in Kumar *et al.*[11,12]. The initial density and temperature fields are taken to be uniform.

3. Control Strategies

Starting with linearized equation for pressure and transverse velocity in inviscid RDT Bertsch *et al.*[13] derive the following relationship between variance of pressure and transverse (dilatational component of velocity) kinetic energy for homogeneous shear flows:

$$F_{equi} \equiv \frac{\overline{u'_2 u'_2}}{\overline{u'_2 u'_2} + \overline{p'p'}/(\bar{p}\gamma\bar{\rho})} \approx \frac{1}{2} \quad \text{at late } St \text{ times}, \tag{7}$$

where we refer to F_{equi} as the equipartition function. In broad-band DNS, equipartition is typically seen beyond St=0.5. Similar partitioning tendency has been observed previously in DNS of decaying isotropic compressible turbulence and of compressible shear turbulence by Sarkar *et al.*[14] and Sarkar *et al.*[15], respectively.

In this study we will highlight the effect of change of Prandtl number on collective behavior. Prandtl number (P_r) is defined as

$$P_r = \frac{\mu c_p}{\kappa_t}, \tag{8}$$

where c_p is the specific heat at constant pressure and κ_t is the thermal conductivity. Prandtl number is the ratio of viscous diffusion rate to thermal diffusion rate. A decrease in Prandtl number enhances the transport

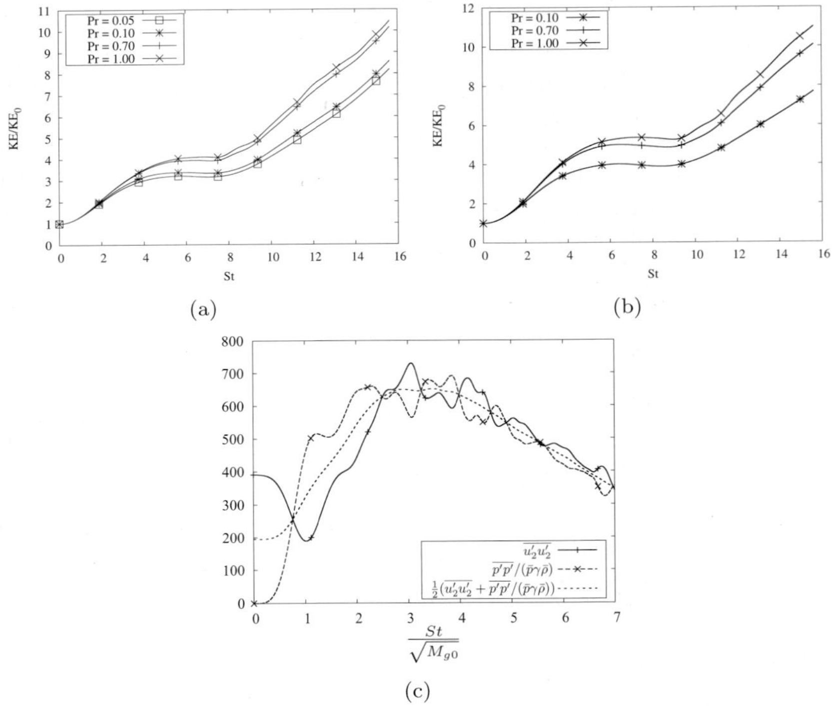

Fig. 3. Effect of Prandtl number $P_r = (0.05, 0.1, 0.7, 1.0)$ on the evolution of kinetic energy in homogeneous shear flows: (a) $M_{g0} = 10, Re_{\lambda 0} = 15$, (b) $M_{g0} = 15, Re_{\lambda 0} = 15$; (c) dilatational kinetic energy $(\overline{u_2'u_2'})$ and potential energy of turbulence $(\overline{p'p'}/(\overline{p'}\gamma\overline{p'}))$ for $M_{g0} = 5, Re_{\lambda 0} = 15$.

of internal energy. Transport of internal energy in real flows can be enhanced by processes like endothermic chemical reactions and thermal non-equilibrium. A reduced Prandtl number in a direct numerical simulation mimics the effect of enhanced transport of internal energy and vice versa. An enhanced transport leads to a reduction in mean and fluctuating internal energy. In gases where temperature and pressure are related through a constitutive relation – *the equation of state* – a reduction in temperature fluctuations manifests in an overall reduction in fluctuating pressure. Due to the equipartition of energy, a reduction in $\overline{p'p'}$ is followed by a reduction in dilatational kinetic energy. u_2' is a good approximation of dilatational component of velocity field in homogeneous shear flows at large shear times[16,17]. Therefore a reduction in dilatational kinetic energy primarily reduces u_2'. Production of turbulent kinetic energy $P = -S\widetilde{u_1'u_2'}$ (see

Eq. 6) is directly dependent on the dilatational component u'_2. A reduction in production leads to a slow-down in the growth of kinetic energy and delays the onset of a fully turbulent flow. If a delay in kinetic energy growth can be demonstrated, such a result would constitute proof of concept of this flow strategy.

3.1. *Results and discussion*

To examine the control concept, we study the effect of Prandtl number on the growth of kinetic energy in homogeneous shear flows. The initial velocity field is random, incompressible, isotropic and satisfies a specified energy spectra $E(\kappa) = A\kappa^4 e^{-B\kappa^2}$. Evolution of kinetic energy in DNS with different Prandtl numbers $P_r = (0.05, 0.1, 0.7, 1.0)$ is shown in Fig. 3 for initial gradient Mach numbers $M_{g0} = 10, 15$ and $Re_{\lambda 0} = 15$. As the Prandtl number is reduced from 0.7 to 0.1, the evolution of kinetic energy undergoes the following changes: (i) The growth of kinetic energy in the case with lower Prandtl number shows an early departure from the 'Burgers growth'. The growth of kinetic energy in Burgers turbulence, which is the 'pressure-released limit', is given by $(k/k_0 = 1 + (St)^2/3)$. (ii) Growth of kinetic energy in Stage-2, which is dominated by acoustic waves, stabilizes at a lower kinetic energy in lower Prandtl number cases. (iii) The onset of Stage-3 in all Prandtl number cases happens at almost same shear time. Although, a closer look shows that Stage-3 begins at slightly later shear time when the Prandtl number is smaller. (iv) The growth rate of kinetic energy in Stage-3 reduces with a reduction in Prandtl number.

We see that, in lower Prandtl number cases, a reduction in the level of kinetic energy in Stage-2 and a slower growth rate in Stage-3 leads to a significantly lower kinetic energy at late shear times.

As discussed earlier, a reduction in Prandtl number leads to a reduction in $\overline{p'p'}$ due to an increased transport of fluctuating internal energy (see Fig. 3(c)). In Stage-1 ($St < 4$) $\overline{p'p'}$ is higher in lower Prandtl numbers which is attributed to the timescale of the energy equation. At initial times (Stage-1) the flow-timescale is much higher compared to the acoustic-timescale which forces the energy equation to be driven by inertia. As the acoustic time-scale becomes comparable to the flow timescale in Stage-2, the transport terms in energy equation become increasingly important and destroy (diffuse) the fluctuations in temperature. This leads to a faster decay of $\overline{p'p'}$ in Stage-2 and Stage-3 in cases with lower Prandtl number. The reduction in $\overline{p'p'}$ is followed by a simultaneous reduction in $\overline{u'_2u'_2}$. A simultaneous reduction in $\overline{u'_2u'_2}$ and $\overline{p'p'}$ is due to the action of pressure-dilatation which exchanges energy between the internal and kinetic modes. The exchange of energy begins only when the timescale of acoustics becomes comparable to that of the shear.

4. Conclusions

We perform DNS for studying the evolution of fluctuations in homogeneous shear flows subject to different Prandtl numbers. We demonstrate a proof-of-concept of thermodynamics-based strategies for controlling the growth of turbulence in high speed compressible shear flows. We show that the equipartition between the potential energy of pressure and dilatational kinetic energy can be exploited for design of novel acoustic-based flow control strategies (see for example[18]).

In homogeneous shear flows, kinetic energy shows a three-stage growth characterized by the timescale of acoustics and flow field. We find that in DNS with reduced Prandtl number, the kinetic energy: (i) is lower in Stage-2, and (ii) has delayed growth in Stage-3. From analysis we find that the reduction of kinetic energy in Stage-2 is primarily due to the weakening of the dilatational modes. The difference between growth of kinetic energy of small and large Prandtl number cases is significant at increasingly early shear times as the dilatational content of the wave increases. Since the growth in Stage-2 is largely dominated by dilatational/acoustic waves, a reduction in Pr reduces kinetic energy in Stage-2.

References

1. J. Sivasubramanian and H. F. Fasel, Numerical investigation of laminar-turbulent transition in a cone boundary layer at Mach 6, **73** (2012).
2. C. Hader and H. F. Fasel, Numerical investigation of porous walls for a mach 6.0 boundary layer using an immersed boundary method, *AIAA*-3081 (2011).
3. J. Sivasubramanian and H. F. Fasel, Transition initiated by a localized disturbance in a hypersonic flat-plate boundary layer, in *AIAA Aerospace Sciences Meeting* (2011).
4. J. Sivasubramanian and H. F. Fasel, Direct numerical simulation of a turbulent spot in a cone boundary-layer at mach 6, in *AIAA 40th Fluid Dynamics Conference* (2010).
5. A. Favre, The equations of compressible turbulent gases, Tech. Rep., DTIC Document (1965).
6. A. Favre, Statistical equations of turbulent gases, *Problems of hydro-*

dynamics and continuum mechanics: Contributions in honor of the sixtieth birthday of Academician LI Sedov, 14th November 1967, 231 (1969).

7. F. Baron, Macro-simulation tridimensionelle d'ecoulements turbulents cisailles, PhD thesis, Universite Pierre at Marie Curie, Paris (1982).
8. U. Schumann, Algorithms for direct numerical simulation of shear-periodic turbulence, in *Ninth International Conference on Numerical Methods in Fluid Dynamics* (1985).
9. T. Gerz, U. Schumann and S. Elghobashi, Direct numerical simulation of stratified homogenous turbulent shear flows, *J. Fluid Mech.* **200**, 563 (1989).
10. J. Kreuzinger and R. Friedrich, DNS of homogeneous shear flow and data analysis for the development of a four-equation turbulence model, *High performance computing in science and engineering*, 59 (2002).
11. G. Kumar, S. S. Girimaji and J. Kerimo, Weno-enhanced gas-kinetic scheme for direct simulations of compressible transition and turbulence, *J. Comput. Phys.* **234**, 499 (2013).
12. G. Kumar, R. L. Bertsch and S. S. Girimaji, Stabilizing action of pressure in homogeneous compressible shear flows: effect of mach number and perturbation obliqueness, *J. Fluid Mech.* **760**, 540 (2014).
13. R. L. Bertsch, S. Suman and S. S. Girimaji, Rapid distortion analysis of high mach number homogeneous shear flows: Characterization of flow-thermodynamics interaction regimes, *Phys. Fluids (1994-present)* **24**, 125106 (2012).
14. S. Sarkar, G. Erlebacher, M. Hussaini and H. Kreiss, The analysis and modelling of dilatational terms in compressible turbulence, *J. Fluid Mech.* **227**, 473 (1991).
15. S. Sarkar, G. Erlebacher and M. Hussaini, Direct simulation of compressible turbulence in a shear flow, *Theo. Comput. Fluid Dyn.* **2**, 291 (1991).
16. G. Blaisdell, G. Coleman and N. Mansour, Rapid distortion theory for compressible homogenous turbulence under isotropic mean strain, *Phys. Fluids* **8**, 2692 (1996).
17. D. Livescu and C. Madnia, Small scale structure of homogeneous turbulent shear flow, *Phys. Fluids* **16**, p. 2864 (2004).
18. J. S. Jewell, R. M. Wagnild, I. A. Leyva, G. V. Candler and J. E. Shepherd, Transition within a hypervelocity boundary layer on a 5-degree half-angle cone in air/CO2 mixtures, in *51st AIAA Aerospace Sciences Meeting*, **0523** (2013).

Damping Numerical Oscillations in Hybrid Solvers through Detection of Gibbs Phenomenon

V. Kalyana Chakravarthy and D. Chakraborty*

Directorate of Computational Dynamics, Defence Research & Development Laboratory Hyderabad, Telangana 500058, India
**E-mail: debasis_cfd@drdl.drdo.in*

A Gibbs phenomenon detector that is useful in damping numerical oscillations in hybrid solvers for compressible turbulence is proposed and tested. It is designed to function in regions away from discontinuities where commonly used discontinuity sensors are ineffective. Preliminary tests indicate that using it alongside a traditional discontinuity sensor in a hybrid code leads to more effective damping of numerical oscillations while leaving the physical oscillations intact.

Keywords: compressible turbulence; hybrid schemes; numerical oscillations.

1. Introduction

Hybrid solvers for simulating compressible turbulence rely on non-dissipative schemes for capturing turbulence while switching to dissipative schemes in the vicinity of discontinuities[1,2]. The use of dissipative schemes should be localized appropriately. The use of dissipative scheme over a broader region around discontinuities leads to undesirable dissipation of physical oscillations while extreme localization leads to numerical oscillations that persist for long distances. If proper localization is not possible, another mechanism is needed to remove oscillations away from discontinuities.

Oscillations away from discontinuities are not unique to hybrid approaches. Hyperviscosity approaches which are often implemented using central compact schemes resort to higher order spectral/compact/dealisaing filtering[3] sometimes in addition to smoothing of hyperviscosity computed using polyharmonic operators to deal with this problem[3,4]. Approaches based on characteristic filtering built using compact schemes[5] also need additional higher order filtering steps unless the base scheme itself is upwinded[6]. Finally, the explicit filtering approaches also rely on two filters[7] one which is ubiquitous and the other localized. Only pure weighted essentially non-oscillatory schemes (WENO) are immune to this problem but they have been deemed[8,9] too dissipative to capture turbulence accurately.

A simple and effective way of dealing with numerical oscillations in hybrid, explicit finite volume schemes by basing the weights on a Gibbs phenomenon detector in addition to a more traditional discontinuity sensor is presented here along with preliminary validation.

2. Numerical Method

An explicit second order Runge-Kutta scheme is used for temporal integration of the governing equations. The net flux on a cell face is computed as follows :

$$F_h = \theta F_c + (1-\theta)F_u \tag{1}$$

F_c and F_u are the centered and upwind fluxes computed, respectively, using fourth order MacCormack and second order RHLL[10] schemes. The MacCormack scheme is apt because it is the simplest non-dissipative scheme that does not need artificial dissipation or special care like the Rhie-Chow correction[11] for preventing checker board type oscillations on non-staggered meshes. Fourth order spatial accuracy is achieved by interpolating fluxes computed at cell centers to faces using stencils proposed by Genin[1]. F_u is computed using the RHLL scheme developed by Nishikawa and Kitamura[10]. It captures shock and contact discontinuities crisply with out suffering from carbuncle phenomenon or the odd-even instability.

The following discontinuity sensor proposed by Genin[1] is used here because it can detect contact lines in addition to shock waves.

$$\begin{aligned} S_{\phi,i} &= \frac{|\phi_{i+1} - 2\phi_i + \phi_{i-1}|}{|\phi_{i+1} - \phi_i| + |\phi_i - \phi_{i-1}|}, if|\phi_{i+1} - 2\phi_i + \phi_{i-1}| \geq \epsilon_\phi \phi_i \\ &= -S_\phi^{th}, otherwise \end{aligned} \tag{2}$$

$$S_{i+1/2} = max(S_{P,i}, S_{P,i+1}, S_{\rho,i}, S_{\rho,i+1}) \tag{3}$$

The constants ϵ_P and ϵ_ρ for pressure and density fields were set to 0.05 and 0.1 while the corresponding threshold values S_P^{th} and S_ρ^{th} equal 0.5 and 0.25 respectively.

Genin[1] used a Heaviside function of smoothness parameter to arrive at the switch (i.e., $\theta_{i+1/2} = 1 - H(S_{i+1/2})$). An alternate way of hybridization is proposed here by detecting Gibbs phenomenon as follows :

$$S'_{i+1/2} = \frac{(\rho_{i+2} - 2\rho_{i+1} + \rho_i)(\rho_{i+1} - 2\rho_i + \rho_{i-1})}{(\rho_{i+1} - \rho_i)^2} \tag{4}$$

$$\theta = min\left[\frac{1}{1+exp(14S)}, exp(min(2.5S', 0.0))\right] \quad (5)$$

Negative values of S' are indicative of the unphysical Gibbs phenomenon. When it is negative, a non-zero weight for the flux split scheme contribution to the overall flux is needed. So, the switching function is now calculated as a function of both S, the shock or contact discontinuity sensor and S', the Gibbs phenomenon detector. As in other alternate approaches[4,8,9,12,13], there are numerical constants involved in this procedure. The values shown in above equation, however, seem to work in all of test cases including some that have been left out here.

3. Results

Four canonical problems are considered here before testing the proposed modification in a large eddy simulation. Physical oscillations are not present in two of the problems while in others, they are generated due to interaction of shocks with spatially varying density or vorticity fields. The physical oscillations have to captured where necessary without generating numerical ones.

This problem involves a Mach 3 shock moving through a sinusoidally varying density field leaving behind physical oscillations. Predictions of the proposed hybrid method on a 400 point mesh (like in previous studies[1,6,12,14]) are compared to a reference high resolution solution in Fig. 1(a) and to a solution computed with weights based on only the discontinuity sensor in Fig. 1(b). The use of Gibbs phenomenon detector seems to suppress not only two points oscillations but also numerical oscillations of higher wavelength that probably arise as the solver responds to Gibbs phenomenon. The physical oscillations, however, remain unaffected.

A two circular dimensional blast wave problem proposed by Toro[15] is chosen as the second test case. This, in a way, is an axisymmetric counterpart of the Sod's shock tube problem and so additional demonstration on the latter is not needed. Further, it involves a contact discontinuity that is not present in the Noh's test problem[8]. The initial density and pressure are both set to unity inside a circle of radius of 0.4, while they equal 0.125 and 0.1 units outside (gas constant is set to unity). A quarter of the problem is simulated for time of 0.25 units by discretizing a unit square domain with a 400x400 mesh[14] and density isolevels are plotted in Fig. 2(a) in order to facilitate comparison with results from a comparative study of Liska and Wendroff[14]. In this study, all schemes retained the axisymmetric nature

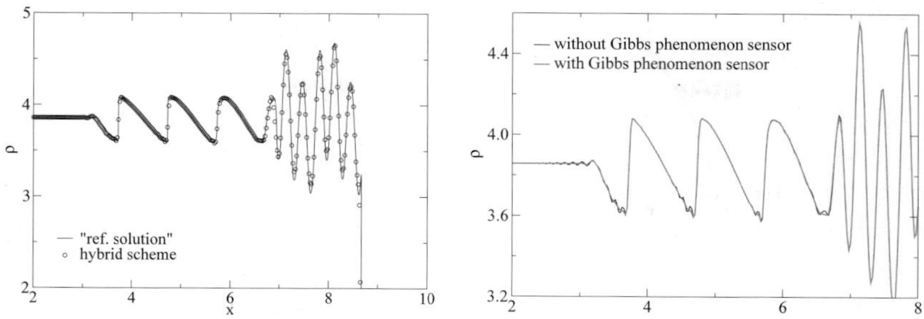

Fig. 1. (a) Comparison of predicted density variation with reference solution (b) Effect of using the Gibbs phenomenon sensor.

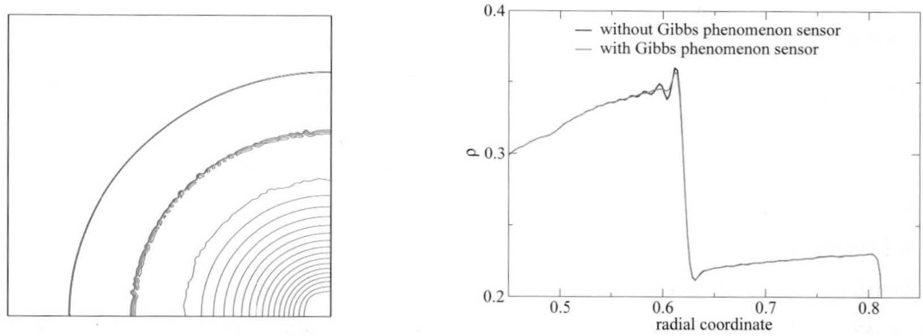

Fig. 2. (a) Predicted density isolevels for the Toro's blast wave problem (b) Effect of using the Gibbs phenomenon sensor on radial density variation.

of the shock wave but not the contact line. An early version of WENO predicted a crisp contact line but was susceptible to numerical instabilities that produced unphysical wiggles when the contact line was aligned to grid directions. High resolution schemes like the piecewise parabolic methods (PPM) also produced unphysical wiggles but with less directional dependency. By comparison, the proposed hybrid scheme predicts a more circular contact line with less directional dependency and smaller unphysical wiggles. Recent WENO schemes can be expected to perform equally well for this problem. The key aspect of this work, however, is a new and simple unphysical sensor. Its effect is illustrated in Fig. 2(b). With out using the unphysical sensor, the numerical oscillations persist away from the discontinuities while using it keeps the post contact line kink in density localized. This kink in density, also a feature of hyperviscosity approaches (as seen in solutions for the Shu-Osher test), though localized, seems unavoidable for now.

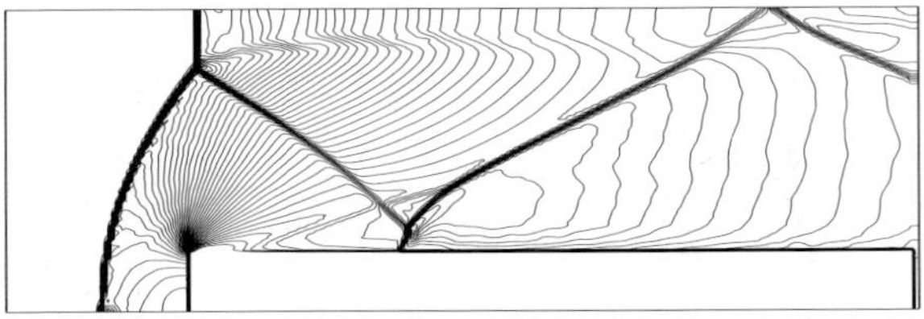

(a) with Gibbs phenomenon detector

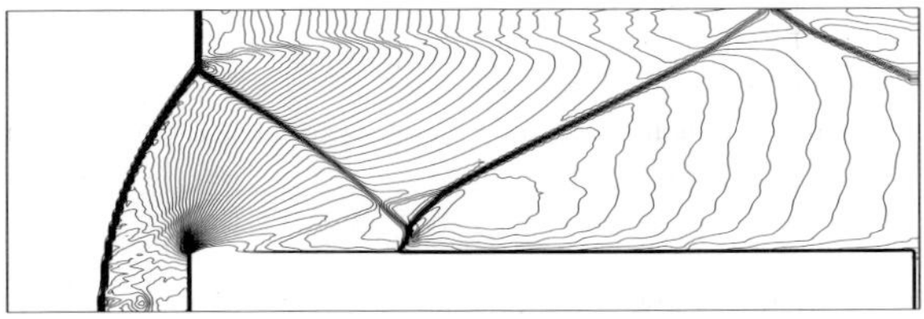

(b) without Gibbs phenomenon detector

Fig. 3. Density isolevels for the Emery test problem.

This flow[16] has many interacting waves and is a good test case for testing hybrid solvers. It involves simulating the response of inviscid Mach 3 flow in a 3x1 sized channel to sudden introduction of a forward facing step at location x=2. Most of the artifacts associated with shock capturing schemes like carbuncle phenomena, odd-even instability would be evident from the density isolevels plot[17]. The density isolevels predicted with and without using the Gibbs phenonemon detector on a 240x80 grid are compared in Fig. 3. With out the Gibbs phenomenon detector, switching generates unphysical oscillations behing the leading shock wave. They are to a large extent suppressed if this detector is activated. It is also to be noted that the two simulations predict identical results around the two slips lines. First one is generated at the shock triple point while the second is generated at the convex corner of the step. There are no unphysical jumps in density field next to the contact lines like in case of the blast wave problem. This problems seems to come up only when the contact lines are normal (and perhaps at a non-zero angle) to the flow direction. If the contact lines are aligned with the flow, there does not seem to be a problem in using the hybrid solvers.

Compressible turbulence involves interaction of shocks with density/entropy fluctuations (as in Shu-Osher problem) and vorticity fields. Shock interaction with a single vortex[5,6] is simulated here on a uniform 250x100 mesh. An isentropic compressible vortex centered initially at

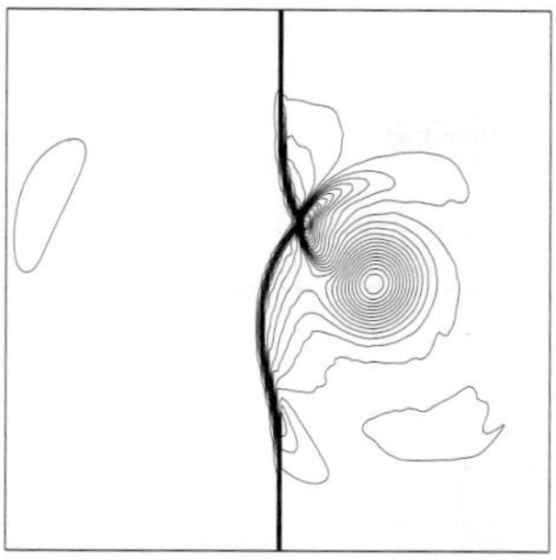

(a) with Gibbs phenomenon detector

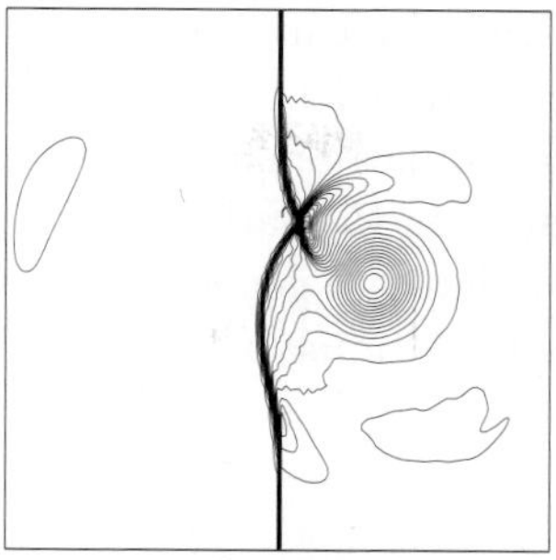

(b) without Gibbs phenomenon detector

Fig. 4. Density isolevels for the shock-vortex interaction problem.

[0.25,0.5] is convected into a stationary Mach 1.1 shock located at at x=1 in a [0,2]x[0,1] domain. The parameters of this test case are available from several previous studies and not repeated here. Comparison of density isolevels in Fig. 4 indicates that post shock oscillations that are generated are effectively suppressed by using the new sensor. Other than these, the density fields are similar to each other and are comparable in terms of captured features to the solutions from earlier studies[5,6]. Lo and coworkers[5] predict a thinner shock by clustering the grid in the axial direction around the shock.

In addition to a separation zone, this flow contains an expansion fan, a slip-line and a shock wave interacting with an unsteady vorticity field. The details of the simulation are provided in Table 1.

Table 1. Parameters for LES of flow over backward facing step.

Parameter	value
Free stream Mach no.	1.5
Total pressure	54psi
Total temperature	300K
Step height(H)	0.25in
Domain size	9.8Hx5.0Hx2.0H
Inflow boundary layer thickness	0.08in
Distance between the inlet and the step location	1.9H
Mesh size	240x140x64

To the extent possible, the LES here are deliberately made identical to one from an earlier study[18] to facilitate comparison. This earlier study used a validated LES solver and a subgrid kinetic energy equation model while a simpler Smagorinsky model[19] is used here. There is another key difference. In the earlier study, flow through a converging-diverging nozzle was simulated to generate the inflow conditions. A compressible version of the digital filtering approach[20] is used here.

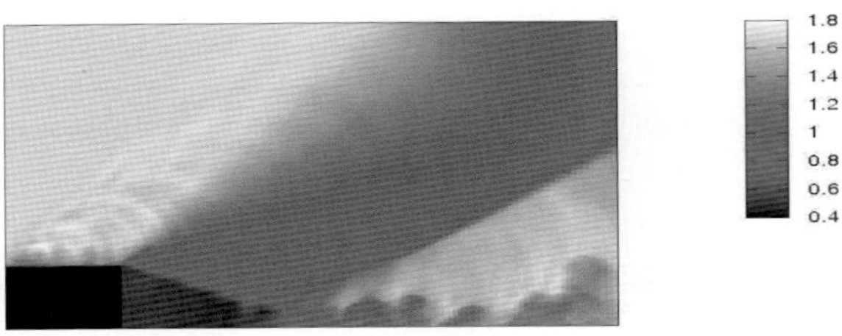

Fig. 5. Density field predicted using LES.

LES performed with and without considering the Gibbs phenomenon detector seem to produce indistinguishable results. The contour plot of instantaneous density field from the former LES is shown in Fig. 5. Unsteady disturbances that propagate along the Mach lines starting at the corner of the inlet and wall boundaries are evident. The transverse velocity at the inlet is set to zero while in reality there is a updraft from the wall. The unsteady fluctuations are also not solutions of the governing equations. These two approximations generate some unphysical disturbances that may affect the flow downstream. The earlier study does not have this problem since a precursor simulation was used the supply the inlet conditions. Other than this artifact, all other features are similar in both studies.

The mean reattachment length turns out to be 2.98H when the Gibbs phenomenon detector is used and 2.97H when it is not. These predictions are close to the value of 3.0H predicted in the earlier study. If uncorrelated random fluctuations are used to specify the inlet turbulence, the reattachment length predicted is about 4.3H. The shear layer instability at the step corner and subsequent breakdown are delayed due to lack of large scale structures in the flow when random noise is used. This delayed reattachment results in a larger recirculation bubble.

The mean velocity predictions from both simulations shown in Fig. 6 very well with those from the reference study. The Gibbs phenomenon detector has no impact probably because the shock is relatively weak and the slip-line is parallel to the flow direction. Other important observation is that the digital filtering approach generates a spectrum of eddies that trigger the shear layer instability as needed.

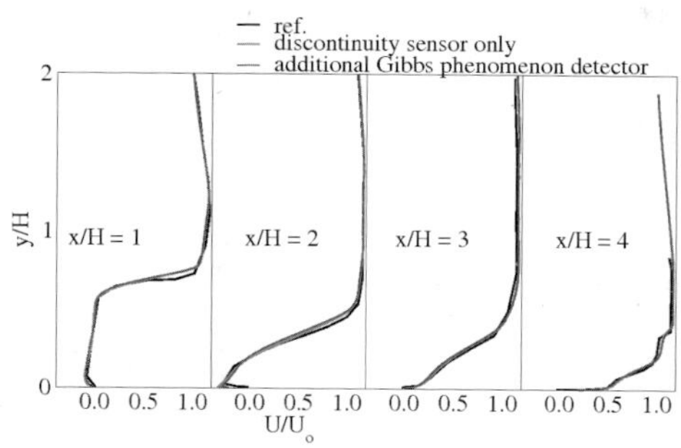

Fig. 6. Predicted mean velocity profiles at various axial locations.

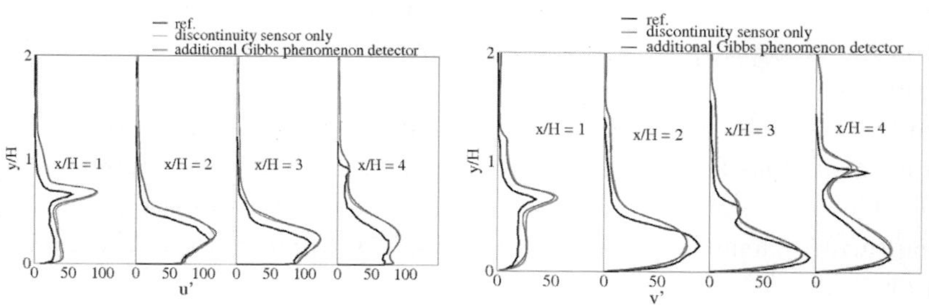

Fig. 7. Predictions of turbulent intensity profiles at various axial locations.

The Gibbs phenomenon detector has no effect on the turbulent intensities as well as seen in Fig. 7. The simulations here overpredict the levels by about 15% compared to the reference study except at the first location. The exact reason for this is not clear. Examination of contour plots indicates the present resolution may be lower than in earlier study. This is, addition to the difference in subgrid modeling, could be a possible reason. The unphysical noise generated at the inflow (see Fig. 5) could also generate higher unsteadiness in the flow as it propagates through the expansion fan at the corner.

4. Conclusions

Inclusion of a simple Gibbs phenomenon detector in the hybridization is effective in suppressing numerical oscillations in preliminary tests. The key feature of this approach is the integrated treatment for numerical oscillations both near and far away from discontinuities. It, however, does not seem to affect the results in the LES test case considered here since the shock is not strong enough and the slip line is parallel to the flow. Further testing is planned.

References

1. F. Genin, Study of compressible turbulent flows in supersonic environment by large eddy simulation, PhD thesis, Georgia Institute of Technology, Atlanta, GA, USA (2009).
2. S. Pirozzoli, Conservative hybrid compact-WENO schemes for shock-turbulence interaction, *J. Comput. Phys.* **178**, 81 (2002).
3. A. W. Cook and W. H. Cabot, Hyperviscosity for shock-turbulence interactions, *J. Comput. Phys.* **203**, 379 (2005).
4. S. Kawai and S. K. Lele, Localized artificial diffusivity scheme for discontinuity capturing on curvilinear meshes, *J. Comput. Phys.* **227**, 9498 (2008).
5. S. C. Lo, G. A. Blaisdell and A. S. Lyrintzis, Higher-order shock capturing schemes for turbulence calculations, *AIAA*-2007-827 (2007).
6. T. Guo-Hua and Y. Xiang-Jiang, A characteristic-based shock-capturing scheme for hyperbolic equations, *J. Comput. Phys.* **225**, 2083 (2007).
7. von Kaenel, L. Kleiser, N. A. Adams and J. B. Vos, Large-eddy simulation of shock-turbulence interaction, *AIAA J.* **42**, 2516 (2004).
8. E. Johnsen, J. Larsson, A. V. Bhagatwala, W. H. Cabot, P. Moin, B. J. Olson, P. S. Rawat, S. K. Shankar, B. Sjogreen, H. C. Yee, X. Zhong and S. K. Lele, Assessment of high resolution methods for numerical simulations of compressible turbulence, *J. Compt. Phys.* **229**, 1213 (2010).
9. S. Pirozzoli, Numerical methods for high-speed flows, *Annu. Rev. Fluid Mech.* **43**, 163 (2011).
10. H. Nishikawa and K. Kitamura, Very simple, carbuncle-free, boundary-layer-resolving, rotated-hybrid Riemann solvers, *J. Compt. Phys.* **227**, 2560 (2008).
11. C. M. Rhie and W. L. Chow, A numerical study of the turbulent flow

past an isolated airfoil with trailing edge separation, *AIAA J.* **21**, 1525 (1983).

12. N. A. Adams and K. Shariff, A high-resolution hybrid compact-ENO scheme for shock-turbulence interaction problems, *J. Comput. Phys.* **127**, 27 (1996).
13. B. Fiorina and S. K. Lele, *An artificial nonlinear diffusivity method for shock-capturing in supersonic reacting flows*, Tech. Rep. 57-70, Annual Research Briefs, Center for Turbulence Research, Stanford University (2005).
14. R. Liska and B. Wendroff, Comparison of several difference schemes on 1D and 2D test problems for the euler equations, *SIAM J. on Sci. Comput.* **25**, 995 (2003).
15. E. F. Toro, *Riemann Solvers and Numerical Methods for Fluid Dynamic*, third edn. Springer-Verlag (2009).
16. A. E. Emery, An evaluation of several differencing methods for inviscid fluid flow problems, *J. Comput. Phys.* **2**, 306 (1968).
17. V. K. Chakravarthy and D. Chakraborty, Modified SLAU2 scheme with enhanced shock stability, *Comput. and Fluids* **100**, 176 (2014).
18. H. Ayyalasomayajula, S. Arunajatesan, C. Kannepalli and N. Sinha, Large eddy simulation of a supersonic flow over backward-facing step for aero-optical analysis, *AIAA*-2006-1416 (2006).
19. J. Smagorinsky, General circulation experiments with the primitive equations, *Monthly Weather Review* **91**, 99 (1963).
20. E. Touber and N. D. Sandham, Large-eddy simulation of low frequency unsteadiness in a turbulent shock-induced separation bubble, *Theo. Comput. Fluid Dyn.* **23**, 79 (2009).

Forward and Inverse 3D Fourier Transforms of a DNS Wavepacket Evolving in a Blasius Boundary Layer

Kean-Lee Kang

NUS Graduate School for Integrative Sciences and Engineering,
National University of Singapore,
Centre for Life Sciences, #05-01,
28 Medical Drive, Singapore 117456

K. S. Yeo*

Department of Mechanical Engineering,
National University of Singapore,
9 Engineering Drive 1, Singapore 117576
**E-mail: mpeyeoks@nus.edu.sg*
Tel: (+65) 65162246

The wavepacket is a model that bears close resemblance to naturally-occurring laminar turbulent transition processes in a boundary layer, but its large number of constituent spectral modes presents a challenge during analysis of the results. In this work, we leverage the extensive flow data available from direct numerical simulations (DNS) to perform a three-dimensional (3D) discrete Fourier transform of a wavepacket in order to investigate the most energetic modes in the flow. Through windowing, it is possible to isolate the wavepacket at different stages of its development, and build up an overall picture of its transition. Furthermore, an inverse transform is applied to subsets of the full spectrum, and this provides insight into the physical structure of its key components.

Keywords: laminar turbulent transition; wavepacket; boundary layer; Fourier transform

1. Introduction

A wave may propagate alone through a boundary layer, or together as a group of waves in a wavepacket.[1–3] The process of wavepacket evolution in a Blasius boundary layer has been widely analyzed as one of several key vehicles for advancing our understanding of laminar turbulent transition in fluids.[4,5] An advantage of the wavepacket model is that it is well able to mimic natural transition processes due to the large number and diversity of modes present in the wavepacket.[6] The downside of this complexity is greater difficulty in analysis and interpretation of results, especially when compared against the classic Tollmien–Schlichting (TS) wave experiments,[1] and modern variants of them.[7] Many theories that describe the interaction and evolution of waves in a boundary layer make assumptions that reduce the number of dimensions of the problem in order to make it tractable and amenable to an analytical solution.[8] Only a few waves at a time are considered, such as the interaction of a two-dimensional (2D) fundamental wave with a symmetric pair of three-dimensional (3D) subharmonic (half the frequency of the fundamental) waves in Craik triads,[9] or a small 3D perturbation riding on a 2D primary disturbance as proposed by Herbert.[10] These theories have achieved much success and held up under experimental verification.

The necessary trade-off resulting from this analytical restriction on the number of modes in the flow is that the overall disturbance is periodic in space and time, and occupies a large spatial footprint. They are unable to replicate the spatial and temporal localization of a wavepacket, for a spectral description of such localization would require a very large number of modes as a consequence of the uncertainty principle.[11] Taking into account a very large number of modes necessitates a computational approach to the problem. In this aspect, much work has been done to apply cutting-edge techniques to decompose a transitional flow into its key physical elements. These techniques include global eigenmodes for linearized dynamics (fully 3D global stability analysis),[12] proper orthogonal decomposition (POD, also known as principal component analysis or singular value decomposition),[8] balanced POD,[13] dynamic mode decomposition[14,15] and nonlinear optimization.[16] In comparison, the fast Fourier transform (FFT) seems to have received somewhat less attention in recent times. However, we believe that it is still a powerful tool for transitional flow analysis, because it rests on a firm mathematical foundation that arbitrary functions can be represented by trigonometric series. Moreover, established theories[9,10] have been founded on the basis of Fourier modes. In fact, the conditions for the convergence of Fourier series representations of functions are very general and practical,[17] so we are able to use discrete Fourier transforms to project a signal onto a basis set that spans the observation window.[18] Additionally, the FFT is computationally efficient,[19] and graphics processing unit (GPU) acceleration is possible.[20]

In this paper, we leverage on these advantages of the FFT to perform 3D spectral analysis on direct numerical simulation (DNS) data of a wavepacket to identify important modes in the flow at different stages of transition. These modes are then isolated and inverse discrete Fourier transforms are applied to recover their spatio-temporal structure. Furthermore, we can quantify their spectral density, which is in turn linked to the physical energy of the system via the famous Plancherel theorem.[21] A survey of past literature indicates that previously published wavepacket Fourier spectra were mostly 2D. The advantage of the 3D $S(\omega, \beta, \alpha)$ spectrum obtained here is that it enables the simultaneous identification of the angular frequency, ω, spanwise wavenumber, β and streamwise wavenumber α of modes in the flow. Previously, the 2D $S(\omega, \beta)$ and $S(\beta, \alpha)$ spectra were obtained separately. The details provided by 3D spectra open the door to the investigation of other important theories. For example, the mode phase speed $c = \omega/\alpha$ can be used to identify phase-locked modes that undergo strong resonant interactions.[22]

2. Methodology

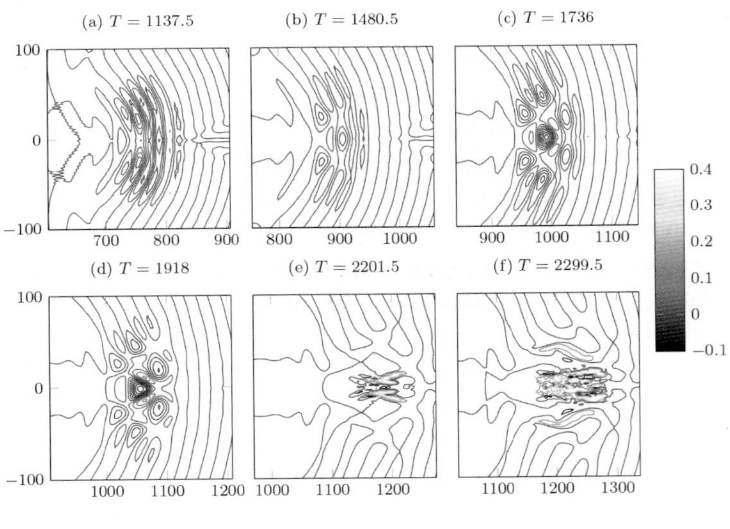

Fig. 1. u-velocity contours of the wavepacket as it propagates downstream through the boundary layer and undergoes laminar-turbulent transition. Note that the axes are X (streamwise) horizontal and Z (spanwise) vertical.

The wavepacket studied in this work is simulated by a published DNS code[23] and closely matches the one used in the simulations of Ref. 5, verified against the experiments of Ref. 4. A single-cycle, wall-normal perturbation velocity sinusoid in a circular disk on the wall in the boundary layer is the initial disturbance of the flow. This perturbation is spatially smoothed by a 2D Gaussian function centred on the disk. Mathematically speaking,

$$v(X,Z,T) = \begin{cases} A\sin(\omega_0 T)\exp\left[-\frac{(X-X_0)^2}{2} - \frac{(Z-Z_0)^2}{2}\right], & 0 \le t \le T_d \\ 0, & \text{otherwise.} \end{cases} \quad (1)$$

X, Y and Z are the streamwise, wall-normal and spanwise cooordinates respectively, all normalized by the displacement thickness at the source $\delta_0 = 2.3182 \times 10^{-3}$ m. u, v and w are the perturbation velocities (on a Blasius base flow) in each direction normalized by the freestream velocity $U_\infty = 6.65$ m/s, $A = 0.0082$ is the amplitude of the disturbance, T is non-dimensional time $T = tU_\infty/\delta_0$ with t measured in seconds, $\omega_0 = 0.063$ is the excitation frequency, $T_d = 2\pi/\omega = 100$ is the disturbance duration corresponding to one full cycle of the sinusoid and $(X_0, Z_0) = (349, 0)$ is the source disk centre located at a Reynolds number of $Re_0 = \delta_0 U_\infty/\nu = 1035$ with kinematic viscosity $\nu = 1.49 \times 10^{-5}$ m^2/s.

Spatio-temporal development of the wavepacket is recorded in a streamwise-spanwise plane over a length of time, at a fixed boundary layer height $y = 0.6\delta$, where δ is the local displacement thickness. The resultant matrix contains streamwise perturbation velocity $u(T, Z, X)$ data that is a function of the non-dimensionalized variables T, Z and X, and this u-matrix can be visualized in slices as shown in Fig. 1 or as an isosurface in Fig. 2(a). The wavepacket evolution begins with an expansion and dispersion from the initial point source in Fig. 1(a). It then begins to acquire more spanwise variations in Fig. 1(b) and a large negative velocity spike at its core in (c). Afterwards, a horseshoe vortex forms between Fig. 1(d) to (e) before breakdown to incipient turbulence in (f). Due to its small spatio-temporal footprint, the wavepacket contains a large number of Fourier modes that interact nonlinearly in a great variety of ways,[24] making it an excellent candidate for the application of 3D Fourier transforms.

Using the methods of Ref. 17, a discrete Fourier transform is applied to $u(T, Z, X)$ data to obtain the spectral density $S(\omega, \beta, \alpha)$, which is a function of the angular frequency ω, spanwise wavenumber β and streamwise wavenumber α as shown in Fig. 2(b). While performing this spectral analysis, the use of windows is necessary to isolate the wavepacket at a particular stage of development and enhance the ability to distinguish a low-energy mode located close to a high-energy mode, such as a small subharmonic seed near a strong fundamental mode. This calls for a window with low side-lobe levels to suppress spectral leakage, and with small worst-case processing loss (the reduction of output signal-to-noise ratio as a result

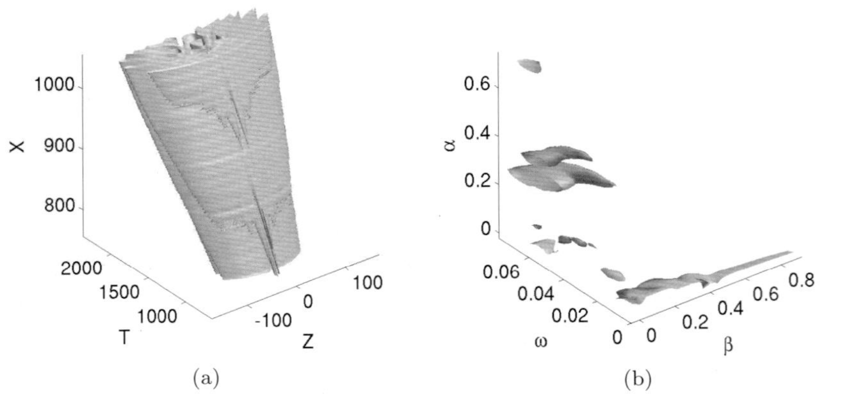

Fig. 2. (a) Wavepacket streamwise perturbation velocity $u(T, Z, X) = 10^{-3}$ isosurface between $X = 755$ and 1057. (b) Wavepacket spectral density $S(\omega, \beta, \alpha) = 2 \cdot 10^{-7}$ isosurface, obtained from a discrete Fourier transform of (a) with Hamming windowing in the T and X dimensions $\Delta T = 1344$ and $\Delta X = 302$, and box windowing in the Z dimension $\Delta Z = 347$, centred at $(T, Z, X) = (1479, 0, 906)$.

of windowing and worst case frequency location). Unfortunately, these are generally conflicting goals, with lower side-lobe levels typically resulting in higher processing loss.[18] The rectangular, Hann, Hamming and Blackman-Harris windows were tested, and the Hamming window was chosen because it provides a good trade-off between these parameters. The Hamming window was applied in the X and T dimensions only, because the wavepacket is self-windowing in the Z-direction due to its limited spanwise extent.

To observe the physical structure of particular modes in the flow, the relevant data subset of the discrete Fourier transform of $u(T, Z, X)$ is Hamming windowed in the three ω, β and α dimensions. All modes outside the subset are set to zero, and an inverse discrete Fourier transform is performed, taking care to include the required conjugate pairs of modes so that the inverse is real.

3. Key Results

For a more detailed view of the 3D spectra in Fig. 2(b), slices can be taken in ω-β planes at various α locations, and these are shown in Fig. 3. Modes of interest are usually local maxima in the spectrum. Some important modes are the fundamental at $(\omega, \beta, \alpha) \approx (0.06, 0, 0.3)$ and the detuned subharmonic mode pair[25,26] at $(\omega, \beta, \alpha) \approx (0.03 \pm 0.008, 0.125, 0.1)$, which

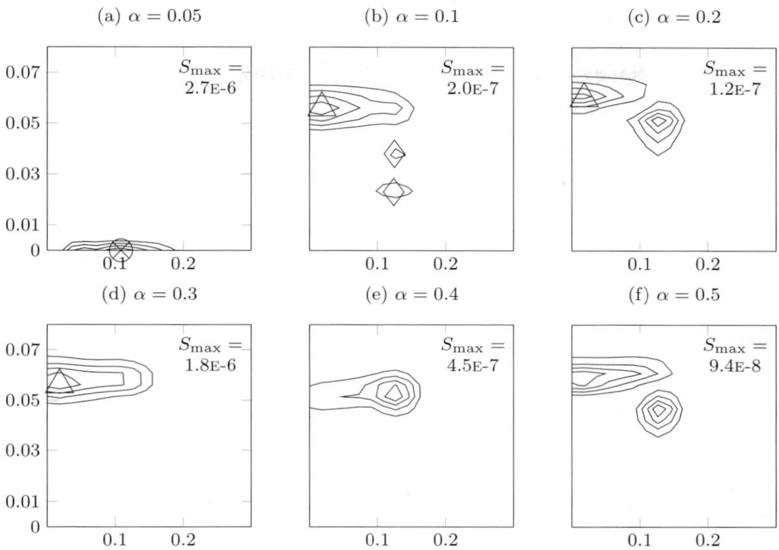

Fig. 3. Slices of the 3D spectra of Fig. 2(b) in ω-β planes. The α value of each slice is indicated above each plot. The fundamental, subharmonic and Klebanoff modes are local maxima in the spectral density, marked by a triangle, diamond and crossed circles respectively. There is a detuned subharmonic mode pair at $\alpha = 0.1$. The contours are uniformly spaced at $\{20\%, 40\%...100\%\}$ of the maximum spectral density $S_{\max}$ in each slice, which is indicated. Note that the axes are β horizontal and ω vertical.

together make up an approximate resonant Craik triad.[9] The other important set of modes are the Klebanoff modes[27,28] with low ω, such as $(\omega, \beta, \alpha) \approx (0.005, 0.1, 0.05)$.

Applying the inverse discrete Fourier transform enables us to uncover the physical structure associated with any subset of modes in the spectrum, and one possible wavepacket decomposition is shown in Fig. 4. The fundamental and detuned subharmonic modes in particular show good agreement with the results of dynamic mode decomposition found in Figure 4 of Ref. 15. Nevertheless, it should be noted that the small spatio-temporal size of the wavepacket at least partially constrains the maximum wavelength or minimum wavenumber of its component modes. The dimensions of the spectral window used also set lower bounds on the frequency and wavenumbers that can be reliably detected, which are 0.005, 0.018 and 0.021 in the ω, β and α dimensions respectively.

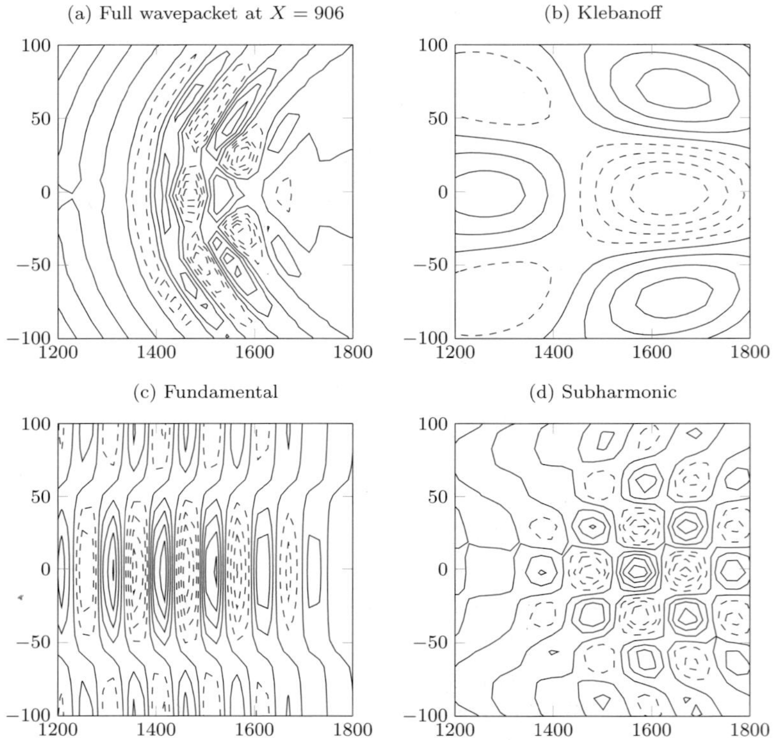

Fig. 4. Normalized perturbation velocity contours ($u/|u|_{\max}$) of the wavepacket centred at $X = 906$ and its key component modes obtained from inverse 3D discrete Fourier transforms of Hamming windowed subsets of the full spectrum. The subsets corresponding to the Klebanoff, fundamental and subharmonic modes cover the ranges (ω, β, α) = (0.005-0.014,0.018-0.072,0.021-0.104), (0.047-0.075,0.018-0.054,0.250-0.354) and (0.014-0.051,0.018-0.145,0.042-0.146) respectively. Solid contour lines denote positive velocity, and dashed lines show negative velocity, with uniformly spaced contour levels $\pm\{0, 0.2, 0.4, ...\}$. Axes are T horizontal and Z vertical.

4. Conclusion

The above sample of key results were performed with a spectral window centred at one intermediate stage of transition. By applying this 3D discrete Fourier transform and inverse technique along a sliding window, we can isolate a transitional wavepacket at many different stages of transition, pinpoint critical dynamics and build up an overall picture of the entire laminar-turbulent transition process.

References

1. G. B. Schubauer and H. K. Skramstad, Laminar boundary-layer oscillations and transition on a flat plate, *J. Res. Natl. Bur. Stand.* **38**, p. 42 (1947).
2. W. O. Criminale and L. S. G. Kovasznay, The growth of localized disturbances in a laminar boundary layer, *J. Fluid Mech.* **14**, 59 (1962).
3. M. Gaster, The development of three-dimensional wave packets in a boundary layer, *J. Fluid Mech.* **32**, 173 (1968).
4. J. Cohen, K. S. Breuer and J. H. Haritonidis, On the evolution of a wave packet in a laminar boundary layer, *J. Fluid Mech.* **225**, 575 (1991).
5. K. S. Yeo, X. Zhao, Z. Y. Wang and K. C. Ng, DNS of wavepacket evolution in a Blasius boundary layer, *J. Fluid Mech.* **652**, 333 (2010).
6. M. A. F. Medeiros and M. Gaster, The production of subharmonic waves in the nonlinear evolution of wavepackets in boundary layers, *J. Fluid Mech.* **399**, 301 (1999).
7. I. B. de Paula, W. Würz, E. Krämer, V. I. Borodulin and Y. S. Kachanov, Weakly nonlinear stages of boundary-layer transition initiated by modulated Tollmien-Schlichting waves, *J. Fluid Mech.* **732**, 571 (2013).
8. P. Holmes, J. Lumley, G. Berkhooz and C. W. Rowley, *Turbulence, Coherent Structures, Dynamical Systems and Symmetry* (Cambridge University Press, 2012).
9. A. D. D. Craik, Non-linear resonant instability in boundary layers, *J. Fluid Mech.* **50**, 393 (1971).
10. T. Herbert, Secondary instability of boundary layers, *Annu. Rev. Fluid Mech.* **20**, 487 (1988).
11. E. M. Stein and R. Shakarchi, *Fourier Analysis: An Introduction* (Princeton University Press, 2003).
12. S. Bagheri, Effects of weak noise on oscillating flows: Linking quality factor, Floquet modes, and Koopman spectrum, *Phys. Fluids* **26** (2014).
13. C. W. Rowley, Model reduction for fluids, using balanced proper orthogonal decomposition, *Intl. J. Bifurcation Chaos* **15**, 997 (2005).
14. C. W. Rowley, I. Mezi, S. Bagheri, P. Schlatter and D. S. Henningson, Spectral analysis of nonlinear flows, *J. Fluid Mech.* **641**, 115 (2009).
15. T. Sayadi, P. Schmid, J. W. Nichols and P. Moin, Reduced-order representation of near-wall structures in the late transitional boundary

layer, *J. Fluid Mech.* **748**, 278 (2014).

16. S. Cherubini, P. De Palma, J.-C. Robinet and A. Bottaro, The minimal seed of turbulent transition in the boundary layer, *J. Fluid Mech.* **689**, 221 (2011).
17. D. E. Newland, *An Introduction to Random Vibrations, Spectral and Wavelet Analysis*, third edn. (Dover Publications, Inc, Mineola, New York, 1993).
18. F. J. Harris, On the use of windows for harmonic analysis with the discrete Fourier transform, *Proc. IEEE* **66**, 51 (1978).
19. S. M. Kay and J. Marple, S. L., Spectrum analysis-a modern perspective, *Proc. IEEE* **69**, 1380 (1981).
20. NVIDIA, CUDA toolkit documentation (2014), `http://docs.nvidia.com/cuda/cufft/index.html`.
21. E. C. Titchmarsh, A contribution to the theory of Fourier transforms, *Proc. London Math. Soc.* **s2-23**, 279 (1925).
22. X. Wu, P. A. Stewart and S. J. Cowley, On the catalytic role of the phase-locked interaction of Tollmien-Schlichting waves in boundary-layer transition, *J. Fluid Mech.* **590**, 265 (2007).
23. Z. Y. Wang, K. S. Yeo and B. C. Khoo, Spatial direct numerical simulation of transitional boundary layer over compliant surfaces, *Comput. Fluids* **34**, 1062 (2005).
24. M. A. F. Medeiros and M. Gaster, The influence of phase on the nonlinear evolution of wavepackets in boundary layers, *J. Fluid Mech.* **397**, 259 (1999).
25. V. I. Borodulin, Y. S. Kachanov and D. B. Koptsev, Experimental study of resonant interactions of instability waves in a self-similar boundary layer with an adverse pressure gradient: I. Tuned resonances, *J. Turbul.* **3**, p. N62 (2002).
26. V. I. Borodulin, Y. S. Kachanov, D. B. Koptsev and A. P. Roschektayev, Experimental study of resonant interactions of instability waves in self-similar boundary layer with an adverse pressure gradient: II. Detuned resonances, *J. Turbul.* **3**, p. N63 (2002).
27. P. S. Klebanoff, K. D. Tidstrom and L. M. Sargent, The three-dimensional nature of boundary-layer instability, *J. Fluid Mech.* **12**, 1 (1962).
28. Y. S. Kachanov, Physical mechanisms of laminar-boundary-layer transition, *Annu. Rev. Fluid Mech.* **26**, 411 (1994).

Reduced Order Modeling by POD of Supercritical Flow Past Circular Cylinder

Parvathi, M.K., Syed Ijlal, Pallavi, G., Sengupta, T. K.*

Department of Aerospace Engineering, IIT Kanpur,
Kanpur, U.P., 208 016, India
**E-mail: tksen@iitk.ac.in*

1. Introduction

A reduced order model (ROM) is developed here using POD[1–3] of DNS data performed for flow past a circular cylinder based on a dynamical system approach[5]. The ROM[4,14] has been developed using first six POD modes by training it with DNS data at the same Reynolds number (Re). It is now feasible to replace the solution of PDE by four pairs of ODE for the amplitude and phase of the instability modes of disturbance vorticity. Recently, we have extended the work by performing a reconstruction of signal at any Re which is not trained with the corresponding DNS data.

The method depends on DNS of 2D flow past the circular cylinder by solving the stream function- vorticity formulation of Navier-Stokes equation (NSE). The solution obtained by using compact scheme for the convective acceleration is shown for a point along the wake centreline ($x/D = 0.504$), with the origin at the centre of the cylinder. In Fig. 1(a), the disturbance vorticity time series is plotted, obtained by subtracting the time average from the instantaneous value of the vorticity from the solution of NSE. The solution grows temporally to begin with, which saturates to a constant value due to nonlinear stabilization. Such an approach of disturbance field to reach a time periodic state from a quiescent condition is described by Stuart-Landau equation[6–8], which holds the predominance of a single mode and the solution is stated to have undergone a Hopf bifurcation. However, Fig. 1(b) clearly show the presence of multiple Hopf bifurcations and dominance of multiple modes, which is also noted by the POD analysis of the vorticity[4,5,9,10]. The POD modes have been classified as (a) the regular or R mode; (b) anomalous mode of first kind or T_1 mode and (c) anomalous mode of second kind or T_2 mode. It has been shown that this characterization[9], in the neighbourhood of primary instability is universal, as noted for two canonical internal and external flows. More flow problems have been studied[11], to show different variations of POD modes.

2. Instability Modes and SLE Equation

As noted in[4,5,10], for any such multi-modal representation of the POD amplitude ($a_m(t)$) and eigenfunction (ϕ_m), we relates those with the instability modes as[11,13]

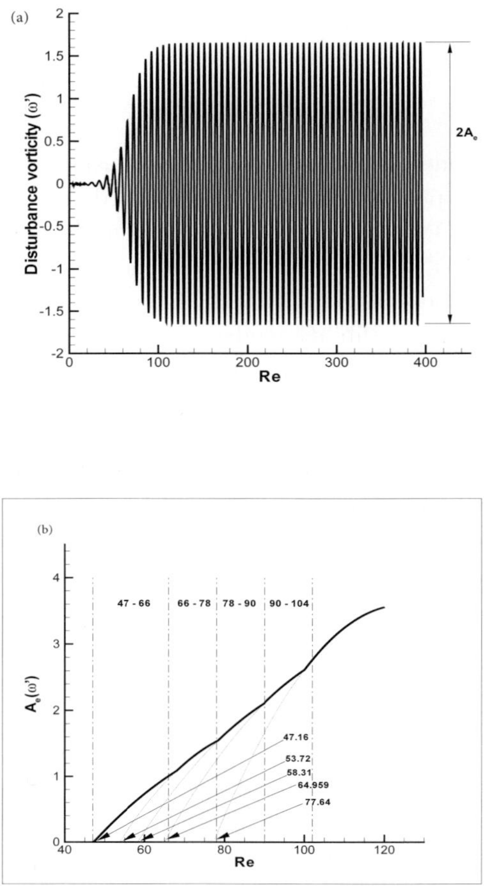

Fig. 1. a) Time evolution of disturbance vorticity computed at the wake centre-line, $x/D = 0.504$ for the Reynolds number, $Re = 80$. b) The equilibrium amplitude of the disturbance vorticity at the same location is plotted as a function of Reynolds number.

$$A_j(t) = \sqrt{\epsilon_j}(a_{2j-1}(t) + ia_{2j}(t)) \quad (1)$$

$$f_j(\vec{X}) = \frac{1}{\sqrt{\epsilon_j}}(\phi_{2j-1}(\vec{X}) - i\phi_{2j}(\vec{X})) \quad (2)$$

where $\epsilon_j = (\lambda_{2j-1}+\lambda_{2j})/\Sigma_{k=1}^{N}(\lambda_k)$. The value of λ_k's are obtained from the POD analysis involving two-point cross correlation of disturbances. This normalization factor is introduced to make the dimension of $A_j(t)$ same as that of vorticity, since ϵ_j is a measure of enstrophy. Multi-modal nature of instability modes have been given by[5] the Stuart-Landau-Eckhaus (SLE) equations,

$$\frac{d|A_j|}{dt} = (\alpha_j)_{real}|A_j| + \Sigma_{k=1}^{M}\ (\beta_{jk})_{real}|A_j||A_k|^2 + \delta(j-2)\gamma_{jm,jd}|A_j|^3 \quad (3)$$

$$\frac{d\theta_j}{dt} = (\alpha_j)_{imag} + \Sigma_{k=1}^{M}\ (\beta_{jk})_{imag}|A_k|^2 \quad (4)$$

In a recent attempt[4,14], dynamical system approach[5] has been modified to account for the phase variation of T_1 modes, which is represented by A_2 here. The idea behind this modification is that this mode displays time variations at two scales: at a slow scale (which can be approximated by RANS procedure and hence called the shift mode[12]) given by A_{2m} and another (A_{2d}) at a rapid scale. This procedure can work only for DNS. Even global linear analysis will fail, as the whole process is dominated by nonlinear interaction, as shown above in Eqs. (3)and (4). Note that A_1 and A_3 are regular modes.

The coefficients of Eqs. (3) and (4) are as given in Table 1, for: $Re = 80,\ 90\ and 100$. Higher the value of the coefficients, weaker is the dependence of that particular mode. For solving the set of equations given by (3), an optimal set of conditions have been obtained following the procedure given in[4,14]. Results of the SLE reconstruction is discussed in the next section. Apart from SLE reconstruction, an alternate approach is also followed to reconstruct the disturbance vorticity via lagrange interpolation using the available DNS data. But the direct lagrange interpolation would not give the correct result as each Re has a different time scale associated with it. This problem is sorted out by scaling the time-scale of each Re with respect to particular Re before interpolation.

3. Results and Discussion

As discussed in the previous section, the SLE reconstruction is performed by the set of ODEs of amplitude and phase alongwith the optimal initial condition. The results are shown for $Re = 150$ of amplitude and phase variation in Figs. 2 and 3, respectively. One notices an accurate match for both the amplitude and phase variation with DNS. Having obtained the amplitude and phase variation from SLE equation, the disturbance vorticity is reconstructed from POD modes by retracing the path. In Fig. 4, reconstructed disturbance vorticity are shown for $Re = 150$. The excellent match with the DNS result shows the consistency of the method. Alternate reconstruction of disturbance vorticity for $Re = 70$ from the DNS data of $Re = 68,\ 72$ and 76 using lagrange interpolation are shown in Fig. 5.

Table 1.1: Amplitude equation coefficient for A_1

Re	α_1	β_{11}	β_{12m}	β_{12d}	β_{13}
80	6.637×10^{-2}	4.239×10^{1}	1.238×10^{2}	-2.012×10^{5}	-4.103×10^{2}
90	7.288×10^{-2}	6.170×10^{1}	2.066×10^{2}	-1.739×10^{5}	-5.070×10^{2}
100	8.450×10^{-2}	-6.735×10^{1}	2.001×10^{2}	-2.065×10^{4}	-1.835×10^{2}

Table 1.2: Amplitude equation coefficient for A_{2m}

Re	α_{2m}	β_{2m1}	β_{2m2m}	β_{2m2d}	β_{2m3}	γ_{2m2d}
80	1.232×10^{-2}	-5.168×10^{2}	-5.664×10^{1}	9.581×10^{5}	7.635×10^{2}	1.470×10^{6}
90	1.950×10^{-2}	-4.770×10^{2}	-1.637×10^{2}	7.844×10^{4}	1.654×10^{3}	1.238×10^{6}
100	5.707×10^{-2}	-1.154×10^{3}	-3.084×10^{2}	1.754×10^{6}	2.409×10^{3}	1.187×10^{6}

Table 1.3: Amplitude equation coefficient for A_{2d}

Re	α_{2d}	β_{2d1}	β_{2d2m}	β_{2d2d}	β_{2d3}	γ_{2d2m}
80	1.525×10^{-1}	5.874×10^{2}	1.404×10^{3}	-9.628×10^{5}	3.754×10^{3}	6.606×10^{-2}
90	1.165×10^{-1}	2.026×10^{2}	-2.650×10^{2}	-3.449×10^{5}	-1.446×10^{3}	1.226
100	1.386×10^{-1}	-7.330×10^{1}	-1.803×10^{2}	-7.398×10^{4}	-4.967×10^{2}	4.325×10^{-1}

Table 1.4: Amplitude equation coefficient for A_3

Re	α_3	β_{31}	β_{32m}	β_{32d}	β_{33}
80	-4.456×10^{-2}	5.219×10^{2}	2.516×10^{2}	-1.110×10^{6}	-1.104×10^{3}
90	9.081×10^{-2}	-1.435×10^{2}	-3.419×10^{2}	-1.890×10^{5}	-4.766×10^{2}
100	8.129×10^{-2}	-6.157×10^{1}	-2.793×10^{2}	1.121×10^{3}	-2.760×10^{2}

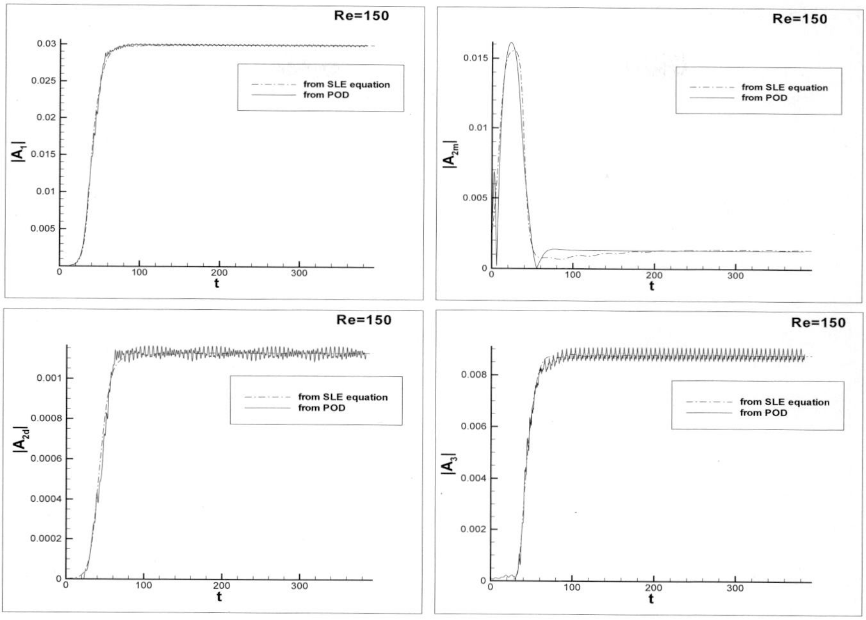

Fig. 2. Reconstructed instability mode amplitude variation with time for Re=150.

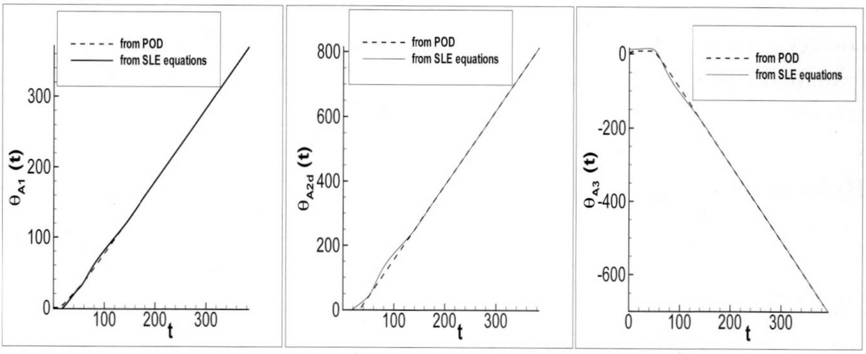

Fig. 3. Reconstructed phase variation with time of the instability modes for Re=150.

4. Conclusions and Future Work

It has been shown that it is possible to replace the arduous task of solving the Navier-Stokes equation to track flow instability and solution bifurcation,

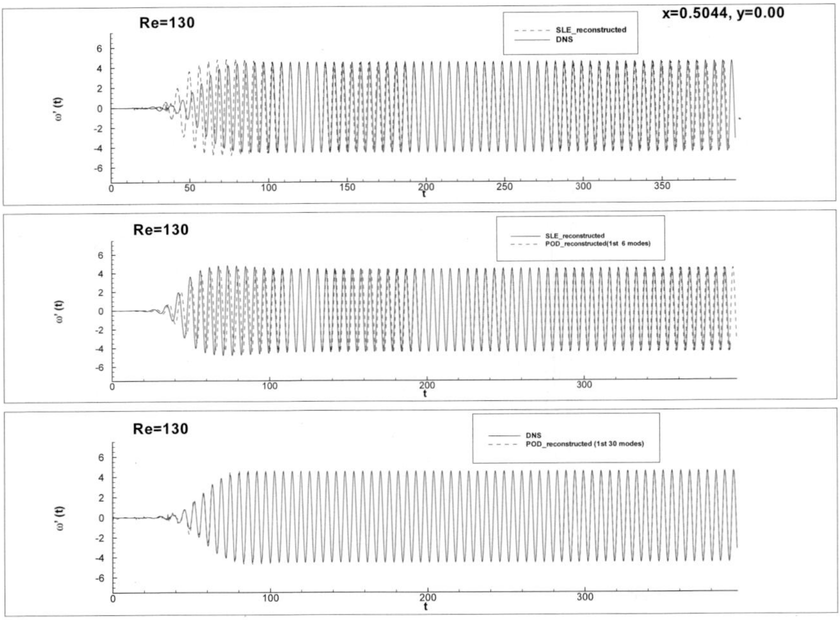

Fig. 4. Disturbance vorticity obtained from DNS reconstructed using the results from the SLE equations for Re=150 at (0.5044,0.0).

by solving a set of ODEs without much effort. While one can get almost correct reconstruction by training the ROM with DNS data, we have now extended the method for any other super-critical Reynolds number without requiring any training from corresponding DNS.

References

1. Kosambi, D. D., Statistics in function space, *J. Indian Math. Soc.*, **7**, 76–88, (1943).
2. Holmes, P., Lumley J. L. and Berkooz G., *Turbulence, Coherent Structures, Dynamical Systems and Symmetry*, (Cambridge University Press, U. K., 1995).
3. Sirovich, L., Turbulence and dynamics of coherent structures. Part I Coherent Structures, Part II Symmetries and Transformations and Part III Dynamics and Scaling, *Q. Appl. Math.*, **45**, 561–590, (1987).
4. Parvathi, M. K., Reduced order modeling of instability for flows using enstrophy-based POD, *M.Tech. Thesis, IIT Kanpur, India*, (2014).
5. Sengupta, T. K., Singh, N. and Suman, V. K., Dynamical system ap-

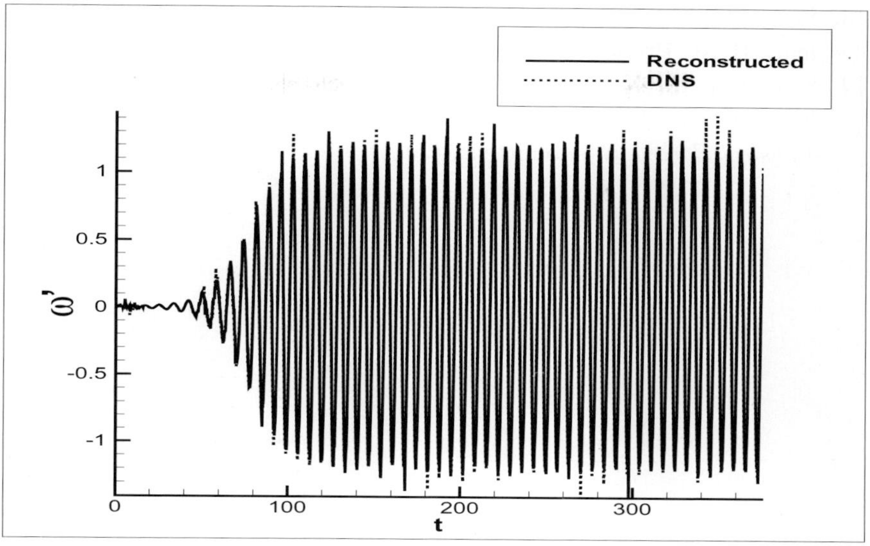

Fig. 5. Reconstructed disturbance vorticity for $Re = 70$ using time scaling interpolation.

proach to instability past flow a circular cylinder, *J. Fluid Mech.*,**656**, 82–115, (2010).

6. Landau, L. D., On the problem of turbulence, *C. R. Acad. Sci.*, USSR, **44**, 311-315, (1944).
7. Eckhaus, W., *Studies in Nonlinear Stability Theory* (Springer Verlag, Berlin, Germany, 1965).
8. Drazin, P. G. and Reid, W. H., *Hydrodynamic Instabilities*, (Cambridge University Press, U.K, 1981).
9. Sengupta, T. K., Singh, N. and Vijay, V. V. S. N., Universal instability modes in internal and external flows, *Computers & Fluids*, **40**, 221–235, (2011).
10. Pallavi, G., Nonlinear instability of flow past a cylinder via POD analysis *M.Tech. Thesis, IIT Kanpur, India*, (2014).
11. Sengupta, T. K., Bhaumik, S. and Bhumkar, Y. G., Nonlinear receptivity and instability studies by POD, *6th AIAA Theoretical Fluid Mechanics Conf., Honululu, Hawaii, USA*, 27–30 June, (2011).
12. Noack, B. R., Afanasiev, K., Morzynski, M. Tadmor, G. and Thiele, F., A hierarchy of low-dimensional models for the transient and post-transient cylinder wake, *J. Fluid Mech.*, **497**, 335–363, (2003).

13. Sengupta, T. K.,*Instabilities of Fluid Flows and Transition to Turbulence* (CRC Press (Taylor and Francis), Florida, USA, 2012).
14. Sengupta, T. K., Haider, S. I., Parvathi, M. K. and Pallavi, G., Enstrophy-based POD for reduced order modeling of flow past cylinder, *Physical Review E.*(submitted), (2015).

Proper Orthogonal Decomposition vs. Fourier Analysis for Extraction of Large-Scale Structures of Thermal Convection

Supriyo Paul

Computational Fluid Dynamics Team, Centre for Development of Advanced Computing Pune,
*Pune 411007, India**
E-mail: supriyop@cdac.in

Mahendra K. Verma

Department of Physics, IIT Kanpur, Kanpur 208016, India

We performed a comparative study of extraction of large-scale flow structures in Rayleigh Bénard convection using proper orthogonal decomposition (POD) and *Fourier analysis.* We show that the free-slip basis functions capture the flow profiles successfully for the no-slip boundary conditions. We observe that the large-scale POD modes capture a larger fraction of total energy than the Fourier modes. However, the Fourier modes capture the rarer flow structures like flow reversals better. The flow profiles of the dominant POD and Fourier modes are quite similar. Our results show that the Fourier analysis provides an attractive alternative to POD analysis for capturing large-scale flow structures.

Keywords: Proper Orthogonal Decomposition; Fourier Analysis; Convection

1. Introduction

In fluid flows, the large-scale structures play a major role in its dynamics. Hence, an identification of such structures is critical for understanding fluid flows. Proper Orthogonal Decomposition (POD)[1–3] is one of the popular methods for this task. In this paper we employ a POD scheme called "snapshot method"[4,5], in which a number of uncorrelated and discrete time snapshots of the flow field are used.

An alternate tool for identifying large-scale structures is Fourier analysis[6], which is relatively easier to compute. Recently, Chandra and Verma[7] showed that the Fourier modes play a critical role in the reversal dynam-

*Present Address: Department of Physics, A.C. College, Jalpaiguri 735101, India. E-mail: supriyopaul@gmail.com

ics of turbulent convection. Note that a large number of low-dimensional models have been constructed using the large-scale Fourier modes. There are several low-dimensional models based on POD modes as well[8,9], but Fourier modes are more amenable for this purpose.

In the present work, we perform a comparative study between POD and Fourier analysis by employing them to Rayleigh-Bénard convection in a 2D box. We will emphasise similarities and dissimilarities between these two diagnostics tools.

2. Large-scale Structures of RBC in a Two-dimensional Box

We simulate *Rayleigh-Bénard convection* (RBC) in a square box of unit dimension. For the velocity field we assume no-slip boundary conditions ($u = v = 0$) on all the walls, and for the temperature field we consider the top and bottom walls to be perfectly conducting, while the side-walls to be insulating. The relevant nondimensionalized equations under Boussinesq approximation for the flow are

$$\partial_t \mathbf{v} + (\mathbf{v} \cdot \nabla)\mathbf{v} = -\nabla P + RaPrT\hat{y} + Pr\nabla^2 \mathbf{v}, \tag{1}$$

$$\partial_t T + (\mathbf{v} \cdot \nabla)T = \nabla^2 T, \tag{2}$$

$$\nabla \cdot \mathbf{v} = 0. \tag{3}$$

where $\mathbf{v} = u\hat{x} + v\hat{y}$ is the velocity field, T is the temperature field, the Rayleigh number Ra is the ratio of the buoyancy term and the nonlinear term, the Prandtl number Pr is the ratio of the kinematic viscosity and the thermal diffusivity, and $\hat{y}$ is the buoyancy direction.

We solve the equations in a unit square box for Prandtl number $Pr = 1$, and Rayleigh number $Ra = 2 \times 10^7$. For this parameter, the flow is turbulent. The flow also exhibits flow reversals, that is, probes near the vertical walls exhibit random reversals of the velocity field. For the simulation we use spectral element code NEK5000[10], and employ 28×28 spectral elements with seventh-order polynomials inside each element. Thus the overall resolution of the simulation is 196×196 grids. The aforementioned resolution is sufficient to resolve the boundary layers. We remark that a two-dimensional RBC in an experiment can be realised when the depth of the setup is much smaller compared to the height and the width; in such systems, the modes along the depth are not generated in a significant manner.

We ran our simulation till the system attains a steady state. During the steady state, we focus our attention on the flow during an interval from thermal diffusive time $t_A = 12.765$ to $t_B = 12.792$, between which a flow reversal is observed. We analyse the dominant flow structures during this interval.

For the POD analysis[4,5] we take 1000 snapshots of the flow in the aforementioned equidistant interval. The simulation data were interpolated on a uniform mesh of (192×192). For the POD analysis, we construct 1000 vectors using the two components of the velocity and the temperature field. Following this, a correlation matrix of these vectors was constructed[4,5]. First ten most energetic POD modes were identified using this correlation matrix[4,5], which contain 97.5% of the total energy. In Fig. 1 we illustrate the energy ratio E_p/E_1, where E_p represents the energy of the p-th POD mode. Clearly, E_p/E_1 decreases sharply with p ($E_1/E_2 \sim 14$). Also, the first POD mode contains 88% of the total energy. The first three POD modes a

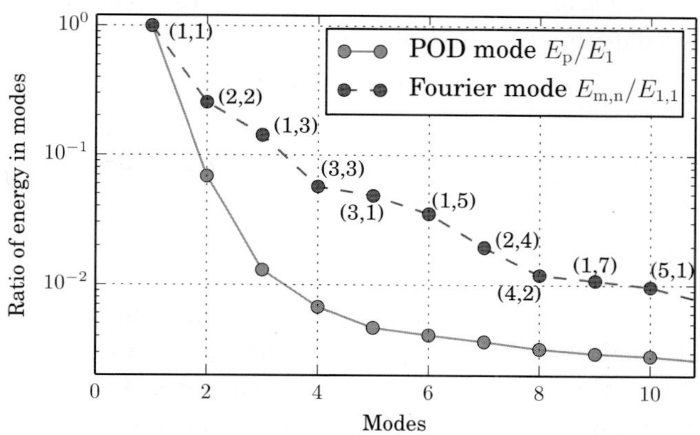

Fig. 1. Ratio of the energies of different POD modes with the 1st POD mode. Also the ratio $E_{m,n}/E_{1,1}$, where $E_{m,n}$ is the energy of the (m,n) Fourier mode.

The flow velocity at all the four walls is zero, i.e., $\mathbf{v} = 0$. A spectral decomposition of the no-slip boundary condition involves Chebyshev polynomials that resolve the boundary layers quite efficiently. Note however that the flow structures in the boundary layers are of "small scales", and they can be ignored while computing the large-scale flow structures. Chandra and Verma[7] observed that the free-slip basis functions defined below

$$u = \sum_{m,n} \hat{u}_{m,n}(\mathbf{k}) \sin(m\pi x) \cos(n\pi y)$$

$$v = \sum_{m,n} \hat{v}_{m,n}(\mathbf{k}) \cos(m\pi x) \sin(n\pi y) \tag{4}$$

capture the large-scale flow structures quite well. Here $k_x = m\pi$, $k_y = n\pi$ with m, n as positive integers, $\hat{u}_{m,n}, \hat{v}_{m,n}$ are the Fourier components of u and v respectively[7]. Note that the free-slip basis functions do not satisfy the no-slip boundary condition. The success of the free-slip basis functions in capturing the large-scale structures is due to the fact that ignored modes at the boundary layers belong to the small-scale structures.

The temperature field satisfies the conducting boundary conditions at the horizontal plates (T=constant), and insulating boundary conditions ($\partial T/\partial x = 0$) at the vertical walls. It is customary to separate T into conducting and convective part, i.e.,

$$T(x, y) = \bar{T}(y) + \theta(x, y) \tag{5}$$

where $\bar{T}(y) = 1 - y$ is the conduction profile for the nondimensionalized system, and θ is the temperature fluctuations over $\bar{T}(y)$. The aforementioned thermal boundary conditions yields:

$$\theta = \sum_{m,n} \hat{\theta}_{m,n}(\mathbf{k}) \cos(m\pi x) \sin(n\pi y) \tag{6}$$

The Fourier transforms can be performed for each snapshot independently, which is one of the main advantages of this analysis, contrary to POD that requires many snapshots. However to study the dynamics and evolution of the flow structures during a reversal, we study 1000 snapshots of the flow. We compute the Fourier modes for each frame [Eqs. (4, 6)]. The energy of the higher wavenumber modes decreases sharply, consistent with the Kolmogorov theory of turbulence. Since our focus is on the large-scale structures, we take the first 10 Fourier modes that contain approximately 93% of the total kinetic energy. In Fig. 1 we plot the ratio $E_{m,n}/E_{1,1}$, where $E_{m,n}$ represents the kinetic energy of the (m, n) Fourier mode. The Fourier analysis reveals that the modes $(1, 1)$, $(2, 2)$, and $(1, 3)$ are the most dominant modes in the flow. The first Fourier mode has 58% of the total energy. The Fourier modes of the flow are shown in right column of Fig. 2 as $F1$-$F3$.

When we compare the two methods for the extraction of large-scale flow structures, we observe that POD modes are more optimal than the Fourier modes. For example, the first 10 POD and first 10 Fourier modes contain respectively 97.5% and 93% of the total energies. Also, the first POD mode contains 88% of the total energy, but the corresponding Fourier mode contains only 58% of the total energy. However, the flow structures of the first three POD modes are distinctly similar in their shape to the first three Fourier modes, but the higher order modes differ. Note however

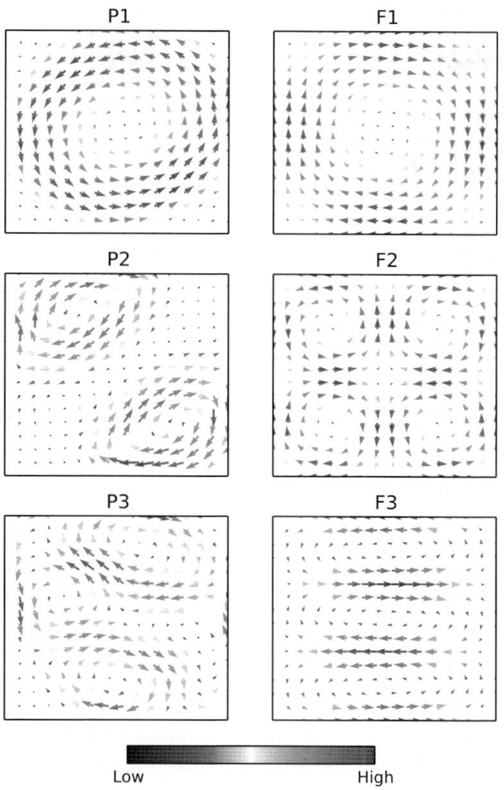

Fig. 2. Plots of the three most energy POD modes (left column) and the corresponding Fourier modes (right column). The top three POD modes are (P1) POD-1, (P2) POD-2, and (P3) POD-3. The top three energetic Fourier modes are (F1) (1,1) mode, (F2) (2,2) mode and (F3) (1,3) mode.

that the directions of the velocity fields of the POD and Fourier modes are anti-correlated, hence the amplitudes of these modes are also opposite to each other (to be discussed below).

The amplitude of a POD mode is computed by projecting the snapshot to the POD mode. The time series of the first five Fourier and POD modes are exhibited in Fig. 3 that shows that the amplitudes of the POD and Fourier modes are anti-correlated to each other. This is consistent with the anti-correlation of the velocity fields for the corresponding POD and Fourier modes (see Fig. 2). During the flow reversal between $t = 12.78$ and $t = 12.785$, the first Fourier mode, as well as the first POD mode, change sign. The second and third modes also show sharp variations during the flow reversal[7]. The fourth and fifth Fourier modes also show noticeable variations during the reversal process, but the corresponding POD modes do not show noticeable variations during this period. This is due to the fact that top three POD modes are most dominant in the flow, while the fourth and the fifth POD modes are quite weak. Thus both Fourier and POD modes provide valuable information on the dynamics of flow reversal, details of which can be found in Sergent and Podvin[8,9], and Chandra and Verma[7].

The reconstruction of the snapshots using the POD and Fourier modes is shown in Fig. 4 where we show the 5-th and 670-th snapshots of the flow. The 5-th snapshot is reproduced accurately by both the POD and Fourier analysis. However for the 670-th snapshot, which depicts the corner-roll flow structure, POD reconstruction is poorer than the Fourier one. The reason for the discrepancy is due to the averaging process of the POD analysis. On the average, the first POD mode is more dominant than the second POD-mode ($E_2/E_1 = 1/14$). For the 670-th snapshot, the second POD mode is the most important mode, but its contribution towards the reconstruction of the snapshot get suppressed by the most dominant first POD mode. On the other hand, Fourier modes correctly reconstructs the 670-th snapshot since each snapshot has its own set of Fourier modes, and the (2,2) mode is more dominant than the (1,1) mode for the 670-th snapshot. Hence Fourier reconstruction is better than the POD reconstruction for dynamic flows.

3. Conclusions and Discussions

In this paper we proposed that the low-wavenumber Fourier modes are good candidates for identifying large-scale flow structures, and they provide an alternative to POD analysis. In support of our argument, we perform a comparative study of the POD and Fourier analysis of 1000 snapshots of the flow profiles in a two-dimensional Rayleigh Bénard convection.

The low wavenumber Fourier modes capture the large-scale structures of the flow quite well. However, the first N POD modes contain more energy than the corresponding Fourier modes. Due to this, a reproduction of a flow pattern requires fewer POD modes than the Fourier modes, but the difference is not very significant. Note however that Fourier modes can be computed for each snapshot separately, but the POD analysis requires a large set of snapshots. A simple estimation shows that the Fourier analysis

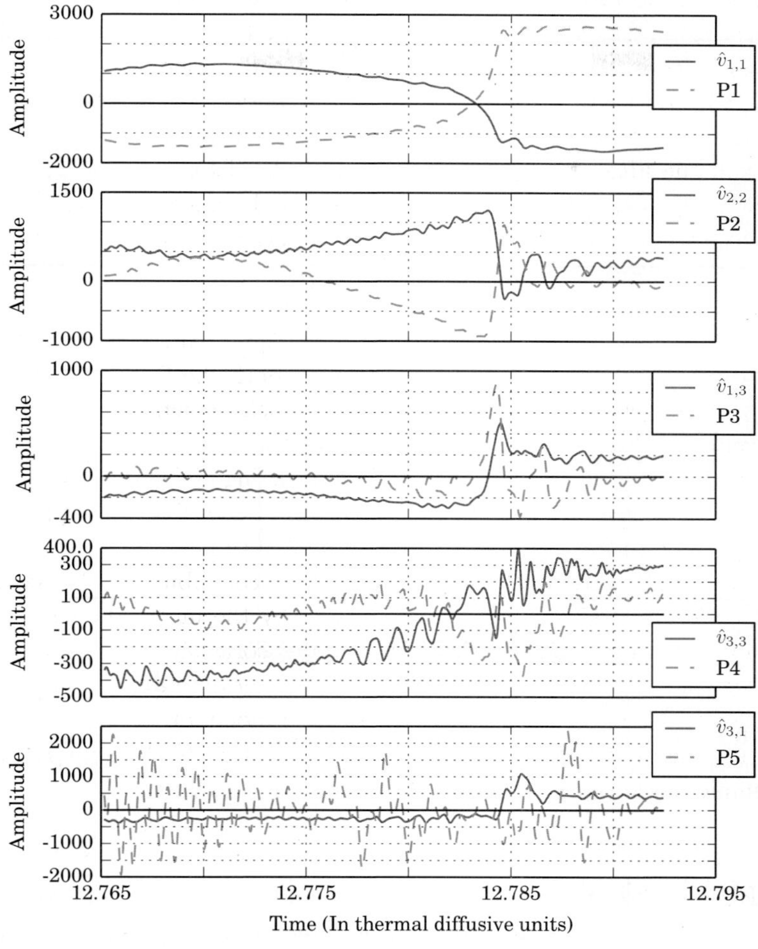

Fig. 3. Plot of the time series of the most energetic five Fourier and POD modes of the vertical velocity v during the reversal. POD coefficients are scaled appropriately for visual clarity.

is computationally less expensive than the POD analysis. This is useful while dealing with large three-dimensional datasets. Moreover, the Fourier modes have simpler visual interpretations than the POD modes.

A major objection to the usage of Fourier analysis is its inapplicability to no-slip boundary conditions. However, the small-scale structures of a boundary layer do not contribute significantly to the large-scale structures.

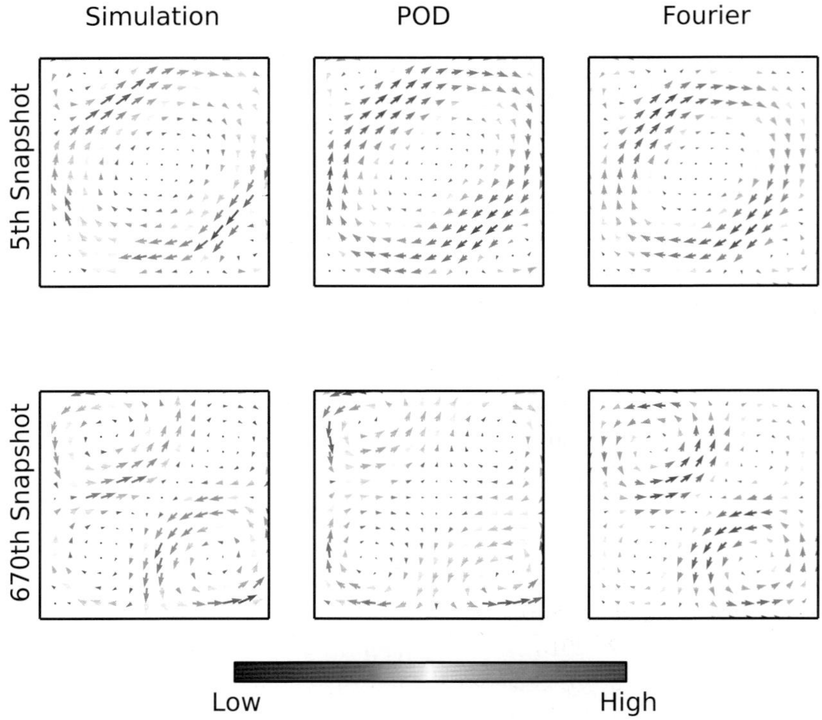

Fig. 4. Reconstruction of the 5th and 670th snapshots using the first ten POD and ten Fourier modes. Both the POD and Fourier reconstruction are reasonably good, but the the 670th snapshot is better reconstructed using the Fourier modes.

The RBC example discussed in this paper and in Chandra and Verma[7] show that free-slip basis can capture the large scale flow structures even under no-slip geometries quite well.

Thus, POD and Fourier analysis have their own advantages and disadvantages. For idealised geometries like a box, Fourier analysis provides an attractive alternative to POD.

Acknowledgment

The authors thank M. Chandra for sharing with us the RBC data. SP thanks C-DAC's CFD group for support and encouragement. The analysis was performed on PARAM-YUVA.

References

1. D. D. Kosambi, *J. Ind. Math. Soc.*, **7**, 76 (1943).
2. J. L. Lumley, "The structure of inhomogeneous turbulent flows," in Atmospheric Turbulence and Radio Wave Propagation, (Nauka, Moscow), p. 166, (1967).
3. G. Berkooz, P. Holmes and J. L. Lumley, *Annu. Rev. Fluid Mech.* **25**, 539 (1993).
4. L. Sirovich, *Q. Appl. Math.* **45**, 561 (1987); L. Sirovich, *Q. Appl. Math.* **45**, 573 (1987); L. Sirovich, *Q. Appl. Math.* **45**, 583 (1987).
5. L. Sirovich, L. Maxey and H. Tarman, "Analysis of turbulent thermal convection", *In Sixth Symposium on Turbulent Shear Flow*, Toulouse, France, p. 68. Springer (1989); A. E. Deane and L. Sirovich, *J. Fluid Mech.* **222**, 231 (1991).
6. M. Lesieur, "Turbulence in Fluids", (Kluwer Academic, Dordrecht), (1990).
7. M. Chandra and M. K. Verma, *Phys. Rev. E* **83**, 067303 (2011); M. Chandra and M. K. Verma, *Phys. Rev. Lett.* **110**, 114503 (2013)
8. B. Podvin and A. Sergent, *Phys. Fluids* **24**, 105106 (2012); *J. Fluid Mech.* **766**, 172 (2015).
9. J. Bailon-Cuba, O. Shishkina, C. Wagner and J. Schumacher, *Phys. Fluids* **24**, 107101 (2012).
10. Paul F. Fischer, James W. Lottes, and Stefan G. Kerkemeier, "nek5000 Web page", http://nek5000.mcs.anl.gov, (2008).

Energy Spectrum and Flux of Buoyancy-Driven Turbulence

Mahendra K. Verma*, Abhishek Kumar+ and Anando G. Chatterjee++

Department of Physics, Indian Institute of Technology Kanpur, Kanpur, 208016, India

E-mail: *mkv@iitk.ac.in, +abhkr@iitk.ac.in and ++anandogc@iitk.ac.in

Using direct numerical simulation we demonstrate that for isotropic stably-stratified flows follow Bolgiano-Obukhov scaling, i.e, the kinetic energy spectrum $E_u(k) \sim k^{-11/5}$, the entropy spectrum $E_\theta(k) \sim k^{-7/5}$, and kinetic energy flux $\Pi_u(k) \sim k^{-4/5}$. This is due to the conversion of kinetic energy to potential energy because of buoyancy. For Rayleigh Bénard convection, we argue that $\Pi_u(k)$ is a nondecreasing function of k due to the positive energy supply rate by buoyancy. Our numerical simulations show that the convection turbulence follows Kolmogorov-like scaling ($E_u(k) \sim k^{-5/3}$) with Kolmogorov's constant approximately 1.4.

Keywords: Startified turbulence; Rayleigh Bénard convection; Direct numerical simulation; Buoyancy-driven turbulence.

1. Introduction

Buoyancy-driven flows are observed in planets, stars, galaxies, as well as in electronic devices and industrial applications. Understanding of such systems are critical for modeling complex fluid flows, e.g., terrestrial atmosphere, weather predictions, planetary boundary layers, etc. One of the most important topic in this field is to study the scaling of spectrum and fluxes of kinetic energy (KE, $u^2/2$), and that of entropy ($\theta^2/2$). Here $\mathbf{u}$ and θ are the velocity and temperature fluctuations[1,2]. In this paper, we will compute these quantities using direct numerical simulations and show that the spectrum differs from Kolmogorov's theory when buoyancy is strong.

The dynamical equations describing the buoyancy-driven flows under the Boussinesq approximation are

$$\frac{\partial \mathbf{u}}{\partial t} + (\mathbf{u} \cdot \nabla)\mathbf{u} = -\frac{\nabla \sigma}{\rho_0} + \alpha g \theta \hat{z} + \nu \nabla^2 \mathbf{u} + \mathbf{f^u}, \tag{1}$$

$$\frac{\partial \theta}{\partial t} + (\mathbf{u} \cdot \nabla)\theta = \frac{d\bar{T}}{dz} u_z + \kappa \nabla^2 \theta, \tag{2}$$

$$\nabla \cdot \mathbf{u} = 0, \tag{3}$$

where $\mathbf{u}$ is the velocity field, θ and σ are the temperature and pressure fluctuations, respectively, with reference to the conduction state, $\hat{z}$ is the buoyancy direction, $\mathbf{f^u}$ is the external force field, d is the characteristic length, and ρ_0, ν, and κ are the fluid's mean density, kinematic viscosity, and thermal diffusivity respectively. Gravity-driven flows exhibit variety of nonlinear phenomena depending on the strength of gravity and the external forcing. We categorise these phenomena in terms of the important parameters of buoyancy-driven turbulence:

$$\text{Brunt Väisälä frequency } N_f = \sqrt{\frac{g}{\rho_0}\left|\frac{d\bar{\rho}}{dz}\right|} = \sqrt{\alpha g\left|\frac{d\bar{T}}{dz}\right|} \tag{4}$$

$$\text{Rayleigh number Ra} = \frac{d^4\alpha g}{\nu\kappa}\left|\frac{d\bar{T}}{dz}\right| \tag{5}$$

$$\text{Froude number Fr} = \frac{u_{\text{rms}}}{dN_f} \tag{6}$$

$$\text{Richardson number Ri} = \frac{\alpha g d^2}{u_{\text{rms}}^2}\left|\frac{d\bar{T}}{dz}\right| = \frac{1}{\text{Fr}^2} \tag{7}$$

$$\text{Reynolds number Re} = \frac{u_{\text{rms}} d}{\nu} \tag{8}$$

$$\text{Prandtl number Pr} = \frac{\nu}{\kappa} \tag{9}$$

where u_{rms} is the rms velocity of flow. The Brunt Väisälä frequency is the frequency of the gravity wave, the Froude number is the ratio of the characteristic fluid velocity and the gravitational wave velocity, and the Richardson number is the ratio of the buoyancy and the nonlinearity $(\mathbf{u} \cdot \nabla)\mathbf{u}$.

The generic behaviour for various range of parameters are:

(1) **Stably stratified flows**: When $d\bar{T}/dz > 0$, the flow is stable and we observe the following behaviour.

   (a) *Internal gravity waves*: When the nonlinearity is weak ($\text{Re} \ll 1$), we obtain internal gravity waves as the solution of the above equations. The frequency of gravity wave is N_f. Weak nonlinearity induces weak turbulence in the gravity waves.

   (b) *Quasi two-dimensional turbulence*: When the nonlinearity becomes significant and gravity too is strong ($\text{Fr} \ll 1$), the flow along the buoyancy direction is suppressed, and it becomes quasi two-dimensional. Such flows have been studied extensively by Lindborg[3,4], Brethouwer *et al.*[5], and others.

(c) *Approximately isotropic turbulence*: When the gravity is relatively small but nonlinearity is strong (Re $\gg 1$), the flow becomes isotropically turbulent. When the Richardson number, which is the ratio of the buoyancy force and the nonlinearity, is of the order of unity, we obtain Bolgiano-Obukhov scaling (to be described below). However when the Richardson number is small, we obtain the usual Kolmogorov's spectrum.

(2) **Convective flows**: When $d\bar{T}/dz < 0$, the flow is intrinsically unstable and it becomes convective. The system exhibits convective rolls and patterns for small nonlinearity, but turbulence for large nonlinearity. We will show below that such flows in the turbulent limit exhibit approximate Kolmogorov's spectrum.

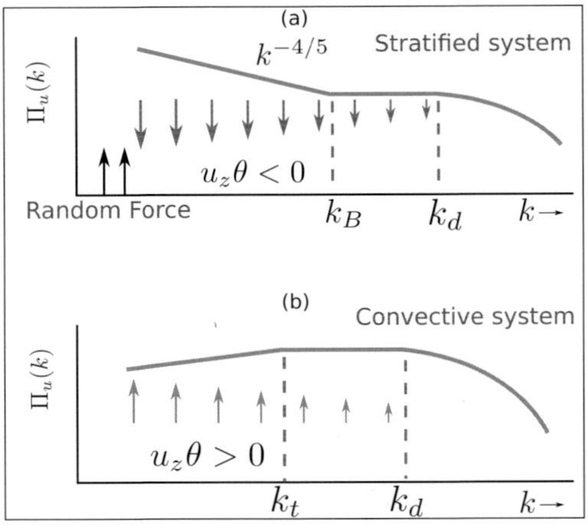

Fig. 1. Schematic diagrams of energy flux $\Pi_u(k)$: (a) In a stably stratified flow, $\Pi_u(k)$ decreases with k due to a negative energy supply rate $\Re\langle u_z(k)\theta^*(k)\rangle$. (b) In a thermally driven flow (e.g., Rayleigh Bénard convection), $\Pi_u(k)$ is a non-decreasing function of k (in inertial range) due to a positive $\Re\langle u_z(k)\theta^*(k)\rangle$. Thus $\Pi_u(k)$ first increases for $k < k_t$, where $F(k) > D(k)$, then $\Pi_u(k) = const$ $k_t < k < k_d$, where $F(k) \approx D(k)$; $\Pi_u(k)$ decreases for $k > k_d$ where $F(k) < D(k)$. Here k_d is the Kolmogorov's dissipation wavenumber. [Figure adapted from Kumar *et al.*[13]]

For stably stratified flows, Bolgiano[6] and Obukhov[7] first proposed a phenomenology, according to which

$$E_u(k) = c_1(\alpha^2 g^2 \epsilon_\theta)^{2/5} k^{-11/5}, \tag{10}$$
$$E_\theta(k) = c_2(\alpha g)^{-2/5} \epsilon_\theta^{4/5} k^{-7/5}, \tag{11}$$

$$\Pi_\theta(k) = \epsilon_\theta = \text{constant}, \tag{12}$$

$$\Pi_u(k) = c_3(\alpha^2 g^2 \epsilon_\theta)^{3/5} k^{-4/5}, \tag{13}$$

where k is the wavenumber, ϵ_θ is the entropy dissipation rate, $E_u(k)$ is the kinetic energy spectrum, $E_\theta(k)$ is the entropy spectrum, $\Pi_u(k)$ is the kinetic energy flux, and $\Pi_\theta(k)$ is the entropy flux, and c_i's are constants. The KE flux of a stably stratified flow is depleted at different length scales due to conversion of kinetic energy to "potential energy"; here $F(k) = \Re\langle u_z(k)\theta^*(k)\rangle < 0$. As a result, $\Pi_u(k)$ decreases with wavenumber (see Fig. 1(a)), and the energy spectrum is steeper than that predicted by Kolmogorov theory ($E(k) \sim k^{-5/3}$). The aforementioned scaling (Eqs. (1-4)) is referred to as *BO* scaling, while the usual Kolmogorov-scaling $k^{-5/3}$ is called Kolmogorov-Obukhov (*KO*) scaling. Note that the aforementioned scaling assumes that the turbulence is fully developed and isotropic. In this paper, we demonstrate that the stably stratified flows exhibit *BO* scaling when buoyancy is strong, and it exhibits *KO* scaling when the buoyancy is weak.

Procaccia and Zeitak[8], L'vov[9], L'vov and Falkovich[10], and Rubinstein[11] proposed that a similar scaling is also applicable to Rayleigh-Bénard convection (RBC) with the assumption that $F(k) < 0$. In this paper, we demonstrate using numerical simulations and phenomenological arguments that $F(k) > 0$ for RBC (see Fig. 1(b)). Consequently $\Pi_u(k)$ would increase with k, and $E_u(k)$ would be either Kolmogorov-like or shallower then $k^{-5/3}$. These observations of non-decreasing $\Pi_u(k)$ contradict the earlier predictions on RBC[8–10], but they are in agreement with the numerical results of Škandera *et al.*[12]. For further details see Kumar *et al.*[13] and Verma *et al.*[14].

2. Numerical Methods

We solve the following non-dimensionalized form of Eqs. (1,2,3):

$$\frac{\partial \mathbf{u}}{\partial t} + (\mathbf{u}\cdot\nabla)\mathbf{u} = -\nabla\sigma + \theta\hat{z} + \sqrt{\frac{\mathrm{Pr}}{\mathrm{Ra}}}\nabla^2\mathbf{u} + \mathbf{f^u}, \tag{14}$$

$$\frac{\partial \theta}{\partial t} + (\mathbf{u}\cdot\nabla)\theta = S u_z + \frac{1}{\sqrt{\mathrm{RaPr}}}\nabla^2\theta, \tag{15}$$

$$\nabla\cdot\mathbf{u} = 0. \tag{16}$$

These equations have been nondimensionalized using d as the length scale, $\sqrt{\alpha g \Delta d}$ as the velocity scale, and Δ as the temperature scale. The parameter $S = (d\bar{T}/dz)/(\Delta/d)$, where Δ is the absolute temperature differences in height d, takes values ± 1. For RBC, the temperature of the top plate is lower than that of bottom one, hence $S = +1$, but for the stably stratified flows, the temperature gradient is opposite, i.e., $S = -1$.

We perform our direct numerical simulation of the above equations using a pseudospectral code Tarang[15] in a three-dimensional periodic box of size $(2\pi)^3$. We use fourth-order Runge-Kutta (RK4) method for time stepping, Courant-Freidricks-Lewey (CFL) condition for computing time step Δt, and 3/2 rule for dealiasing. We apply a random force to the flow in the band $2 \leq k \leq 4$ using the scheme of Kimura and Herring[16] to obtain a steady turbulent flow.

For RBC, we performed a simulation in unit box with 512^3 grid. For the horizontal plates, we employ a free-slip boundary condition for the velocity field and a conducting boundary condition for the temperature field. For the vertical walls, we apply a periodic boundary condition for both fields. Note that for RBC $\mathbf{f^u} = 0$ since the buoyancy makes the flow turbulent by itself.

3. Results and Discussions

3.1. *Stably stratified turbulence*

Table 1. Parameters of our numerical simulations for stably stratified turbulence (SST) and Rayleigh Bénard convection (RBC): flow type, grid size, Rayleigh number Ra, Richardson number Ri, Froude number Fr, Reynolds number Re, kinetic energy dissipation rate ϵ_u, thermal dissipation rate ϵ_θ, and averaged Δt. We choose Prandtl number Pr = 1 for both the runs.

Flow Type	Grid	Ra	Ri	Fr	Re	ϵ_u	ϵ_θ	Δt
SST	1024^3	5×10^3	0.01	1.4	649	114	150	3.5×10^{-6}
RBC	512^3	10^7	16	NA	790	8.8×10^{-3}	10^{-3}	6.2×10^{-4}

We perform large-resolution simulations for Pr = 1 and Rayleigh number Ra $= 5 \times 10^3$. We report our results when the flow has reached a steady state. We find that the Richardson number Ri = 0.01. Thus buoyancy is comparable to the nonlinear term in our simulation. The other parameters of our runs are listed in Table 1.

We compute the KE and entropy spectra and fluxes for the steady-state data of the above run. Figures 2(a, b) indicate that the *BO* scaling fits with the numerical data better than the *KO* scaling. The KE and entropy

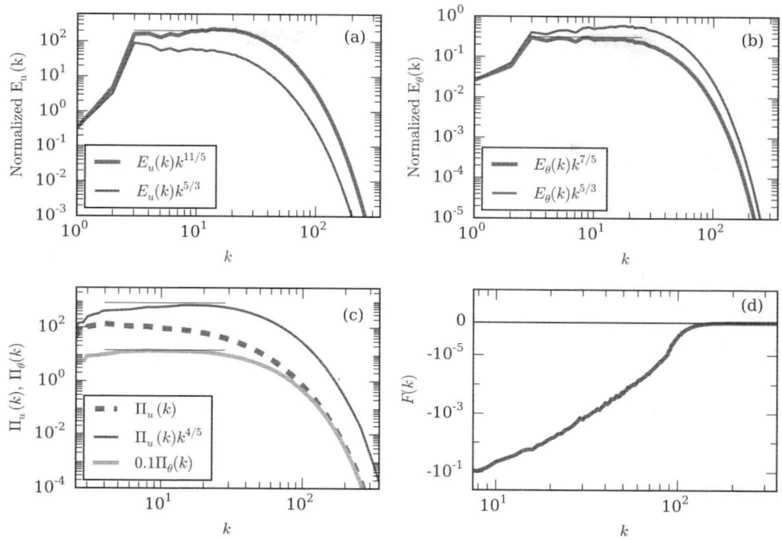

Fig. 2. For stably stratified turbulence (SST) with Pr $= 1$ and Ri $= 0.01$, plots of (a) normalized KE spectra, (b) normalized entropy spectra for Bolgiano-Obukhov (*BO*) and Kolmogorov-Obukhov (*KO*) scaling, (c) KE flux $\Pi_u(k)$, normalized KE flux $\Pi_u(k)k^{4/5}$, and entropy flux $\Pi_\theta(k)$, and (d) negative $F(k)$ for stably stratified flow.

fluxes are plotted in Fig. 2(c). Clearly, the KE flux $\Pi_u(k)$ is positive, and it decreases approximately as $k^{-4/5}$, consistent with the *BO* predictions [Eq. (13)]. The entropy flux Π_θ is a constant, and the energy supply rate by buoyancy, $F(k) = \Re\langle u_z(k)\theta^*(k)\rangle$, is negative (see Fig. 2(d)). These results show that the *BO* scaling is valid for stably stratified flows for Ri $= O(1)$. We remark that the flow is approximately isotropic.

3.2. *Rayleigh Bénard convection with* Pr $= 1$

We performed a numerical simulation of RBC with for Pr $= 1$ and Rayleigh number Ra $= 10^7$ (see Table 1). We employ free-slip boundary condition for our simulation. We report the spectra and fluxes of kinetic energy and entropy $(\theta^2/2)$ after the flow has reached a steady state.

In Fig. 3(a), we plot the normalized KE spectra for the *BO* and the *KO* scaling. The plots indicate that the *KO* scaling fits better than the *BO* scaling for a narrow band of wave numbers (the shaded region, $15 < k < 40$). In Fig. 3(b), we plot the entropy spectrum that exhibits a dual branch, with the upper branch scaling as k^{-2}, whose origin is related to the

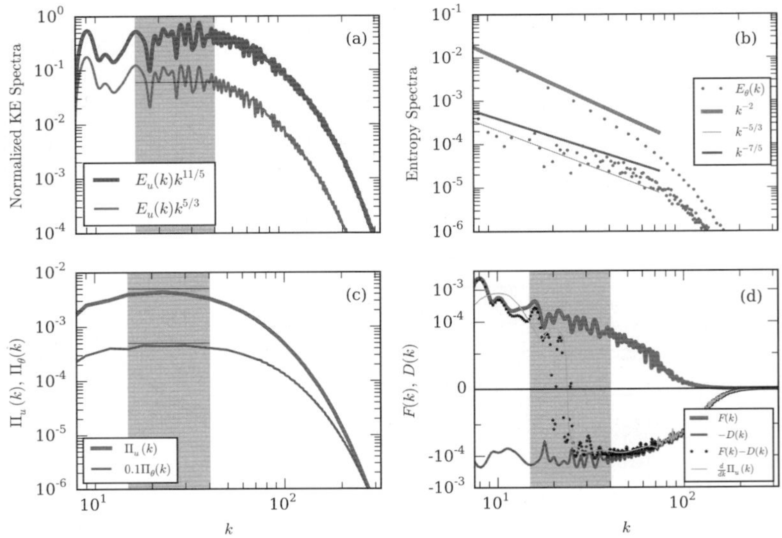

Fig. 3. For convective turbulence (RBC) with Pr $= 1$ and Ra $= 10^7$, plots of (a) normalized KE spectra for (BO) and (KO) scaling; (b) entropy spectrum that exhibits a dual branch. The upper branch of $E_\theta(k)$ matches with k^{-2} quite well, while the lower part is fluctuating. (c) KE flux $\Pi_u(k)$ and entropy flux $\Pi_\theta(k)$; (d) $F(k)$, $D(k)$, $F(k) - D(k)$, and $d\Pi_u(k)/dk$. Shadow region shows the inertial range.

observation that $\hat{\theta}(0, 0, 2n) \approx -1/(2n\pi)$, where $\hat{\theta}$ is the Fourier transform of the temperature fluctuations[17]. The KE and entropy fluxes, shown in Fig. 3(c), demonstrate that the fluxes are constant, consistent with the KO scaling.

For RBC we observe that $F(k) > 0$, as shown in Fig. 3(d). In Fig. 3(d), we also plot $D(k)$, $F(k) - D(k)$, and $d\Pi_u(k)/dk$. For the wavenumbers band $15 < k < 40$, $F(k) \approx D(k)$, hence $d\Pi_u(k)/dk \approx 0$ or $\Pi_u(k) \approx const.$ This constant KE flux yields $E_u(k) \sim k^{-5/3}$. Note that the aforementioned Kolmogorov's $-5/3$ spectrum is not due to a constant energy flux, but due to a cancellation of the energy supply rate with the dissipation rate. For our run, the Reynolds number Re ≈ 790, which is moderate. For Re ≈ 790, the Kolmogorov's length is $\eta = \mathrm{Re}^{-3/4}L \approx L/133$, where $L = 1$ is the size of the box. Consequently the dissipation starts at the length $l_d = 60\eta \approx L/2$ or at the wavenumber of $k_d = 2\pi/l_d \approx 4\pi$, which is consistent with our results exhibited in Fig. 3.

The aforementioned arguments show that the behaviour of $F(k)$ and

$\Pi_u(k)$ for the stably stratified flow and RBC are quite different[13]. Our numerical results are consistent with our flux-based arguments, but they differs from those of Procaccia and Zeitak[8], L'vov[9], and L'vov and Falkovich[10].

We also compute Kolmogorov's constant for the above run. We find that in the inertial range $\Pi_u \approx \epsilon_u \approx 8.8 \times 10^{-3}$, and $E_u(k)k^{5/3} \approx 0.06$. Therefore,

$$K_{Ko} = \frac{E_u(k)k^{5/3}}{\Pi_u^{2/3}} \approx 1.4, \tag{17}$$

which is in agreement with the theoretical and experimental estimates.

The Richardson number for the RBC run is approximately 16, which is related to the Péclet number Pe as following:

$$\begin{aligned} \mathrm{Ri} &= \frac{\alpha g d (d\bar{T}/dz)}{u^2/d} \\ &= \frac{\alpha g d^4 (d\bar{T}/dz)}{\nu\kappa} \frac{\nu}{\kappa} \frac{\kappa^2}{u^2 d^2} \\ &= \frac{\mathrm{RaPr}}{\mathrm{Pe}^2}. \end{aligned} \tag{18}$$

Verma *et al.*[18] showed that

$$\mathrm{Pe} \approx 0.24\sqrt{\mathrm{RaPr}}. \tag{19}$$

Hence $\mathrm{Ri} \approx 1/0.24^2 \approx 16$, which matches very well with our numerical results.

3.3. *Rayleigh Bénard convection for small and large* Pr

For RBC, the turbulence behaviour for very small or very large Prandtl number differs significantly from that with moderate Prandtl numbers ($\mathrm{Pr} \sim 1$). For zero and small Prandtl number, Mishra and Verma[17] showed that the temperature fluctuations and hence the buoyancy are concentrated at low wavenumbers. This feature makes Kolmogorov phenomenology applicable for zero and small Prandtl numbers. Therefore the kinetic energy spectra for zero and small Prandtl number are Kolmogorov-like.

For infinite and large Prandtl numbers, Pandey *et al.*[19] showed that the viscous dissipation dominate the nonlinear term $\mathbf{u} \cdot \nabla \mathbf{u}$. Consequently, the kinetic energy spectrum scales $k^{-13/3}$. The entropy spectrum exhibits dual branch (k^{-2} and another branch) for all Prandtl number flows.

In summary, RBC turbulence with small to moderate Prandtl numbers exhibit Kolmogorov scaling, while that with large and infinite Prandtl number exhibit $E_u(k) \sim k^{-13/3}$.

4. Summary and Conclusions

In summary, we analyse stably-stratified and convective turbulence using numerical simulations. For the stably-stratified flows, we restrict ourselves to approximate isotropic regime that occurs for $\mathrm{Fr} \gtrsim 1$ and $\mathrm{Ri} \lesssim 1$. For such stably stratified flows, our numerical results on the energy spectrum, the energy flux, and the energy supply rate by buoyancy reveal that they exhibit Bolgiano-Obukhov BO scaling. We show that the energy supply rate due to buoyancy, $F(k)$, is negative, hence the kinetic energy is transferred to the potential energy, which is dissipated.

We remark that the stably stratified flows become quasi two-dimensional when $\mathrm{Fr} \ll 1$. The turbulent behaviour of such flows is very different, and it is being actively studied[3,4].

For turbulence in RBC with moderate Prandtl number, we show that the kinetic energy flux is a non-decreasing function of k (since $F(k) > 0$), and the energy spectrum of KE cannot be steeper than $k^{-5/3}$ in the bulk. We observe $k^{-5/3}$ energy spectrum for kinetic energy. The flux of kinetic energy and entropy remain approximately constant here. We estimate Kolmogorov's constant to be around 1.4, which is in a very good agreement with theoretical and numerical results. When combined with earlier work on RBC, convective turbulence with small to moderate Prandt numbers exhibit Kolmogorov scaling, while that with large and infinite Prandtl number exhibit $E_u(k) \sim k^{-13/3}$.

5. Acknowledgments

We thank K. R. Sreenivasan, Anindya Chatterjee, Pankaj Mishra, and Mani Chandra for useful comments and suggestions, and T. K. Sengupta and the organizers for hosting the wonderful and interactive IUTAM meeting. We also thanks the editors of the proceedings for the useful suggestions on the paper. Our numerical simulations were performed at Centre for Development of Advanced Computing (CDAC) and IBM Blue Gene P "Shaheen" at KAUST supercomputing laboratory, Saudi Arabia. This work was supported by a research grant SERB/F/3279/2013-14 from Science and Engineering Research Board, India.

References

1. E. D. Siggia, *Ann. Rev. Fluid Mech.* **26**, 137 (1994).
2. D. Lohse and K. Q. Xia, *Ann. Rev. Fluid Mech.* **42**, 335 (2010).
3. E. Lindborg, *Geo. Res. Lett.* **32**, L01809 (2005).
4. E. Lindborg, *J. Fluid Mech.* **550**, 207 (2006).
5. G. Brethouwer, P. Billlant, E. Lindborg, and J.-M. Chomaz *J. Fluid Mech.* **585**, 343 (2007).
6. R. Bolgiano, *J. Geophys. Res.* **64**, 2226 (1959).
7. A. N. Obukhov, *Dokl. Akad. Nauk SSSR* **125**, 1246 (1959).
8. I. Procaccia and R. Zeitak, *Phys. Rev. Lett.* **62**, 2128 (1989).
9. V. S. L'vov, *Phys. Rev. Lett.* **67**, 687 (1991).
10. V. S. L'vov and G. E. Falkovich, *Physica D* **57**, 85 (1992).
11. R. Rubinstein, Icqmp-94-8, cmott-94-2, *NASA Tech. Memo.* (1994).
12. D. Škandera, A. Busse, and W. C. Müller, *High Performance Computing in Science and Engineering, (Springer, Berlin), Part IV*, 387 (2008).
13. A. Kumar, A. G. Chatterjee and M. K. Verma, *Phys. Rev. E* **90**, 023016 (2014).
14. M. K. Verma, A. Kumar, and A. G. Chatterjee, *Physics Focus* **25(1)**, 45 (2015).
15. M. K. Verma, A. G. Chatterjee, K. S. Reddy, R. K. Yadav, S. Paul, M. Chandra, and R. Samtaney, *Pramana* **81**, 617 (2013).
16. Y. Kimura and J. R. Herring, *J. Fluid Mech.* **698**, 19 (2012).
17. P. K. Mishra and M. K. Verma, *Phys. Rev. E* **81**, 056316 (2010).
18. M. K. Verma, P. K. Mishra, A. Pandey, and S. Paul *Phys. Rev. E* **85**, 016310 (2012).
19. A. Pandey, M. K. Verma, and P. K. Mishra *Phys. Rev. E* **89**, 023006 (2014).

DNS of a Buoyant Turbulent Cloud under Rapid Rotation

Avishek Ranjan* and P. A. Davidson

Department of Engineering, University of Cambridge, Cambridge, CB2 1PZ, UK
**E-mail: ar606@cam.ac.uk*

We conduct the DNS of a buoyant turbulent cloud under rapid rotation in a 512^3 periodic box at low Rossby numbers (Ro). The initial condition chosen is relevant to the equatorial plane of the Earth's inhomogeneous outer core. A random but uniform distribution of buoyant blobs of different sizes is centrally placed in the box. It is observed that columnar structures form and grow towards the box boundary. Using helicity as a diagnostic, we confirm that these structures are formed by inertial waves which arise from the buoyant cloud. A near perfect helicity segregation of negative in the north ($+z$) and positive in the south ($-z$) exists at $Ro = 0.01$ but not for $Ro = 0.1$. Moreover, at $Ro = 0.1$, there is a significant distortion in the buoyancy field which vanishes at $Ro = 0.01$. These observations are attributed to a wave-induced mean flow at higher Ro. The kinetic energy distribution shows oscillations in the buoyant cloud region. A POD analysis of horizontal velocity field indicates that these are stationary inertial waves of frequency 2Ω.

Keywords: Rotating turbulence; helicity; buoyant cloud; inertial waves

1. Introduction

Turbulence under rapid rotation exists in several geophysical flows, such as in the atmosphere, the oceans and even the core of the Earth. In most of these situations, the turbulence is primarily inhomogeneous. An example of such a case is a localized patch of turbulence with quiescent fluid on either side, called 'a turbulent cloud'. Recently, there have been some investigations of such freely decaying turbulent clouds under rotation [1,2] and stratification [3]. In this study, we consider the evolution of an unbounded turbulent cloud composed of buoyant blobs, 'a buoyant cloud', under rapid rotation. This initial condition is relevant to the equatorial plane in the outer core of the Earth where there is a high radial heat flux reported to be a source of inertial waves [4]. For the purpose of our study, we consider only the Coriolis and buoyancy forces and ignore other forces such as those due to magnetic field and density stratification. It is important to note that rotation and gravity are perpendicular to each other as is the case at the equator, i.e. $\boldsymbol{\Omega} = \Omega\hat{\boldsymbol{e}}_z$ and $\boldsymbol{g} = -g\hat{\boldsymbol{e}}_y$.

The relative strength of the inertia and Coriolis forces is characterized by the Rossby number, $Ro = u/2\Omega l$, where u and l are characteristic velocity and length scales, respectively. The momentum conservation in a rotating reference frame for an inviscid Boussinesq flow at low Ro (so that advection is small) can be written as

$$\frac{\partial \boldsymbol{u}}{\partial t} = -\frac{1}{\rho}\nabla\tilde{p} + 2\boldsymbol{u} \times \boldsymbol{\Omega} + c\boldsymbol{g}, \tag{1}$$

where $\tilde{p}$ is the modified pressure which includes the centrifugal force, $c = \rho^{'}/\rho$ and $\rho^{'}$ is the density perturbation. The vorticity equation, corresponding to Eq. (1), is

$$\frac{\partial \boldsymbol{\omega}}{\partial t} = 2(\boldsymbol{\Omega} \cdot \nabla)\boldsymbol{u} + \nabla c \times \boldsymbol{g}, \tag{2}$$

from which we can derive,

$$\frac{\partial^2}{\partial t^2}(\nabla^2 \boldsymbol{u}) + 4(\boldsymbol{\Omega} \cdot \nabla)^2 \boldsymbol{u} = 2(\boldsymbol{\Omega} \cdot \nabla)(\boldsymbol{g} \times \nabla c). \tag{3}$$

If there is no buoyancy, this equation supports internal waves of the form $\boldsymbol{u} = \hat{\boldsymbol{u}} \exp\left[\iota(\boldsymbol{k} \cdot \boldsymbol{x} - \varpi t)\right]$, called inertial waves, of frequency $\varpi = \pm 2(\boldsymbol{k} \cdot \boldsymbol{\Omega})/k$ and group velocity,

$$\boldsymbol{C_g} = \pm\frac{2\Omega}{k}(\hat{\boldsymbol{e}}_\Omega - \cos\theta\hat{\boldsymbol{e}}_k). \tag{4}$$

where θ is the angle between $\boldsymbol{\Omega}$ and $\boldsymbol{k}$. The low frequency inertial waves (for which $(\boldsymbol{k} \cdot \boldsymbol{\Omega}) \approx 0$) travel parallel to $\boldsymbol{\Omega}$ and with the highest group speed, $C_g = 2\Omega/k$. These waves are helical, and posses helicity ($h = \boldsymbol{u} \cdot \boldsymbol{\omega}$); the waves with negative helicity propagate upwards (i.e. along $+z$), and those with positive helicity propagate downwards[5] (along $-z$).

It may be inferred from Eq. (3) that buoyancy in the right hand side acts as a source of waves. It has been observed that inertial waves spontaneously emerge from a localized turbulent cloud under rotation and form columnar structures[1,2]. In the Earth's liquid outer core, although there is no clear separation of scales such as in fully developed turbulence, it is possible that the same linear wave propagation results in columnar convection cells as observed in dynamo simulations, otherwise attributed to Ekman pumping. To investigate this, we conduct direct numerical simulation (DNS) of a buoyant cloud under rotation.

2. Methodology

The pseudo-spectral DNS code used in this study (based on discrete Fourier transform) solves the full momentum equation in rotating frame (i.e. Eq. (1) plus non-linear and viscous terms) in a 512^3 periodic box of size 2π. Time-advancement is done using second-order predictor-corrector scheme; an integrating factor to account for rotation[6] is incorporated in the code. The de-aliasing of non-linear terms is carried out using phase-shifting and all modes greater than $\sqrt{2}N/3$ (where $N = 512$) are truncated.

2.1. *Initial condition*

In this study, the initial condition for DNS is a cloud of buoyant Gaussian blobs ($b_i \sim \Sigma \exp -[(x - X_i)^2 + (y - Y_i)^2 + (z - Z_i)^2]/\delta_i^2$) distributed randomly and uniformly in the center of the periodic box as shown in Fig. 1(a), where $b = cg$, δ_i is the size and (X_i, Y_i, Z_i) is the center of the blob i. A random distribution of Z_i (in a short range about the center) and X_i, Y_i (space filling in horizontal) creates a buoyant cloud with a vertical thickness $\sim 0.11 \times 2\pi$ (see Fig. 1(a) for example). Table 1 lists the two cases considered in the present study and other relevant parameters. Under quasi-steady conditions, Eq. (2) reduces to

$$2\Omega \frac{\partial u_z}{\partial z} \approx -g \frac{\partial c}{\partial x} = -\frac{\partial b}{\partial x}. \tag{5}$$

The input initial velocity field is zero in the entire domain. For a blob of size δ, using Eq. (5) we can write $u_z \sim b/\Omega$ which further implies that $Ro \sim b/2\Omega^2\delta$. The Ω is thus decided based on the Ro, the strength and characteristic size of the buoyancy field. The first time step calculation is based on the group speed of inertial waves, i.e. $C_g \sim 2\Omega\delta$ with a small Courant number of 0.02 to ensure adequate resolution. The simulation is carried out at a low Ro so that the buoyancy time scale is much larger than the rotation time scale which is important to observe inertial wave dynamics. The simulations are stopped after a very short time interval (less than or equal to one eddy turn over time) so as to minimize the effect of a periodic boundary.

3. Results and Discussion

3.1. *R0.1*

The DNS results are visualized in terms of the iso-surfaces of buoyancy b, z-component of velocity u_z, and iso-surfaces of u_z^2 coloured by helicity. For

Table 1. DNS cases considered in the present study.

Case	Ro	δ (% of box size)	Number of blobs
R0.1	0.1	2-5	1000
R0.01	0.01	1-2	10000

the case R0.1, Fig. 1 shows the buoyancy field at $\Omega t = 0, 4, 8$. Figures 2 (a-e) show the iso-surfaces of u_z at $\Omega t = 1, 2, 4, 6, 8$, respectively. Alternating pair of columnar cyclones and anticyclones form in the buoyant cloud and grow towards the box boundary. Moreover, these occur in a plane perpendicular to gravity, i.e. the XZ plane. Furthermore, we show that these columnar structures are low-frequency inertial wave packets. Figure 3 shows the iso-surfaces of ${u_z}^2/{u_z}^2_{max} = 0.1$ coloured by helicity at various times. Most columnar structures above (north) the buoyant cloud have negative helicity, while those below (south) the cloud have positive helicity, a property typical of inertial waves (refer §1). It can be noted in Fig. 1 that the buoyancy field gets distorted at $\Omega t = 8$ (Fig. 1(c)). These distortions occur at the positions from where the columnar structures emerge vertically. Similar distortion was observed earlier for a single buoyant Gaussian eddy[7] under rotation but was not attributed to inertial waves. In order to get a quantitative idea of the inertial wave propagation, we plot the total positive and negative helicity in all horizontal (XY) planes in $1 < Nz < 512$ shown in Fig. 4. It is observed that helicity produced in the buoyant cloud is symmetric upto $\Omega t = 2$. However, surprisingly so, it becomes asymmetric thereafter, i.e. more positive helicity than negative helicity is produced which propagates downwards ($-z$). To examine this, we consider the plane-averaged energy ($1/2u^2$) in Fig. 4(b). It is observed that, unlike helicity, the kinetic energy profile is vertically symmetric. Kinetic energy is generated by buoyancy in the cloud and carried away by inertial waves. However, it is interesting to see the kinetic energy profile oscillate inside the buoyant cloud. For instance, the peaks at $\Omega t = 3, 6$ are at lower kinetic energy than at $\Omega t = 2, 4$ respectively. To examine the above observations in more detail, we proceed to a different initial condition with greater number of smaller blobs (for enhanced homogeneity), at ten times lower Ro (note that the Ro in Earth's core is much lower $\sim 10^{-6}$).

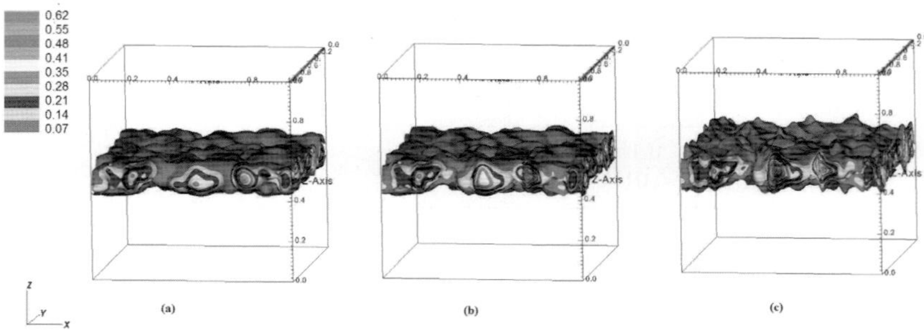

Fig. 1. R0.1: Iso-surfaces of buoyancy b at Ωt = (a) 0 (b) 4 (c) 8.

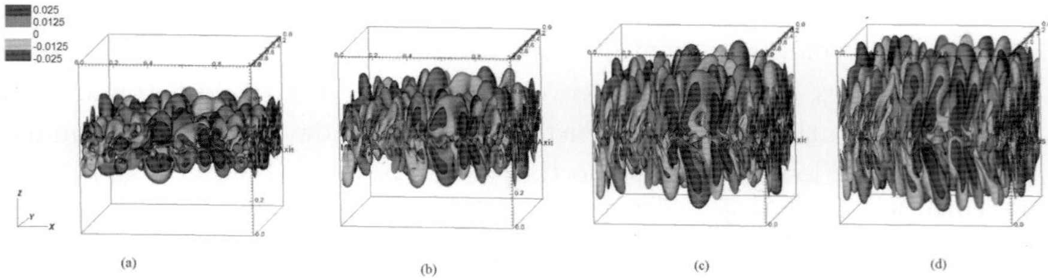

Fig. 2. R0.1: Iso-surfaces of u_z at Ωt = (a) 2 (b) 4 (c) 6 (d) 8 [note that the thresholds are lower for (a), (b) than for (c)-(d)].

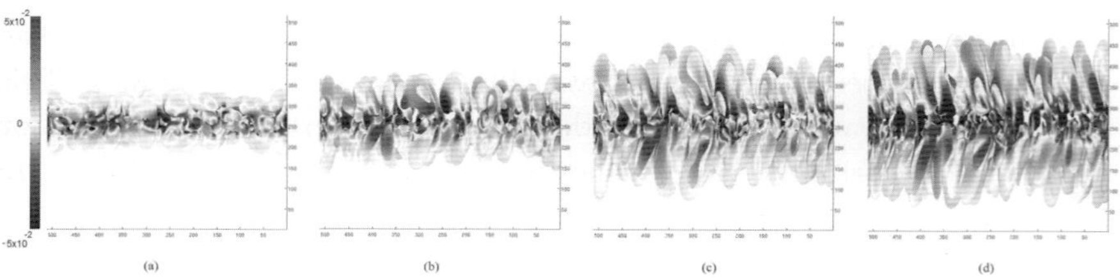

Fig. 3. R0.1: Iso-surfaces of ${u_z}^2/u_{z\,max}^2 = 0.1$ coloured by helicity (dark is positive, light negative) at $\Omega t = (a)2(b)4(c)6(d)8$. The axes are Nx, Nz respectively, i.e. grid points along X and Z axes respectively, where $Nz = (z/2\pi)512$; $Nx = (x/2\pi)512$.

3.2. *R0.01*

The buoyancy fields for the case R0.01 at $\Omega t = 0, 10$ are shown in Fig. 5. It is evident that the field remains roughly unchanged with no distortion that was observed for R0.1 (Fig. 1). We, therefore, conclude that the distortion is due to a wave-induced mean flow in the buoyancy field which is higher at higher Ro, and also leads to slightly-bent columnar structures for R0.1 as

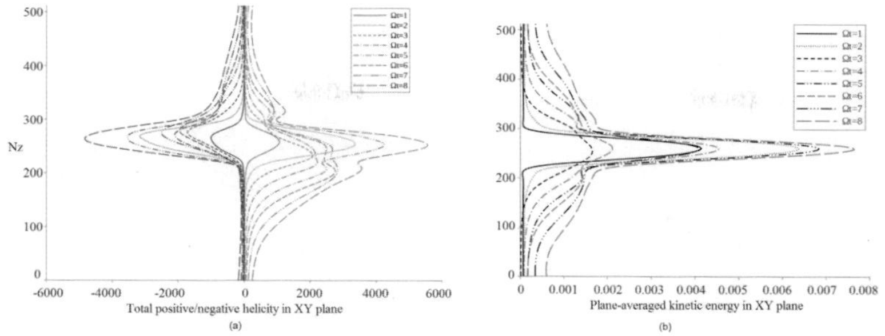

Fig. 4. R0.1: (a) Total positive and negative helicity (b) plane-averaged kinetic energy in horizontal planes, $1 < Nz < 512$.

see in Fig. 2 (the corresponding structures at R0.01, shown in Fig. 6, are nearly erect). The corresponding iso-surfaces of normalized u_z^2 are shown in Fig. 7. A near-perfect distribution of negative/positive helicity in the north/south is observed, which was not seen in R0.1. This is also evident by considering the quantitative vertical distribution of helicity in Fig. 8(a) which shows clear symmetry in helicity segregation by inertial waves. It is likely that a distortion in the buoyancy field causes the cyclones and anticyclones (with opposite sign of u_z) to overlap which results in a helicity of the opposite sign between the cyclone-anticyclone pair (as seen for R0.1).

The vertical distribution of kinetic energy is shown in Fig. 8(b). The oscillations in the buoyant cloud region are still present at $Ro = 0.01$. This shows that these are not linked to the Ro or the distortion in the buoyancy field. We suspect that these are caused by inertial waves of maximum frequency, $\pm\, 2\Omega$ (These waves are stationary and do not propagate any kinetic energy as their group velocity is zero, see Eq. (4) in §1). To ascertain this, we conduct a proper orthogonal decomposition (POD) analysis.

3.3. *POD analysis*

POD, also called principle-component analysis, is a popular mathematical tool[8] used in fluid mechanics to identify the most dominant flow structures present in an unsteady flow. In POD, the velocity field is expressed in terms of a linear combination of deterministic functions. Unlike the Fourier decomposition, POD modes are derived from the flow field itself, making it more physical. POD is also very efficient as the maximum energy of the flow is captured by the first few eigenvalues arranged in descending order

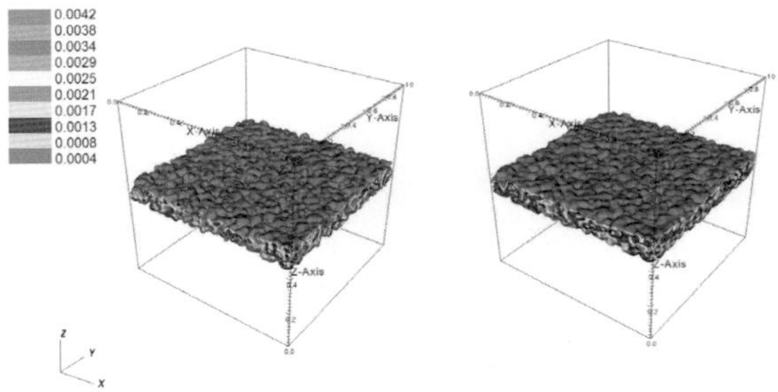

Fig. 5. R0.01: Buoyancy fields at $\Omega t = 0$ (left) and $\Omega t = 10$ (right). Image from [8].

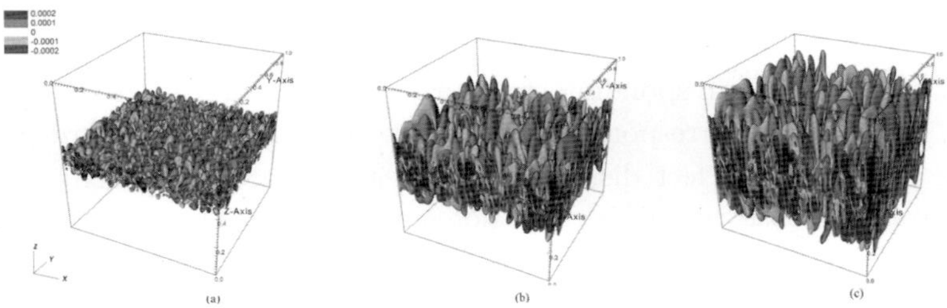

Fig. 6. R0.01: Iso-surfaces of u_z at $\Omega t =$ (a) 1 (b) 4 (c) 8. Image from [8].

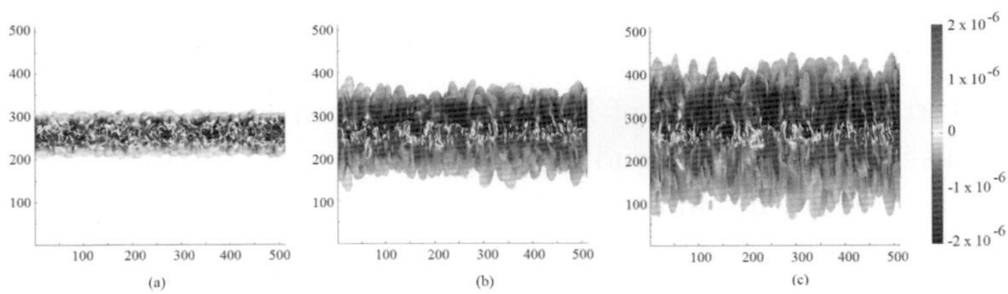

Fig. 7. R0.01: Iso-surfaces of ${u_z}^2/u_{z\,max}^2 = 0.025$ coloured by helicity (dark is positive, light negative) at $\Omega t =$ (a) 1 (b) 4 (c) 8. Image from [8].

and in the associated eigenmodes. Being a linear decomposition, it is ideal to study inertial waves. For discrete datasets, the implementation of POD is conveniently done through singular value decomposition (SVD)[9]. For our purpose, we consider the central vertical (XZ) plane of velocity fields u_x, u_y (we expect the 2Ω oscillations in horizontal direction) obtained from the DNS of R0.01. The input to the POD analysis is planar snapshots in time $\Omega t = 0.9 - 13.6$ (steps of $\Omega t = 0.1$) for both u_x, and u_y fields. The output is in terms of the 50 spatial and temporal modes. The first spatial mode (corresponding to the highest eigen value) shows the most dominant

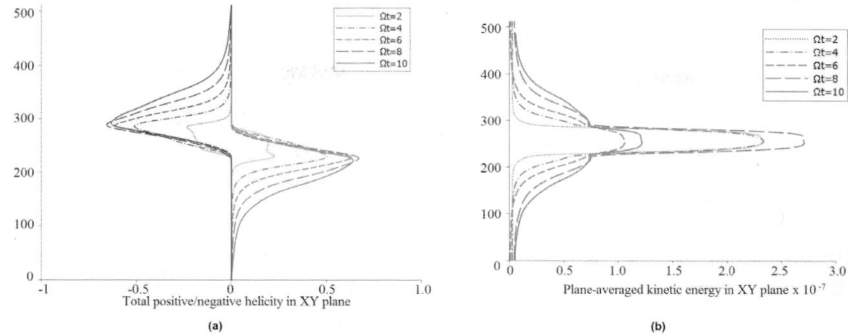

Fig. 8. R0.01: (a) Total positive and negative helicity (b) plane-averaged energy in horizontal planes. (a) is taken from [8].

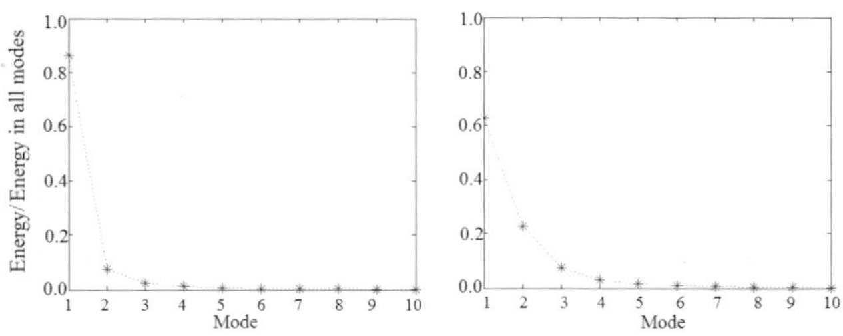

Fig. 9. Energy in a POD mode normalized by total energy in all 50 POD modes obtained from planar DNS data of u_x (left) field and u_y (right) field (central XZ plane).

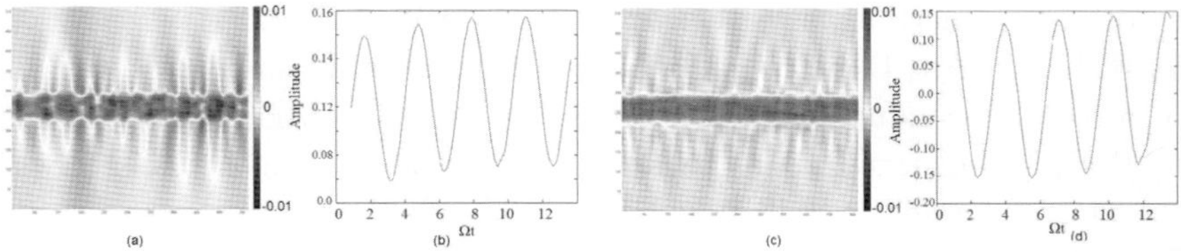

Fig. 10. First (a) spatial (b) temporal mode obtained from u_x field; second (c) spatial (d) temporal mode obtained from u_y field. Note that the first temporal mode obtained from u_x peaks at $\Omega t = 1.6$, 4.8, 7.9 and 11.0, while the second mode from u_y peaks at $\Omega t = 3.9$, 7.1, and 10.3 (at these times, the temporal modes looks exactly like the central XZ plane of the corresponding velocity component, not shown here).

flow structures, the second shows the next dominant ones, and so on. The temporal mode shows the time co-ordinates at which the corresponding spatial mode was more or less dominant. In our POD results, we find that the first 5 modes have almost all the energy (shown in Fig. 9) and thus capture almost all the information in the flow. The first spatial mode for u_x and the second mode for u_y are shown in Fig. 10(a, c), whereas the corresponding temporal modes are shown in Fig. 10(b, d) (We plot these modes only as they show oscillations). The expected time period of 2Ω oscillations is $1/(2\Omega/2\pi) = \pi/\Omega$ which comes out to be π if time is in Ωt units. This is exactly the time period that is observed in these temporal modes. This analysis proves beyond doubt that the oscillations in kinetic energy seen earlier (Figs. 4(b), 8(b)) are due to stationary inertial waves. The other dominant spatial modes not shown here, such as the first mode obtained from u_y, u_z and second mode from u_x, show tall columnar structures formed by the low-frequency inertial waves, as is expected.

The authors are grateful to Prof P K Yeung (Georgia Tech, US) for sharing DNS code.

References

1. P. A. Davidson, P. J. Staplehurst and S. B. Dalziel, On the evolution of eddies in a rapidly rotating system, *J. Fluid. Mech.* **557**, 135 (2006).
2. A. Ranjan and P. A. Davidson, Evolution of a turbulent cloud under rotation, *J. Fluid. Mech.* **756**, 488 (2014).
3. A. Maffioli, P. A. Davidson, S. B. Dalziel and N. Swaminathan, The evolution of a stratified turbulent cloud, *J. Fluid. Mech.* **739**, p. 229 (2014).
4. P. A. Davidson, The dynamics and scaling laws of planetary dynamos driven by inertial waves, *Geophys. J. Int.* **198**, 1832 (2014).
5. H. K. Moffatt, Dynamo action associated with random inertial waves in a rotating conducting fluid, *J. Fluid. Mech.* **44**, p. 705 (1970).
6. P. K. Yeung and Y. Zhou, Numerical study of rotating turbulence with external forcing, *Phy. Fluids* **10**, p. 2895 (1998).
7. M. G. St. Pierre, On the local nature of turbulence in earth's outer core, *Geophys. Astrophy. Fluid Dyn.* **83**, p. 293 (1996).
8. D. D. Kosambi, Statistics in function space, *J. Ind. Math. Soc.* **7**, p. 76 (1943).
9. L. Sirovich, Turbulence and the dynamics of coherent structures part i: coherent structures, *Q. App. Math.* **45**, p. 561 (1987).

Numerical Simulation of Shock-Bubble Interaction using High Order Upwind Schemes

Abhishek Kundu* and Sudipta De**
CSIR-Central Mechanical Engineering Research Institute,
Durgapur 713209, India
**Research Scholar, AcSIR-CMERI,*
E-mail: abhishekkundu45@gmail.com
***s_de@cmeri.res.in*

We compute the evolution of a helium bubble in ambient air after a shock of Mach 1.22 passes over it. A thirteenth order upwind scheme is employed for this purpose, in the framework of the AUSM$^+$ algorithm. The basic features of this class of problems have been captured, and the present numerical results match well with earlier experimental and numerical findings. In addition, the high spectral accuracy and low numerical dissipation of the explicit Euler solver reveals features of the deformed bubble – particularly after the formation of small-scale rolled-up vortices due to the Kelvin-Helmholtz instability – that are difficult to detect with low order upwind schemes.

Keywords: Shock-bubble interaction; Richtmyer-Meshkov instability; high resolution upwind scheme; AUSM$^+$ algorithm; Euler equations.

1. Introduction

Passage of a shock wave over a gas bubble present in another gas generates patterns at the interface separating the two gases – a phenomenon known as the Richtmyer-Meshkov instability (RMI) [1] – named after Richtmyer [2], who first presented a rigorous theoretical and numerical analysis of this subject, and Meshkov [3], who experimentally confirmed the predictions of Richtmyer. Initial amplification of perturbations is due to the baroclinic torque term in the equation of vorticity – which represents vorticity generation from misalignment of density and pressure gradients. A density gradient is always present at the interface of two different gases. The pressure gradient across a passing shock wave would generally have a different direction compared to the density gradient, resulting in the RMI. After the initial growth of disturbances, Kelvin-Helmholtz (KH) instability takes over, creating a layer of small scale rolled-up structures and intense turbulent mixing at the interface. Numerical simulation of this problem therefore requires application of schemes with high accuracy in the wavenumber space. The presence of shock waves also means that one has to apply extra artificial diffusion locally in the vicinity of shocks. This can be achieved explicitly, as in the Jameson-Schmidt-Turkel scheme[4], or through flux limiters[5] designed to convert a large-stencil interpolation formula to a form limited to one or two cells. In this paper we follow the latter approach. It is also possible to take a weighted sum of different interpolation schemes to construct the numerical flux, as in the WENO class of schemes[6]. But the WENO schemes are computationally expensive, and may even require further hybridization with compact schemes[7], increasing their complexity further. It has already been demonstrated for the one dimensional wave equation that application of explicit artificial diffusion can produce results similar to those from more complex WENO schemes[8]. In the present paper, we assess the performance of a thirteenth order upwind scheme in the AUSM^{+}[9] framework for shock-bubble interaction.

An important application of the RMI is in promoting 'mixing between fuel and oxidizer in supersonic and hypersonic air-breathing engines'[1]. Before we apply our high resolution methodology in more complex reactive solvers, we need to carefully estimate its capability for a class of problems where high accuracy is required together with effective limiting of fluxes. We have already tested our numerical procedure for the Schardin's problem[10], and the present paper tests it for a gas mixture in multiple-gamma formulation, as outlined by Marquina and Mulet[12].

2. Problem Definition and Numerical Set-up

The RMI was experimentally studied by Haas and Sturtevant[13] for cylindrical and spherical gas bubbles. The cylindrical bubble was formed by circulating a test gas (helium) within a $0.5\mu m$ thin nitrocellulose membrane supported at the ends by thicker Pyrex windows. A shock wave was then passed through this cylindrical gas volume, and the resulting deformation of the gas volume was observed. This problem was numerically investigated by Quirk and Karni[11] and later by Marquina and Mulet[12]. Our computational domain is the same as that in Marquina and Mulet. It is a 445×44.5 rectangle, as shown in Fig. 1. The Mach 1.22 shock moves from right to left, interacting with the cylindrical bubble. Only the upper half of the bubble is considered in the computational domain, with symmetry conditions at the centreline. The right boundary is the inflow, where we apply Riemann invariant based boundary condition. The outflow boundary at the left uses extrapolation based on local one-dimensional inviscid (LODI) relations[14]. However, the time of travel of the bubble during which we observe it does not seem to require use of complex boundary conditions and a simple extrapolation of variables is likely to serve the purpose equally well[11]. The top boundary is a free-slip wall. The MPI based parallel solver solves the Euler equations on a 8000×800 grid, using a time step which is between 0.002 to 0.005. A four-stage Runge-Kutta scheme is used for time integration[15]. We solve the Euler equations on a two dimensional Cartesian grid. To handle the two-component gas mixture, a separate equation is solved for the mass fraction of helium. The complete procedure is documented by Marquina *et al.*[12] in details.

The thirteenth order scheme that has been used is given by

$$\begin{aligned} u_{j+\frac{1}{2},L} = \frac{1}{360360}(&30u_{j-6} - 425u_{j-5} + 2851u_{j-4} - 12164u_{j-3} \\ &+37886u_{j-2} - 97249u_{j-1} + 263111u_j + 211631u_{j+1} \\ &-58639u_{j+2} + 16436u_{j+3} - 3584u_{j+4} + 511u_{j+5} - 35u_{j+6}) \end{aligned} \tag{1}$$

The above scheme constructs the cell-face fluxes in the AUSM^+ algorithm. A similar form can be written for the right state. The numerical dissipation associated with this scheme is active only in the high wavenumber range, and this makes the scheme suitable for tracking the evolution of small scale vortices. The amplification factor and the group velocity[16] of the scheme are displayed in Fig. 2. Very accurate compact schemes for solving unsteady flow phenomena have appeared in recent literature[17], and these may be utilized to solve the present problem in future.

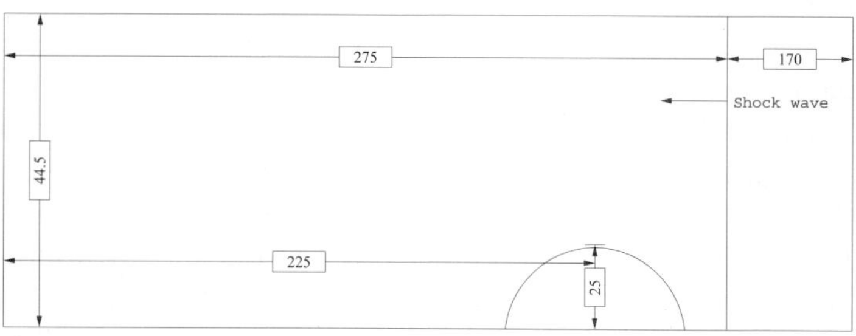

Fig. 1. Schematic of the computational domain (not to scale). All dimensions are in mm.

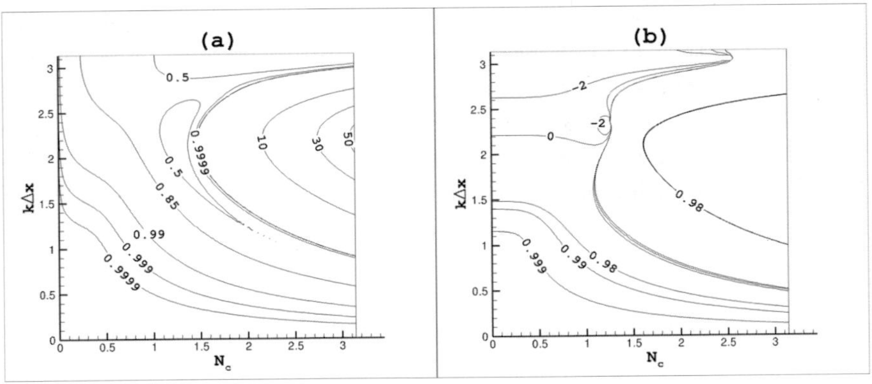

Fig. 2. The (a) amplification factor and (b) group velocity of the 13th order scheme.

3. Results and Discussion

The sequence of events after the passage of the shock is shown in Fig. 3. After the shock impinges on the upstream end of the bubble, a reflected wave travels backwards, and a transmitted wave passes through the bubble, compressing it longitudinally in the process. At t = 53 μs (frame (a) in the figure), the transmitted shock is seen leaving the bubble at its downstream end and retransmitted to the ambient air. The reflected shock is also seen behind the upstream (right) edge of the bubble. The part of the incident shock that does not interact with the bubble remains connected with the bubble for a longer duration. Immediately behind the transmitted shock, two secondary transmitted shocks are also observed, as noticed in[13]. Initial circular shape of the bubble does not remain circular for long, and as observed in the experiment of Haas and Sturtevant, deforms into a kidney shaped structure, shown in Fig. 3, frame (e). With time, the length of the bubble increases, which is a characteristic feature of a lighter gas bubble in a heavier medium[18]. Along the centreline, ambient air pushes the upstream side of the bubble towards the downstream side even after the departure of the shock, forming a jet-like structure. This results in both the ends of the bubble finally connecting to each other (frames (g) and (h)), which nearly divides the originally cylindrical bubble into two halves. This penetration of the helium bubble by surrounding air is nothing but the 'spike' of the RMI[13]. This spike rolls-up due to the KH instability at later stages[1]. In Fig. 4 we observe the bubble at t = 556.326 μs. The small-scale vortices visible at the bubble-air interface are the ones that are caused by the KH instability. This roll-up is more prominent at the upstream side of the bubble - most of these structures being at the lower end where the ambient air is about to penetrate the bubble. At this point of time, the vorticity on the rolled-up structures at the interface is mostly negative. Streaks of positive vorticity are found surrounding the negative small-scale vortices. Positive vorticity is also seen deposited at the downstream side of the bubble. This vorticity plays an important role in the 'turbulent mixing zone' (TMZ)[1] that forms later.

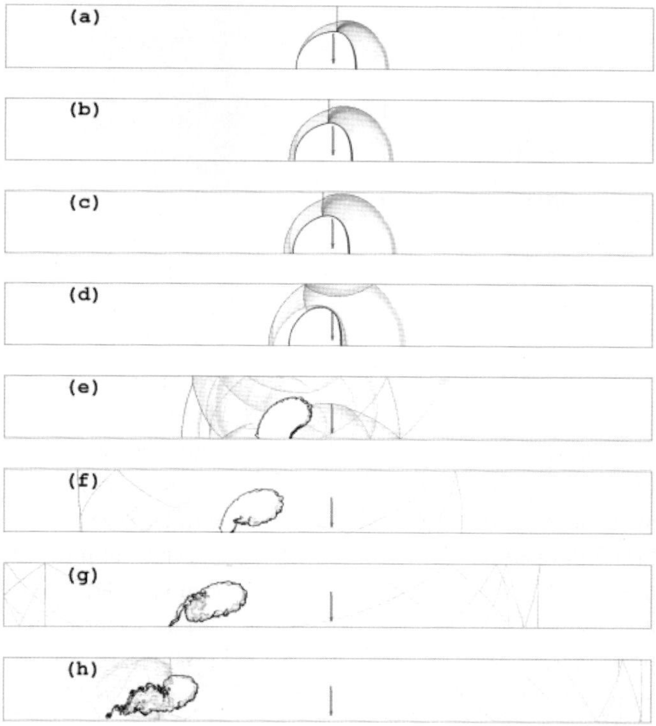

Fig. 3. Numerical schlieren[12] of shock-bubble interaction, starting from the instant when the shock touches the upstream (right) edge of the bubble: (a) 53 μs, (b) 66 μs, (c) 75 μs, (d) 102 μs, (e) 260 μs, (f) 445 μs, (g) 674 μs, (h) 983 μs. Computed using the AUSM$^+$ algorithm with a 13th order upwind scheme. The arrow indicates the initial location of the bubble centre.

In the later stages of the deformation of a spherical helium bubble in air with an incident shock at Mach 1.45, Ranjan *et al.*[19] noted formation of a pair of vortex rings after the 'inversion' (penetration) of the bubble

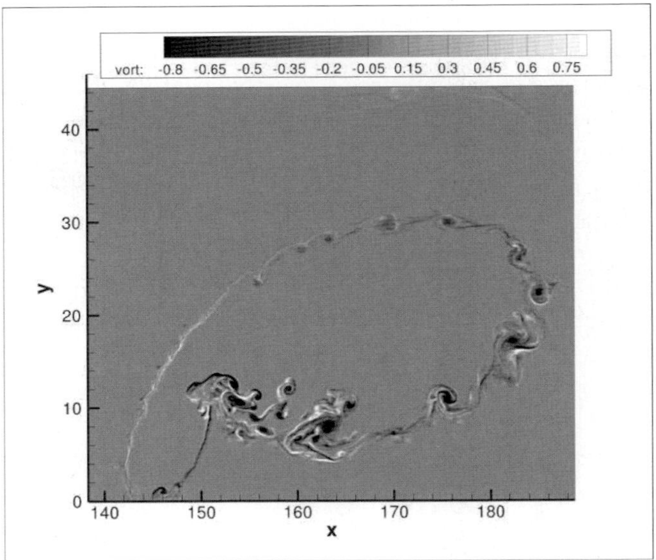

Fig. 4. Vorticity at $t = 556.326\mu s$: Kelvin-Helmholtz instability generated roll-up at the interface. 32 contour levels have been plotted between -0.8 and 0.8 at an increment of 0.05, excluding 0.

by surrounding air. The trailing ring fell behind the primary ring ahead (downstream) of it, causing axial elongation of the bubble. This was an experimental observation. Numerically, no distinct vortex ring could be seen in the trailing 'lobe', which had 'pure, unmixed helium' in it. We are not computing with a pure helium bubble. In our case it is a cylindrical bubble of helium-air mixture deformed by a Mach 1.22 shock. Despite these differences, we look into the later stages of development for similar phenomena.

Figure 5 shows the mass fraction of helium at t = 983 μs. At this stage the further division of the bubble initiates. At the downstream side, penetration of outside air is more prominent, with local pockets of helium-rich gas. The upstream 'lobe', on the other hand, has more of the original helium-air mixture. This is similar to the numerical observation of Ranjan *et al.*, as mentioned above. Figure 6 shows the bubble at a slightly more advanced time, when the downstream side of the bubble has allowed more air to come in, the upstream side still maintaining the nearly pure original mixture. Only at the interface of the upstream side one notices presence of outside air. This confirms the numerical observation of Ranjan *et al.*[19].

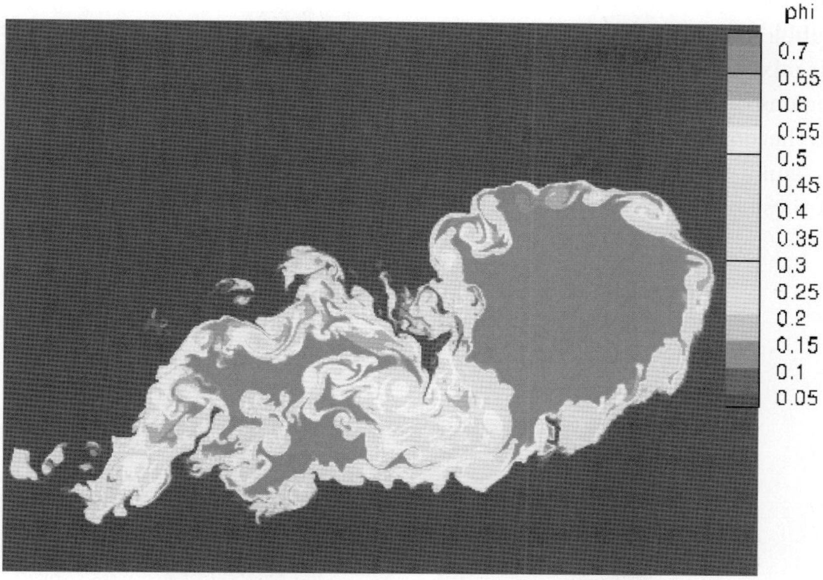

Fig. 5. Mass fraction (ϕ) of helium at $t = 983\mu s$. The relatively undisturbed upstream part of the bubble (ϕ close to 0.72) has initial gas mixture within it, which is helium contaminated by 28% air.

Though from experimental observations it is generally thought that two separate vortices form, these are nothing but interface vorticity entrained by the jet that penetrates the bubble from its upstream side. It appears that after penetrating the bubble almost completely, the air in the jet is pulled outward. At this point the vorticity at the interface has a chance to enter the bubble interior, the downstream side of which fills up with these small-scale vortices through filamentation and mixing. Figure 7 shows the corresponding vorticity contours.

4. Concluding Remarks

An Euler solver has been developed using a thirteenth order upwind scheme that computes the evolution of a helium bubble (helium-air mixture with 72% helium) after it interacts with a passing shock wave. The basic flow features at different stages of evolution noted in earlier experimental and computational studies have been identified. The high resolution nature of the scheme reveals more features at an advanced stage when the bubble is almost divided into upper and lower halves. The downstream side of the bubble is seen to have greater mixing of outside air with the original helium-air mixture forming the bubble. Instead of formation of two distinct vortical regions on each side of the centreline at late stages, we notice a single one at the downstream side which is more of a turbulent mixing zone formed by small-scale vortices of mainly clockwise vorticity. Careful examination of these results would allow us to characterize the bubble in greater details.

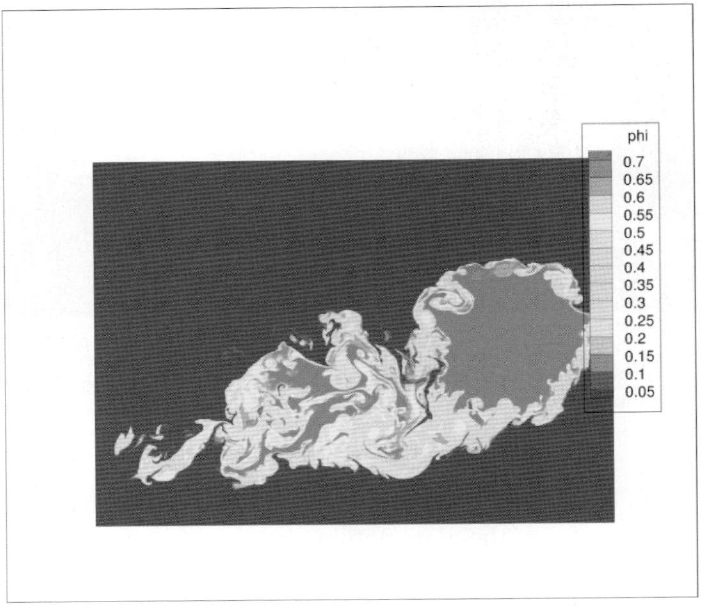

Fig. 6. Mass fraction (ϕ) of helium at $t = 1052.43\mu s$. The relatively undisturbed upstream part of the bubble (ϕ close to 0.72) has initial gas mixture within it, which is helium contaminated by 28% air.

References

1. M. Brouillette, The Richtmyer-Meshkov instability, *Ann. Rev. Fluid Mech.*, **34**, 445 (2002).
2. R. D. Richtmyer, Taylor instability in shock acceleration of compressible fluids, *Comm. Pure Appl. Math.*, **8**, 297 (1960).
3. E. E. Meshkov, Instability of the interface of two gases accelerated by a shock wave, *Sov. Fluid Dyn.*, **4**, 101 (1969).

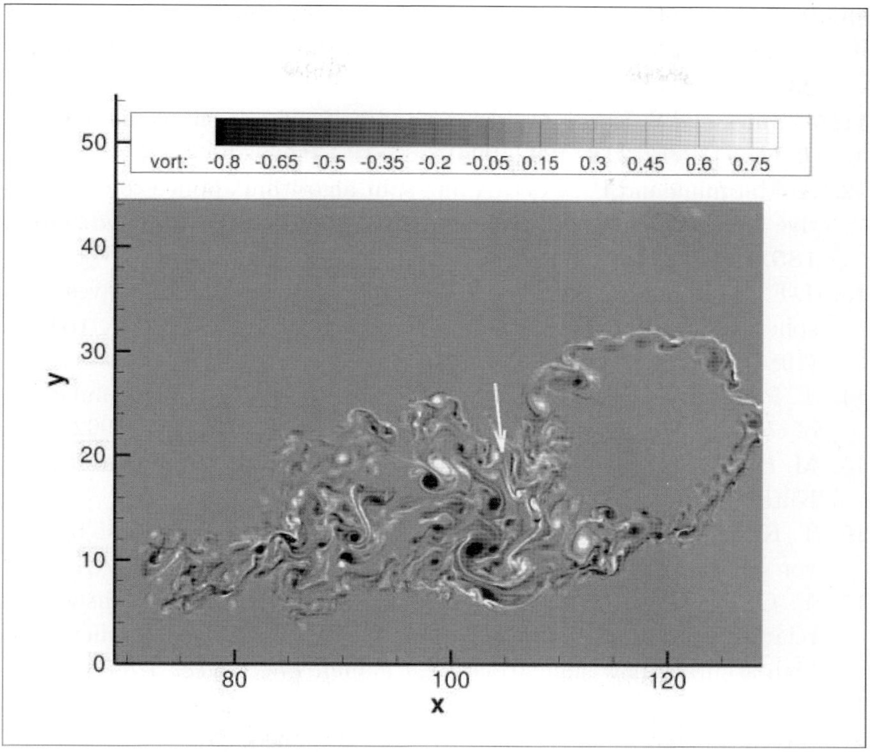

Fig. 7. Vorticity at $t = 1052.43\mu s$.

4. A. Jameson, W. Schmidt and E. Turkel, Numerical simulation of the Euler equations by finite volume methods using Runge-Kutta time-stepping schemes, *AIAA Paper 1981-1259, AIAA 14th fluid and plasma dynamic conference* (Palo Alto, California, 1981).
5. C. Hirsch, *Numerical Computation of Internal and External Flows, Volume 2* (John Wiley, Chichester, West Sussex, England, 1995).
6. G. S. Jiang and C.-W. Shu, Efficient implementation of weighted ENO schemes, *J. Comput. Phys.*, **126**, 202 (1996).
7. S. Pirozzoli, Conservative hybrid compact-WENO schemes for shock-turbulence interaction, *J. Comput. Phys.*, **178**, 81 (2002).
8. T. K. Sengupta, G. Ganerwal and A. Dipankar, High accuracy compact schemes and Gibbs' phenomenon, *J. Sci. Comput.*, **21** (3), 253 (2004).
9. M.-S. Liou, A sequel to AUSM: AUSM$^+$, *J. Comput. Phys.*, **129**, 364 (1996).

10. P. Halder, S. De, K. P. Sinhamahapatra and N. Singh, Numerical simulation of shock-vortex interaction in Schardin's problem, *Shock Waves*, **23**, 495 (2013).
11. J. Quirk and S. Karni, On the dynamics of a shock-bubble interaction, *J. Fluid Mech.*, **318**, 129 (1996).
12. A. Marquina and P. Mulet, A flux-split algorithm applied to conservative models for multicomponent compressible flows, *J. Comput. Phys.*, **185**, 120 (2003).
13. J.-F. Haas and B. Sturtevant, Interaction of weak shock waves with spherical and cylindrical gas inhomogeneities, *J. Fluid Mech.*, **181**, 41 (1987).
14. T. Poinsot and S. K. Lele, Boundary conditions for direct simulations of compressible viscous flows, *J. Comput. Phys.*, **101**, 104 (1992).
15. M. H. Carpenter and C. A. Kennedy, Fourth-order 2N-storage Runge-Kutta schemes, NASA TM 109112 (1994).
16. T. K. Sengupta, A. Dipankar and P. Sagaut, Error dynamics: beyond von Neumann analysis, *J. Comput. Phys.*, **226**, 1211 (2007).
17. Y. G. Bhumkar, T. K. Sengupta and T. W. H. Sheu, A dispersion relation preserving optimized upwind compact difference scheme for high accuracy flow simulations, *J. Comput. Phys.*, **278**, 378 (2014).
18. G. Layes, G. Jourdan and L. Houas, Experimental investigation of the shock wave interaction with a spherical gas inhomogeneity, *Phys. Fluids*, **17**, 028103 (2005).
19. D. Ranjan, J. H. Niederhaus, J. G. Oakley, M. H. Anderson, R. Bonazza and A. Greenough, Shock-bubble interactions: features of divergent shock-refraction geometry observed in experiments and simulations. *Phys. Fluids*, **20**, 036101 (2008).

Rayleigh-Taylor Instability of a Miscible Fluid at the Interface: Direct Numerical Simulation

*Ashish Bhole[a], Soumyo Sengupta[b], Aditi Sengupta[c], K. S. Shruti[d], Nidhi Sharma[a]

[a]*Department of Aerospace Engineering, IIT Kanpur, Kanpur, U.P., 208 016, India, *E-mail: ashishbhole07@gmail.com*

[b]*Department of Mechanical and Aerospace Engineering, OSU, Columbus, Ohio, USA, E-mail: soumyoss08@gmail.com*

[c]*Department of Mechanical Engineering, MIT Manipal, Manipal, Karnataka, 576 104, India, E-mail: dolaaditi@gmail.com*

[d]*Department of Space Engineering & Rocketry, BIT Mesra, Ranchi, Jharkhand, 835 215, India, E-mail: ks.aero91@gmail.com*

In the present study, a novel methodology of 2D Direct Numerical Simulation (DNS), incorporating non-Boussinesq, inhomogeneous formulation with non-periodic boundary conditions is used to study the Rayleigh-Taylor Instability (RTI) of miscible fluid at the interface. The evolution of RTI is shown with density contours and mixing length. The system considered here is completely isolated and hence entropy evolution of the system reported here is relevant from non-equilibrium thermodynamics point of view.

Keywords: Rayleigh-Taylor instability, Buoyancy driven thermal instability, DNS, Inhomogeneous flow

1. Introduction

RTI is a flow instability triggered when a heavier fluid rests on top of a lighter fluid to experience an unsettling moment. Traditionally, RTI has been studied using an inviscid mechanism[1] and the corresponding viscous mechanism has been reported in[2–4]. Two fluids at different temperatures experience an instability at the plane interface by which these try to acquire a least combined potential energy. A nondimensional Atwood number[5] (A) has been defined as $A = (\rho_u - \rho_l)/(\rho_u + \rho_l)$, where ρ is the density and subscripts u and l refer to upper and lower fluids. For our case, $A = 0.1056$ corresponding to a temperature difference $\Delta T = T_l - T_u$ of 70 K. The novelty of the present study is that we have circumvented the use of Boussinesq approximation by conducting 2D DNS of RTI using compressible Navier-Stokes equations. Mikaelian[6] has noted *direct numerical simulations are the best way to assess the validity of Boussinesq approximation.* The dispersion relation preserving (DRP) property of the numerical method[8] adopted, $OUCS3$ compact scheme for spatial discretization and optimized three-stage Runge-Kutta method for time integration, has helped to incorporate the high accuracy required to compute instabilities affected by high wavenumbers. It is well-known[2,9] that 2D perturbations are more potent in creating RTI and Kelvin-Helmholtz instability, compared to 3D perturbations in the inviscid framework. This prompted Youngs[10] to perform 2D simulation,using the MAC method to simulate introduced multiple-wavelength perturbations. This justifies the present study of 2D problem with accurate numerical methods and fine resolution to understand RTI. Global spectral analysis of the numerical methods employed has been performed, details of which are available[8]. The system considered here is a 2D rectangular box, shown in Fig. 1, consisting of air at different temperatures separated by an interface placed at half the height of the box, which is impulsively removed at $t = 0$. To model RTI, we have taken the colder fluid (heavier) to rest atop the hotter (lighter) fluid, the horizontal barrier creating the interface is non conducting. The problem of interface tracking has been bypassed by considering air in the two halves. The walls of the box are adiabatic and hence the system is completely isolated thermodynamically. The impulsive removal of the barrier creates the initial perturbation at the junction of the interface with the side walls, required to trigger RTI. A good match between experiment and computation will be realized, if the initial condition in experiments represent an impulsive start.

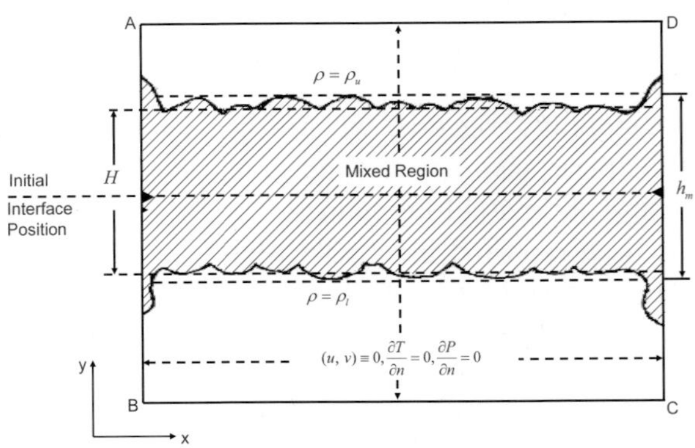

Fig. 1. Schematic of the computational domain.

2. Problem Formulation

The governing equations of motion are the 2D unsteady, compressible Navier-Stokes equation which conserve mass, momentum and energy[11] . The density gradients existing in the flow will be accounted for using this approach. A set of four nonlinear partial differential equations for the fluid have been expressed in divergence form, the details are available in[11,12] as

$$\frac{\partial \hat{Q}}{\partial t^*} + \frac{\partial \hat{E}}{\partial x^*} + \frac{\partial \hat{F}}{\partial y^*} = \frac{\partial \hat{E}_v}{\partial x^*} + \frac{\partial \hat{F}_v}{\partial y^*} \tag{1}$$

where the conserved variables are given by

$$\hat{Q} = \begin{bmatrix} \rho^* \\ \rho^* u^* \\ \rho^* v^* \\ \rho^* e_t^* \end{bmatrix} \tag{2}$$

The convective flux variables $\hat{E}$ and $\hat{F}$ are given as

$$\hat{E} = \begin{bmatrix} \rho^* u^* \\ \rho^* u^{*2} + p^* + (\rho^* - \rho_s) g_x x \\ \rho^* u^* v^* \\ (\rho^* e_t^* + p^*) u^* \end{bmatrix} \tag{3}$$

$$\hat{F} = \begin{bmatrix} \rho^* v^* \\ \rho^* u^* v^* \\ \rho^* v^{*2} + p^* + (\rho^* - \rho_s) g_y y \\ (\rho^* e_t^* + p^*) v^* \end{bmatrix} \tag{4}$$

and the viscous flux vectors $\hat{E}_v$ and $\hat{F}_v$ are given as

$$\hat{E}_v = \begin{bmatrix} 0 \\ \tau_{xx}^* \\ \tau_{xy}^* \\ u^* \tau_{xx}^* + v^* \tau_{xy}^* - q_x^* \end{bmatrix} \tag{5}$$

$$\hat{F}_v = \begin{bmatrix} 0 \\ \tau_{yx}^* \\ \tau_{yy}^* \\ u^* \tau_{yx}^* + v^* \tau_{yy}^* - q_y^* \end{bmatrix} \tag{6}$$

The reference density is taken as $\rho_s = 1.225 kg/m^3$ and with $\Delta T = 70K$, the temperatures on the top and bottom of the interface at the onset $(t = 0)$

are $T_u = 298K$ and $T_l = 368K$, respectively. The densities of air above and below the interface are obtained at $t = 0$, in terms of reference density (ρ_s) respectively, as $\rho_u = \rho_s$ and $\rho_l = \rho_s \frac{T_u}{T_l}$. The similitude is expressed in terms of the non-dimensional parameters, Reynolds number (Re), Prandtl number (Pr), Froude number (Fr), Eckert number (Ec) and Gay-Lussac number (Ga) given by,

$$Re = \frac{\rho_s U_s L}{\mu_s}, \qquad Pr = \frac{\mu_s C_p}{\kappa}, \qquad Fr = \frac{U_s^2}{gL}, \qquad Ec = \frac{U_s^2}{C_p \Delta T}, \qquad Ga = \frac{\Delta T}{T_s} \tag{7}$$

where Froude and Eckert numbers represent the ratio of inertial force to gravitational force and enthalpy difference, respectively. The Gay-Lussac number or thermal expansion number relates the temperature difference from a reference temperature (which is taken as $T_s = 298K$) which becomes relevant here as RTI is instigated by a temperature difference between air taken in the two halves of the box. The reference velocity (U_s) is taken as 20 m/sec, which is used to non-dimensionalize the u and v components of velocity, and define Reynolds number which is used as 10^7. Other reference values used in the present simulations are given by, μ_s, C_p and κ, which have been calculated from $Re = 10^7$, $Ec = 0.01$ and $Pr = 0.7$, respectively. Similarly, reference pressure and Mach number are calculated from ρ_s, T_s and with Ec and ΔT.

2.1. *Boundary and initial conditions*

The flow is computed in a domain given by $0 \le x \le 60$ and $0 \le y \le 30$, discretized with uniformly spaced 1000 points in the x-direction and 600 points in the y-direction.

Along the walls (AB, BC, CD, AD) for $x \ge 0$ and $y \ge 0$, no-slip adiabatic conditions are imposed at the walls, which makes the flow in the box completely isolated from the surrounding. Such a thermodynamical system will yield solution with which one can study RTI in all essential details from linear to nonlinear stages, as well as estimating rotationality and entropy generation without the need to use Boussinesq approximation. The corresponding non-dimensional wall boundary conditions are given by:

$$u = 0;\ v = 0;\ \frac{\partial \theta}{\partial n} = 0 \text{ and } \frac{\partial p}{\partial n} = 0$$

where $\theta = \frac{T^* - T_s}{\Delta T}$ and $p = \frac{p^* - p_s}{\rho_s U_s^2}$.

In simulating the flow, the heavier air on top chamber at $t = 0$ is taken as $T = T_s$ (298K), so that $\theta = 0$ for the colder air and $\theta = 1$ for the warmer air at the bottom. For such a discontinuous θ distribution, the interface nondimensional temperature at $t = 0$ is given as 0.5. The pressure boundary condition reflects the low Mach number incompressible flow limit for which a thin shear layer approximation is assumed on the walls.

3. Results and Discussion

The commencement of RTI is due to the baroclinic generation of vorticity as a consequence of the misalignment of pressure and density gradients. The baroclinic source term is $-\frac{1}{\rho^2}\nabla\rho \times \nabla p$

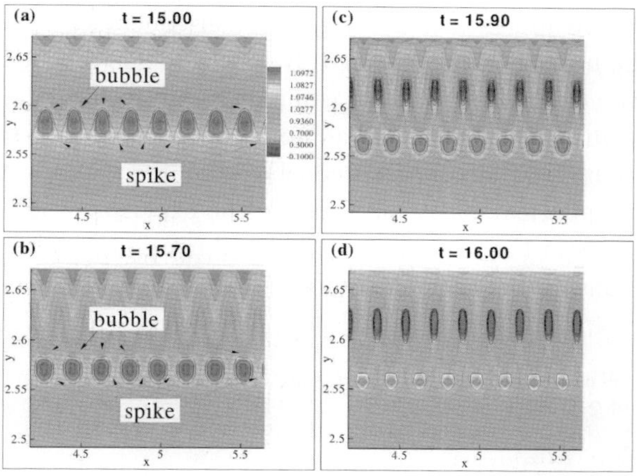

Fig. 2. Density contours illustrating the exact mechanism of propagation of bubbles and spikes as fronts.

The vorticity is concentrated at the junction of the side-walls with the interface at $t = 0$ as a result of the presence of large density gradients. Once vorticity is created baroclinically, it nonlinearly self-advects leading to interface becoming increasingly convoluted in space and hence leading to violent instabilities. In the present 2D simulations, the flow is not constrained to be periodic in x-direction, contrary to that used in other studies[3,4] and thus the viscous effects originating from the side walls dictate the flow evolution during RTI which is what one also observes in actual experiments[14]. Lawrie[15] sums up the situation by noting that *small initial perturbations become accentuated due to the baroclinic vorticity they supply to themselves, and they organise into self-advected structures known as 'bubbles' if they contain relatively less dense fluid and are rising upwards, and 'spikes' if they are more dense... While the bubbles and spikes remain coherent structures, the growth rate of the instability decreases - as predicted by the theoretical models balancing buoyancy and drag.*

In the density contours of Fig. 2 one notices the migration of bubbles upward in a leapfrog motion where newer bubbles appear above the spikes while the older bubbles weaken and eventually disappear. Here the denser fluid penetrates into the lighter fluid assisted by gravity, as evident by the spike formations at the early stages of RTI.

From the density contours shown in Fig. 3, one clearly notes the perturbations to originate at the side walls and at the same time disturbances increase in size in the interior with bubble formation. This process is accompanied by the appearance of spikes and consequently the mixing region. By $t = 30$, the nonlinear stage is already reached. For the later time evolution, the distinct phases are almost completely mixed over the full domain. In Fig. 4, visualizations from Read's[14] experiment have been shown so that the present numerical simulations can be validated. One can clearly observe the same physical events occuring in the experiment as well as in the numerical results in Fig. 3, at different non-dimensional times. Thus, there exists a relationship between the numerical data and real experiments.

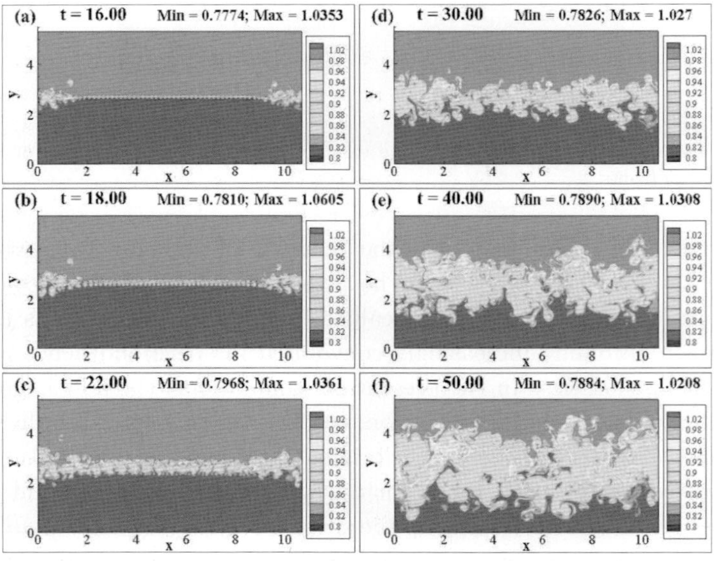

Fig. 3. Density contours at indicated times: (a) to (c) showing bubble and spike formation and (d) to (f) showing mixing region.

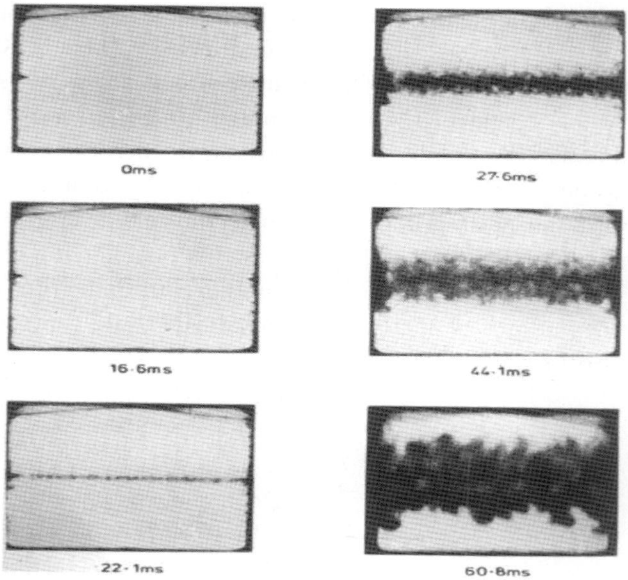

Fig. 4. Flow visualizations reported in Read[14] of RTI at the interface of two fluids of different density, in the incompressible limit achieved by constant acceleration.

3.1. *Mixing length and Reynolds number in the flow*

DNS addresses the issues arising with evolving length scales which change the equivalent Reynolds number characterizing the flow. In reference[3], a Reynolds number ($Re_H = H\dot{H}/\nu$) was defined where it reached a maximum value of 32,000. H is defined in terms of a mole fraction (X) for the heavy fluid with $X = (\rho - \rho_l)/(\rho_u - \rho_l)$. With homogeneity in x-direction, an equivalent ensemble average was defined by averaging in the horizontal direction: $\langle X \rangle = \frac{1}{L}\int_0^L X(x, y_0, t)dx$. The length scale H, was defined[3] by $0.01 \leq \langle X \rangle \leq 0.99$. In the simulation, a significantly higher value of Re_H can be reached, due to the fact that an inhomogeneous, non-periodic and compressible flow formulation has been used for which we have chosen a nominal Reynolds number as 10^7. In periodic flows, there are no side-wall vortices to affect the flow dynamics. In Fig. 5, the evolution of the length scale (H), velocity scale (dH/dt) and Re_H are shown as functions of time. Up to $t \approx 12$, the mixing region seems almost constant with very marginal growth (identified with the linear regime of RTI growth) and is the time beyond which the bubbles and spikes make their appearances simultaneously. Later on, we will note that these 'bubbles' and 'spikes' formation are synonymous with formation of rows of vortices. Once these structures appear, the growth of the mixing region is very rapid till it covers the whole height of the box, as noted in the top frame of Fig. 5. If this mixing length is truly ensemble averaged, then the variation of H with t would be distinctly smoother, like the one shown in frame (a). The resultant variation of H is reflected also in the middle frame where the absolute value of dH/dt is plotted as a function of time. Rapid variations of $|dH/dt|$ result in larger value of Re_H as shown in frame (c). This shows the ability of the present method to simulate RTI and capture very fine length and time scales.

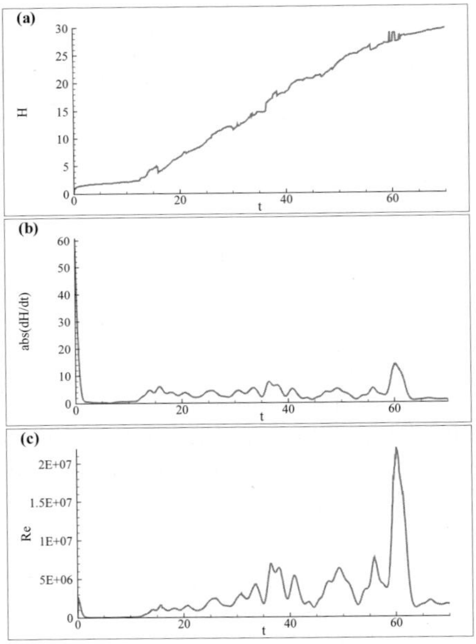

Fig. 5. Evolution of visual width H, velocity dH/dt and Re_H as functions of time.

3.2. *Entropy creation during RTI*

One can estimate the specific entropy (s_{mn}) change in a cell identified by its central co-ordinate by (x_m, y_n) during a time step of Δt for a perfect gas, with a temperature dependent relation of specific heat (c_p), as $\Delta s_{mn} =$

$s(x_m, y_n, t+dt) - s(x_m, y_n, t)$ obtained from

$$\Delta s_{mn}(t) = (a-R)[\text{Ln}\frac{T(x_m, y_n, t+\Delta t)}{T(x_m, y_n, t)}]+b[T(x_m, y_n, t+\Delta t)-T(x_m, y_n, t)]+$$

$$\tfrac{c}{2}[T^2(x_m, y_n, t+\Delta t) - T^2(x_m, y_n, t)] + \tfrac{d}{3}[T^3(x_m, y_n, t+\Delta t) - T^3(x_m, y_n, t)]$$

$$-R\,\text{Ln}\frac{\rho(x_m, y_n, t+\Delta t)}{\rho(x_m, y_n, t)} \tag{8}$$

where R is the specific gas constant for dry air (287 J/kgK), $a = 970.6491713 J/kgK$, $b = 0.06792127 J/kgK$,$c = 1.65814917 \times 10^{-4} J/kgK$,$d = -6.7886740 \times 10^{-8} J/kgK$, details are provided in Appendix A10[7]. In Fig. 6, events occurring during RTI are visualized in terms of entropy which is evaluated at discrete points for a $3000x1800$ grid using Eq. (8). Points above the interface $(i = 1501, j = 901)$ (top frame), at the interface $(i = 1501, j = 900)$ (middle frame) and below the interface $(i = 1501, j = 899)$ (bottom frame) are considered. At $t = 6.2$, a marked dip in the entropy is observed which is associated with the formation and development of bubble and spikes. One notes that during flow instabilities, coherent structures form and this appearance of orderliness leads to decrease in change of entropy. During the flow process and its instability, the system does not progress through a succession of equilibrium states rather the dominance of non equilibrium thermodynamics is observed. In Fig. 6, one also notices that the value of entropy is fluctuating with time and it attains negative values at the interface and below the interface. Although, the entropy at a location in the domain can increase, it can do so only by decreasing the entropy at another location by the same or greater amount. In fact, Prigogine[13] noted *If the system is perturbed, the entropy production will increase, but the system reacts by coming back to the minimum value of the entropy production.* It is also acknowledged[13] that *non-equilibrium may be a source of order.* The non periodic nature of the boundary conditions does not allow the system to reach a state of thermodynamic equilibrium and eventually the system settles down to a state of "least dissipation".

4. Conclusions

Present inhomogeneous simulations match qualitatively very well with the experimental flow visualization in[5,14,15]. as shown in Figs. 3 and 4. The role of inhomogeneity brings its effects by not requiring any initial perturbations, as the evanescent vortices on the side-walls convect and diffuse.

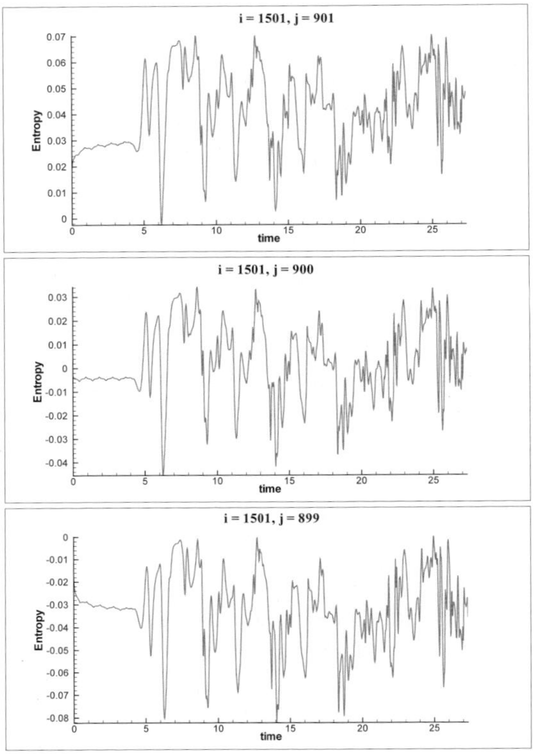

Fig. 6. Change in entropy evaluated above the interface ($i = 1501, j = 901$), at the interface ($i = 1501, j = 900$ and below the interface ($i = 1501, j = 899$) for a $3000x1800$ grid.

This sets in motion the RTI all along the interface. Very soon the nonlinear phase of RTI is attained and one observes the creation of bubbles and spikes with vigorous mixing setting in and by $t = 200$, the process almost completes. The present study of RTI in a truly inhomogeneous system has yielded valuable insight on this fundamental fluid dynamical problem and in the near future we will extend the present study for 3D flows.

References

1. P. G. Drazin & W. H. Reid, *Hydrodynamic Stability.* Cambridge Univ. Press, U.K. (1981)
2. S. Chandrasekhar, *Hydrodynamic and Hydromagnetic Stability.* Oxford Univ. Press, UK (1961)

3. W. H. Cabot & A. W. Cook, Reynolds number effects on Rayleigh-Taylor instability with implications for type la supernovae. *Nature*, **2**, 562–568.(2006)
4. A. W. Cook, W. Cabot & P. L.Miller, The mixing transition in Rayleigh-Taylor instability. *J. Fluid Mech.*, **511**, 333–362. (2004)
5. A. G. W. Lawrie & S. B. Dalziel, Rayleigh-Taylor mixing in an otherwise stable stratification. *J. Fluid Mech.* **688**, 507–527. (2011)
6. K. O. Mikaelin, Boussinesq approximation for Rayleigh-Taylor and Richtmyer-Meshkov instabilities. *Phys. Fluids*, **26**, 054103. (2014)
7. M. A. Saad 1997 *Thermodynamics Principles and Practice.* Prentice-Hall Inc.,USA
8. T. K. Sengupta 2013 *High Accuracy Computing Methods: Fluid Flows and Wave Phenomenon.* Cambridge Univ. Press, USA.
9. T. K. Sengupta 2012 *Instabilities of Flows and Transition to Turbulence.* CRC Press, Taylor and Francis Group, Florida, USA.
10. D. L. Youngs, Numerical Simulation of turbulent mixing by Rayleigh-Taylor instability *Physica D. Nonlinear Phenomena* , **12**, 32–44. (1984)
11. T. K. Sengupta, A. Bhole & N.A. Sreejith, Direct numerical simulation of 2D transonic flows around airfoils. *Comput. Fluids*, **88**, 19–37. (2013)
12. K. A. Hoffmann & S.T. Chiang, *Computational Fluid Dynamics, Vol. II.* Engineering Education Systems, Kansas, U.S.A.(1998)
13. Ilya Prigogine, Time, Structure and Fluctuations *Nobel Lecture in Chemistry*, 8 December (1977)
14. K. I. Read, Experimental investigation of turbulent mixing by Rayleigh-Taylor instability. *Physica D*, **12**, 45–58. (1984)
15. A. G. W. Lawrie, Rayleigh-Taylor mixing: confinement by stratification and geometry. PhD thesis, DAMTP, University of Cambridge, UK. (2009)

A High Resolution Differential Filter for Large Eddy Simulation on Unstructured Grids for High-Order Methods

M. Najafiyazdi*, S. Nadarajah and L. Mongeau

Department of Mechanical Engineering, McGill University, Montreal, QC, H3V OC3, Canada
** E-mail: mostafa.najafiyazdi@mail.mcgill.ca*
www.mcgill.ca

The applicability of a high resolution low-pass differential spatial filter for unstructured grids was investigated for typical high-order one-dimensional Finite Element methods. The goal was to investigate the accuracy and stability of the filter when applied on high-order elements. The one-dimensional differential filter was discretized based on Lagrange and Hierarchical Legendre base functions. Manufactured solutions were used to demonstrate the performance of the filter on two one-dimensional test cases for elements of order up to six. The results suggest that the proposed filter is effective for explicit filtering in Large Eddy Simulations using high-order methods.

Keywords: Higher-Order Methods; Differential Filter; Lagrange bases; Hierarchical Legendre bases; Large Eddy Simulation.

1. Introduction

The basic idea of Large Eddy Simulations (LES) is to decompose the flow variables into the sum of a large-scale resolved component, and a small-scale subgrid component, through the application of a low-pass, spatial filter either implicitly or explicitly.

Explicit filters have been successfully applied on structured grids using various schemes[1–6]. Attempts were made to extend the notion of explicit filtering to unstructured grids using weighted sums of neighbouring node values[7,8]. These methods were shown to be highly dependent on the local grid topology and perform poorly in the presence of skewed elements[9]. Differential filters offer an alternate approach for LES. A differential filter was proposed by Najafi-Yazdi *et al.*[10] as an extension of Germano's work[11].

In the present study, the same differential filter was discretized using a Finite Element Model (FEM) for high-order elements. Derivations and analyses were performed for one-dimensional elements of order up to six.

Two different families of base functions for high-order elements were investigated, namely Lagrange polynomials and Hierarchical Legendre polynomials. The goal was to find the most suitable base functions for minimal dissipation and numerical stability at high orders.

This paper is organized as follows. The least squares Galerkin discretization of the filter is provided in Sec. 2 for Lagrange and Hierarchical Legendre bases. The transfer functions for internal and end solution points are compared with each other. In Sec. 3 numerical examples are used to illustrate the performance of the discretized filter on high-order elements using Lagrange base functions. Conclusions are drawn in Sec. 4.

2. Discretization on High-Order Elements

The differential filter equation

$$\overline{\phi} + \frac{\partial}{\partial \xi_j}\left(\alpha^2 \frac{\partial}{\partial \xi_j}\overline{\phi}\right) = \phi + \frac{\partial}{\partial \xi_j}\left(\beta^2 \frac{\partial}{\partial \xi_j}\phi\right), \tag{1}$$

where α and β are two filter parameters, is discretized for an arbitrary non-boundary node using a least squares Galerkin formulation. The use of different quadrature points changes the coefficient of the shape functions and subsequently the behaviour of the filter at different wavenumbers.

2.1. *Lagrange base*

The shape functions based on Lagrange polynomials on a reference element are defined as

$$N_q = \frac{1}{a_q}\prod_{\substack{k=0\\k\neq q}}^{P}(\xi - \xi_k) \quad \text{for all } q = 0, 1, \cdots, P, \tag{2}$$

where a_q is a coefficient to ensure that N_q is unity at the solution point q of the element.

For a P_2 element, the final discretized filter equation for an end solution point, SP_1 in Fig. 1, yields

$$\begin{aligned}\alpha_{f_1}\overline{\phi}_{i-1} + \alpha_{f_{\frac{1}{2}}}\overline{\phi}_{i-\frac{1}{2}} + \overline{\phi}_i + \alpha_{f_{1/2}}\overline{\phi}_{i+1/2} + \alpha_{f_1}\overline{\phi}_{i+1} = \\ a\phi_i + \frac{b}{2}(\phi_{i+1/2} + \phi_{i-1/2}) + \frac{c}{2}(\phi_{i+1} + \phi_{i-1}).\end{aligned} \tag{3}$$

The coefficients α_{f_1} , $\alpha_{f_{1/2}}$, a, b , and c depend on the two parameters α

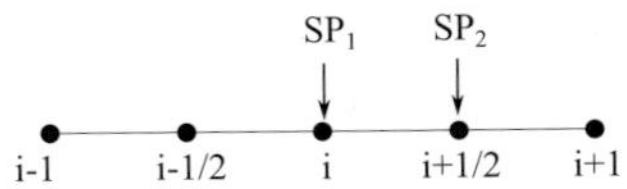

Fig. 1. One-dimensional P_2-element grid.

and β, and are defined as

$$\alpha_{f_1} = \frac{1/15 + 8/3\alpha^2}{4/15 - 14/3\alpha^2}, \alpha_{f_{1/2}} = \frac{1/30+2/3\alpha^2}{14/3\alpha^2-4/15}, \tag{4}$$

$$a = \frac{7/3\beta^2 - 2/15}{7/3\alpha^2 - 2/15}, \quad b = -\frac{1/30+1/2\beta^2}{7/3\alpha^2-2/15}, \quad c = \frac{1/15 + 8/3\beta^2}{7/3\alpha^2 - 2/15}. \tag{5}$$

The indices $i \pm 1$ and $i \pm 1/2$ correspond to the solution points on neighbour elements as shown in Fig. 1. Using von Neumann analysis, the transfer function of the discretized differential filter for the end points of an element is obtained as

$$T(kh) = \frac{a + b\cos(kh/2) + c\cos(kh)}{1 + 2\alpha_{f_1}\cos(kh) + 2\alpha_{f_{1/2}}\cos(kh/2)}, \tag{6}$$

where kh is the non-dimensional wavenumber such that $kh \in [0, P\pi]$ where P is the order of the element. For an internal solution point, SP_2, a similar discretized equation can be derived. The transfer function for the internal solution point is obtained as

$$T(kh) = \frac{\tilde{a} + \tilde{b}\cos(kh/2)}{1 + 2\tilde{\alpha}_{f_{1/2}}\cos(kh/2)}, \tag{7}$$

where $\tilde{a}$, $\tilde{\alpha_{f_{1/2}}}$ and $\tilde{b}$ are the coefficients obtained for the internal solution point as

$$\tilde{a} = \frac{16\beta/3 - 8/15}{16\alpha/3 - 8/15}, \tilde{b} = -\frac{8\beta/3 + 1/15}{16\alpha/3 - 8/15}, \tilde{\alpha}_{f_{1/2}} = -\frac{8\alpha/3 + 1/15}{16\alpha/3 - 8/15}. \tag{8}$$

In order to force the transfer function to vanish at grid cut-off, i.e. $k_g h = P\pi$, where P is the order of the element, the numerator of the transfer functions in Eqs. 6 and 7 should be zero. These constraints for SP_1 and SP_2 yield $\beta^2 = \frac{1}{160}$ and $\tilde{\beta}^2 = \frac{3}{80}$.

The transfer functions for elements of up to sixth order are shown in Fig. 2. All transfer functions have characteristics similar to those of the original filter proposed by Najafi-Yazdi *et al.*[10]: (*i*) the transfer functions are close to unity at low to moderate wavenumbers; (*ii*) the transfer functions vanish at grid cut-off to ensure complete energy content removal. A 6th order element, however, yields amplification, i.e. $|T| > 1$, over a range of wavenumbers due to the discretization scheme.

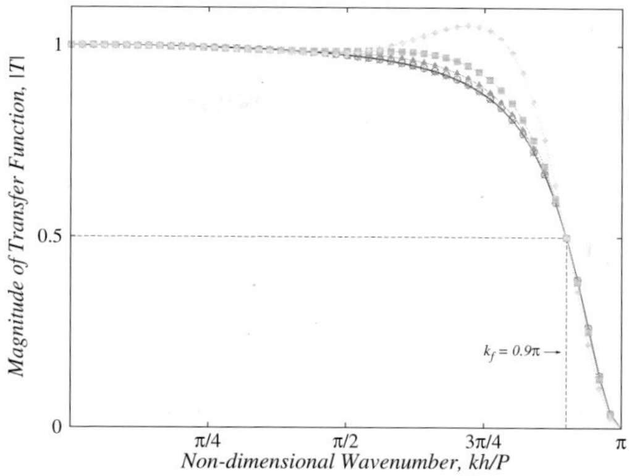

Fig. 2. Transfer functions of the end solution points of (solid) P_1, (star) P_2, (circle) P_3, (triangle) P_4, (square) P_5, and (diamond) P_6 elements.

2.2. *Hierarchical Legendre base*

The shape functions based on Hierarchical Legendre polynomials on a reference element are defined as

$$N_{-1}(\xi) = \frac{1-\xi}{2} \ , \ N_1(\xi) = \tfrac{1+\xi}{2} \ , \ N_q(\xi) = \frac{L_q(\xi) - L_{q-2}(\xi)}{\sqrt{2(2q-1)}} \ , \tag{9}$$

where L_q is the Legendre polynomial of order q. The scalar is calculated within such an element as

$$\phi = C_{-1}N_{-1} + C_1 N_1 + \sum_{q=2}^{P} C_q N_q \ . \tag{10}$$

The coefficients C_q are calculated based on U_{-1}, U_1, and U_q for $q = 2, \cdots, P$ such that $U_q(\xi)$ is unity at the qth quadrature point and zero at all other points. Therefore, the choice of quadrature points affects the values of these coefficients and the discretization of the differential filter. For the scope of this work, two sets of quadrature points were considered: 1) equi-distributed; and 2) Gauss-Lobatto points. For a 2nd order element, the coefficients were obtained as

$$C_{-1} = \phi_{-1} \ , \ C_1 = \phi_1 \ , \ C_2 = \frac{4}{\sqrt{6}}\left(\frac{\phi_1 + \phi_{-1}}{2} - \phi_2\right) \ . \tag{11}$$

The discretized filter equation for an end-point, SP_1, of a 2nd order element using hierarchical base functions yields

$$\begin{aligned}\overline{\phi}_i + \alpha_{f_1}(\overline{\phi}_{i+1} + \overline{\phi}_{i-1}) + \alpha_{f_{1/2}}(\overline{\phi}_{i+1/2} + \overline{\phi}_{i-1/2}) = \\ a\phi_i + b(\phi_{i-1/2} + \phi_{i+1/2}) + c(\phi_{i-1} + \phi_{i+1}) \ ,\end{aligned} \tag{12}$$

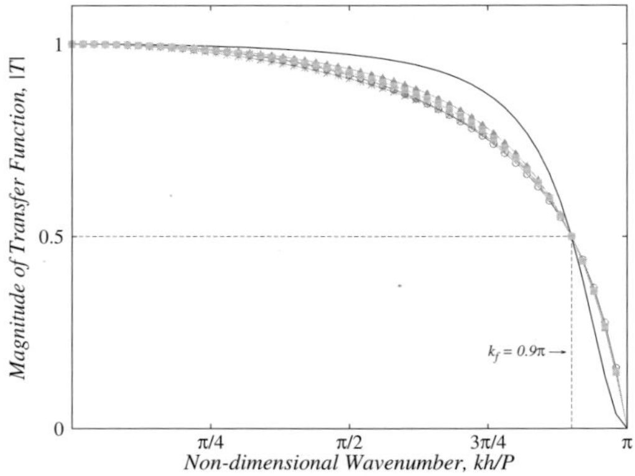

Fig. 3. Transfer functions of the end solution point with Hierarchical Legendre bases on Gauss-Lobatto points for an element of (solid) 2nd, (star) 3rd, (circle) 4th, (triangle) 5th, and (square) 6th order.

where

$$\alpha_{f_{1/2}} = -\frac{20\alpha + 2}{35\alpha - 8}, \alpha_{f_1} = \frac{5\alpha + 2}{70\alpha - 16}, \tag{13}$$

$$a = \frac{35\beta - 8}{35\alpha - 8} b = -\frac{20\beta + 2}{35\alpha - 8} c = \frac{5\beta + 2}{70\alpha - 16}, \tag{14}$$

The transfer function in the spectral domain is given with $\beta^2 = \frac{1}{40}$ as

$$T(kh) = \frac{1 + 2\alpha_{f_{1/2}} \cos(kh/2) + 2\alpha_{f_1} \cos(kh)}{a + 2b\cos(kh/2) + 2c\cos(kh)} . \tag{15}$$

For equi-distributed solution points, the shape functions are the same as those obtained from Lagrange polynomials. However, Gauss-Lobatto solution points result in a different discretization, and therefore a different transfer function for the discretized differential filter. Figure 3 shows the filter transfer function for SP_1 for 2nd to 6th order elements using Gauss-Lobatto solution points. The results show that stabilized discretized differential filters of higher orders are achievable with a Hierarchical Legendre-basis discretization using Gauss-Lobatto solution points. But this is obtained at the cost of more dissipation at medium wavenumbers.

For an internal solution point, the discretization stencil is generally asymmetric resulting in some dispersive properties. Figures 4(a) and (b) show the magnitude and phase of the filter transfer function for solution points SP_1 through SP_5 of a 5th order element. Dispersion errors remain very low for internal solution points despite the asymmetry of their numerical stencil. Moreover, the magnitude of the filter transfer function is

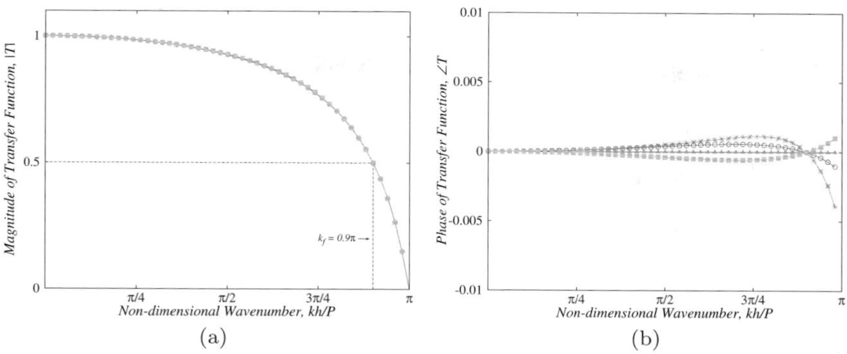

Fig. 4. (a) Magnitude and (b) phase of transfer functions for a 5th order 1D element with Hierarchical Legendre bases on Gauss-Lobatto points for (solid) SP_1, (star) SP_2, (circle) SP_3, (square) SP_4, and (diamond) SP_5 solution points.

similar for all internal solution points at all wavenumbers.

2.3. *Comparisons*

Figure 5 compares the transfer functions of an end solution point of a 6th order element with Lagrange bases vs Hierarchical Legendre bases on Gauss-Lobatto points. The filter cut-off is set to $k_{1/2} = 0.85\pi$. It can be seen that the filter transfer function on Lagrange-basis elements may overshoot unity. This can cause numerical instability, and negative effects on the turbulence dynamics by energy injection at moderate wavenumbers. This implies that the FEM discretization of the filter with Lagrange polynomials for high-order elements is prone to numerical instabilities. Transfer functions for Hierarchical Legendre bases show more stability at the cost of more dissipation at moderate wavenumbers.

These results call for a more detailed study of the role of quadrature points on the performance of the proposed differential filter. Further, analytical studies are required to investigate if discretization based on Hierarchical Legendre polynomials can lead to a stable formulation.

3. Numerical Examples

Two one-dimensional numerical examples are used to study the performance of the differential filter on high-order elements: a noisy sinusoidal pulse, and a square pulse signal. In both examples Lagrange base polynomials were used as base functions for the ease of implementation. The

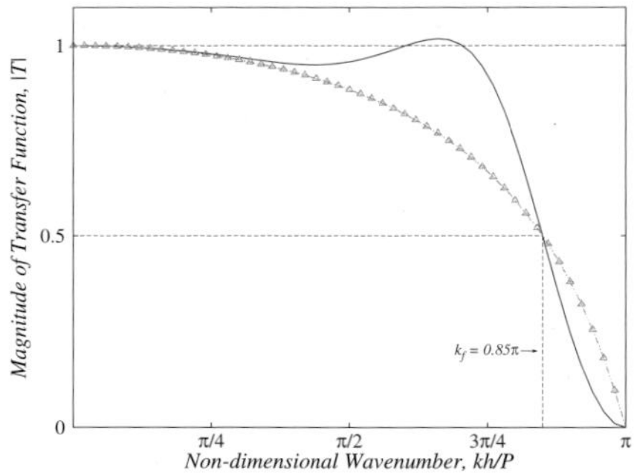

Fig. 5. Transfer functions of the end solution point of a 6th order 1D element with, (solid) Lagrange bases, and (triangle) Hierarchical Legendre bases on Gauss-Lobatto points.

order of the elements and the filter cut-off were chosen such that the filter transfer function is always less than unity to ensure numerical stability.

3.1. *Noisy sinusoidal pulse*

This test case has been adopted from the original work of Najafi-Yazdi *et al.*[10,12] in order to validate that the filter for high-order elements can be designed such that perfect attenuation is achieved at the grid cut-off wavenumber. A one-dimensional signal of the form

$$u(\tilde{x}) = e^{-5\tilde{x}^2}\left[2\sin(2\pi\tilde{x}) + 3\sin(4\pi\tilde{x}) + 0.5\sin(2\pi/3\tilde{x})\right] \tag{16}$$

was considered for $x \in [-1, 1]$ where $\tilde{x} = 20x - 10$. A sawtooth wave of amplitude 0.5 was added to the scalar field u. For simulations using different element orders, a fixed number of degrees of freedom, i.e. DoF = 121, was used.

For all element orders between one and five, the filter completely removed the sawtooth wave while preserving low wavenumbers. Figure 6(a) compares the filtered signal with the noise free signal on different grids.

3.2. *Square signal*

To demonstrate the effectiveness of the filter on a solution with a continuous spectrum of wavenumbers, a single square pulse was used.

In Fig. 6(b) the unfiltered spectrum is compared with the one after filtering. It can be seen that filtering preserved the magnitudes of low to moderate wavenumbers while wavenumbers close to grid cut-off are attenuated such that at grid cut-off the wave magnitude vanishes.

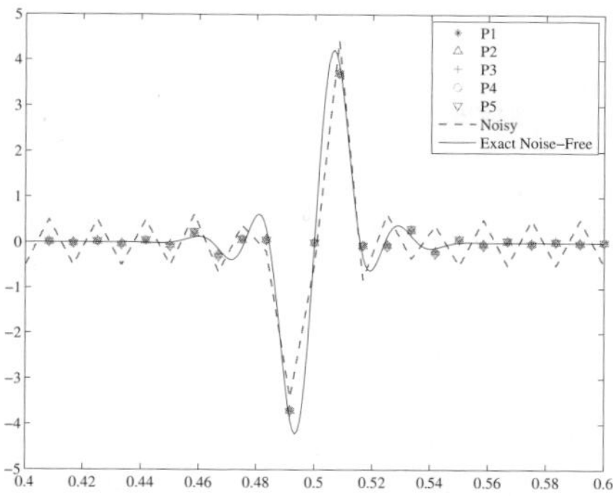

(a)

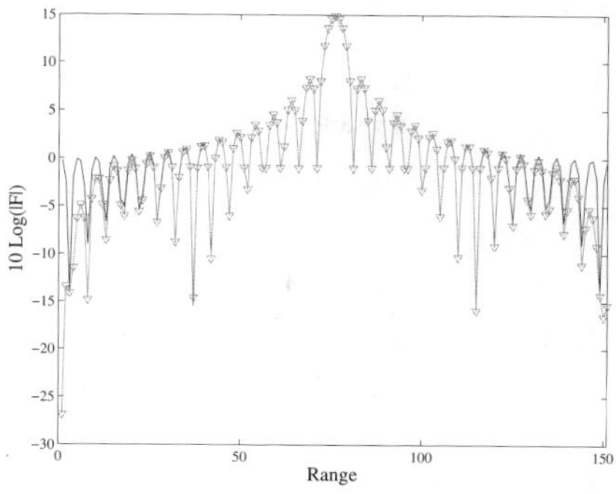

(b)

Fig. 6. Comparing (a) filtered vs unfiltered noisy sinusoidal pulse; and (b) frequency spectra of (solid) the unfiltered and (triangle) the filtered square pulse.

4. Conclusions

The high resolution low-pass differential filter of Najafi-Yazdi *et al.*[10] was extended for higher order elements. The discrete form for one-dimensional high order elements was derived using FEM. Two different sets of basis functions were investigated, namely Lagrange and Hierarchical Legendre polynomials. For the Hierarchical Legendre polynomials Gauss-Lobatto quadrature points were used.

It was shown that Lagrange-base discretization is prone to cause numerical instabilities due to amplifications over a range of wavenumbers. The filter discretized with Hierarchical Legendre bases on Gauss-Lobatto points did not demonstrate overshoots and displayed more dissipation at moderate wavenumbers and less dissipation at high wavenumbers. Further investigations for the stability of this discretization scheme are required.

Two numerical test cases were presented to show the effectiveness of the filter on high-order elements. It was shown that a sawtooth wave was completely removed from a sinusoidal signal while energy contents at other wavenumbers where mostly preserved. Application of the filter on a square pulse resulted in the energy content near grid cut-off to drop exponentially, which implies the possibility of pseudo-spectral behaviour for the differential filter at high orders. The results presented in this work suggest that the proposed filter is promising for explicit filtering in LES with high-order methods. Future work will include a more detailed study of the effect of dissipation and dispersive properties of the filter on numerical solutions.

References

1. S. K. Lele, Compact finite difference schemes with spectral-like resolution, *J. Comput. Phys.* **103**, 16 (1992).
2. M. R. Visbal and D. P. Rizzetta, Large-Eddy Simulation on Curvilinear Grids Using Compact Differencing and Filtering Schemes, *J. Fluids Engg.* **124**, 836 (2002).
3. D. P. Rizzetta, M. R. Visbal and G. A. Blaisdell, A time-implicit high-order compact di erencing and ÿltering scheme for large-eddy simulation, *Int. J. Numer. Meth. Fluids* , 665 (2003).
4. A. Uzun, J. Bin and M. Y. Hussaini, High-fidelity numerical simulation of a chevron nozzle jet flow, *Int. J. Aeroacoustics* **10**, 531 (2011).
5. C. Bogey and C. Bailly, Computation of a high Reynolds number jet and its radiated noise using large eddy simulation based on explicit filtering, *Comput. Fluids* **35**, 1344 (2006).
6. D. Fauconnier, C. Bogey and E. Dick, On the performance of relaxation filtering for large-eddy simulation, *J. Turb.* **14**, 22 (2013).
7. A. L. Marsden, O. V. Vasilyev and P. Moin, Construction of commutative filters for LES on unstructured meshes, *J. Comput. Phys.* **175**, 584 (2002).
8. A. Haselbacher and O. V. Vasilyev, Commutative discrete filtering on unstructured grids based on least-squares techniques, *J. Comput. Phys.* **187**, 197 (2003).
9. S. T. Bose, P. Moin and F. Ham, Explicitly filtered large eddy simulation on unstructured grids, *Annual Research Briefs (CTR)*, 87 (2011).
10. A. Najafi-Yazdi, M. Najafiyazdi and L. Mongeau, A high resolution low-pass differential filter for large eddy simulations on unstructured grids, in *Proc. 7th AIAA Theoretical Fluid Mechanics Conference*, Atlanta, US, (2014).
11. M. Germano, Differential filters for the large eddy numerical simulation of turbulent flows, *Phys. Fluids* **29**, 1755 (1986).
12. A. Najafi-Yazdi, M. Najafi-Yazdi and L. Mongeau, A high resolution differential filter for large eddy simulation: Toward explicit filtering on unstructured grids, *J. Comput. Phys.* **292**, 272 (2015).

A Critical Assessment of Simluations for Transitional and Turbulent Flows

Tapan K Sengupta

High Performance Computing Laboratory, Dept. of Aerospace Engg.,
Indian Institute of Technology Kanpur,
Kanpur 208 016, India.
E-mail: tksen@iitk.ac.in, tapansg@gmail.com

In the symposium we have seen a very wide range of simulations for transitional and turbulent flows. In writing this epilogue, the author takes the responsibility of identifying some common elements which add to success of desired DNS and LES. Of course, the single common element of any such activities is to be aware of numerical dispersion relation and how it relates to physical dispersion relation. We also investigate suitability of a specific combined implicit-explicit (IMEX) time integration method in performing DNS, which has been used to explore the so-called bypass transition. While IMEX methods have often been used, yet its properties and potentials are not assessed. Specifically, error analysis by solving a model equation that mimics some of the physical processes in DNS is not performed. A model equation with exact solution allows tracking of error with time for IMEX time integration methods, with either explicit or implicit spatial discretization. This analysis is motivated by recent DNS of flow transition over a flat plate reported in Bhaumik and Sengupta (Phys. Rev. E, **89**, 043018, 2014) and Sayadi *et al.* (J. Fluid Mech., **724**, 480-509, 2013), both trying to mimic the classical experimental results on flow transition in Schubauer and Skramstad (J. Aero. Sci., **14**, 69, 1947). These two simulations show distinctly different computed solutions, yet the average flow quantities show remarkable similarity. In Bhaumik and Sengupta (2014), explicit dispersion relation preserving time integration method has been used everywhere in the domain. In contrast, in Sayadi *et al.* (2013), implicit time integration method is used very close to the plate, while explicit time integration method is used in rest of the domain. Such splitting of domain into explicit and implicit regions causes spurious internal reflection of the signal at the interface between these regions, creating unphysical, spurious and upstream-propagating q-waves which considerably contaminates solution. This is investigated in detail along with other issues of high resolution computing here to show suitable methods for DNS.

Keywords: DNS; LES; Transitional Flow; Turbulent flow; Error analysis; q-waves; Compact schemes; IMEX method; Anisotropic wave propagation; Round-off error

1. Introduction

Direct numerical simulation (DNS) is attempted by many investigators for canonical problems. Despite its widespread use as a tool of scientific computing, yet there are no known accepted definition of what constitutes DNS. It is not adequate to state that not using any model(s) of transition and turbulence in simulating high Reynolds number flows qualifies as DNS. In that case, all laminar flow computations should have been considered as DNS. It is understood in fluid mechanics context, that simulating flow fields with wide-band energy and enstrophy spectrum without models constitute DNS. In a generic simulation of dynamical system governing space-time dynamics, this definition can be broadened to include simulations of wide-band spectral events without any numerical artifacts. Such numerical artifacts could be wide ranging from altering the governing dynamical equation; trying to study so called free vibration problems by studying forced vibration; adding spurious dispersion and dissipation of numerical methods, so much so that the physical dispersion relation (if it can be identified in the first place) is poorly followed by adopted numerical dispersion relation and adding other sources of numerical errors. In this context, the main issue is one of evaluating the numerical dispersion relation. It is somewhat ironic that the celebrated error analysis of von Neumann used all over the world today, and which remained classified during the second world war years due to its supposed importance, has been the main reason, which has spawned misunderstanding of numerical dispersion relation. This has been corrected by the author's group and reported in Refs. 33 and 35, following an early lead in Ref. 14. This will be amply demonstrated by a simple test case and rudimentary metrics identified for correct error analysis. This may provide another qualitative description of what constitute DNS. In short, DNS is one, in which dispersion relation preserving (DRP) method is used and which minimizes all known sources of error. This has been supported by Prof. Sheu and Prof. Deville in the panel discussion that DRP schemes have to be established as the principal tool for DNS and LES.

Thus it appears that not obtaining the correct numerical dispersion relation is responsible for the slow growth of scientific computing, apart from nonlinearity of the governing differential equation. However, computers are used to solve nonlinear problems to understand the physics of the problem. So the inability to solve nonlinear problems in closed form provide opportunity for scientific computing to bring the second wave of development in mathematical physics. Establishment of calculus brought the first wave of scientific knowledge creation. So it is going to be very interesting to be traveling with the second wave! Computing and continuum mechanics is again the crucible where ideas and applications are fine-grained and analyzed for further development. However, we will keep our attention focused mainly on fluid dynamical problems in this section.

The second-most important issue of scientific computing related to inability of handling non-periodic problems. Lack of analysis tools for numerical methods has somewhat delayed progress so far. The community (specially who are more mathematically inclined) has spent a lot of time to study so-called periodic problems (note the warning of Prof. Deville in the panel discussion in this context). A global spectral analysis (GSA) has appeared in literature over more than a decade from the author's group (see Refs. 28 and 36 for complete details) for full domain analysis of non-periodic problems. It is equally relevant to note that GSA can accommodate all types of discretization schemes and nothing escapes its scrutiny, be it a finite difference, finite volume, finite element or Fourier spectral method – all such schemes have been analyzed[29,32]. The existence of such a tool should be viewed as a great advantage for all sentient researchers in the field.

Of course GSA requires adopting a model equation for calibration and the one adopted here is the linear wave equation or one-dimensional (1D) convection equation. The reason for the choice of this model equation needs some elaboration. In the context of the symposium theme, we gathered to discuss about transitional and turbulent flow. The governing Navier-Stokes equation (NSE) is a statement of dynamical equilibrium between mainly two processes: convection and diffusion; apart from imposed pressure gradient and unsteadiness. For high Reynolds number transitional and turbulent flows, convection mainly dictates the dynamics. So from that perspective alone, adopting linear convection equation appears as the lowest common denominator. It is not relevant to distinguish between 1D and multi-dimensional convection, as it has been amply demonstrated in Ref. 39, as to how GSA can be extended to multi-dimensions and related errors due to anisotropy of numerical methods can also be studied. The second reason for adopting linear wave equation is due to the perspective that in fluid dynamic studies, we place a body in a given surrounding and find out how disturbances created by a body propagates. Thus, it is a problem of signal propagation. If we adopt the description of Fourier-Laplace transform to the disturbance field, denoted by the unknown u here, then it can be represented in 1D description by

$$u(x,t) = \iint U(k,\omega_0)\, e^{i(kx-\omega_0 t)}\, dk\, d\omega_0 \tag{1.1}$$

Thus the disturbance field is ensemble of wave components with characteristic number (k,ω_0) in the continuum, as defined mathematically. This applies equally well, if the dynamical system consist of waves or eddies. Thus, adoption of 1D convection equation seems relevant and is given by

$$\frac{\partial u}{\partial t} + c\frac{\partial u}{\partial x} = 0 \tag{1.2}$$

This is the simplest possible description of a space-time dependent problem. Using the representation of Eq. (1.1), one notes that this PDE transforms to an algebraic relation

$$\omega_0 = kc \tag{1.3}$$

This is the physical dispersion relation. We must hasten to emphasize that dispersion relation is not simply the governing differential equation in spectral plane. In some problems, like surface gravity waves, the governing differential equation can be simply elliptic PDE, yet the boundary condition helps one relate wavenumber and circular frequency, i.e., the dispersion relation. Thus in a general connotation, physical dispersion relation is a consequence of governing differential equation (for hyperbolic PDEs) and/ or boundary condition(s). The latter waves are termed as dispersive wave, as their origin may not be related to a hyperbolic PDE. Thus, 1D convection equation helps one relate the space-time dependence of the problem clearly.

One of the problems of computing, relates to many well-known methods' inability to satisfy Eq. (1.3) in a numerical framework. This fact was not helped by the celebrated von Neumann analysis[6,7], which assumed that the numerical error of Eq. (1.2) to follow the same governing equation. That is, if the error is defined as the difference between exact (u) and numerical solution (u_N) as, $e(x,t) = u(x,t) - u_N(x,t)$, then the error propagation according to von Neumann is given by

$$\frac{\partial e}{\partial t} + c\frac{\partial e}{\partial x} = 0 \tag{1.4}$$

This intuitive expectation that signal and error are governed by same equation originated from the thinking that the governing equation is linear. This also assumes that c is kept constant during computation. Now in discrete computing, evaluation of a spatial derivative is given by an equivalent representation in a hybrid representation by

$$\frac{\partial u}{\partial x} = \int ik_{eq}U(k,t)e^{ikx}dk \tag{1.5}$$

Thus in Eq. (1.5), Fourier-Laplace amplitude is multiplied by an equivalent wavenumber k_{eq}, instead of the actual wavenumber, k, that one would do in Fourier spectral method. Thus, the resolution of numerical derivative is often described by the quotient k_{eq}/k plotted as a function of nondimensional wavenumber kh, as in Refs. 19 and 28. This along with the observation that c in Eq. (1.2) remains the same in computations led to the numerical dispersion written as

$$\omega_N = k_{eq}c \tag{1.6}$$

where Ω_N is the numerical circular frequency. This may seem logical, but leads to erroneous numerical group velocity as,

$$V_{gN} = \frac{d\omega_N}{dk} = c\frac{dk_{eq}}{dk}$$

As group velocity gives the energy propagation speed, this numerical group velocity implies that this quantity is independent of time integration scheme and solely determined by spatial discretization scheme. This apparent contradiction clearly indicates that the numerical dispersion relation in Eq. (1.6) cannot be right. Unfortunately this error has continued unabated with attendant consequences in Ref. 42. This is despite the fact that Haltiner[14] has indicated that the numerical phase speed is different from the physical phase speed, $c_N \neq c$. No one paid particular attention to this innocuous observation, except the author's group in Refs. 33 and 35, who were prompted to identify the correct numerical dispersion as

$$\omega_N = kc_N \tag{1.7}$$

where k remains the independent variable of the problem and space-time discretization method determines, $c_N = c_N(k)$ and Eq. (1.7) determines the numerical circular frequency. The correct numerical group velocity is determined by[28,33,35],

$$V_{gN} = c_N + k\frac{c_N}{dk} \tag{1.8}$$

Following the development of correct numerical dispersion relation, the wrong error propagation was corrected and reported in Refs. 28 and 35 to obtain the correct error dynamics given in the following section. By now we have understood the reason for adopting 1D convection equation to calibrate numerical method. It is for this reason in Refs. 20 and 28, this equation has been extensively used to develop the subject. We need to realize that this equation provides the necessary condition for accepting a numerical method for solving NSE. If a numerical method cannot satisfactorily solve Eq. (1.4), then there are no reasons to believe that method will work for solving NSE. However, a method which solves Eq. (1.4) efficiently and accurately, does not guarantee that the same will be good for NSE. That is why we term it as a necessary condition, but not sufficient for DNS.

Finally, additional justification is made for which we use this model equation, for which we owe it to the wisdom of Scottish philosopher David Hume, who made a poignant assumption about new knowledge is established, not necessarily by method of induction. This is known as the theory of black swan. This is described next.

2. One-Dimensional Convection Equation: Theory of Black Swan and Error Dynamics

Till the time European settlers came to Australia, it was an accepted norm in Europe to consider all swans to be white. However, upon finding black swans in the new continent, Hume advanced his observation on flaw of trying to prove a theorem by method of induction. Stated in simple language, it was noted that observation of thousands of white swans does not make all swans white, specially if one sees an additional white swan. However, if the additional swan turns out to be black, then the central proposition of all swans being white becomes unfounded. With this analogy in perspective, the authors' group approached the problem, of wrong numerical dispersion relation in Eq. (1.6) and the theme of error always following the same linear dynamics of the signal, by finding a counter-example of central tenet of von Neumann error and stability analysis. If one can find one linear system, for which error does not follow signal, then we disprove von Neumann stability analysis by the analogy with theory of black swan. This turned out to be case with 1D convection equation in Ref. 35, details of which can be found in Ref. 28 also.

The numerical solution $u_N(x_j, t^n)$ can be expressed by the hybrid representation at the jth node of a uniform grid with spacing h as, $u_N(x_j, t^n) = \int U(kh, t^n)\, e^{ikx_j} dk$. Here, k is the wavenumber. Generally speaking, literature on implicit and explicit time integration methods do not list advantages and disadvantages by quantifying properties responsible to control numerical error. For example, one defines a numerical amplification factor in GSA by $G_j = U(k, t^{n+1})/U(k, t^n)$[28,35]. The effects of different numerical parameters are readily available from the error dynamics obtained by GSA, which is given for the 1D convection-diffusion equation as[35]

$$\frac{\partial e}{\partial t}+c\frac{\partial e}{\partial x} = -c\left[1-\frac{c_N}{c}\right]\frac{\partial \bar{u}_N}{\partial x} - \int \frac{dc_N}{dk}\left[\int ik' A_0\,[|G|]^{t/\Delta t}\, e^{ik'(x-c_N t)} dk'\right] dk$$

$$- \int \frac{\text{Ln}\,|G|}{\Delta t} A_0\,[|G|]^{t/\Delta t}\, e^{ik(x-c_N t)}\, dk \qquad (2.1)$$

Here, c_N and V_{gN} are the phase speed and the group velocity of the numerically computed signal. Thus on comparing Eqs. (1.4) and (2.1), we find two new problems: (i) In discrete computing, one cannot solve 1D convection with zero error, for which an exact solution is already known. This is due to the fact that, a priori, one cannot say that a space-time discretization method will always ensure $|G| \equiv 1$ and thus suffer error due to the last term on the right hand side of Eq. (2.1). There are implicit methods like Crank-Nicolson (CN) method[7] which satisfy this condition and this error is avoided. We also note that for this model non-dispersive, non-dissipative equation, one must have $|G| = 1$ to satisfy non-dissipative nature. This prompts us to state that any numerical method must be neutrally stable, otherwise it will fail this test, even with numerical methods which are proven to be stable! However, even if we have a perfectly neutrally stable method, all discrete computations suffers dispersion error via the numerical phase speed being non-constant, unlike that is given in the original problem. This prompts us to state the uncertainty principle of computing, which points to the fact that complete error-free solution cannot be even ensured for this model equation.

(ii) Also in discrete computation of a system governed by linear dynamics, error and signal do not follow the same dynamics. There are forcing terms on the right hand side of Eq. (2.1). This was erroneously assumed to be zero in von Neumann's theory, which was accepted as given from simple theoretical consideration only. This also warns us that computing is different from theory and one should be alert for all sources of errors.

3. In Search of Accuracy: Explicit and Implicit Methods

We have already noted that the resolution of spatial discretization is often reported in plots of k_{eq}/k versus kh, where h is the uniform spacing between two successive points. The ordinate k_{eq}/k is ostensibly chosen[19,20,28] with the implicit belief that the Fourier Spectral method[5,8,13,15] provides the maximum resolution and ideally this quotient should be equal to one for all length scales. This expectation tends to deceive at times! Fourier spectral method used in conjunction with explicit multi-stage Runge-Kutta time integration may not always be successful. For example, the reported DNS of homogeneous turbulence in Refs. 10, 11 and 47 has used Fourier-Spectral method for spatial discretization with two-stage Runge-Kutta method (RK2), referred to as RK2-FS method, which has actually mild numerical instability for finite CFL number ($N_c = c\Delta t/h$, where Δt is the time step chosen for computing Eq. (1.4)). Thus the computational parameters chosen for comparison of different discretization schemes are kh and N_c, which in fact gives us the grid spacing (h) and time step (Δt) as the controlling parameters. Figure 1, reproduced from Ref. 29 clearly shows the attribute of numerical instability of RK2-FS scheme.

Despite the fact that Fourier spectral method has the limited applicability for periodic problems (as in homogeneous isotropic turbulence), this is still favored due to its better resolution. However, for space-time dependent governing equations, performance of combined space-time dis-

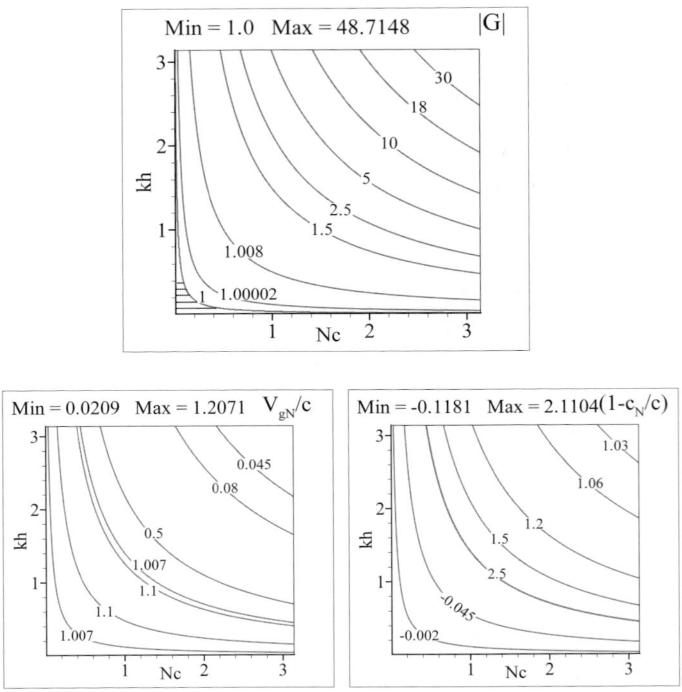

Fig. 1. Results of GSA for RK2-FS method, with —G— showing the numerical amplification factor, V_{gN}/c as the non-dimensional dispersion property of the numerical method and $(1 - c_N/c)$ as the phase error.

cretization scheme is more relevant. The $|G|$-contours for RK2-FS method graphically show this effect of instability in the top frame of Fig. 1, i.e., $|G| > 1$ in the entire (N_c, kh)-plane, except the very narrow shaded region near the origin. It is noted that the dispersion indicated by V_{gN}/c contours and the phase error indicated by $(1 - c_N/c)$ represent, for the chosen value of $N_c = 0.5$ in Ref. 47 are unacceptably high for RK2-FS method. Numerical group velocity is more than 10% from the actual value for high wavenumber components. Similar error is noted for phase error at higher wavenumber. We display computation of Eq. (1.4) for this N_c will produce solution to blow up in Fig. 2, even when the CFL number was reduced more than five times than that was used in Ref. 47 for solving NSE. Why the homogeneous turbulence simulation in Ref. 47 did not blow up for such a high N_c value is due to use of hyperviscosity in the method, which damps higher wavenumber components. This remains as the onus of the authors to show that this is indeed so for the model Eq. (1.4). Adding hyperviscosity is tantamount to modifying the dynamics of the system. At least for Eq. (1.4), one should demonstrate the effect of hyperviscosity on the numerical properties by GSA for the periodic problem.

We also note that in reporting the results of homogeneous turbulence simulation in Ref. 47, the authors did not state if the turbulence was initiated by initial perturbation or the turbulence is initiated due to present numerical instabilities at some unknown scales. If that is indeed the case, i.e., *creation of turbulence by numerical noise can have different transitional flow extent and the integrated load over aerodynamic bodies will be method-dependent in such cases. Ideally in computing such flows, one must compute the equilibrium flow first and then subject such equilibrium flow to realistic ambient disturbances.*[29]. For the flow in Ref. 47 is in itself an idealization for the canonical flow, so there is no question of *realistic turbulence*, even though some researchers refer to flow in a wind tunnel before the test section – after the contraction – as the flow which is similar to the homogeneous turbulence.

Referring back to the numerical properties shown in Figs. 1 and 2, are we ready to consign spectral method completely? This question is worth asking with respect to the comparative numerical properties shown in Fig. 3. Properties of three stage Runge-Kutta (RK3) method used in conjunction with spectral method (RK3-FS) are shown side by side with RK2-FS method in this figure. One aspect immediately stands out that RK3-FS has some region in (N_c, kh)-plane where the method is neutrally stable for Eq. (1.4). The figure caption points out the obvious conclusion that unlike

RK2-FS method, the wave-packet computed using RK3-FS method will be attenuated for $N_c = 05$, the value used in Ref. 47. It will also be the case for the lowered N_c value used in Fig. 2. One can hope for achieving correct solution of Eq. (1.4) if N_c is vanishingly small and kh restricted below one. However, the value for N_c with RK2-FS method will have to be even smaller with much smaller kh.

We also note that though RK2-FS method requires two-thirds the com-

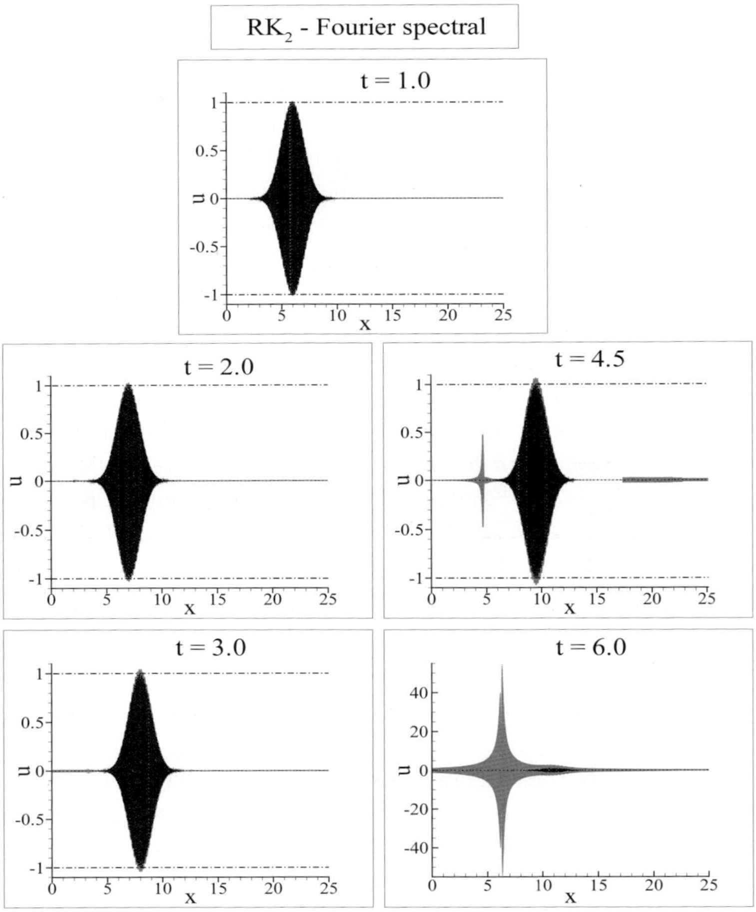

Fig. 2. Propagation of computed wave-packet for Eq. (1.4) with $c = 0.1$, $k_0h = 0.8$ and $N_c = 0.0983$ using RK2-FS method. In all the frames, exact solution is shown by dotted lines. Where k_0 is the central wavenumber of the wave-packet chosen. It is noted that the actual wave-packet has finite band-width around this central value.

putational effort needed for RK3-FS method, there are other potential benefits for the latter method, as noted in Fig. 3, where the dispersion and phase errors are noted to be lower by an order of magnitude. It is for these reasons one should explore higher order temporal discretization schemes.

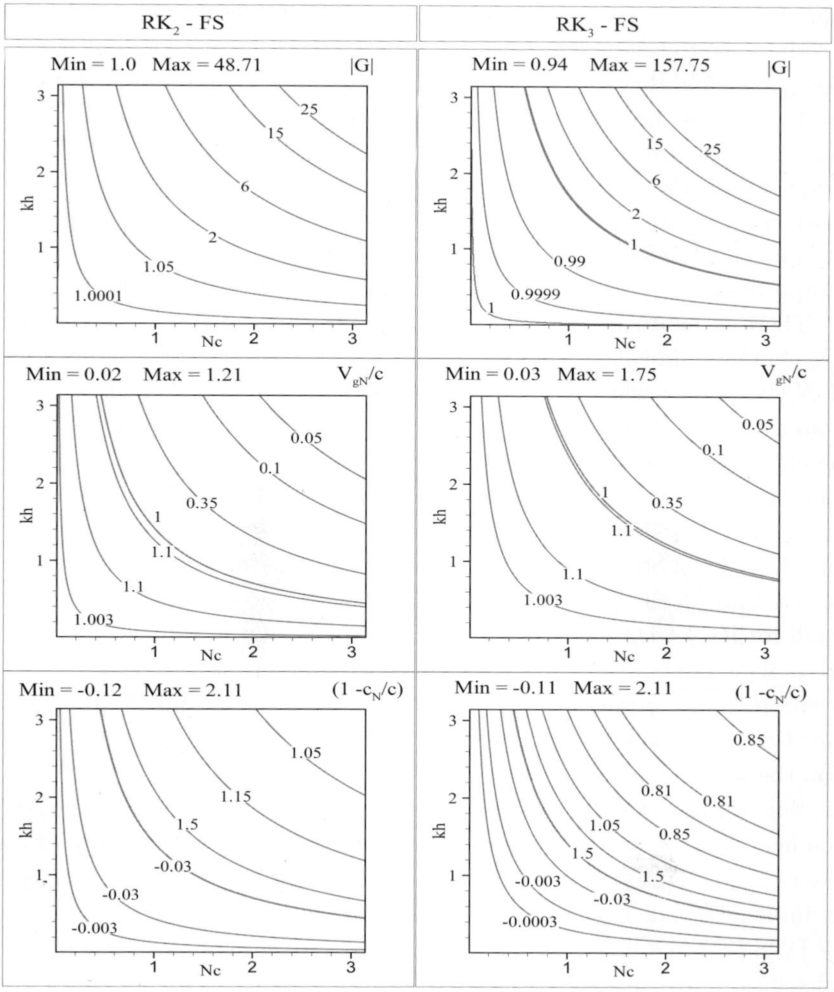

Fig. 3. Comparative results of GSA for RK2-FS and RK3-FS method showing —G—, V_{gN}/c and $(1 - c_N/C)$ contours in (N_c, kh)-plane. Note that the instability of RK2-FS method disappears in RK3-FS method. However, the solution of Eq. (1.4) will be highly attenuated for $N_c = 05$. It will also be attenuated for $N_c = 0.0983$ that was used in Fig. 2.

However, as noted earlier that we should restrict ourselves to only two-time level methods, as more than two-level methods will invoke spurious numerical modes[28,33]. This is seen more graphically in Fig. 4, where four stage Runge-Kutta method is used with second order central difference scheme (CD2) and FS methods of spatial discretization. One can see that the dispersion and phase errors of RK4-FS methods are far superior over that of RK3-FS method, even the neutral stability region is also significantly more, enabling one to take larger values of N_c with RK4-FS method, but nothing like $N_c = 0.5$ in Ref. 47 or even the value $N_c = 0.0983$ used in Fig. 2, for the propagation of the wave-packet will be useful. If we compute with smaller N_c values, then it is possible to use up the full range of kh prescribed by Nyquist criterion. Comparatively speaking, RK4-FS method allows taking much larger value of N_c, as compared with RK3-FS method so that the computational efforts will be lower, even though one uses an extra stage of computation.

There is another reason for displaying the two methods compared in Fig. 4. One immediately notes that RK4-CD2 method has a larger range of N_c values for which the full Nyquist range: $(kh)_{max} = \pi$, is available from the point of view of neutral stability. Of course, one has the potential benefit of using RK4-CD2 method of solving non-periodic problems. However, there are two principal reasons for which this will never be practiced: First, the dispersion and dissipation errors are orders of magnitude higher for RK4-CD2 method, as compared to RK4-FS method. Secondly, one notices a limit on the value of useful range of kh for which dispersion error is small for RK4-CD2 method; but more importantly the value of $kh = \pi/2$, above which the numerical group velocity becomes negative. This situation is said to create q-waves[32] above $kh > \pi/2$, which propagates in the wrong direction, as opposed to the exact solution of Eq. (1.4) indicates. We caution the reader that the presence of q-waves, of however small scale it may be, fluid dynamics of system will be altered qualitatively in an unphysical manner. The q-wave is manifestation of extreme dispersion error of any space-time discretization scheme and its effect being present in the range of useful length scale makes the solution useless. Thus, another rule of thumb for DNS is stated here: Do not allow q-waves to contaminate your numerical solution. We also note that Fourier spectral methods do not suffer from q-waves. However, FS methods only work for periodic problems. Are there spatial discretization methods with superior properties than CD2 method and which does not have the limitations of spectral methods?

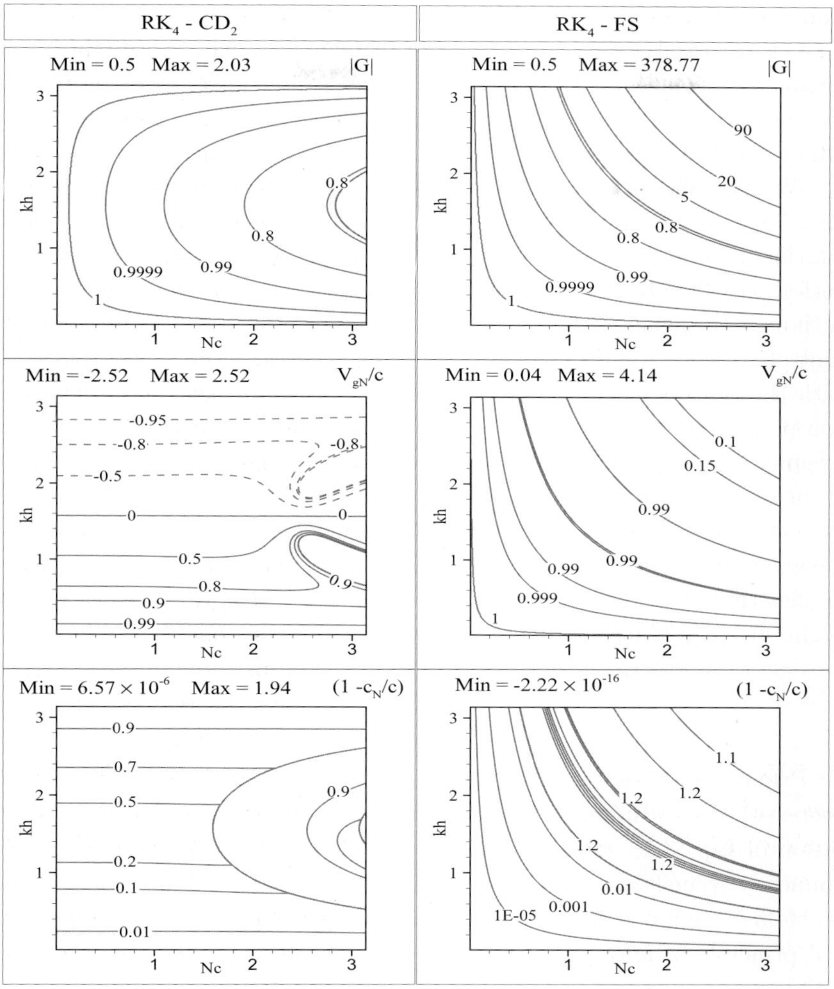

Fig. 4. Comparative results of GSA for RK4-CD2 and RK4-FS method showing —G—, V_{gN}/c and $(1 - c_N/C)$ contours in (N_c, kh)-plane. Note that the instability of RK4-CD2 method has larger region of neutral stability as compared to RK4-FS method. However, the dispersion and phase error property clearly indicates that why RK4-FS will be preferred over RK4-CD2 method for the solution of Eq. (1.4)

3.1. *Implicit spatial discretization methods*

There are indeed the compact difference schemes, which have near-spectral accuracy and have been used for direct simulation of practical flows in fluid mechanics and aerodynamics extensively. These methods with some care can compute acoustic problems starting from first principles. Just for the sake of introduction, we introduce the readers with one such method from a class of optimal upwind compact schemes[36,37]. We will focus on OUCS3 scheme, which is notionally only second order accurate, but provides resolution which is one among the best in practice, in the literature.

We show the properties of the compact scheme OUCS3 introduced in Ref. 36 for a non periodic problem and here we show the scheme for an interior node and other nodes at and close to the boundary. The GSA is performed when RK4 time integration strategy is used for solving Eq. (1.4). Although the schemes have upwind as a descriptor, yet it is in the users' control to work with a central or upwind version. Even when one works with the central scheme, there is inherent bias in all compact schemes for non-periodic problems due to the need of using one-sided boundary closure scheme. For the OUCS3 scheme, this was avoided in a symmetrized version of the scheme in Ref. 9, known as symmetrized OUCS3 or SOUCS3 scheme. First, the stencil for the basic OUCS3 scheme along with boundary stencils are given in a uniform grid with spacing h. The use of uniform grid spacing implies that such compact schemes work in the transformed plane, instead of the physical plane. The interior stencil is given by

$$\alpha_{j-1}\, u'_{j-1} + \alpha_j\, u'_j + \alpha_{j+1}\, u'_{j+1} = \frac{1}{h}\sum_{k=-2}^{k=2} q_k\, u_{j+k} \tag{3.1}$$

Primed quantities represent numerical derivatives obtained by simultaneous solution of linear algebraic equations, which are obtained by application of Eq. (3.1) at interior nodes. Furthermore, one requires one-sided boundary stencils at nodes $j = 1, 2$ and $N-1, N$, for solving non periodic problems. Such one-sided boundary closure schemes can violate the physical principle of information propagation. While obtaining solution using high accuracy compact schemes, inherent bias of these stencils can also accentuate the above mentioned problem of violating physical principle. Explicit boundary closure stencils are given for $j = 1$ and $j = 2$ as

$$u'_1 = \frac{1}{2h}(-3u_1 + 4u_2 - u_3) \tag{3.2}$$

$$u'_2 = \frac{1}{h}\left[\left(\frac{2\gamma_2}{3} - \frac{1}{3}\right) u_1 - \left(\frac{8\gamma_2}{3} + \frac{1}{2}\right) u_2 + (4\gamma_2 + 1)\, u_3 - \left(\frac{8\gamma_2}{3} + \frac{1}{6}\right) u_4 + \frac{2\gamma_2}{3} u_5\right] \tag{3.3}$$

where $\gamma_2 = -0.025$ is a parameter used to achieve better numerical properties of neutral stability, lesser phase and dispersion errors for the solution of Eq. (1.4). Similarly, one can write boundary closure schemes for $j = N$ and $j = N - 1$ using another parameter, $\gamma_{N-1} = 0.09$ by

$$u'_N = \frac{1}{2h}(3u_N - 4u_{N-1} + u_{N-2}) \tag{3.4}$$

$$u'_{N-1} = -\frac{1}{h}\left[\left(\frac{2\gamma_{N-1}}{3} - \frac{1}{3}\right) u_N - \left(\frac{8\gamma_{N-1}}{3} + \frac{1}{2}\right) u_{N-1} + (4\gamma_{N-1} + 1)\right.$$
$$\left. u_{N-2} - \left(\frac{8\gamma_{N-1}}{3} + \frac{1}{6}\right) u_{N-3} + \frac{2\gamma_{N-1}}{3} u_{N-4}\right] \tag{3.5}$$

To achieve desired numerical properties, coefficients in Eq. (3.1) are upwinded[28,36]: $\alpha_{j\pm1} = 0.3793894912 \pm \eta/60$, $\alpha_j = 1$, $q_{\pm2} = \pm0.183205192/4 + \eta/300$, $q_{\pm1} = \pm1.57557379/2 + \eta/30$ and $q_0 = -11\eta/150$, where $\eta = -2.0$ is the upwind parameter for the interior stencil. Undesirable bias of the above OUCS3 scheme is readily apparent, as one notices that the stencils for $j = 1$, 2 and $j = N$, $(N-1)$ are antisymmetric. This problem of lack of symmetry was removed using a procedure in Ref. 9, which results in identical properties for all nodes located symmetrically about the central node.

To dispel the myth of order of schemes, we note the comparison of scale resolution in evaluating first derivative in Fig. 5(a) by two different methods, both of which are second order accurate schemes. Both the displayed methods in this frame being of same order, yet the scale resolution of SOUCS3 is order of magnitude better than CD2 scheme. There is of course the obvious difference between CD2 and OUCS3 schemes, the former being an explicit scheme, while the latter is an implicit scheme.

To account for error contributing terms, we have compared $|G|$- and V_{gN}/c-contours between CD2-CN and SOUCS3-RK4 methods, in Figs. 5(b) to 5(e). In Fig. 5(b), one observes neutral stability over the complete (N_c, kh)-plane for the CD2-CN method, with respect to the third term on the right hand side of Eq. (2.1). However, V_{gN}/c-contours for CD2-CN method show less dispersion only very close to the origin in Fig. 5(c), i.e., only for very small time steps and for very large length scales. Higher wavenumber components will not only propagate erroneously, but any wavenumber above $kh = \pi/2$, will move as the q-waves[32] and will be a major source of error altering the physical flow. Corresponding properties for the SOUCS3-RK4 scheme are shown in frames 5(d) and 5(e). Figure

5(d) shows $|G_j|$-contours and indicates a very small range of N_c where neutral stability prevails. However, significant improvement is noted for V_{gN}/c-contours in Fig. 5(e), which makes SOUCS3-RK4 method preferable over CD2-CN method. Also for SOUCS3-RK4 scheme, we have a

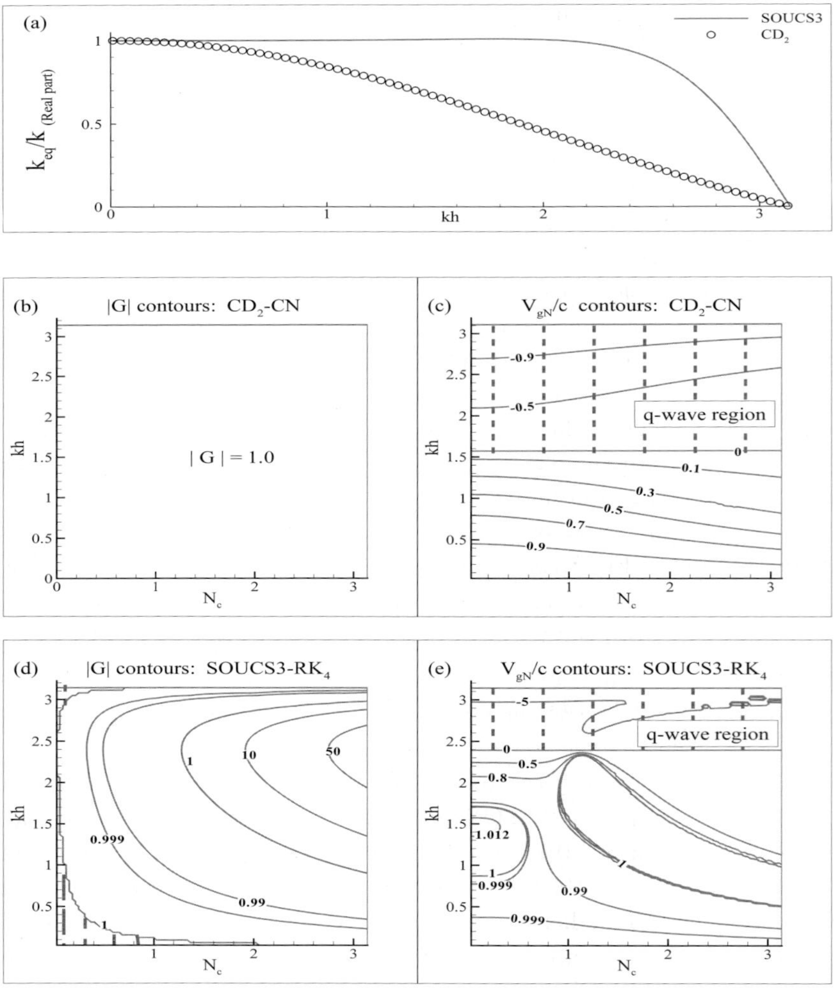

Fig. 5. Comparison of numerical amplification factor ($|G|$) and normalized numerical group velocity (V_{gN}/c) contours for the SOUCS3-RK4 method with CD2-CN method for solving Eq. (1.4). Top frame of the figure shows the scale resolution provided by the SOUCS3 scheme as compared to CD2 scheme. The properties are obtained using 101 points.

smaller q-wave region at significantly higher wavenumbers, as compared to CD2-CN method.

There are significant progresses made in developing high accuracy DRP compact schemes[38]. Any scheme which has numerical and physical dispersion relations identical over a significant region in (N_c, kh)-plane and which is also neutrally stable in the same region is called a DRP scheme. The DRP compact scheme in Ref. 38 has been used to solve many practical inhomogeneous flows and is termed the OCRK3 scheme. This method has been used in recent times for the DNS of inhomogeneous flows from the receptivity to fully developed turbulent stage and reported[30,31]. Computed fully developed two-dimensional turbulent flow in these references displayed k^{-3} spectrum for the energy, as has been shown theoretically by Batchelor[2] and Kraichnan and Montgomery[18].

In Fig. 6, corresponding numerical properties of OCRK3 scheme are shown and one notices the presence of a range of N_c up to 0.21, for which the scheme is neutrally stable, for all resolved values of kh. The contours also reveal that this scheme remains stable up to $N_c = 0.99$. In the lower right frame of Fig. 6, contours of $(1 - c_N/c)$ or normalized numerical phase speed have been plotted and one notes significant improvement in the shaded region admitting error only up to 0.2%. In the bottom left frame, normalized numerical group velocity contours are plotted and the shaded region indicate the values of kh and N_c for which the error is less than 0.5% for the solution of Eq. (1.4). Once again the comparison between OCRK3 scheme with RK2-FS scheme establishes that the latter method is inherently flawed, on both counts of numerical instability and dispersion/phase error properties.

In Fig. 7, we note the solution of Eq. (1.4) for the propagation of a wave-packet given by

$$u(x, t = 0) = e^{-\alpha(x-x_0)^2} \cos(k_o(x - x_o)) \tag{3.6}$$

We have chosen $\alpha = 0.5$ with center of packet at $x_0 = 5$ in the computational domain $0 \leq x \leq 100$ with 32,768 equi-spaced points with grid-spacing of $\Delta x = 0.00305$. For this case, convection speed is chosen as $c = 0.1$ and the non dimensional central wavenumber of $k_o h = 0.8$ with a time step of $\Delta t = 0.003$ are used. For the chosen parameters, corresponding value of CFL number is $N_c = 0.09830$. For the RK2-FS method shown in Fig. 2. One can notice the formation of Gibbs' phenomenon[37] at upstream side after $t = 4.5$, which gives rise to spurious oscillations on the upstream side and this phenomenon becomes more dominating at later times, as can be seen at $t = 6.0$. Thereafter the numerical result are so laced with errors that these are not shown. The error is due to poor phase and dispersion properties of RK2-FS method as shown in Fig. 1. Numerical results corresponding to OCRK3 scheme are shown in Fig. 7 at various time instants with the exact solution shown again by dotted lines. This case is computed up to $t = 400$ and beyond. Due to good phase and dispersion properties of OCRK3 scheme, a good match with the exact solution is possible without any dissipation, phase and dispersion errors even up to $t = 400$. This is due to the fact that for the chosen values of numerical parameters, OCRK3 method is neutrally stable with negligible phase and dispersion error - a feature of the DRP method. As q-waves are formed for $kh > 2.4$, one needs to reduce h so that the essential dynamics is kept within this

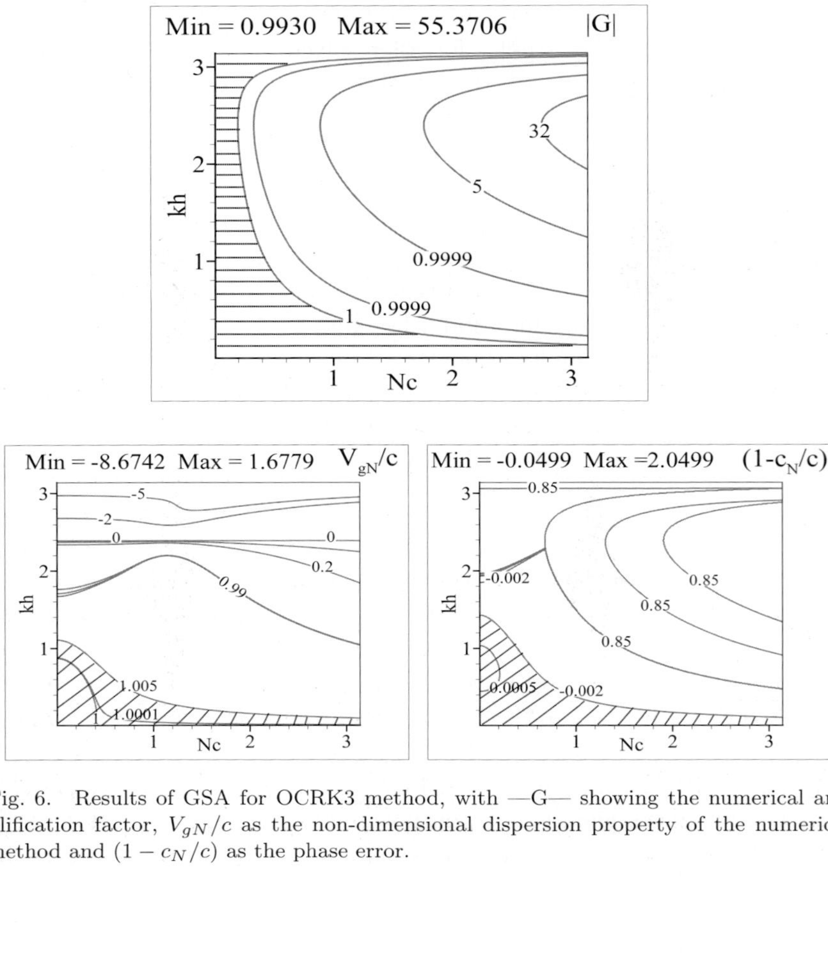

Fig. 6. Results of GSA for OCRK3 method, with —G— showing the numerical amplification factor, V_{gN}/c as the non-dimensional dispersion property of the numerical method and $(1 - c_N/c)$ as the phase error.

range of kh. Other staggered compact schemes will be described by which 3D transitional flow simulation has been reported in Ref. 3. Having seen the existing possibilities of spatial discretization for DNS and LES, it is necessary to discuss about time integration methods.

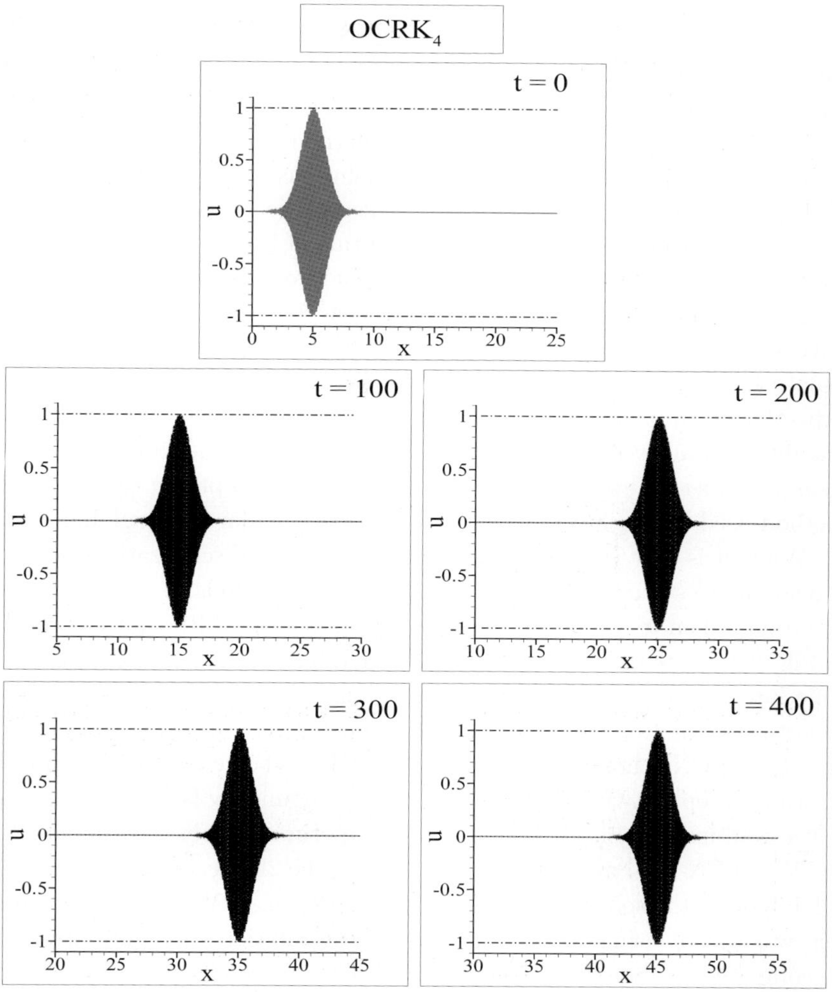

Fig. 7. Propagation of computed wave-packet for Eq. (1.4) with $c = 0.1$, $k_0h = 0.8$ and $N_c = 0.0983$ using OCRK3 method. In all the frames, exact solution is shown by dotted lines. Where k_0 is the central wavenumber of the wave-packet chosen. It is noted that the actual wave-packet has finite band-width around this central value.

4. Explicit and Implicit Methods for Time Integration

We already have noted that for explicit time integration schemes increasing the order amounted to increase in accuracy, in terms of the numerical properties which are responsible for error in Eq. (2.1). One of the advantages of implicit methods like Crank-Nicolson method is their unconditional stability ($|G_j| = 1$) which from numerical instability point of view, allows one to take larger time step, as shown in Fig. 5 for CD2-CN method used to solve Eq. (1.4). Implicit methods obtain unknowns at the current time level by solving simultaneous linear algebraic equations by direct or iterative methods, which makes these methods computationally intensive. While this relates to computational cost of implicit methods, there are practical cases of solving flow problems where implicit solutions suffer tremendously due to dispersion error, as given by first and second terms of Eq. (2.1).

Figure 8 taken from Ref. 40 shows solution of Eq. (1.4) over a long time with wave-packet as the initial condition, for two different time steps given by $N_c = 0.14545$ and 1.4545. This has been done purposely to show the adverse effects of dispersion. On the contrary, many researchers[25,26,46] have been using CN-CD4 method partially in the domain for solving NSE. One expects that implicit methods due to their inherent unconditional neutral stability will allow taking very large time steps. But the results in Fig. 8, clearly shows that even for small N_c the error is unacceptable and this is the method used by the authors[21,25,26,46] in these references for solving NSE.

While it is now well-known that implicit spatial discretization methods provide more accuracy as compared to explicit methods, it is not well understood, when it comes to temporal discretizations, although the results in Fig. 8 indicate inadequacy of CN-CD4 method. In a recent study, role of time integrations in computing a transitional wall bounded flow has been reported[40]. In Ref. 40 using CD4 method for spatial discretization[25,26], specifically CN scheme is compared with RK4 scheme of time integration to study receptivity of zero pressure gradient boundary layer to monochromatic wall excitation, in simulating the experimental observations in Ref. 27. When CN-CD4 method is used in solving the 2D receptivity problem by solving NSE, the solution blew up after a finite time. When the same problem was solved using RK4-CD4 method, the solution was quite accurate, as compared to that obtained by OUCS3-RK4 method.

Despite this observations for convection-diffusion problems, researchers often use implicit-explicit (IMEX) methods, for which the convection parts are handled using higher accuracy explicit methods, while the diffusion part is computed implicitly with unconditional stability for any time step chosen. This is the principal reason for many researchers[1,12,16,22,24,26] to use IMEX methods by combining explicit and implicit methods to solve problems which involve stiff source terms, relaxation and diffusion terms in governing equations.

In solving NSE, this has led to proposing projection methods[17,23] which treat convection terms explicitly and diffusion terms implicitly. For solv-

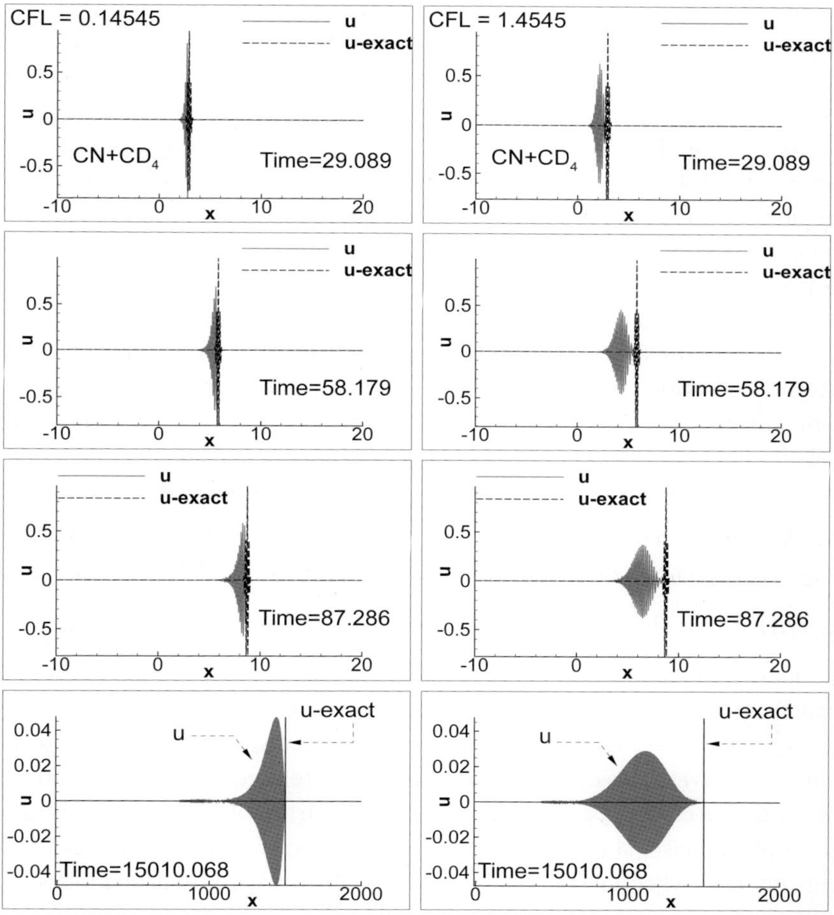

Fig. 8. Propagation of computed wave-packet following Eq. (1.4) with $N_c = 0.14545$ (left) and $N_c = 1.4545$ (right) using CN-CD4 method. In all the frames, exact solution is shown by dotted lines. It is noted that the maximum amplitude of signal appears to continually reduce due to dispersion, as this method satisfies $|G| = 1$ for all combinations of kh and N_c.

ing convection-diffusion problems by spectral and finite difference schemes, IMEX methods have been studied in Ref. 1, where the authors experienced difficulties with multi-grid method due to weak decay of high frequency modes which caused poor convergence. Spectral collocation methods experience aliasing and thereby creating high frequency modes and encounter numerical difficulties due to same reasons. However with such widespread usages, there are no quantitative guidelines available as given here, due to lack of systematic study to characterize numerical methods.

In Ref. 3 high accuracy staggered compact scheme (OSCS), with optimized three stage Runge-Kutta method (ORK3) has shown the onset of 3D transition also associated with the formation of spatio-temporal wave-front (STWF), as it was shown earlier for 2D transition[30]. Simulation reported in Ref. 26 for 3D transition, the authors failed to even capture STWF, using CN-CD4 method used very close to the plate, while in the rest of the domain, RK3-CD4 method has been used. Use of implicit method near the wall is supposed to allow relaxed time step, where fine grid is used to resolve large flow-gradients. However, the dispersion error of implicit method is not confined close to the wall. We show here that the interface of implicit and explicit region, very close to the wall, is the site of partial internal reflections, whose role is to continually generate error-packets with high wavenumber creating q-waves and thus changing the flow physically. This is responsible for the failure by the authors in Ref. 25 and 26 to even trace the STWF, as it is shown in Ref. 3 to be caused by very low levels of perturbation.

5. IMEX Method for Solving Transitional Flow

To qualify the above statement, we first investigate 1D and 2D convection equations using IMEX method, which requires an overlap region, as shown in Fig. 9(a). The behavior of the interface between these two parts of the domain is investigated and shown that the interface gives rise to internal reflections as error-packets, destroying the accuracy of the solution. Such q-waves are a consequence of extreme dispersion error, as shown by GSA[28,32]. This internal reflection is explained by Fig. 9(b), where we have plotted the numerical dispersion relation for CN-CD4 and CN-OUCS3 schemes. For example, for the CN-CD4 scheme if the signal/ error corresponds to the point P_2, then the reflected signal/ error will also have the same numerical circular frequency (ω_N) and hence the reflected signal will correspond to the point R_2 in this figure. As P_2 corresponds to a lower value of kh, it is readily apparent that the reflected packet would correspond to point R_2 with larger value of wavenumber, nearer to the Nyquist limit ($kh = \pi$). Some result was generalized in Ref. 43 while investigating CD2 spatial discretization, which is verified by GSA to hold true only for CD2 spatial discretization scheme. The general expression comes out of numerical dispersion relation

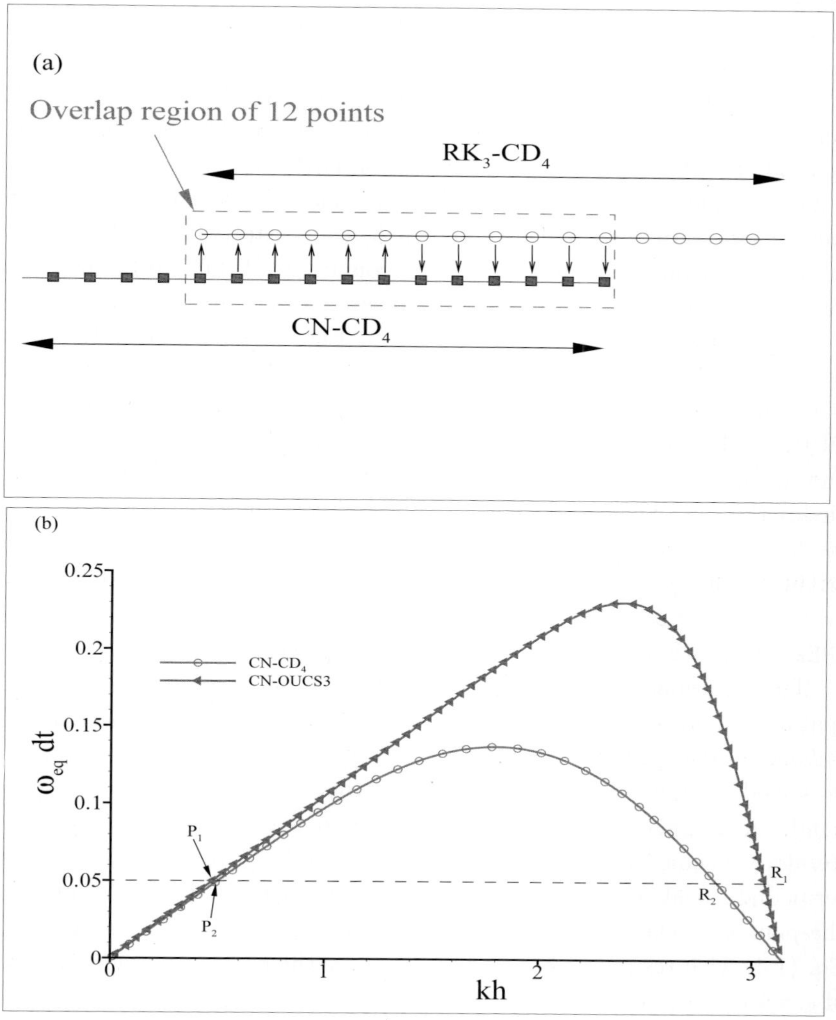

Fig. 9. (a) Schematic diagram of overlap region with arrowheads indicating data transfer during time integration. (b) Numerical dispersion relation for CN-CD4 and CN-OUCS3 methods, used here to explain internal reflection at sub-domain boundaries.

in Eq. (1.6) and is explained in Fig. 9(b).

For the IMEX method used for analysis here, CN scheme is used for implicit time integration, and for explicit time integration RK3 scheme is used – both with CD4 scheme for spatial discretization. This combination of methods are also used in Refs. 25 and 26. Fifty overlap points are used to solve Eq. (1.4) and corresponding error propagation is shown in Figs. 10(a-b). No effects in the order of error is noted, even for reduced twelve overlap points taken (results not shown), with error dynamics showing identical propagation properties with two peaks.

The CD4 spatial scheme has lower spectral resolution and poorer dispersion properties, as compared to compact schemes. To compare IMEX method of[25,26,46] with other high resolution DRP methods, OUCS3 scheme is also considered here with CN and RK3 time integration schemes. Compact schemes are not symmetric and their use in parallel computing brings in additional problems at sub-domain boundaries[34]. The error evolution for this method is plotted for the indicated times in Figs. 10(c-d). At $t = 9$ in Figs. 10(a-b) and at $t = 10$ in Figs. 10(c-d), one can see the error level at the peak M is of the order of 10^{-5}. Increase in number of overlap points, do not show any perceptible change in error levels for OUCS3 scheme. The peak M is due to dispersion error and peak P is due to internal reflection at the mid-point of the overlap region (when M passes through it). In between these two peaks there is an enhanced error level E which is absent in CD4 spatial discretization scheme and it's origin is also investigated here. Time variation of error for this case is available at `http://spectral.iitk.ac.in/hpcl/sites/default/files/IMEX_Animation/1D_OUCS3_50_pts_overlap.gif`.

The additional peak at E for OUCS3 method is not present for CD4 spatial discretization case. For CD4 method the numerical solution of pentadiagonal matrix equation arises due to the use of CN-scheme, which can be solved *exactly* with very low error. For the compact scheme on the other hand, such *exact* solver is not available and we have used Bi-CGSTAB iterative scheme[44] with a tolerance of $\epsilon = 10^{-8}$. It is noted that the intermediate peak E also appears for CN-CD4 method, if instead of using the pentadiagonal solver, one uses Bi-CGSTAB iterative solver in solving Eq. (1.4) with results shown in Fig. 11(a) for the particular tolerance level of $\epsilon = 10^{-8}$.

To illustrate the source of this additional error peak, we consider IMEX time integration schemes with CD4 spatial scheme. For CN-CD4 scheme the resulting penta-diagonal matrix is inverted by a direct solver and using Bi-CGSTAB iterative method with tolerance $\epsilon = 10^{-8}$. The error for both these cases are also shown in Fig. 11(a) at $t = 180$. One notes from Fig. 11(a) that while higher error at the intermediate peak E is present for results obtained using Bi-CGSTAB solver, such an additional peak of enhanced error is absent for results obtained by direct solver. When the tolerance level of the Bi-CGSTAB solver is tightened to 10^{-16}, such an intermediate peak at E is seen to be absent and error levels in both cases are seen to coincide, as shown clearly in Fig. 11(b).

In most cases of solving PDE's by IMEX methods, one would not have direct methods and instead would be forced to use iterative solver. For the cases shown in Fig. 11, when the error peak M passes through the mid-point of the overlap region, the number of iterations required for $\epsilon = 10^{-16}$ increases to six, from three iterations required for $\epsilon = 10^{-8}$. While this may appear as insignificant increase in computational efforts, the situation becomes adverse when a signal reflects back from the domain boundaries,

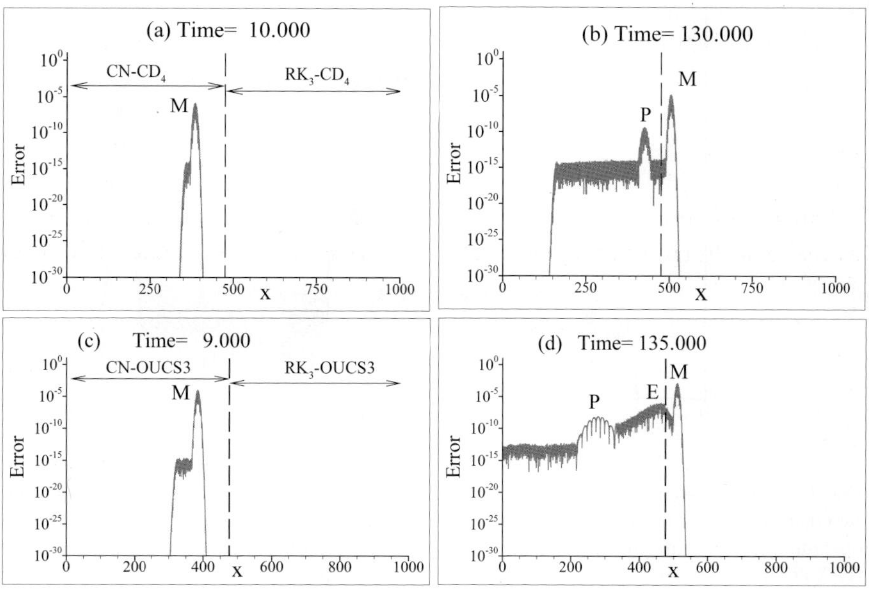

Fig. 10. (a-b) Time evolution of error for the case of fifty points overlap solving 1D convection equation using CD4 scheme along with IMEX time integration scheme (CN and RK3 methods for implicit and explicit part of the domain). (c-d) Error for the case of fifty points overlap for the solution of 1D convection equation using OUCS3 scheme along with IMEX time integration scheme.

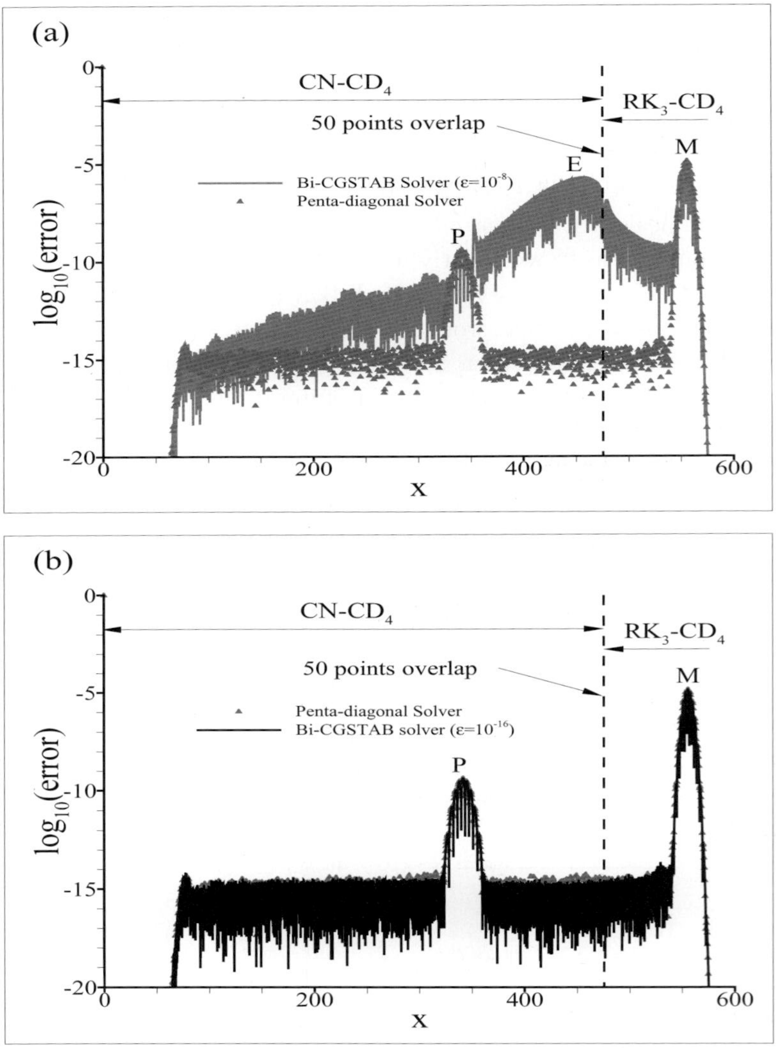

Fig. 11. Error distribution showing the peaks of the solution of 1D convection equation by CD4 discretization for linear algebraic solution. (a) Comparison of error distribution obtained by direct method (pentadiagonal solver) and Bi-CGSTAB iterative solver with the tolerance, $\epsilon = 10^{-8}$ (b) Comparison of error distribution obtained by direct method and Bi-CGSTAB iterative solver with the tolerance, $\epsilon = 10^{-16}$.

with orders of magnitude increase in number of iterations. After each such reflections from domain boundaries, the signal will also be partially reflected at the interface of the explicit and implicit time integration sub-domains.

Each of which will contribute to enhanced error and subsequent increase in number of iterations for iterative solver. It is noted that in Refs. 25, 26 and 46, the implicit domain size is so small that such series of total and partial reflections from domain and sub-domain boundaries will cause significantly increased error, swamping the actual signal. This was the reason that the authors did not obtain STWF at all in the solution of NSE. There is an additional effect for the case of DNS reported in Refs. 25 and 26, where the implicit region is confined to a region very close to the solid wall. Spurious internal reflections create error-packets which move towards the wall and upon reflection from there, will make the iterative solver for the implicit zone even slower significantly. If the error is reduced associated with E (by tightening the tolerance level), then there will still be the additional problem of created q-waves at the interface. These q-waves will numerically excite the flow causing what the authors have termed *bypass transition.*

5.1. *Two-dimensional linear convection equation*

To calibrate numerical schemes for solving NSE, an appropriate model equation is represented by multi-dimensional wave propagation. We confine attention to 2D wave propagation, with its attendant issues of q-waves, which is an added error source[32] augmented by anisotropic propagation of signal if the grid spacing in both directions are not homogeneous[39]. Here specifically a wave-packet propagating at an angle is investigated, as in Ref. 39. The model 2D linear convection equation which is used in such investigations is given as,

$$\frac{\partial u}{\partial t} + c_x \frac{\partial u}{\partial x} + c_y \frac{\partial u}{\partial y} = 0 \tag{5.1}$$

where $c_x = c\cos\theta$; $c_y = c\sin\theta$ with θ as the angle of propagation.

Numerical properties of 2D wave propagation problem by CN-CD4 and RK3-CD4 schemes are shown in Fig. 12. The properties of these schemes shown in Fig. 12 are obtained following the procedure in Ref. 39. Here, h_x and h_y are the uniform grid spacing in x and y-directions, respectively. Similarly, k_x and k_y are the wavenumber components in x- and y-directions, respectively. It is well known[39] that the grid aspect ratio has to be such that the signal is aligned with the wave propagation direction along the lines joining the node for higher accuracy. The CFL numbers in x and y-directions are Nc_x and Nc_y and are related by $Nc_y = Nc_x \left[\frac{\tan\theta}{\lambda}\right]$, where $\lambda = \frac{h_y}{h_x}$ is the grid-aspect ratio. If we choose $Nc_x = Nc_y$, then we have to take $h_x = h_y$, as we have taken $\theta = 45^0$ as the wave propagation direction.

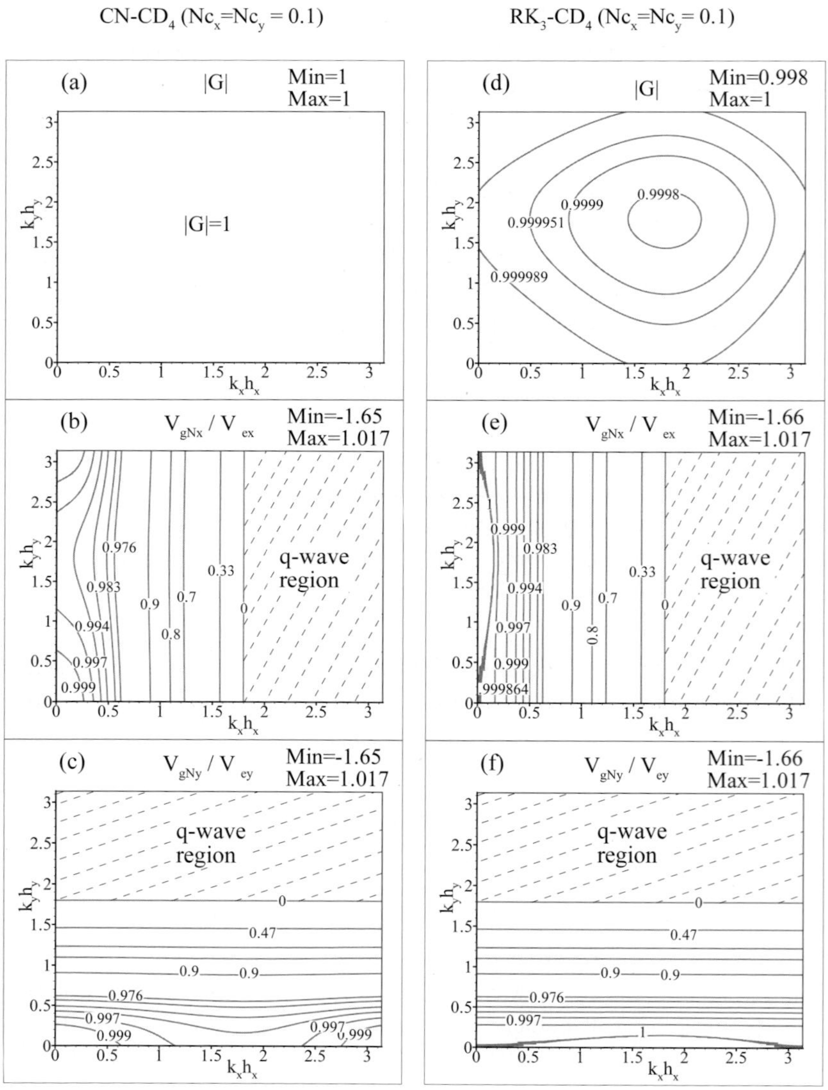

Fig. 12. Numerical properties of CN-CD4 and RK3-CD4 schemes: (a) and (d) Numerical amplification factor; (b) and (e) normalized numerical group velocity in x-direction and (c) and (f) normalized numerical group velocity in y-direction.

In Fig. 12, we have chosen $Nc_x = Nc_y = 0.1$ and plotted the properties in (k_xh_x,k_yh_y)-plane by showing the numerical properties of CN-CD4 and RK3-CD4 schemes on left and right frames, respectively. For CN-CD4 scheme, $|G| = 1$ for all combinations of k_yh_y and k_xh_x representing neutral stability over the entire resolvable spectral domain. This is due to implicit nature of the time-integration scheme. In frame (b), normalized V_{gN_x}-contours (normalized with respect to the exact group velocity component V_{ex}) are plotted. One observes the DRP region from this frame where $0.999 \leq V_{gN_x}/V_{ex} \leq 1.001$. This condition is satisfied in the spectral region where $k_xh_x \leq 0.28$ and $k_yh_y \leq 0.6$. Beyond $k_xh_x = 1.8$, presence of q-wave can be clearly noted, which is indicated with dashed lines. Similarly, frame (c) shows the non-dimensional group velocity component (normalized with respect to V_{ey}) in the y-direction. Here, presence of q-wave region is noted above $k_yh_y = 1.8$. For RK3-CD4 scheme also, one notes $|G|$ to be very close to 1. In Figs. 12(e) and 12(f), we plot V_{gN_x}/V_{ex} and V_{gN_y}/V_{ey} for RK3-CD4 scheme. Comparing Figs. 12(e, f) with corresponding plots in Figs. 12(c, d), one notes significant improvement in the extent of the DRP region for RK3-CD4 scheme over CN-CD4 scheme. This once again proves the superiority of the explicit RK3-CD4 scheme over implicit CN-CD4 scheme. Thus, the latter may have unconditional neutral stability (as a consequence the third term on the right hand side is zero in Eq. (2.1)), yet its dispersion error is significantly poorer as compared to the explicit RK3-CD4 scheme. This, coupled with the fact that explicit scheme is less expensive, suggests that computing in the full domain by explicit scheme will be preferred as compared to fully implicit treatment or even adopting IMEX schemes.

In Fig. 13, the numerical properties of the CN-OUCS3 and RK3-OUCS3 schemes are shown by plotting $|G|$ and normalized group velocity components in (k_xh_x, k_yh_y)-plane. The CN-OUCS3 scheme has the desired neutral numerical stability as noted in frame (a) over the entire resolvable spectral domain. The RK3-OUCS3 scheme is also near-neutral over the entire range of (k_xh_x, k_yh_y) and has similar stability property. As compared to RK3-CD4 scheme, group velocity property is even better with marginal deficiency centered at higher values of $k_xh_x = 2.5 = k_yh_y$. However more importantly, the normalized group velocity components show significantly improved dispersion properties with increased values of k_xh_x and k_yh_y above which q-waves are observed. These properties have been derived following the procedure reported in Ref. 39.

The initial condition for solving this equation is a circular Gaussian

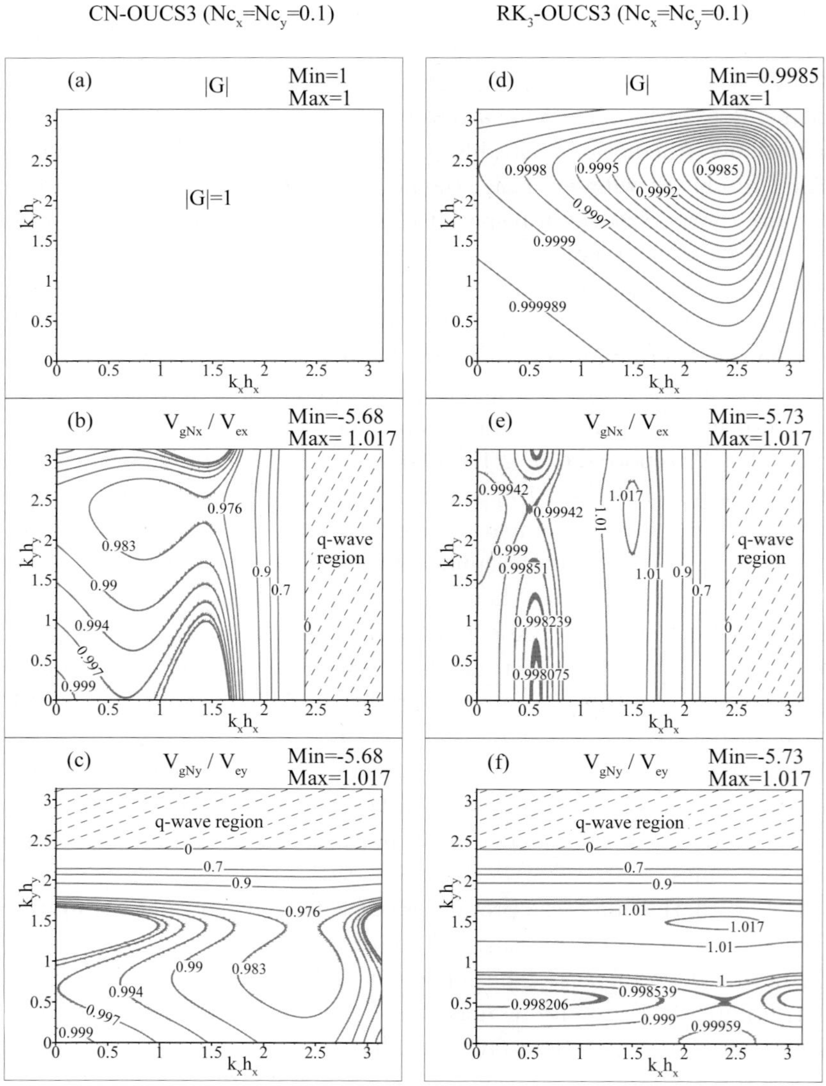

Fig. 13. Numerical properties of CN-OUCS3 and RK3-OUCS3 schemes: (a) and (d) Numerical amplification factor; (b) and (e) normalized numerical group velocity in x-direction and (c) and (f) normalized numerical group velocity in y-direction.

patch given by

$$u = e^{-\alpha((x-x_0)^2+(y-y_0)^2)} \sin\left[k_0\sqrt{(x-x_0)^2+(y-y_0)^2}\right]$$

where a dominant wavenumber k_0 and α, decide the width of the wave-packet and are chosen as 1 and 500 here.

For solving Eq. (5.1) by IMEX method with CD4 discretization, the value of θ is chosen as 45^0 and c as $0.2\sqrt{2}$. The CN-CD4 scheme is used in the implicit zone and RK3-CD4 scheme in the explicit zone. The Bi-CGSTAB solver with $\epsilon = 10^{-8}$ as the tolerance criterion is used in the implicit zone. We have used an uniform grid of 100 points along x- and y-directions for the reported simulations. The splitting of the domain into implicit and explicit zones is performed in x-directions only with twenty four points overlap. Also, solution of Eq. (5.1) is performed with OUCS3 scheme for comparison using identical values of all parameters. The improved numerical properties of CN-OUCS3 and RK3-OUCS3 schemes also show up when the same initial condition, in a square domain of size 10×10 is used for the reported error contours in Fig. 6. For this case initial location of the patch is chosen at $x_0 = 5.6$ and $y_0 = 5.6$.

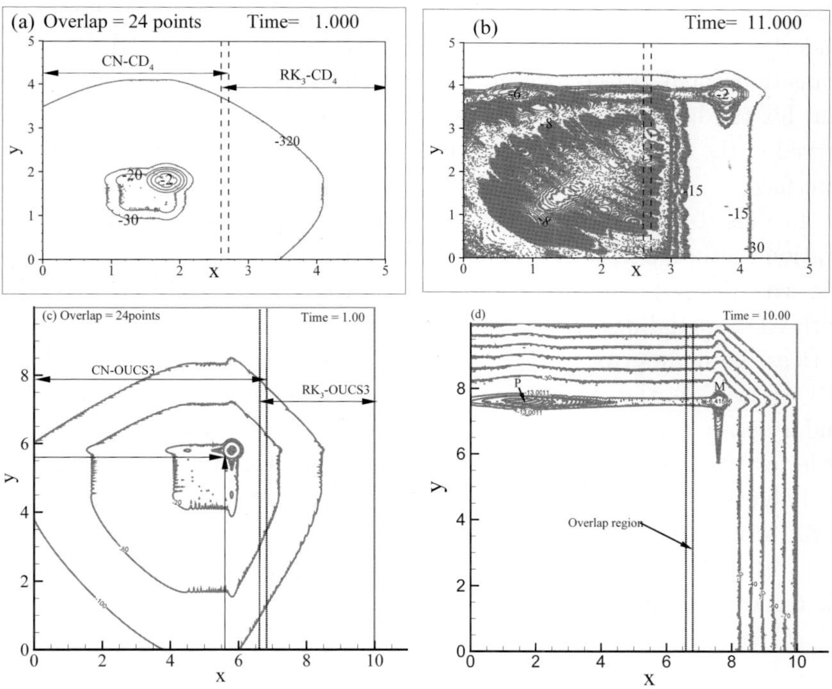

Fig. 14. Comparison of time evolution of error for the case of twenty four overlap points for the solution of 2D convection equation using CD_4 and OUCS3 spatial discretization in frames (a)-(b) and (c)-(d), respectively.

In Figs. 14(a-b), we have plotted the distribution of error in the computed domain at the indicated discrete times for IMEX scheme with CD4 for spatial discretization. In the frames, the error contours are plotted showing the index n, such that the error is represented as 10^n. One notices the error level at the center of the error-packet (M) to be of the order of 10^{-2}, surrounded by errors of lower level. In frame (b), the main error-packet is centered about the mid-point of the overlap region, which is followed by a region inside the rectangular contour, which is at round-off error level (10^{-15}). This error is initiated by the q-waves, when the signal crosses the overlap region. The spurious error component is noted to travel in opposite direction with respect to the exact solution's direction of propagation. Once the error-packet along with the signal (M) crosses the overlap region, additional elongated error-packets (P) are created, which move in the negative x-direction, at the y-location corresponding to the center of the main error packet (M). This horizontal error-packets move upstream with respect to the main error-packet. In solving NSE, these error-packets (P) will behave like convected vortices propagating upstream and constantly buffeting the underlying shear-layer, something akin to the unphysical scenario of bypass transition noted in Refs. 25, 26 and 46. These spurious error-packets grow and hit the boundary at later time. There are hardly any differences observed in the error levels, if we increase the number of overlap points at the interface.

In Figs. 14(c-d), error of the computed solution using IMEX time integration method and OUCS3 spatial discretization scheme is shown. Here the error levels due to dispersion are significantly lower than IMEX scheme with CD4 spatial discretization, as evident from the frames. Here also one notices the main error packet (M) along with the secondary error-packets (P) created, as the signal transits through the overlap region. The secondary error packets caused by the internal reflection display values of the order of 10^{-6}. Time variation of error for both CD4 and OUCS3 cases are available at `http://spectral.iitk.ac.in/hpcl/sites/default/files/IMEX_Animation/2D_CD4_12_pts_overlap.gif` and `http://spectral.iitk.ac.in/hpcl/sites/default/files/IMEX_Animation/2D_OUCS3_24_pts_overlap.gif`.

6. 3D Receptivity of Boundary Layer to Wall Excitation

Error-packet creation by IMEX scheme at the interface of implicit and explicit region can cause wrong flow simulation, which is noticed when two simulations, with and without using the IMEX scheme, are compared. The cause of transition to turbulence for low level of background disturbances was noted as the creation of STWF in Ref. 3, which originate via a linear mechanism due to flow instability. However in Refs. 25 and 26, the authors took a very small domain and apart from employing a wall exciter, *random* excitation was added at the inflow and no STWF was reported, as has been reported in the other publication[3]. It is noted that for 2D transition also the authors[30] reported presence of the STWF. Thus, STWF is ubiquitous for flow instability and transition process.

To explore the role played by the STWF for a 3D disturbance field, NSE has been solved using DRP methods[3,28]. Schematic of the computational domain used here is shown in Fig. 15(a), with the shear layer excited by a Gaussian circular patch (GCP) exciter vibrated time-harmonically, which is periodic in the spanwise direction, with a period of z_{max}. Such an excitation creates planar and oblique 3D TS wave-packets, depending upon the geometry of the exciter.

Imposed wall-normal velocity caused by the exciter is given by,

$$v_{wall}(x, z) = \alpha_1 A_m(x, z) H(t) e^{i\omega_0 t}$$

where α_1 and ω_0 denote the amplitude and the frequency of excitation. Here, $H(t)$ is the Heaviside function needed for the impulsive start of excitation and $Maximum(|A_m(x, z)|) = 1$. The amplitude of the GCP exciter varies as $A_m = (1 + \cos(\pi r/r_{max}))/2$, where r_{max} is the radius of the circular patch. Periodicity in spanwise direction implies existence of exciters at a periodic interval of z_{max}. Significantly longer domain is used to allow full nonlinear evolution of the STWF in the computational domain, as compared to that used in Refs. 25 and 26. Also, the present computations do not use any *random* wide-band excitation and use two-level explicit time integration method ORK3, which does not have spurious numerical modes[28,33]. Optimized staggered compact scheme OSCS[3] is used for spatial discretization and superior rotational form of velocity-vorticity formulation[4] is used for computations.

The perspective view of the streamwise disturbance velocity (u_d) is shown in the (x, z)-plane, in Fig. 15(b) at $t = 15$ for GCP exciter, with $\alpha_1 = 0.01$ and $z_{max} = 1$, which is taken from[3]. The non-dimensional frequency is $F = 0.5 \times 10^{-4}$, where $F = 2\pi\nu f/U_\infty^2$; ν is the kinematic viscosity and f is the frequency of excitation in Hertz. Figure 7(b) identifies the three elements of response, as was shown earlier[30] for the creation of 2D turbulence. Following the local solution, planar and a pair of oblique TS wave-packets are seen in Fig. 15(b). For this case, leading the disturbance structure is the STWF, which grows rapidly and is the actual precursor of transition to turbulence for this case[3].

The computed results[3] are the DNS for 3D disturbance flow field caused by the GCP excitation, which takes the flow from receptivity to the turbulent spot generation stage. This is compared to the solution reported earlier[26]. While in Sayadi *et al.*[26], no STWF was noted, yet the wall skin friction shown in Fig. 16, seem to match the time-averaged empirical data extremely well. We note that the authors[25,26] used *random noise*,

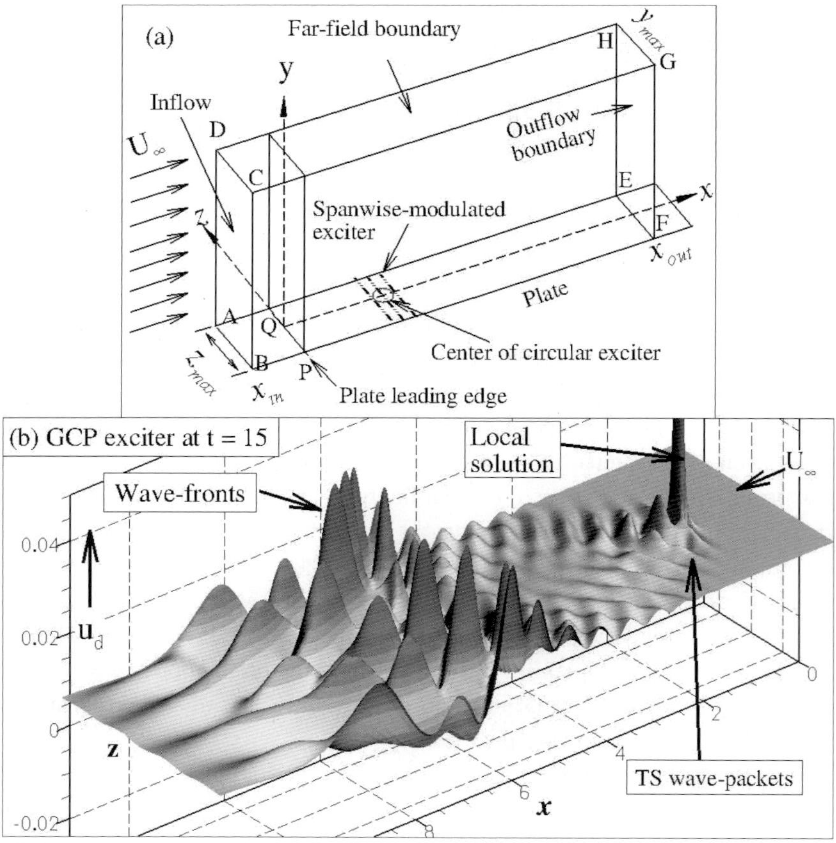

Fig. 15. (a) Schematic diagram of a computational domain shown for Gaussian circular patch (GCP) exciter with center located at x_{ex}. (b) Perspective plot of streamwise disturbance velocity at a height $y/L = 0.00215$ for $t = 15$ after onset of GCP excitation, for the case with $x_{ex} = 1.5$ and nondimensional excitation frequency, $F = 5 \times 10^{-5}$ for a one percent excitation amplitude and Reynolds number based on L is $Re = 10^5$.

apart from wall excitation for their IMEX method, which caused additionally upstream propagating error-packet to convect very close to the wall. Such complex excitation pattern does not show a clear solution as shown in Fig. 15[3], where one notices a local solution; oblique Tollmien-Schlichting waves and the STWF. Presence of STWF is a clear indication of clean computation without numerical contaminations, as is the case reported in[25,26]. Such convoluted numerical process has been termed as *bypass transition.* In contrast, the other solution[3] is typical of K-type transition reported experimentally[27].

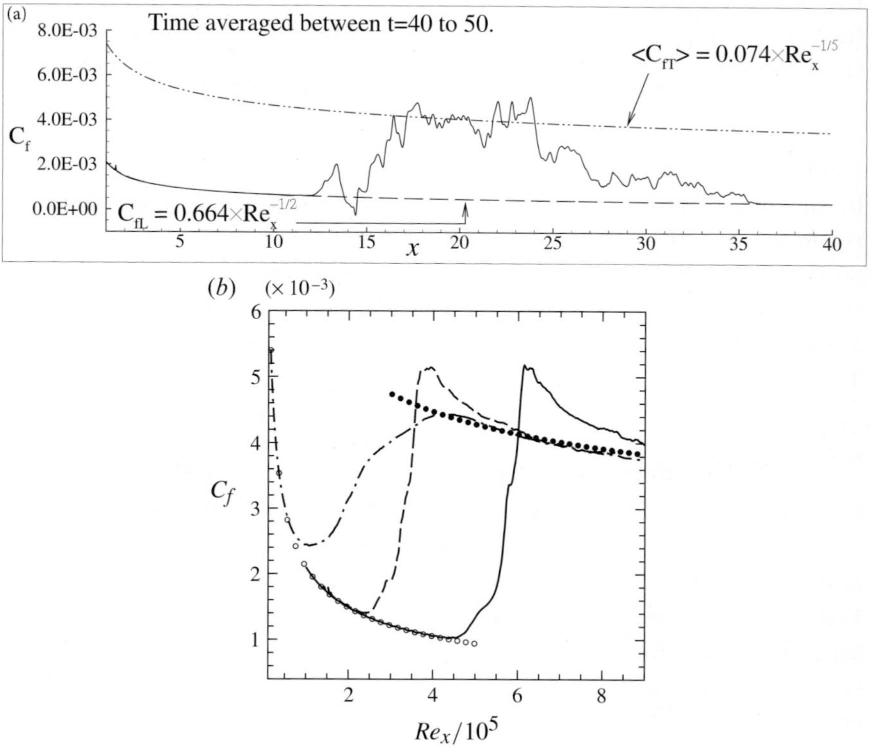

Fig. 16. Comparison of computed time-averaged wall skin friction by (a) the ORK3 method in Bhaumik & Sengupta (PRE, 2014) and (b) the IMEX method in Sayadi *et al.* (JFM, 2013).

6.1. *Random excitation and dispersion relation in transitional and turbulent flows*

In early researches on DNS of transitional flow (see for example the work of Refs. 25, 26 and 41), so-called *random noise* has been used to excite the ribbon-induced experimental results for transition in Schubauer and Skramstad[27]. Here in reviewing this aspect on transitional flow, one may ask the relevant question: In a finite domain discretization with finite spacing, is it possible to simulate and use truly white or random noise?

Unfortunately the answer is "no"! For uniform spacing (h), the maximum wavenumber resolvable is given by the Nyquist criterion: $k_{max} = \pi/h$. For finite number of points used in the domain, it is impossible to go to the limit of $h \rightarrow 0$, so that $k_{max} \rightarrow \infty$. Now if k_{max} remains bounded, the spatial scales remain bounded. Couple this with the dispersion relation (physical and numerical) that relates wavenumber with circular frequency, one comes out with the eventual conclusion: Having established the finite bandwidth of spatial scale, the bandwidth of temporal scale is also limited by the dispersion relation. Once this is understood, it is realized that such *random noise* does not excite all the scales - as expected from white noise. So the system will not necessarily settle down to natural mode of excitation. That furthermore leaves us with the question: Do we have always a dispersion relation for nonlinear transitional flows and turbulent flows? This is a question without an unambiguous answer at this point in time.

7. Summary and Conclusion

Here, we have analyzed the numerical properties, evolution and propagation of numerical errors for general space-time discretization schemes by identifying the correct dispersion relation of the model 1D convection equation. With such a model equation, it is possible to show the special attributes of scientific computing, i,e, even for a linear system, signal and error do not share same dynamics. For the 1D convection, this analysis reveals that the error is governed by forced dynamics, with various sources of error acting as the forcing. Thus, in scientific computing it is paramount to find out major sources of errors. While the computing community is fixated with the simple aspects of scale resolution (the so-called k_{eq}/k expression) and numerical instability (most of the time, without considering the time integration scheme), it is more important to quantify dispersion and phase error also. In fact, the instability analysis of von Neumann is shown to be flawed and a fresh approach reveals the correct answer in solving the model

1D convection equation. It is quantitatively shown that the error is created by numerical stability/ instability; dispersion error and phase error as the primary reasons affecting high accuracy computing. It is noted that one requires neutrally stable methods with correct numerical phase speed and group velocity.

We also argue why 1D convection equation is relevant for high performance computing of transitional and turbulent flow. We have now at our disposal a full domain global spectral analysis tool for any discretization scheme. The main irritant is found to be dispersion error for high accuracy computing. In its extreme form, dispersion error can create upstream propagating error packets and this is termed as the q-waves.

This type of combined analysis for space-time discretization tells us some aspects of prevalent computing. We have focused on three examples highlighting three aspects of computing. In the first, we make the case for near-spectral accuracy compact schemes for non-periodic problems. Even for periodic problem employing Fourier spectral method, we suggest the more appropriate time discretization schemes being two-time level multi-stage schemes, e.g. the Runge-Kutta methods should be above second order. The RK2 method is unstable, while RK3 method being stable can also cause error. It is noted from an example of solving Navier-Stokes equation for the 2D receptivity to wall excitation problem, that RK4 method in conjunction with CD4 scheme for space provides adequate accuracy, as efficiently as a compact scheme used with RK4 method.

We focus on a new source of error in combined implicit-explicit (IMEX) time integration method, which is gaining popularity in many branches of scientific computing. As an example, we investigate the *bypass transition* reported[25,26,46], where the numerically caused transition is due to severe error caused at the interface of explicit and implicit parts of the domain by creation of propagating error-packets by partial internal reflection. While the implicit part of the time-integration is performed by Crank-Nicholson (CN) scheme, the explicit part is carried out by RK3 method. The analysis of creation and propagation of errors have been performed by solving 1D and 2D model convection equations, which admit exact solution and excellently mimics the physical convection process. We have used CD4 and OUCS3 schemes for spatial discretization of the convective derivatives in these equations, for comparison of two such presentations in the symposium by an explicit time integration method and another method that proposes *bypass transition* using IMEX time integration scheme. In the latter, one has to adopt an overlap region between the implicit and explicit parts of the computational domain[21,25]. Schematic of such an overlap between implicit and explicit zones are shown in Fig. 9(a). It has been shown in Figs. 10 and 11 by solving 1D convection equation, that when the error-packet associated with the actual signal passes through such overlap zone, it creates additional spurious high-wavenumber error-packets of comparable magnitude which propagates upstream of the original signal. Use of iterative solvers to solve the resulting linear algebraic equations for the implicit part of the domain for IMEX method also generates additional error-packets, those are also of similar orders of magnitude as that of the original error-packet.

In Figs. 12 and 13, we compare the numerical properties of the CN-CD4 scheme with RK3-CD4 scheme and CN-OUCS3 scheme with RK3-OUCS3 scheme. It is shown that RK3-scheme displays lesser error as compared to the CN-scheme used with CD4 spatial discretization because of its better dispersion properties. In Fig. 14, we show the evolution and propagation of error for model 2D convection equation to show that the passage of error-packet corresponding to the original signal through the overlap region creates additional elongated error-packets, similar to the 1D case shown in Figs. 10 and 11. This newly generated error-packets travel in the opposite direction with respect to the exact solution's direction of propagation.

For performing DNS of flows past a flat plate[25,26], the implicit zone is restricted to a confined space close to the wall[21,25]. Therefore, such error-packets of significant amplitude produce additional artificial forcing term, which is responsible for early numerical transition of the flow which has been termed as *bypass transition*. In contrast, explicit ORK3 time integration method along with high resolution OSCS scheme is used[3], where spatio-temporal wave-front (STWF) has been established as the physical precursor of transition for flows past a flat-plate with low background disturbance level. In Refs. 25 and 26, detection of STWF and its role in causing flow transition was not found because of the IMEX method which generates significant error-packets close to the wall that prevents creation of STWF in the first place. While the other reason is due to inadequate size of the computational domain and in this small domain numerical turbulence was created further by the use of *random or white noise*, along with the monochromatic wall-excitation. Yet, the match of the computed skin friction coefficient with empirical data for the transitional flow is exceedingly striking. Indeed if it is possible to reach the turbulent state numerically quickly in a very small domain without going through the formal steps of physical transition for low amplitude of excitation, then it opens up the interesting possibility in turbulence research.

References

1. U. M. Ascher, S. J. Ruuth and B. T. R. Wetton, Implicit-explicit methods for time-dependent partial differential equations, *SIAM J. Num. Anal.*, **32**, 797–823 (1995).
2. G. K. Batchelor, Computation of the energy spectrum in homogeneous two-dimensional decaying turbulence, *Phys. Fluids*, **12**, (suppl. II), 233–239 (1969).
3. S. Bhaumik and T. K. Sengupta, Precursor of transition to turbulence: Spatiotemporal wave front. *Phys. Rev. E*, **89**, 043018–1 (2014).
4. S. Bhaumik and T. K. Sengupta, A new velocity-vorticity formulation for direct numerical simulation of 3D transitional and turbulent flows, *J. Comput. Phys.*, **284**, 230–260 (2015).
5. C. Canuto, M. Y. Hussaini, A. Quarteroni and T. A. Zang, Spectral Methods in Fluid Dynamics, Springer Verlag, New York, USA (1987).
6. J. G. Charney, R. Fjørtoft and J. von Neumann, Numerical integration of the barotropic vorticity equation, Tellus, **2**(4), 237–254 (1950).
7. J. Crank and P. Nicolson, A practical method for numerical evaluation of solutions of partial differential equations of the heat conduction type, *Proc. Camb. Phil. Soc.*, **43**, 50–67 (1947).
8. M. O. Deville, P. F. Fischer and E. H. Mund, High Order Methods for Incompressible Fluid Flow, Cambridge Univ. Press, Cambridge, UK (2002).
9. A. Dipankar and T. K. Sengupta, Symmetrized compact scheme for receptivity study of 2D transitional channel flow, *J. Comput. Phys.*, **215**, 245–273 (2006).
10. D. A. Donzis, P. K. Yueng and K. R. Sreenivasan, Dissipation and enstrophy in isotropic turbulence: resolution effects and scaling in direct numerical simulation, *Phys. Fluids*, **20**, 045108 (2008).
11. D. A. Donzis and P. K. Yueng, Resolution effects and scaling in numerical simulation of passive scalar mixing in turbulence, Physica D: Nonlinear Phenomena, **239**(14), 1278–1287 (2010).
12. F. X. Giraldo, J. F. Kelly and E. M. Constantinescu, Implicit-explicit formulations of a three dimensional nonhydrostatic unified model of the atmosphere (NUMA), SIAM *J. Sci. Comp. (SISC)* (2013).
13. D. Gottlieb and S. A. Orszag, Numerical Analysis of Spectral Methods: Theory and Application, SIAM-CBMS, Philadelphia, USA (1977).
14. G. J. Haltiner and R. T. Williams, Numerical Prediction and Dynamics Meteorology, 2nd edn., John Wiley & Sons., USA (1980).

15. G. E. Karniadakis and S. J. Sherwin, Spectral/hp Element Methods for CFD, Oxford Univ. Press, Oxford, UK (1999).
16. A. Kanevsky, M. H. Carpenter, D. Gottlieb and J. S. Hesthaven, Application of implicit-explicit high order Runge-Kutta methods to discontinuous-Galerkin schemes, *J. Comp. Phys.*, **225**, 1753–1781 (2007).
17. J. Kim and P. Moin, Application of a fractional-step method to incompressible Navier-Stokes equation, *J. Comp. Phys.*, **58**, 308–322 (1985).
18. R. Kraichnan and D. Montgomery, Two-dimensional turbulence, *Reports in Progress in Physics*, **43**, 547–619 (1980).
19. S. K. Lele, Compact finite difference schemes with spectral like resolution, *J. Comp. Phys.*, **103**, 16–42 (1992).
20. H. Lomax, T. H. Pulliam and D. W. Zingg, Fundamentals of CFD, Springer Verlag, Berlin, Germany (2002).
21. S. Nagarajan, Leading edge effects of bypass transition, Ph. D. Thesis, Dept. of Mech. Engineering, Stanford University, (2004).
22. P.-O. Persson, High-order LES simulations using Implicit-Explicit Runge-Kutta Schemes, Proc. of the 49th AIAA Aerospace Sciences Meeting and Exhibit *AIAA* 2011-684 (2011).
23. C. Pozrikidis, Introduction to Theoretical and Computational Fluid Dynamics, Oxford Univ. Press, U. K. (2011).
24. S. J. Ruuth, Implicit-explicit methods for reaction-diffusion problems in pattern formation, *J. Math. Biology*, **34**, 148–176 (1995).
25. T. Sayadi, Numerical simulation of controlled transition to developed turbulence in a zero-pressure-gradient flat-plate boundary layer, Ph. D. Thesis, Dept. of Mech. Engineering, Stanford University (2012).
26. T. Sayadi, C. W. Hamman and P. Moin, Direct numerical simulation of complete H-type and K-type transitions with implications for the dynamics of turbulent boundary layers, *J. Fluid Mech.*, **724**, 480–509 (2013).
27. G. B. Schubauer and H. K. Skramstad, Laminar boundary layer oscillations and the stability of laminar flow, *J. Aero. Sci.*, **14**, 69–78 (1947).
28. T. K. Sengupta, High Accuracy Computing Methods: Fluid flows and Wave Phenomena, Cambridge University Press, USA (2013).
29. T. K. Sengupta, Theoretical and Computational Aerodynamics, John Wiley and Sons, Chichester, U. K. (2015).
30. T. K. Sengupta and S. Bhaumik, Onset of turbulence from the receptivity stage of fluid flows. *Phys. Rev. Lett.*, 154501–5 (2011).

31. T. K. Sengupta, S. Bhaumik and Y. G. Bhumkar, Direct numerical simulation of two-dimensional wall-bounded turbulent flows from receptivity stage, *Phys. Rev. E*, **85**, 026308 (2012).
32. T. K. Sengupta, Y. Bhumkar, M. K. Rajpoot, V. K. Suman and S. Saurabh, Spurious waves in discrete computation of wave phenomena and flow problems, *App. Math. Comput.*, **218**, 9035–9065 (2012).
33. T. K. Sengupta and A. Dipankar, A comparative study of time advancement methods for solving Navier-Stokes equations, *J. Sci. Comp.*, **21**, 225–250 (2004).
34. T. K. Sengupta, A. Dipankar and A. K. Rao, A new compact scheme for parallel computing using domain decomposition, *J. Comp. Phys.*, **220**, 654–677 (2007).
35. T. K. Sengupta, A. Dipankar and P. Sagaut, Error dynamics: Beyond von Neumann analysis, *J. Comp. Phys.*, **226**, 1211–1218 (2007).
36. T. K. Sengupta, G. Ganeriwal and S. De, Analysis of central and upwind compact schemes, *J. Comp. Phys.*, **192**(2), 677–694 (2003).
37. T. K. Sengupta, G. Ganeriwal and A. Dipankar, High accuracy compact schemes and Gibb's phenomenon, *J. Sci. Compt.*, **21**(3), 253–268 (2004).
38. T. K. Sengupta, M. K. Rajpoot and Y. G. Bhumkar, Space-time discretizing optimal DRP schemes for flow and wave propagation problems, *Comput. Fluids*, **47**(1), 144–154 (2011).
39. T. K. Sengupta, M. K. Rajpoot, S. Saurabh and V. V. S. N. Vijay, Analysis of anisotropy of numerical wave solutions by high accuracy method, *J. Comp. Phys.*, **230**, 27–60 (2011).
40. T. K. Sengupta, V. K. Sathyanarayanan, M. Akhil and M. Sriramakrishnan, Role of time integration in computing transitional flows caused by wall excitation, *J. Sci. Comp.* DOI 10.1007/s10915-014-9967-1 (2014).
41. P. Spalart and N. Yang, Numerical simulation of ribbon-induced transition in Blasius flow, *J. Fluid Mech.*, **178**, 345–365 (1987).
42. C. K. W. Tam and J. C. Webb, Dispersion-relation-preserving finite difference schemes for computational acoustics, *J. Comp. Phys.*, **107**, 262–281 (1993).
43. L. N. Trefethen, Group velocity in finite difference schemes, SIAM *Reviews*, **24**(2), 113–136 (1982).
44. H. A. Van der Vorst, Bi-CGSTAB: A fast and smoothly converging variant of Bi-CG for the solution of non-symmetric linear systems, SIAM *J. Sci. Stat. Comput.*, **12**, 631–644 (1992).

45. R. Vichnevetsky and J. B. Bowles, Fourier Analysis of Numerical Approximations of Hyperbolic Equations, SIAM Pub., Philadelphia, USA (1982).
46. X. Wu and P. Moin, Direct numerical simulation of turbulence in a nominally zero-pressure-gradient flat-plate boundary layer, *J. Fluid Mech.*, **630**, 5–41 (2009).
47. P. K. Yueng, D. A. Donzis and K. R. Sreenivasan, Dissipation, enstrophy and pressure statistics in simulation at high Reynolds number, *J. Fluid Mech.*, **700**, 5–15 (2012).

IUTAM Symposium on "Advances in Computation, Modeling and Control of Transitional and Turbulent Flows"

Panel Discussion on

Status and Future of High Performance Computing Related to Transition and Turbulence

Panel members: Prof. T. K. Sengupta, Dr. P. Spalart, Prof. D. Gaitonde, Prof. J. Jimenez, Prof. M. Deville, Prof. Sanjiva K. Lele, Prof. Tony Sheu, Prof. Peter Davidson

A panel discussion was organized on the final day of the symposium to take stock of the proceedings and to take views of experts on status and future of HPC related to transition and turbulence. The panel consisted of Prof. T. K. Sengupta (IIT Kanpur), Dr. P. Spalart (The Boeing Company), Prof. D. Gaitonde (Ohio State University), Prof. J. Jimenez (Politecnica de Madrid, Spain), Prof. M. Deville (EPFL, Switzerland), Prof. S. K. Lele (Stanford University), Prof. T. W. H. Sheu (NTU, Taiwan) and Prof. P. A. Davidson (Cambridge University) and was moderated by Dr. Vidyadhar Mudkavi (CSIR NAL).

Prof. Sengupta made the following opening remarks and set tone for the subsequent discussions.

The main point raised was whether there exist metrics for HPC and, if so, whether the same are quantifiable to suggest if computational results obtained are acceptable and correct. This is intimately related to the other question: What really constitutes high accuracy computing? Is high order computing synonymous to high accuracy computing? Or, are there fundamental differences? He elaborated that filtering is implicit in all computing and it must be recognized and taken care of. A high order scheme does not necessarily guarantee high accuracy if the associated errors are not quantified and eliminated/reduced. Thus, there is a need for canonical problems, which must be solved by any code without any ambiguity, before one claims validity of solutions produced for complex problems using such codes. He also probed the connectivity between experiments and HPC both with attendant limitations. He specifically enunciated difference between dispersion and diffusion which is often used interchangeably that is incorrect. One must get space-time discretization correct with regard to numerical dispersion and reflection. He also alluded to perils of "me-too" syndrome, which, in his opinion, is taking computing in the wrong direction. In support of performing canonical problems first, he quoted Einstein: "Any intelligent fool can make things bigger and more violent. It takes a touch of genius and a lot of courage to go in the opposite direction."

He made a strong case for spatio-temporal analysis, as was done for a 1D convection equation, resulting in replacement of traditional von Neuman error analysis. This is especially true, as some recent results show different routes to turbulence due to limitations of different computational procedures, obscuring any physical insight when turbulence is created by forcible artificial numerical errors. He laid emphasis on developing neutrally stable codes as opposed to so-called stable algorithms without affecting the underlying physics.

Dr. Spalart was of the opinion that in order to serve the engineering community there is a need to solve complex 3D problems. As such, adaptive grids will play an important role in future. Numerical schemes developed for 1-D may work, but these need to work for 3D problems as well. He said that the RANS codes are here to stay from an engineering perspective.

Prof. Gaitonde said that people are beginning to compute which extend experiments. In many cases, experiments are not even possible and the only route is through computations. He observed that there aren't many one billion-cell calculations. He placed emphasis on scalability and role of artificial intelligence in handling large amounts of data associated with large simulations. He also said that one cannot anymore be mere subject expert. There is greater need to collaborate. One now needs software and hardware knowledge; capability in numerical analysis; parallel algorithms etc., to develop effective HPC codes.

Prof. Jimenez said that "experiments" are being performed in computers for the last ten years. He was aware of many billion-cell computations and also claimed that many perform 40/50 billion cell computations and these are becoming common place. He also said that we are not allowed to speculate anymore theoretically alone as one can go ahead and test the theory using computing. He also mentioned that canonical flows are as important as other flows which are now computed with equal rigor. In this context, he however, noted that a mouse can be felt and understood better, than an "elephant" by proverbial blind men. The problem arises when the size is large and different portions of the problem seem unconnected.

This is where big data research can help computations and interpretations. Individuals with their limited capacity may not be in a position to interpret the results of large problems.

Prof. Deville said he has seen wide range of techniques in the landscape of spatial discretization techniques such as finite difference, finite volume, finite element, spectral element and so on. But one needs to ask what should be the right temporal discretization. He quoted Richard Hamming to lay emphasis on the point he was making: "The purpose of computing is insight, not numbers." He said, however, that he liked to re-interpret Hamming and would rather say that the purpose of computing is numbers up to certain degree of accuracy that gives different insights. As such he was in favor of high order methods. He was of the opinion that all problems in the world are bounded, even the dynamo. Therefore, one must not do any periodic flows, as they do not represent real problems. He also touched upon important aspect of influence of programming languages. He said new programming languages help save human effort, which can be used for better things.

Prof. Lele interpreted HPC to mean LES/DNS with specific emphasis on aeroacoustics. He said that the main drivers for aeroacoustics should let one decide what to capture and what to model. In this connection, there is no room for brute force approach, implying solving the problem from first principle. It is important to understand what compromises we end up making while modeling. In aeroacoustics simulations, the near field must be captured using computations, while the far field should be modeled using acoustic analogy. He pointed out that one must be aware upfront as to where one is operating, as enunciated by Prof. Sengupta in terms of various forms of errors. He discussed the example of crackle noise simulation in supersonic jet using about 207 million nodes. Simulations help discover new physical phenomenon, which in turn, help us model better. Modeling is, in general, hard. Engineering simulations require multi-levels of fidelity such as RANS, Hybrid-LES, LES, DNS etc. It is important to define benchmark problems. Blind workshops play crucial role in this regard.

Touching upon advances in computer hardware, he highlighted hyper-threading, memory hierarchy, accelerators etc. which play significant roles as newer machines are dominated by larger number of smaller CPU cores. As the race to exa-scale unfolds, more challenges to computing community will arise. Training in computational science and handling big data for better models offer challenges, as well as opportunities. He said that he did not believe in end of theory proposition with emergence of artificial intelligence.

Exa-scale and related issues open a second golden age in computing.

Prof. Sheu, while appreciating the high standard of the papers presented at the symposium, held a view that steady state simulations dominate most practical applications and, as such, mostly second order accuracy is sufficient for such activities. Flow instability, bifurcation of solutions, transitions and small Reynolds number problems are new important areas of application. He felt that dispersion relation preserving, high order schemes are required. He called upon young researchers to practice programming in CUDA as it can speed up by order of 70 for Euler solutions and 40 for Navier-Stokes solutions.

Prof. Davidson strongly felt that there is a need to look at fundamental problems which he has been himself asking for last fifteen years. He particularly mentioned about his discussions with Prof. Pirozzoli who asked him those same questions regarding turbulence and associated asymptotic theories and Prof. Davidson felt that those questions still remain relevant.

Prof. Pirozzoli said that he has been solving turbulent flows for last ten years and has not seen self-similar cascades in his simulations. Thus, he advocated that we continue to ask fundamental questions and not only worry about big scale DNS.

There was a question from Prof. Baig about grid sensitivity which evoked many responses from the panelists. Prof. Sengupta did not share his full-hearted support for such exercises, as refining a grid alone can result in completely different numerical properties and hence simulation results. In this context, he quoted the example of simple flow past a cylinder where even today there is no unanimity on critical Reynolds number, as this depends on the number of points considered using the same method. Prof. Lele said that grid refinement should accompany change in time step also, so that the results are self-consistent with different grids. This was echoed also by Prof. Sheu that numerical properties are important and he highlighted the importance of dispersion relation preserving methods. On this aspect he and Prof. Sengupta share identical views.

In summary, while the opening remarks of Prof. Sengupta tried to bring focus on what really constitutes high performance computing, it was clear that the experiences of different researches is widely different as there appears to be philosophical differences in the use of HPC for solving flow problems in nature. It is evident, therefore, that there is an urgent need to deliberate and accept a common definition for HPC in this complex field. General feeling emerged during the discussion that fundamental problems and problems involving big data should continue reinforcing each other.

The allegorical feel for the mouse (canonical problems) should continue by blind men (researchers), alongside attempts to feel the elephant (turbulence and other complex problems involving big data) too.

When is DNS a DNS: a brief comment

Katepalli R. Sreenivasan
Professor of Physics; Senior Vice Provost, NYU, USA
E-mail: katepalli.sreenivasan@nyu.edu

DNS in turbulence started some forty-five years ago with very low Reynolds number flows: the low computational power available at the time wouldn't allow anything more ambitious. The first DNS of homogeneous and isotropic turbulence was made on a 32^3 periodic box[1]. The largest box now is 8192^3 in size[2]: the Reynolds number of this simulation is larger than those realized in grid flows of most large wind tunnels[3]. This is an impressive quantitative change brought about by the enormous increase in computing power. Such huge *quantitative* changes sometimes result in the need for a *qualitative* change in outlook, and, in this instance, that outlook concerns the grid resolution.

In the past, the gold standard in DNS was to compute with a grid resolution equal to the mean Kolmogorov scale $\eta = (v^3/ < \epsilon >)^{1/4}$, where v is the kinematic viscosity, ϵ represents the energy dissipation rate and the angular brackets indicate the spatial average. This seemed perfectly reasonable given the widely held view that η represents the smallest dynamical scale in turbulent flows. Most calculations have actually cut corners by resolving no better than a few multiples of $\eta, 2$ or 3 being the commonly chosen factor.

At high Reynolds numbers, huge fluctuations kick in — e.g., in dissipation, in squared vorticity, etc. When one has huge fluctuations, tens of thousands larger than the mean (see[2]), one also has much smaller local Kolmogorov scales (as can be seen by just replacing the mean dissipation rate in the definition of η by its instantaneous value). This makes scales much smaller than η dynamically relevant[4] — and the smallness factor becomes smaller with increasing Reynolds number: because of extreme intermittency, events whose amplitudes are small on the average can be rather large whenever and wherever they occur (which, of course, will happen rarely). So the effective viscous scale shrinks in relation to η, which makes it necessary to employ a finer grid to resolve it. Thus, while a grid that resolves η is fine enough at low Reynolds numbers (most simulation take some liberty even with this, as already pointed out), it falls short of the requirement of a true DNS at much higher Reynolds numbers.

No one disputes that fluctuations of scales smaller than η exist but most have the reaction that there is no need to take them seriously because of a huge spectral roll-off in the dissipative range. That is certainly not adequate as just mentioned. And the problem shows up as more prominently the higher the Reynolds number we compute, and more consequential it becomes — especially if one is interested theoretically in tails of various distributions[5] or, in practice, in topics such as combustion[6].

How much of a resolution, then, should a proper DNS aim for? This depends on the Reynolds number and the quantity one wants to resolve well[7,8]. The theoretical estimate at the highest Reynolds number is that the smallest scale is of the order $\eta x\ Re^{-1/4}$, but it is less stringent in practice 9 — nevertheless more stringent than merely resolving η. What it does in practice is that the Reynolds numbers one estimates for a given computing power, from standard formulae using the mean Kolmogorov scale η are optimistic and open themselves up to various questions on the reliability of results. And there is much more to be learnt by accurate simulation of modest Reynolds number flows than by resorting to poorly resolved flows at high Reynolds numbers 7.

References

1. S. A. Orszagand G. S.Patterson,—it Phys.Rev.Lett. **28**, 76 (1972).
2. P. K. Yeung, X. M. Zhai and K. R. Sreenivasan, Proc. Nat. Acad. Sci. (submitted, 2015).
3. M. Sinhuber, E. Bodenschatz and G. P. Bewley, *Phys. Rev. Lett.* **114**, 034501 (2015).
4. V. Yakhot and K. R. Sreenivasan, *J. Stat. Phys.* **121**, 823 (2005).
5. J. Schumacher, K. R. Sreenivasan, and P. K. Yeung, *J. Fluid Mech.*531, 113 (2005).
6. K. R. Sreenivasan, Flow, Turbulence and Combustion **72**, 115 (2004).
7. J. Schumacher, K. R. Sreenivasan and V. Yakhot, *New J. Phys.* **9**, 1 (2007).
8. J. Schumacher, J. D. Scheel, D. Krasnov, D. A. Donzis, V. Yakhot, and K. R. Sreenivasan, Proc. Nat. Acad. Sci. USA, 111, 10961 (2014).
9. D. A. Donzis, P. K. Yeungand KR Sreenivasan, *Phys. Fluids* **20**, 045108 (2008).